Publications of the

Carnegie Endowment for International Peace

Division of Economics and History

ECONOMIC AND SOCIAL HISTORY OF THE WORLD WAR

JAMES T. SHOTWELL, LL.D.

General Editor

Rumanian Series

DAVID MITRANY, Ph.D.

Editor

THE LAND & THE PEASANT IN RUMANIA

THE WAR AND AGRARIAN REFORM (1917–21)

BY

DAVID MITRANY
PH.D. (London), B.Sc. (Econ. London)

LONDON: HUMPHREY MILFORD
OXFORD UNIVERSITY PRESS
NEW HAVEN: YALE UNIVERSITY PRESS
1930

Printed in Great Britain

TO THE MEMORY

OF

PROFESSOR L. T. HOBHOUSE

1864–1929

EDITOR'S PREFACE

In the autumn of 1914, when the scientific study of the effects of war upon modern life passed suddenly from theory to history, the Division of Economics and History of the Carnegie Endowment for International Peace proposed to adjust the programme of its researches to the new and altered problems which the War presented. The existing programme, which had been prepared as the result of a conference of economists held at Berne in 1911, and which dealt with the facts then at hand, had just begun to show the quality of its contributions; but for many reasons it could no longer be followed out. A plan was therefore drawn up at the request of the Director of the Division, in which it was proposed, by means of an historical survey, to attempt to measure the economic cost of the War and the displacement which it was causing in the processes of civilization. Such an 'Economic and Social History of the World War', it was felt, if undertaken by men of judicial temper and adequate training, might ultimately, by reason of its scientific obligations to truth, furnish data for the forming of sound public opinion, and thus contribute fundamentally towards the aims of an institution dedicated to the cause of international peace.

The need for such an analysis, conceived and executed in the spirit of historical research, was increasingly obvious as the War developed, releasing complex forces of national life not only for the vast process of destruction but also for the stimulation of new capacities for production. This new economic activity, which under normal conditions of peace might have been a gain to society, and the surprising capacity exhibited by the belligerent nations for enduring long and increasing loss—often while presenting the outward semblance of new prosperity—made necessary a reconsideration of the whole field of war economics. A double obligation was therefore placed upon the Division of Economics and History. It was obliged to concentrate its work upon the problem thus presented, and to study it as a whole; in other words, to apply to it the tests and disciplines of history. Just as the War itself was a single event, though penetrating by seemingly unconnected ways to the remotest parts of the world, so the analysis of it must be developed according to a plan at once all-embracing and yet adjustable to the practical limits of the available data.

During the actual progress of the War, however, the execution of this plan for a scientific and objective study of war economics proved impossible in any large and authoritative way. Incidental studies and surveys of portions of the field could be made and were made under the direction of the Division, but it was impossible to undertake a general history for obvious reasons. In the first place, an authoritative statement of the resources of belligerents bore directly on the conduct of armies in the field. The result was to remove as far as possible from scrutiny those data of the economic life of the countries at war which would ordinarily, in time of peace, be readily available for investigation. In addition to this difficulty of consulting documents, collaborators competent to deal with them were for the most part called into national service in the belligerent countries and so were unavailable for research. The plan for a war history was therefore postponed until conditions should arise which would make possible not only access to essential documents but also the co-operation of economists, historians, and men of affairs in the nations chiefly concerned, whose joint work would not be misunderstood either in purpose or in content.

Upon the termination of the War the Endowment once more took up the original plan, and it was found with but slight modification to be applicable to the situation. Work was begun in the summer and autumn of 1919. In the first place a final conference of the Advisory Board of Economists of the Division of Economics and History was held in Paris, which limited itself to planning a series of short preliminary surveys of special fields. Since, however, the purely preliminary character of such studies was further emphasized by the fact that they were directed more especially towards those problems which were then fronting Europe as questions of urgency, it was considered best not to treat them as part of the general survey but rather as of contemporary value in the period of war settlement. It was clear that not only could no general programme be laid down *a priori* by this conference as a whole, but that a new and more highly specialized research organization than that already existing would be needed to undertake the Economic and Social History of the War, one based more upon national grounds in the first instance and less upon purely international co-operation. Until the facts of national history could be ascertained, it would be impossible to proceed with comparative analysis ; and the different national histories were themselves of almost baffling intricacy and variety. Consequently the former European Committee of Research was dissolved, and in its place it was decided to erect an Editorial Board in each of the larger countries and to nominate special editors in the smaller ones, who should concentrate, for the present at least, upon their own economic and social war history.

The nomination of these boards by the General Editor was the first step taken in every country where the work has begun. And if any justification was needed for the plan of the Endowment, it at once may be found in the lists of those, distinguished in scholarship or in public affairs, who have accepted the responsibility of editorship. This responsibility is by no means light, involving, as it does, the adaptation of the general editorial plan to the varying demands of national circumstances or methods of work ; and the measure of success attained is due to the generous and earnest co-operation of those in charge in each country.

Once the editorial organization was established there could be little doubt as to the first step which should be taken in each instance toward the actual preparation of the history. Without documents there can be no history. The essential records of the War, local as well as central, have therefore to be preserved and to be made available for research in so far as is compatible with public interest. But this archival task is a very great one, belonging of right to the governments and other owners of historical sources and not to the historian or economist who proposes to use them. It is an obligation of ownership ; for all such documents are public trust. The collaborators on this section of the war history, therefore, working within their own field as researchers, could only survey the situation as they found it and report their findings in the form of guides or manuals ; and perhaps, by stimulating a comparison of methods, help to further the adoption of those found to be most practical. In every country, therefore, this was the point of departure for actual work ; although special monographs have not been written in every instance.

This first stage of the work upon the war history, dealing with little more than the externals of archives, seemed for a while to exhaust the possibilities of research. And had the plan of the history been limited to research based upon official documents little more could have been done, for once documents have been labelled 'secret' few government officials can be found with sufficient courage or initiative to break open the seal. Thus vast masses of source material essential for the historian were effectively placed beyond his reach, although much of it was quite harmless from any point of view. While war conditions thus continued to hamper research, and were likely to do so for many years to come, some alternative had to be found.

Fortunately, such an alternative was at hand in the narrative, amply supported by documentary evidence, of those who had played some part in the conduct of affairs during the War, or who, as close observers in privileged positions, were able to record from first- or at least second-hand knowledge the economic history of different phases of the Great War, and of its effect upon society. Thus a series of monographs was planned consisting for the most part of unofficial yet authoritative statements, descriptive or historical, which may best be described as about half-way between memoirs and blue-books. These monographs make up the main body of the work assigned so far. They are not limited to contemporary, war-time studies; for the economic history of the war must deal with a longer period than that of the actual fighting. It must cover the years of 'deflation' as well, at least sufficiently to secure some fairer measure of the economic displacement than is possible in purely contemporary judgements.

With this phase of the work the editorial problems assumed a new aspect. The series of monographs had to be planned primarily with regard to the availability of contributors, rather than of source material as in the case of most histories; for the contributors themselves controlled the sources. This in turn involved a new attitude towards those two ideals which historians have sought to emphasize, consistency and objectivity. In order to bring out the chief contribution of each writer it was impossible to keep within narrowly logical outlines; facts would have to be repeated in different settings and seen from different angles, and sections included which do not lie within the strict limits of history; and absolute objectivity could not be obtained in every part. Under the stress of controversy or apology, partial views would here and there find their expression. But these views are in some instances an intrinsic part of the history itself, contemporary measurements of facts as significant as the facts with which they deal. Moreover, the work as a whole is planned to furnish its own corrective; and where it does not, others will.

In addition to this monographic treatment of source material, a number of studies by specialists is already in preparation, dealing with technical or limited subjects, historical or statistical. These monographs also partake to some extent of the nature of first-hand material, registering as they do the data of history close enough to the source to permit verification in ways impossible later. But they also belong to that constructive process by which history passes from analysis to synthesis. The process is a long and difficult one, however, and work upon it has only just begun. To quote an apt characterization, in the first stages of a history like this one is only ‘picking cotton’. The tangled threads of events have still to be woven into the pattern of history; and for this creative and constructive work different plans and organizations may be needed.

In a work which is the product of so complex and varied co-operation as this, it is impossible to indicate in any but a most general way the apportionment of responsibility of editors and authors for the contents of the different monographs. For the plan of the History as a whole and its effective execution the General Editor is responsible; but the arrangement of the detailed programmes of study has been largely the work of the different Editorial Boards and divisional Editors, who have also read the manuscripts prepared under their direction. The acceptance of a monograph in this series, however, does not commit the editors to the opinions or conclusions of the authors. Like other editors, they are asked to vouch for the scientific merit, the appropriateness and usefulness of the volumes admitted to the series; but the authors are naturally free to make their individual contributions in their own way. In like manner the publication of the monographs does not commit the Endowment to agreement with any specific conclusions which may be expressed therein. The responsibility of the Endowment is to History itself—an obligation not to avoid but to secure and preserve variant narratives and points of view, in so far as they are essential for the understanding of the War as a whole.

The present volume calls for special editorial comment. There are two ways to deal with the problems covered by the Economic and Social History of the World War. There is, on the one hand, the close analysis of the immediate effects, the treatment of events in their own setting without regard to their historical antecedents or their effects upon the subsequent developments of economic or social life; on the other hand, there is the purely historical treatment which places the events of the War in the long perspective of national development and follows the story through slow and complicated processes of recovery. The Economic and Social History of the World War contains both types of analysis; the strictly limited study of war-time phenomena, and the surveys which present not only the detail of contemporary happenings but an interpretation of their meaning. This study is of the latter type. And the fact that it is largely a pioneer exploration in an area as yet not well defined in either theory or practice has made it necessary to describe the historical background of the problem in almost as great detail as the problem itself. Mr. Mitrany has in a truly scientific spirit attempted to deal with the agrarian problem of south-eastern Europe as it has shaped itself under varying pressures of war and politics. It is a large canvas which has here been filled with the figures of those inarticulate masses of men and women to whose fate the historian has been as often indifferent as the politician has been unjust. At last, in this volume, the Rumanian peasant speaks for himself to the whole world. And his plea for social, economic, and political justice is, in its very nature, a force which makes for peace. There is, therefore, an added reason for including in this series the volume which makes this appeal intelligible to English readers.

J. T. S.

AUTHOR'S PREFACE

The agrarian reform described in this volume was a result of the World War, in the same sense as was the Russian Revolution. The slow action of centuries created the conditions which made such an outcome possible, and the shock of the Great War shattered the social structure of eastern Europe at its weakest joint. In the potency of their contribution to that effect War and history perhaps had an equal share. Inevitably, however, there is more to be told about the slow work of history than about the stark blow of the War. For the proper understanding of the reform this study, therefore, had to go beyond the limits of the War years. Its real meaning could not have been made clear without projecting the reform against its historical background, especially as writers on south-eastern Europe have been so engrossed with its politics hitherto as to give scant attention to social history. For the same reason the study had to take in a number of post-war years; the application of the reform itself took a number of years to complete, and its effects, to be measured with any solidity, had to be observed over as long a period as possible. This method imposed itself—paradoxical though it may sound—just because in this case the effect of the War was so deep. Where the War merely caused some transient body to be set up, like the Allied Shipping Board, or some existing practice to be temporarily suspended, like the parliamentary control of expenditure, its effect was simple, direct and co-extensive in time, and could be placed nicely within the framework of the War. But such limits would have been altogether artificial and distorting for the description of an effect which has revolutionized the whole social progress of a nation.

The point is still more evident when one considers not effects but causes of war. Special war boards and exceptional war rules have no place in the latter category at all. They are merely adjuncts in the conduct of war, but are neither produced by, nor do they revise, the issues which underlie the conflict. Not so with the social foundations of States. They are both the warp, as causes, and the weft, as effects, from which is woven the progress of the nations through war and peace. If it is certain that without the War those sweeping social changes would not have occurred, who could tell if the War itself would have happened, were eastern Europe to have had a different social history?

There are two aspects of this study on which it might be as well to say a word here. One is the mood of the historical part of the narrative. The picture drawn in that part is so dark, almost without any half-lights, that it may possibly strike the general reader as biassed. Yet it is just as likely, I fear, that those who have been spectators of the unbridled manner in which the Rumanian peasant's power of work and gentleness of temper were abused before the War, may rather reproach me with appearing to make light of a great wrong, by speaking of it in too measured tones.

The second point concerns the statistical and sociological material used in this study. In a letter to his friend Suvorin, written in the 'nineties, Chekhov complained that 'in Russia there is a terrible dearth of facts and a terrible abundance of speculations of all sorts'. Since that time the zemstvo workers have provided Russia with the most elaborate agrarian statistics and social studies of any country, the United States excepted. But Chekhov's remark applies with distressing accuracy to Rumania. Her agrarian and social statistics have never yet been taken seriously by those responsible for the budget of the statistical services. I am conscious, therefore, of many shortcomings, which might have been made good if the means at the disposal of the officials concerned had been equal to their ability and goodwill.

Research for this study having stretched over a number of years, with several spells of work in the field, it will be evident that I have received information and assistance from many more people than I could publicly thank here. The kindness of members of the former Rumanian Government—M. Vintilă Brătianu, M. I. Duca, M. C. Argentoianu, and M. G. Cipăianu—as well as of the present Minister of Agriculture, M. I. Mihalache, made it possible for me to obtain help from departments already overburdened with work. I owe immense gratitude for their unstinted help to my friend General Radu Rosetti, son of the distinguished historian of the agrarian question, to Dr. Gr. Antipa, Director of the Natural Science Museum, and to Professor G. Ionescu-Siseşti, Director of the Institute of Agronomic Research. I am deeply indebted to the indefatigable general secretary of the Rumanian Academy, Professor I. Bianu, and to the staff of the Academy's reading room, for much courtesy and help; and likewise to Dr. Gheron Netta, Director of the Rumanian Economic Institute, and his assistants. I have received much encouragement from my friend Professor Dimitrie Gusti, President of the Rumanian Social Institute. Professor Al. Nasta, Director of the Central Resettlement Office, and his assistant M. N. Chiţoiu, have given me all the aid in their power. I have received also most willing help from MM. P. Rosiade, I. Lupan, and E. Filotti, of the Ministry of Agriculture; and from MM. E. Marian and F. Kémeny, technical experts to the 'Sămânţa' company. My friend M. Emanoil Bucuţa never failed to reply to a letter and to give all the help he could, as did also my friend M. Aureliu Popescu, now Commercial Attaché in Paris, and M. I. Coler, editorial secretary of the *Adevĕrul*.

In this country I have received from Mr. C. S. Orwin, Director of the Oxford Institute of Agricultural Economics, valuable advice for the solution of certain difficulties in the translation of technical terms, and he has also been kind enough to read in proof the chapters on Production. Dr. G. Pavlovsky has prepared the Index.

D. M.

KINGSTON BLOUNT, OXFORD.
December 1929.

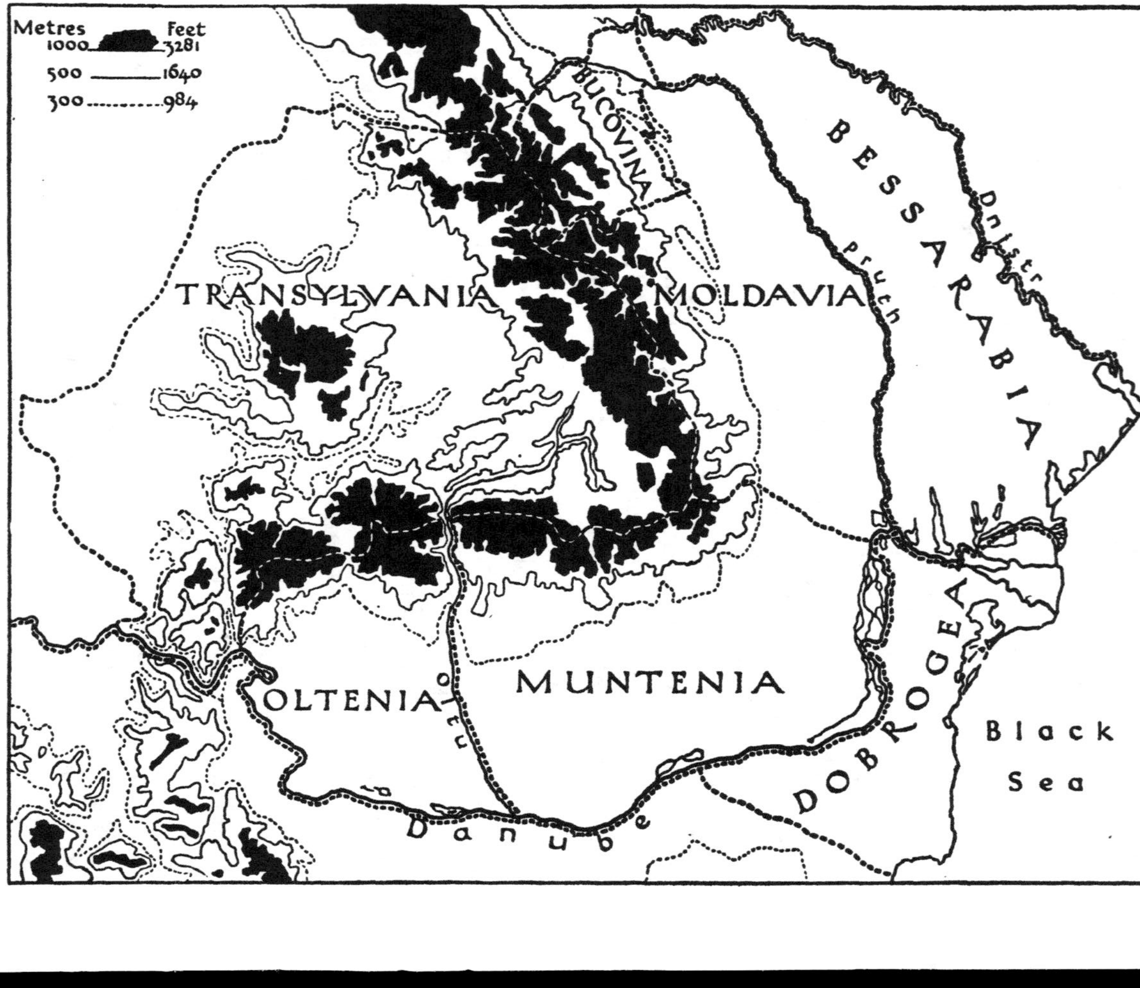
Metres
Feet
1000
3281
500
1640
300
984
BUCOVINA
BESSARABIA
Dnistr
Pruth
TRANSYLVANIA
MOLDAVIA
MUNTENIA
OLTENIA
Oltu
DOBROGEA
Danube
Black Sea

CONTENTS

INTRODUCTION: THE EFFECT OF THE WORLD WAR UPON THE AGRARIAN STRUCTURE OF EUROPE . xxiii

PART I

THE AGRARIAN PROBLEM IN RUMANIAN HISTORY

I. SOCIAL FREEDOM AND POLITICAL DEPENDENCE
1. The Middle Ages: Land and People 4
The rulers 5
The ruled 6
Land tenure 7
2. Foundation of the Principalities 8
Relations between princes and cnezi 10
The emergence of an upper class 11
3. The beginnings of serfdom 13
The loss of freedom 14
The increase in servitudes 17
The limitation of the right to land 21
The rape of yeoman land 22

II. POLITICAL FREEDOM AND SOCIAL DEPENDENCE
1. The Organic Statutes 25
The Moldavian statute 27
The Muntenian statute 31
2. Revising the organic statutes 34
3. The condition of the peasantry on the eve of national independence 38

III. NATIONAL INDEPENDENCE AND THE EMANCIPATION OF THE PEASANTS
1. The new factors 42
2. The reform of 1864 47
The nature of the reform 50
The background of the reform 52
3. Practical effects of the reform 57

IV. THE NEW STATE AND THE NEW SERFDOM (1864–1917) 63
1. Land policy after 1864: 66
first period 66
second period 69
2. The laws on agricultural contracts 73
3. The Peasant's burdens 81
4. Wars and revolutions 85

PART II

THE NEW LAND REFORM

V. LAND LAWS: POLITICAL HISTORY 95
Fixing the basis of the reform 101
Legislative evolution of the proposals 104
The peasantist interlude 112
The second expropriation 115

VI. LAND LAWS: THE NATURE OF REFORM, GENERAL LEGISLATIVE PROVISIONS 120
1. The expropriation:
The first 122
The second 129
The mechanism 131
2. The resettlement: 136
The distribution of holdings 136
The mechanism of resettlement 140
Communal grazings and woodlands: grazings . . 141
Forests 144
Colonization 147
3. The problem of compensation 148
The expropriation price 151
The resettlement price 156
4. Measures for the protection of production 158
The associations of resettlement 158
The cultivation of new holdings 160
5. Measures for safeguarding peasant property . . . 160
Consolidation 160
The maintenance of peasant holdings 161
Inheritance 161
Sale and mortgage of holdings 162
The restriction of large property 164
6. Variations in the several land laws 164
The main differences 168
Nationalist tendencies 175

PART III

THE APPLICATION OF THE REFORM AND ITS RESULTS

VII. THE EFFECTS OF THE REFORM ON THE DISTRIBUTION OF LAND PROPERTY
1. The change in the Old Kingdom 185
The expropriation 187
The resettlement 191
Criticisms of application 194
2. The effects of reform in the new provinces 200
Bessarabia 200
Bucovina 204

Transylvania 206
The change in the distribution of land among the various nationalities 211
3. Summary of the effect of the reform on the distribution of land property 220

VIII. THE EFFECTS OF THE REFORM ON THE ORGANIZATION OF FARMING 228
1. Changes in the scale of agricultural exploitations 232
2. Changes in the means of production 252

IX. THE EFFECTS OF THE REFORM ON PRODUCTION
1. The change in the distribution of crops 284
2. Present position and the future of wheat and maize . . 295
Wheat 295
Maize 303
3. The effect of the reform on production 306
The cultivated area 308
Production 312
The new provinces 325
4. The effect of the reform on prices and export 338
Export 338
Value and prices 348

X. THE EFFECTS OF THE REFORM ON RURAL ECONOMY
1. The effects on live stock 357
2. The effect on rural industries 365
Domestic industries 365
Agricultural industries 371
3. The effect on the co-operative movement 373
History of the movement 373
Legal status of co-operation 378
Credit co-operatives 381
Federals of popular banks 389
Central of popular banks 390
Consumers' co-operatives and co-operatives of producers . 393
Co-operatives of production 396
Agricultural co-operation 398
Landholding societies 399
Land purchase societies 400
Agricultural societies 402
Co-operation in the new provinces 403
The State and co-operation 406
The new co-operative code 409

XI. THE EFFECTS OF THE STATE'S ECONOMIC AND FINANCIAL POLICY UPON THE WORKING OF THE REFORM. 414
1. The resources of agriculture 416
Working capital as affected by reform 416
The supply of agricultural credit 421

2. The burdens of agriculture 427
Taxation 427
Tariffs 432
Transport 441
3. Mercantilism or agrarianism? 444

XII. THE SOCIAL AND POLITICAL EFFECTS OF THE REFORM. 460
1. Social effects on urban life and institutions 463
2. Social effects in the village 470
The economic situation of the peasantry 475
Material conditions of life 481
(*a*) Housing 482
(*b*) Alimentation 486
(*c*) Alcoholism 490
3. Health and vital statistics 495
4. Cultural and psychological changes 509
(*a*) Education 509
Elementary education 510
Professional education 515
Adult education 520
(*b*) Crime and conflict 525
(*c*) Changing psychology, as seen in religious and social outlook 528
Religion 529
Social psychology and class division 531
5. The Political effects of the reform 537
The time and temper of the nationalist middle-class . 540
The peasant revival 546
The peasant movement 553

CONCLUSION 567

APPENDIX I. Traditional forms of joint landholding . . . 583
II. Métayage in rumanian agriculture . . . 588
III. The financial accounts of the reform . . . 592

BIBLIOGRAPHY 594

INDEX 612

MAPS

Physical map of Rumania xviii

The effect of the reform on the distribution of arable property *at end*

EXPLANATORY NOTES

1. *Geographical Names.*

Geographical names are used here in accordance with Rumanian custom. The province generally called Wallachia is always referred to in Rumanian as *Muntenia*, and this form is adopted in the following pages.

Muntenia included Oltenia. Moldavia included Bucovina until 1774, and Bessarabia until 1812, and again between 1856 and 1878.

Dobrogea is here used in its Rumanian form, and not in the more habitual corrupt forms, Dobrudsha or Dobrudja, with the Turkish termination. It was acquired by Rumania in 1878, and its southern part, known as the Quadrilateral, from Bulgaria, in 1913.

For the sake of convenience, all the new territory beyond the Carpathians, acquired by Rumania after the War, is referred to as Transylvania, though it includes part of the Banat (of Temesvár) as well as the provinces known as Crişana and Maramureş.

2. *Technical Terms.*

The reader who may have occasion to refer to the original sources will find in them two terms which do not appear in these pages, and which have no equivalent in English usage:

Improprietărire, in Rumanian, designates the opposite operation to expropriation, i. e. the parcelling out among the peasants of the land taken over from the landowners; both operations were carried out by the State, on a national scale. To render the word into English as 'impropriation' would have been to give an altogether new sense to an old term: 'to impropriate' meaning in English 'to put the possessions of the Church into the hands of laicks' (Dr. Johnson). Therefore, *improprietărire* has been rendered here as 'resettlement', by analogy with the 'closer settlement' used to describe the action of public authorities which acquire large or medium-sized farms and split them into smallholdings.

Comasare, in Rumanian, describes an operation which has no parallel in English agrarian history, namely, the gathering together, by exchange or reallotment, of a peasant holding which had consisted of several inclosed but widely separated fields (and not, be it noted, the transition from strip farming in open fields to 'inclosed' farming). *Comasare* is not necessarily accompanied by an increase in the size of the holdings, and therefore 'consolidation' appeared the most suitable way of describing it.

3. *Pronunciation.*

Below are given the English equivalents of accents and cedillas used in Rumanian orthography:

â, *î* = have no equivalent in English—slightly sharper than *ă*.
ă, *ĕ* = like the *e* in fath*e*r.
ş = *sh* (as in *sh*out).
ţ = *ts* or *tz* (as in *ts*ar).
c = *tsh*, before *i* and *e* (but = *k* before *a*, *o*, *u* and consonants.)
g = *dj* (as in gentle) when followed by *e* or *i*, otherwise as in garden.
u = *oo* in mood.

4. *Weights and Measures.*

Quintals and *tons* in the following pages are always used in their metric values;
Chilă (from the Turkish *kilè*) is an old corn measure, equal to 679·268 litres in Muntenia and 430 litres in Moldavia;
falce (from *falcem*—literally the quantity of grass mown, in a given space of time) = 80 *prăjini* = 1·43 ha. (approximately one and a half hectares);
pogon (from the Russian) = 5012 sq. m. (approximately half a hectare);
desyatin (Russian) = 1·09 hectares = 2·7 acres;
jugar (from *jugerum*), used in Austria and Hungary (*katastral joch*) = 5754 sq. m. (slightly over 1·3 acres);
hectare = 2·471 acres (approximately 2·5 acres).

INTRODUCTION

THE EFFECT OF THE WORLD WAR UPON THE AGRARIAN STRUCTURE OF EUROPE

AMONG the various social effects of the Great War the downfall of the class of large landowners has been the most outstanding on the Continent. No other effect compares with that either in intensity or extent. It has been active throughout the Continent, and in principle has penetrated even into Great Britain—the last stronghold of 'landlordism'. But in Britain no peasants stood ready to push it over the borderline between program and policy.

As a mere dispossession of propertied people the phenomenon has not been limited to the land. After the War wealth changed hands on an enormous scale, and in all the fields of economic activity. Whole sections of the population, for instance, were ruined by the collapse of the German currency. Similarly with the depreciation of the Lancashire cotton shares. Dr. Hilferding, the financial expert of the German Socialists, said a few years ago that 'never before has expropriation gone to the lengths to which capitalists are driving it now; never has the sanctity of contract and property been so desecrated as during and after the War. In the hands of the capitalists this has led to a gigantic accumulation of private wealth, and even of power over public wealth.'[1] Nevertheless, this vast and merciless transfer of wealth in the industrial West has been taken for granted by the very critics who seemed profoundly shocked at a parallel process in the agrarian East.

It is true that the two processes, though alike in kind, differed in more than one respect. In the West, wealth changed hands mainly among individuals of the same class, by speculation and keen competition; both these methods, in the capitalist system, being perfectly legitimate means towards attaining the perfectly lawful end of piling up wealth. In the East, the transfer of

[1] Speech at the International Socialist Congress. (*Hamburger Echo,* May 25, 1923).

wealth was from one class to another, and it was more or less forcibly imposed either by the State or by revolution. And while in the West the event led merely to a fresh concentration of wealth, in the agrarian countries the result was a wholesale division of the main source of wealth among those who had little or nothing. Moreover, in industry and finance the event changed the beneficiaries without in any way affecting the system, whereas in the agrarian East the division of large property meant a change, likewise, from large-scale to peasant farming. Finally, while the expropriation which Dr. Hilferding deplored goes on day by day on stock-exchanges and in the markets, the War having merely sharpened its action, so that luck or shrewdness might at any moment turn the scales in the victims' favour, the agrarian expropriation has been sudden and final, leaving those who suffered under it without any prospect of appeal.

On the European Continent wealth in the form of land has become rare and may soon be unknown. The War has acted as a sharp dissolvent of a state of things wherein many elements, economic and social, were already working a change. Our ideas on property—perhaps the most stubborn of all social canons—have altered considerably during the last half-century. 'There is no doubt', wrote Duguit, 'that the view which regards property as a subjective right is being replaced by a conception which regards it as a social function.'[1] Theoretically this change of outlook made greater strides with regard to land ownership than other forms of property. Social philosophers were continually pointing out how inadmissible it was that land, which was not created by man's efforts yet was the primary source of his existence, should be owned despotically. In practice, however, the new view was more extensively applied in the industrial field, because the somewhat brutal features of the Industrial Revolution called for increased public control, and the organized pressure of the workers helped to secure it.

In agriculture, this philosophical evolution was reinforced later on by practical needs. The crisis caused by the expansion of oversea corn-growing gave fresh support to Proudhon's plea

[1] *Le Droit Social, le Droit Individuel et les Transformations de l'État.* Paris, 1908.

that a man who tilled a piece of land should own it. Only intensive agriculture could extract from Europe's old soil returns capable of withstanding oversea competition; and intensive agriculture on a large scale proved unremunerative with paid labour. Hence, for the sake of production the trend of European agriculture during the last fifty years has been towards the division of large estates. Even among Socialists those who had specialized in agrarian problems, like Dr. David or M. Vandervelde, reverted to Proudhon's distinction between freehold and possession, and advocated that in the Socialist program the demand for the nationalization of the land should be coupled with provisions for its individual exploitation.

The view, therefore, that large estates should be divided into smallholdings formed the leitmotiv of the agrarian reforms passed in Europe in pre-war years. But such reforms were few and feeble. In most European countries, more especially in the East, the landed classes retained sufficient political influence to obtain favours from the State by means of which they could continue to profit from extensive cultivation and at the same time ward off the pressure of the landless peasants. But the War severed that ancillary connexion between large landowners and State, and as a result the flood-tide of agrarian reform, held up by political devices, was freed. Redistribution of political power, it must be noted, and in some cases revolution, preluded the reforms; even in victorious countries, in spite of the fact that military victory often strengthens the power of the ruling classes, they had to be adopted.

There is no doubt that in certain agrarian regions the change received impetus from the fact that the bulk of the landowners were not of the same nationality as the mass of the peasants. But the results were not very different where the landed class was autochthonous, as in Russia and old Rumania. The character of the reforms was determined not by nationality but by social relationship. However consonant with economic needs, philosophic creeds, or at times nationalist prejudices they may happen to be, they are firstly a social phenomenon. They mark the fall of the landlords, and out of the social cataclysm—caused primarily by the Great War—the triumphant emergence of the peasants.

The various reforms passed on the Continent after the War represent different stages in these two results. At one extreme is Russia. That country has done away with large property altogether, and with no compensation whatever to former owners. Then, most of the countries bordering on Russia, as well as Yugoslavia and Czechoslovakia, have acted on a similar principle, dividing the greater part of the large estates among the peasants; in their case the former owners received compensation, but always less than the actual value of the land. Finally, in central and western Europe—in Austria, Hungary, Germany, &c.—the reforms have aimed merely at facilitating the expansion of peasant farming; they have changed, that is, the details but not the basis of the existing agrarian organization, and the land has been purchased for the peasants at current prices.

In pointing out these differences Professor Max Sering explains that the first group of reforms followed the Russian model, while the second followed the European.[1] If this is meant to imply that the reforms were moulded by the more or less 'civilized' outlook of their makers, the remark fails to reveal the real cause of the difference. The range of each reform was determined by the state of the agrarian system which it had to correct. In central and western Europe, where preferential measures in favour of the large owners had to some extent deflected the action of economic factors, relatively mild reforms sufficed to redress that legislative bias. But in eastern Europe the landowners had retained a feudal hold on the social and political life of the region, and only reforms of revolutionary dimensions could bring that state of things up to the level of the more advanced part of the Continent. The reforms had to retrieve ground, in a greater or lesser degree, in the measure in which the progress of agrarian conditions had been retarded. In eastern Europe their task was nothing less than to complete at long last the demolition of feudalism begun in the West by the French Revolution.

The main post-war reforms, therefore, are in direct line with the great nineteenth-century measures which emancipated the

[1] Introduction to *The Agrarian Revolution in Europe*. (In Russian.)

peasants. The two groups form part of a continuous movement, only now completed; yet between the two are differences and even contrasts which serve to reveal the startling change which the position of the peasant has undergone in the interval. The earlier reforms were in the main the achievement of the new Liberalism, which was moved to work for the emancipation of the peasants by factors which were transforming the outlook and life of that period. Their humanitarian philosophy deprecated all restrictions on personal liberty; Constitutional government demanded the equality of all citizens before the law; and the new economic doctrine required freedom of movement for labour and capital alike. But in our time the new reforms have been carried through single-handed by the peasants themselves, running rather contrary to prevailing social and economic tendencies, and taking no account certainly of economic conditions. One of the chief results which the middle-class expected from the earlier emancipation of the peasants was the release of a supply of labour and of cheap foodstuffs for the expanding industrial cities. The new reforms, which have given so many peasants a fresh chance on the land, tend to reduce the supply of labour, and this just in those less developed countries which aspire to create an industry of their own. Moreover, formerly the towns were glad to have the villages behind them in their fight against the strongholds of feudalism, as in 1848 for example; but because the present reforms swing the political pendulum back towards the country-side, the former alliance has given way to acute antagonism between country and towns.

The two groups of reforms show equally strong contrasts in character as in background. The main features of the nineteenth-century measures were, for the peasants, of a negative character. They were freed from feudal servitudes, but they had to pay compensation either in a lump sum or in taxes, and also they generally lost some of the land which they had formerly farmed, as, for instance, in Eastern Prussia, where a large rural proletariat only came into being after the reforms—an effect which gradually had to be remedied by successive measures for closer settlement. Quite otherwise in the twentieth century. Whatever privileges the landlords had enjoyed have been utterly swept away, without any compensation, while the peasants have received most of the land on very easy terms; everywhere the new reforms have raised the peasants' status and standard of living. But while under the earlier measures the peasants were the losers, real benefits were reaped by the landlords. Their estates were freed from servitudes in a period of expanding corn-trade and rising land values, transactions in land were made free and the way opened for the concentration of landed property. The new reforms have well-nigh ruined the landowners—latifundia have been destroyed altogether—and the new laws contain provisions which restrict the sale of land and bar the way to any remaking of large private estates.

The gulf which separates the two related groups of reforms appears very clearly when one compares the more extreme incidents in each of them. M. Leroy-Beaulieu observed as early as 1898 that 'a Russian Revolution might be the greatest historical event since the French Revolution, representing in a way its counterpart at the other end of Europe'.[1] It is very true that the Russian Revolution, which released the spring of the new land reforms, is the eastern counterpart and complement of the French Revolution. But the span of time which separates the two events explains the great change in their positive effects. Both of them are milestones in the social progress of Europe, marking the successive breakdown of feudalism in the West and in the East. But the one took place on the threshold of the Industrial Revolution, with its impetuous unchaining of the forces of production and trade. The other stands at the beginning of a Social Revolution, which will as impetuously press forward the problem of distribution. The one, though helped by the masses, was in the service of the urban middle-class; during the French Revolution land was sold to the peasants merely for securing revenue, and it was divided into smallholdings only as an afterthought and as an adjunct in the party struggle. But the Russian counterpart has turned out a revolution for the peasants, while the share of the apparently dominant urban class is secondary and incidental. Nothing conclusive could be proved by comparing the professed Jacobin individualism of

[1] *L'Empire des Tsars et les Russes.* Paris, 1897–8, vol. ii. p. 624.

the first with the professed Bolshevik communism of the second. As a French writer has pointed out, the Jacobins, who claimed to defend property and threatened with death those who propagated the 'agrarian law', that is the Communists, had in fact requisitioned, confiscated, and expropriated.[1] The Bolsheviks proclaimed the extermination of individual property, but in effect they have contributed to its enormous expansion and consolidation in Russia and in the neighbouring agrarian countries.

All things considered, therefore, the French Revolution in the West and the Russian Revolution in the East are two vastly dissimilar specimens of the same genus. The old worlds which both demolished were alike, but there is little in common between the new worlds to which each has given birth. Both put an end to aristocratic rule, maintained with the revenue of feudal estates. But the first prepared the way for the rise of the capitalist middle-class, whereas the second heralds the political predominance of the working-class. The new movement having taken rise in agrarian regions, the difference between the two events is best seen in their effect on the peasant. The first did no more for the peasant than to release him from his servile fetters. But the second has placed upon his shoulders the mantle of power.

In Rumania it was not until 1918 that the right of the peasant to till his own fields first was formally recognized, through the law for compulsory labour passed in that year. Until then, the mass of the peasants had in practice remained bound to the landlords. Throughout the evolution of the political régime, from Turkish suzerainty to national independence and from autocratic principalities to Constitutional kingdom, the nature of the peasant servitudes had remained the same; only the form in which they were imposed had varied, to fit the legal system of each period. The creation of the Rumanian Principalities found the villagers as free joint-holders of the village lands, burdened with no other duties than that of giving the village headman one-tenth of the produce and three days' service in the year. In time, however, that yeoman status was more and more

[1] A. Mathiez, *Le Bolchévisme et le Jacobinisme.* Paris, 1920, p. 14.

encroached upon, until the peasants were pressed into serfdom. But their decadence did not follow from the political trials which befell the whole region after the coming of the Turks. On the contrary, each downward step in the peasants' social status corresponded rather to some moment of recovery in the country's political status.

The first prince to turn the mass of the peasants into villeins, at the end of the sixteenth century (in Muntenia), was Mihaiu the Brave, who also equipped the first professional army, and, with its help, roused the Rumanian provinces to a last flicker of independence before they finally succumbed to the Turks. Whereas the first to decree the formal abolition of serfdom, about the middle of the eighteenth century, was one of the ill-reputed Phanariote princes, Constantin Mavrocordat. During the long period of subjection to the Turks the peasants shared in the general misery, but their trials were due to chronic lawlessness, and to sporadic abuses by the landlords, and not to any formal depression of their class. Throughout those centuries their land rights remained intact. Their servitudes increased only slightly, as the landlords were kept in check by the ruling powers. But step by step, as the provinces escaped from Turkish rule, the peasants were reduced to such a state of serfdom as they had never known before. The first acts to prelude the end of Turkish rule and its replacement by a Russian protectorate (the Treaties of Kutshuk Kainardji, 1772, and of Iassy, 1792), and the return of native princes (Convention of 1802), were followed by a severe increase in the peasants' labour dues and for the first time by the restriction, especially, of their rights to land. When the provinces recovered their political autonomy (Convention of Ackermann, 1826, and Treaty of Adrianople, 1829) there was immediately a fresh increase in labour dues and other servitudes, and a further restriction of the peasants' land rights; and this being the first occasion on which the boiars forming the national *divan* were allowed to legislate, they used it to transform their usufructuary title to the land into one of full ownership. When in the 'sixties of the last century the new State acquired independence, the peasants were formally emancipated, in compliance with the injunction of the Great Powers; but the holdings of the peasants were again reduced, to a point where they no longer offered sufficient means of subsistence, while labour servitudes were promptly reintroduced in practice by means of the laws on agricultural contracts. At the same time the landlords were released of all their obligations towards the peasants, and many of them were allowed to retain land which was by ancient right the portion of the villagers. If the fate of the peasantry be judged not by legal texts, but by the moral and physical conditions of life on the land, then there is no doubt that the peasants reached the worst degree of economic and social bondage after their emancipation, under national government and· the formal rule of Constitutional liberties. The elementary conditions for the real emancipation of the peasants were created only as a result of the War, after the Russian Revolution, by the agrarian and electoral reforms proclaimed in 1917 and enacted after the armistice.

The case of the Rumanian peasants, because it happens to be peculiarly flagrant, illustrates very well the difference between the two categories of rural reforms. The emancipation of the peasants in 1864, which in fact left them in every respect worse off than they were before, but which opened the way for the enrichment of the landlords, belonged to the group of measures through which *laissez faire* was introduced into agriculture in the first half of the nineteenth century. The new reform—the special subject of this study—belongs to the group initiated by the War and the Russian Revolution, and its effect has been to oust the landlords and to leave the peasants in control of agriculture.

The specific features of the Rumanian case also help to prove the second premiss—that the mainspring of the new reforms was political and that it was released by the Great War. The nineteenth-centry reforms were in the West part and parcel of the transformation of rural economics, but they were everywhere pressed by the monarchies, which felt the need of curbing the power of the landlords. Even in Russia. But in Rumania, after the wilting away of the Turkish domination the landed class *was* the central power. There was no middle-class to counter the will of the landlords, nor a ruler with enough authority to check it, as the position of the new foreign dynasty was insecure. Hence with every release from foreign control the strength of the landlords increased, and the burdens which they laid upon the peasants increased in the same degree. It was this political supremacy of the landed class which made possible the continuance of a servile system until well on in the twentieth century; and it was the snapping of this supremacy by the War which made way for the new reform. The War had not otherwise dislocated the country's economic and social organization: the whole structure of Rumanian society having remained as it was before, there was no organic reason for the breaking up of its feudal-agrarian foundations had its chief beneficiaries been able to preserve them. The anxious efforts which they made to re-entrench themselves on the heights of State after the reform, in order to retain as administrators the control they had lost as landlords, prove how abnormal had been the granting of 'land and liberty' in 1917.

From whatever angle, general or particular, one regards the post-war land reforms, it is their social aspect which stands out, as a revolutionary change which will leave its mark upon the future history of Europe. Nor is that outcome of the World War perhaps without a wider political significance. Eastern Europe has had an inordinate share of invasions and wars and revolutions. This explains in a large measure the economic and social backwardness of the region. Yet it is possible that the present forcible redress, caused by the stress of yet another and fearful war, may extract some good for the world at large from the fact of its having been retarded so long, if what Professor Seignobos said in 1919 proves as true in the future as it has proved in the past. Writing on the 'Downfall of Aristocracy in Eastern Europe' he said: 'We seek guarantees against a return of the war spirit. What régime is more pacific than a democracy of peasant proprietors! Since the world began, no such community has ever desired or prepared or commenced a war.'

PART I

THE AGRARIAN PROBLEM IN RUMANIAN HISTORY

CHAPTER I

SOCIAL FREEDOM AND POLITICAL DEPENDENCE

As a nation, the Rumanians have suffered more than their normal share of warring and duress, but those frequent political storm-clouds have not been without a measure of silver-lining to them. More than once their shadows have given shelter to popular institutions which might have faded beneath a brighter political sky; popular customs and rights have somehow survived through ages when rulers were wont not to tolerate them, because in those times either rulers could not exist in the Rumanian regions, or had more than their work cut out to keep themselves above the flood of invaders. One can illustrate the paradox with the marvellous fate of the Bessarabian peasants, in our own time. About a century ago, Russia tore that province away from the body of the Rumanian Principality of Moldavia and set about turning it into a loyal dependency. It had a hard life under the wilful rule of the Tsars; but because these princes would not let any ray of light penetrate to the masses, either by way of self-government or of education, a whole century of alien discipline has left the life of the peasants untouched. They now return to Rumania as after a long winter sleep, with all the ways which their ancestors took with them in 1812—with, as they still call it, their 'Moldavian' language and customs and traditions. And this is but the most recent instance of the curiously double-sided fate which has been the lot of the Rumanian people during fifteen centuries and more.

From the time when the Romans, bending before the onrush of the barbarians, withdrew their legions from the region between the Danube and the Carpathians, abandoning what had been 'Dacia Felix', about A. D. 270, its inhabitants fell victims to an unbroken chain of alien invasions and intrusions; more of them, probably, and more varied than those which have passed over any other people of Europe. Yet in spite of that, or, as I believe, largely because of that, the mass of the people was able to preserve its customs and ways to an amazing degree. The distinctive costume which the Rumanian peasant wears every day is still identical, from headwear to footgear, with that which common people, as shown on the bas-reliefs of Trajan's column, had worn when the Romans conquered the region eighteen centuries ago. The Romance language which they talk has survived, as a mystery island in the Slav ocean which surrounds it, and which in the political and religious and other fields has more than once submerged the whole region. And so it has been with many customs of the soil and with the ancient rights of those who tilled it. As long as the land was a prey to alien invaders and oppressors, its rulers were hard pressed to preserve their estate; they could hardly find means to try to increase it at the expense of the masses, nor could they take the risk of rousing their enmity by such an attempt. In the general misfortunes of the country the Rumanian peasant, no doubt, has had to bear a heavy share of misery; yet he, at any rate, kept the freedom of his own person and his rights to the use of the land in times when in the civilized West his fellow peasants were being pressed into serfdom. Not till the beginning of the nineteenth century, when the Rumanians began to enter upon the road to national independence, was the peasantry formally deprived of its title to the land and of the right to move on it freely. But this subjection, which could not come before, had fortunately come too late. The system was misbegotten when everywhere the towers of feudalism were tottering or were being successfully stormed. For a while it was able to vegetate in that uncongenial period, but not to spread solid roots; and it broke down, therefore, after the Great War, more speedily and completely than elsewhere under the pressure of the reawakened masses.

The Middle Ages. Land and people. As soon as the Roman legions had abandoned the forts and entrenchments which protected the eastern confines of the Empire, the barbarian hosts broke through the gap between the Black Sea and the Carpathians and flooded the plains which stretched below the mountains to the Danube. These invaders seemed as innumerable as they were insatiable. Wherever they passed little was left above ground of whatever they could consume or carry away with them; and anything else that stood in their way merely roused their passion for destroying, as they passed onwards or retreated in their expeditions. Nothing was safe from them except that which was too lowly and weak to tempt the greed or pique the temper of these wild nomad warriors.

Walls or princely power could not stem the invasion; they only drew upon them the blows of the invaders. And so the people of the region lost the art of living in cities, as well as that of organizing in greater numbers for better government, during the many centuries in which the barbarian invasions continued. Public and private life in former Dacia was reduced to very humble and simple forms, as a means of natural self-protection. The old political hierarchy and organs faded away, and so did the inner articulations and the outer boundaries of the country. Its sorely harassed and badly reduced inhabitants lived in small disjointed communities more like tribal groups than anything else, except that they retained a sense of their kinship and of common attachment to a vast and vague *Ţara Românească* (Rumanian land). The organization of each village was equally simple. Life and wealth of all being equally exposed they all had to stand together; there was no room for stiff hierarchical forms. Money economy did not exist; the division of labour was rudimentary; and in general economic and social life was bound to be of the simplest when the village had always to hold itself ready to fly with children and chattels before some approaching invader. In the absence of elaborate and active central authorities requiring considerable supplies and men for their maintenance, the heads of the villages did not acquire till later on in the period the power to oppress and to punish which elsewhere the executors of more exacting masters derived from their functions.

The Rulers. Under the restrained and fragile rule of the *domni* each village was in effect a self-governing unit, a *judeţie,* in charge of a *judeţ* (*judex*) the most active authority of the time.[1] The *judeţ* united in his person all the functions of a feudal lord: he led the men of his *judeţie* in war; he judged among them in peace; he gathered taxes for the ruler, or later for the passing

[1] The *judeţi* were themselves divided into three classes: (1) the *cnezi,* descendants of the old noble caste, held the charge by right of heritage, and their children had an equal right to rule over a proportionate number of peasants; (2) the *juzi,* elected or appointed, held the office temporarily; (3) the *vatamani* probably were bailiffs appointed by *cnezi* who had jurisdiction over more than one *judeţie.*

barbarian masters. In return for these functions, besides being himself exempted from taxes, he was entitled to one-tenth of the produce raised by the villagers and he could claim from each peasant three days' labour in the year. He also had the sole right to mill and to sell strong beverages. But with this his privileges and prerogatives ended—far short of the point reached by those of feudal lords in the West.

The Ruled. The chief difference from the West lay in the status of the peasantry. The peasants were altogether free in their persons and in their possessions. They could move whenever and wherever they liked, tied in no way to the soil or to the *județ*, except by the duty of paying him a reward in labour and kind for services rendered. There was already, however, an exception to that general rule. Most of the villages had a class of settlers called *vecini* (neighbours), who had probably been made prisoners in war. These people, unlike the native peasants, were bound to the land, and they had also to labour many more days than the others for the *județ*. Originally this servile class was very small, but it appears to have grown in time, misfortune causing some of the free peasants to barter their freedom away for a measure of protection or reward, and thus to become *vecini* tied to the land. It was at worst only a mild form of serfdom; it was mitigated in effect and extension by the custom which passed on its burdens only to the male children (perhaps because the orignal *vecini* were all men, captured in battle, but their wives free native women); and in the all-important matter of a right to use the land the *vecini* enjoyed exactly the same rights as the free peasants.[1]

[1] The origin and character of the status of *vecin* is still in dispute. For a more recent summary of the various arguments, see G. N. Leon, *Istoria Economiei Publice la Români*, Bucarest, 1924, pp. 57–72, 98–110, and 115–20. M. Leon divides the Rumanian peasantry into three categories: (1) *moşneni* or *răzeşi*, who owned land and were free in their persons, paying only tithe and a limited labour due; (2) *clăcaşi*, who were free, but had no land of their own and worked 3–24 days in return for land; (3) *rumâni* or *vecini*, who were serfs with or without land, owing service without limit; they could be sold with or without the estate on which they lived, but serfdom did not include their women and children, in which respect they stood better than serfs elsewhere. This remained the social structure till the reform of Mavrocordat (1746–9).—M. A. D. Xenopol, in his article on the *History of the Agrarian Question*, points out that when a serf purchased his freedom, the customary formula said that he 'was purchasing himself free of *rumânie* with all the land he possessed, in return for a payment'. Judicial decisions likewise show that the land of a *rumân* passed into

Land Tenure. The tenure of land was regulated by ancient customs forming a body of unwritten 'Rumanian law' (*jus valachorum*). Each village disposed of certain stretches of land which together formed the *hotar* (boundary) and were the common patrimony of the village as a whole. The bulk of the *hotar* consisted of meadows, grazing and woodland, and these were used jointly by the whole village. The arable land was divided into a number of equal strips, called *jireabie* in Moldavia and *delniţe* in Muntenia, decreasing in size as one travelled from the plains to the hills, each household being entitled to one of these strips. Grazing land was common. The *judeţ* may have been entitled to more than one strip, but there is no evidence that there was any reserved 'demesne'. Neither had the *domn* any superior title to the land; his rights consisted in a claim to the tithe, augmented by voluntary gifts on festive occasions. There was sufficient land with which to endow new households and the strips were indivisible. House, garden, plantation, and arable land were regarded as the private possession of the individual householder, passing in heritage to the youngest son.[1]

Such, very broadly, was the structure of early Rumanian society, exclusively rural and egalitarian. The bulk of the people were of the same origin and led the same life; there was but the merest class division, based less on power and wealth than on a rudimentary division of functions; and as long as the supreme hope was to escape from the dangers of the time with bare existence, all were united by the same elementary interest, which forbade rivalry and strife. Nor was there any reason to compete for possessions. As far as the ordinary needs of life were concerned, the peasants could have lacked

the hands of the winning side, and, therefore, that the serf did not lose his land when he sold his personal freedom.

[1] Professor Iorga adduced philological evidence, in a paper submitted to the International Congress of Agriculture (Bucarest, June 1929), to show that the oldest agrarian régime among the Rumanian people was that of an agricultural community. There is in the Rumanian language no word of Latin origin to indicate a boundary, other than *margine*, which refers to geographical and not to legal limits. The Latin words referring to individual possession, *finis* and even *limes*, are completely lost. Their meaning is contained in two words of alien origin: *graniţă* (from the German *Grenze*), which now serves to indicate the limits of the State, and *hotar* (from the Magyar), generally used in reference to limits of property. The word *câmp* does not mean the determined estate of a private owner, but merely the arable soil under cultivation. (*Evolution of the Rural Question*, p. 2.)

little if anything at all. They had a right practically to the whole of the land and to the whole produce of their labour. Their superiors in the village only took whatever land was left over after the needs of the householders had been satisfied, and they could not have aspired to more, nor did they till much on their own, for there was no market for the surplus; and what they received in kind from the villagers amply satisfied their personal needs. At the same time, the tithe of corn and hay, and the three days' labour in the year, must have rested lightly on the shoulders of that free and frugal peasantry.

However terrible, then, life must have been between the Danube and the Carpathians under the constant dread of barbarian invaders, there was in that external instability a peculiar safeguard for the personal and economic freedom of the peasants against possible internal oppressors. All efforts to build up power or possessions were so unpromising as not to be worth while. And it therefore happened that the old sterling customs persisted unspoilt as long as, for these reasons, there were no scribes to record them on parchment and no functionaries to dictate their enforcement. Agrarian discords began to germinate below the Carpathians only when quieter times dawned for the region and the inhabitants could come together to found the Rumanian Principalities.

Foundation of the Principalities. When the tide of barbarian invasions receded, two vassals of the Hungarian king crossed the mountains and gathering together the broken ties between the isolated rural groups, established the Principalities of Moldavia and Muntenia, in the thirteenth and fourteenth centuries respectively. Whether it was the lure of old transmuted memories which set these people moving from their homes, or whether they were simply wandering in search of a more fruitful soil on which to settle, one does not know; but it was in any case not a military expedition and all the circumstances of that achievement point to a peaceful progress.

Neither of the two founders came with great followings, and the Moldavian prince set up his rule against the will of his Hungarian suzerain, so that they had to rely upon the friendly welcome of the established populations rather than upon the prowess of whatever armed men they had in their train. It was a gathering of closely kindred clans rather than a conquest of the weaker by the stronger, and popular tradition has recorded that by calling the event pithily the *descălecare* (the dismounting). The new men crossed the Carpathians by one of the high passes, travelled to where the valley opened into the plain, where they came upon some of the larger villages, and having proclaimed their purpose they dismounted from their horses and settled down among the people they had come to rally. The knowledge that the first Rumanian States were created in such peaceful ways is important because it explains why all the essential customs of the people, though merely traditional and unwritten, were left untouched by the new political masters. Of these customs, evidently none concerned the people so much as their right to the land, a right which remained unaffected by the historical events that were taking place.

No written law concerning these rights, or any others, existed either before or immediately after the foundation of the two principalities. Their continuation is proved thereby at least in a negative way, for any change would no doubt have been decreed in writing, as were other matters henceforward. The first written documents belong to the end of the fourteenth century in Moldavia, and the beginning of the fifteenth in Muntenia; they were mainly concerned with confirming existing rights of *judeţie*, now a vassalage under a firmer ruler, which it was therefore well to have ratified by him. Later, written judgements of the princes, dealing with cases in which land belonging to one village had been encroached upon by another, show that it was always the people of the injured village, and not its *judeţ*, who pleaded the claim against the usurper. And there is still better proof to be found as to the persistence of the old land customs in the oldest known written laws of the Rumanian people, the *pravile* of Vasile Lupu and Mateiu Bassarab, neither of which know anything of land disputes between peasants, individually or jointly, and *judeţ*, but deal in regard to land tenure only with boundary disputes between villages as a whole. There is indeed no record of any dispute between villagers and *judeţ* concerning the right to the use of the land as long as the period of native rule lasted. Up to the end of the seventeenth century the *hotar* must invariably have been regarded as belonging jointly to the village community.

In the meantime, however, as we shall presently see, the idea that the title of the *judeţ* to the *hotar* was superior to that of the mass of the villagers, had been gaining ground, and it had penetrated also into the two old laws of Vasile Lupu and Mateiu Bassarab. But it was as yet strictly limited by the right of each household to a piece of land, and by the strict rule, contained in these laws, that the man who tilled the soil with his labour and his seed was bound to give a tenth of the produce to him who gave the land, but never more than that. Land tenure, in short, continued to rest on the principle that all the inhabitants had an equal right to the use of the soil, as well as to the bulk of what they produced with their labour. But with the founding of the two States new and more complex political and social conditions were created which began to eat into that just ancestral system. The migrants from Hungary had brought with them knowledge of the feudal customs which prevailed in the country from which they came; and, in addition, Byzantine influence was introducing in Muntenia the conception of property as embodied in the old Roman law. Upon the prevailing relations, based on equal and independent rights between the local magnates and the peasants, there were now being grafted the relations based on personal service and liegedom between those magnates and the central rulers—a feudalized superstructure upon a patriarchal base—and in time the new was bound to affect what remained of the old.

Relations between princes and cnezi. Altogether, the documents of the fifteenth, sixteenth, and seventeenth centuries show that formally the foundation of the principalities in no way affected the traditional Rumanian law. As far as the letter of the law was concerned, *cnezi* and peasants kept the status and the relations in which they had formerly stood to each other. But in reality the natural background of their relations had inevitably shifted as soon as central political rulers had come upon the scene. If it took some time before the effect of this wrought itself out upon the status of the peasants, on that of the *cnezi* it reacted at once. The centre of gravity of their class moved from the village to the court. The confirmation of their existing rights and, especially, the granting of new privileges were henceforward in the hand of the princes; while, in their turn, the princes had to depend for pomp and power upon the service of the upper class. A new sun had risen, but its light and warmth were intercepted by the satellites, whose shadow alone spread ever wider over the villages.

Whether the princes brought with them knowledge of western feudal custom or not, it was in any case only natural that if the rights of existing communities over their *hotare* were respected, such land as had not yet been appropriated should come to be regarded as within the lordship of the prince. Likewise, those villages which stood under a temporary *județ* were henceforward considered as falling within the jurisdiction of the princes, as *sate*[1] *domnești* (princely villages). So that at least two categories of *județ* no longer owed their status to hereditary rights or to the trust of the village, but to the favour of the princes. Henceforward it was the prince who granted the *județie* over the *sate domnești*, for a certain period or for life, or, more rarely, as an hereditary gift, to those who had served him or pleased him. And it was the prince who from the wastes so far unappropriated occasionally allotted to individuals or to monasteries fresh *hotare* with the right to set up new villages and to exercise hereditary *județie* over them. These new villages usually were known by the name of the founder—satul Albeștilor, satul Negreștilor, &c., which is the origin of the present Albești, Negrești, &c.—and originally their *județie* could not be alienated from the family of the founder. It was, therefore, clearly a privilege rather than a property; and the princely *urice* (deeds of gift) explicitly prove, indeed, that they gave away the *județie* of the village but not the ownership of its *hotar*.

Emergence of upper class. With nothing altered, therefore, in the outward status and legal rights of the *județ*, subtle changes were nevertheless transforming them into an upper class with interests and claims and customs increasingly diverging from

[1] *sat*=village (in Albanian *fsat*) is of Thracian origin in the opinion of O. Densușianu, but M. Bogrea traces it back to the Latin *fossatum*.

those of the common people. At first, perhaps, those changes were more psychological than material. With the creation of central authorities, dispensing justice and favours, the *judeţ* had to look to the princes and no longer to the people for advancement; they began to leave the village, seeking functions at Court, for which there was ever-growing opportunity. Wars and the frequent changes of rulers offered chances for profitable adventures or intrigues. Contact with foreign peoples led to changes in customs and dress, and thereby to more obvious differences from the common people of the land. One can well see, moreover, how the building of a new village, held from the prince and named after its founder, must have given him and his family a proprietary sense, if not a right, over the settlement. Similarly, though such gifts of *judeţie* were relatively few, the confirmation of existing hereditary rights were many, and as these deeds came to be recorded in writing, while the rights of the peasants remained customary, this also helped to give a more solemn ring to the titles of the upper class; and claims passed on merely by word of mouth could more easily be denied or garbled, by those who held power, than rights laid down in black on white. Altogether, the new life gave the upper class fresh ambitions and new powers, which they felt less reluctant to abuse as their ties with the villagers loosened.

Some of these abuses were sporadic, the successful tricks of individuals for their own profit and to the loss of a particular village. In some cases, e. g., they obtained the prince's sanction for selling their *judeţie*, or, what was more serious, part of the village land which had not yet been put into cultivation. But these were isolated wrongs, and their effects not oppressive, as land was still plentiful and its value altogether depended on the people settled on it, who had to be conciliated. A real agrarian problem began when the new ruling faction set about extending its wealth and privileges, as a class, at the expense of the peasantry. The various functions known as *boierii* (Russian—*boyar*; Slavonic—*bolyar*), created by the new rule, were a monopoly of these men, who thus came to be known as the *boiar* class. On the other hand, their segregation as a land-owning upper-class was furthered by the uneasy early life of the two

Principalities. Invasions were now rarer, though not ended, but wars for the throne drained many a village of its population. Peasants from less fertile lands would move to such emptied villages; and great efforts were also made to attract foreign settlers for repopulating old villages or for building up new ones, the princes usually exempting these new settlements, called for this reason *slobozii* (=freedoms; Slav=*svobodi*, Bulg.=*sloboden*), of all taxes, for a number of years. These foreign immigrants were settled always with the status of *vecini*; moreover, they came from regions densely populated and more developed economically, where in consequence the rule of the landlord was heavier, and the Rumanian landlords now treated them likewise. It was inevitable that gradually that sterner treatment should be extended to the native inhabitants.[1] Land was held in these new or repopulated villages by the same rules as in the old, but the newcomers not unlikely looked upon the *județ*, from whom each received his part, as a dispenser of strange possessions, with power to give or to refuse—in short as a master, and so they began to know him as the *stăpân* (=master; Slav.=*stopanu*). Socially and politically, therefore, the boiars were drawing away from the village; and the same circumstances which brought this about also welded them into an upper class with privileges to defend and increasing needs to satisfy.

The beginnings of serfdom. One can give here only the briefest sketch of the motives which impelled the boiar class to covet the land and the liberty of the peasants, and of the means by which they succeeded in grasping them. It was not a short and sudden process. On the contrary, it spread over several centuries. The boiars continuously nibbled at the patrimony of the villagers as their own needs and chances slowly rose, during the despondent period of Turkish domination; but it was only after the repulse of the Turks, when the country enjoyed fresh security and when more active relations with the West gave value to its surplus of produce—then only was a final determined assault made upon

[1] St. Antim, *Chestia Țărănească*, p. 96. On the other hand, colonies of Rumanian peasants were found in Galicia governed according to the *jus valachorum*, which in Poland constituted a privileged régime. (Rosetti, *Pentru Ce . . .* p. 73.)

the land and labour of the peasantry, well on in the nineteenth century.

It has been customary with writers on this subject to see two periods in the action which reduced the Rumanian peasantry to serfdom. The dividing line, on this view, was formed, roughly, by the Turkish conquest of the provinces, the peasants retaining most of their old rights that side of the line, when strong native princes still ruled over the land, but gradually losing them after the line was crossed and government passed into alien hands and decayed. This reading of events would be justified in so far as the peasants were the likeliest prey for the fabulously corrupt administrators which, under Turkish rule, dropped like locusts upon the land. When everyday life was one long abuse it must have been hardest for the peasant, for he had little to spare and could not make good his loss from any one lowlier than himself. But these were the accidentals of the time rather than of his standing, and in any case not the springs of his fall in status. For these one must look to the circumstances which shaped the country's social evolution, such as those we have described as creating an upper class with characteristic ambitions and powers. The first formal inroad upon the independent status of the peasantry followed indeed a material step in the organizing of the central power, on its military side, and it was made when that power reached its highest glory, before the Turks finally crushed it.

The loss of freedom. In three highly instructive papers read by General R. Rosetti before the Rumanian Academy,[1] he has shown that in the second half of the fifteenth century the armies were based on the general duty of all those who owned land, or had a right to use it, to share in the defence of the country; and, further, that the ordnance services were of the simplest, because the peasants were obliged to bring their own arms and food with them whenever the alarm was raised. In other words, in a time of continuous warfare the country's defence was based on a *levée en masse*, that is, on the goodwill of the peasants and on their ability to keep themselves and their horses provisioned; and this

[1] MEMOIRS OF THE HISTORICAL SECTION, Series III, tom. iv, Mem. 9, 1925; *Studies on the Manner in which War was made by Ştefan the Great*, 1454–1504.

presumed that they had to be allowed if not the whole at least a wide margin of their traditional economic and social independence. This military system was radically changed, however, by Mihaiu the Brave (of Muntenia, 1593–1601). Partly because of the need of opposing a more disciplined force to the increasing pressure of the Turks, and partly because of the roving ambitions of this warrior-prince, Mihaiu the Brave for the first time equipped a standing professional army. In this the trained and better equipped *cnezi* and *judeţi* no doubt played an important part. On the other hand, most of the peasants were no longer needed as willing soldiers, but they were needed the more as compliant labourers, as the central power now required considerable supplies for its military establishment, as well as for the payment of the heavy annual tribute to the Sultan. And so it was Mihaiu the Brave—the only Rumanian prince to have achieved the feat of bringing all the Rumanian lands under his sceptre—who at the height of his power bound all the peasants to the land; though, according to M. Phillipide, Mihaiu merely generalized and gave legal sanction to a custom which was already widespread. He decreed that all those who were settled on land whose *judeţie* belonged to another individual, which was the general rule, should be *vecini*; and then it was that the name of *rumân*, by which the peasant was commonly known in Muntenia, came to mean the same as *vecin*. In Muntenia, serfdom became the normal status of most of the peasants.

The conditions of the peasantry have not always been identical in the two provinces. Serfdom spread more rapidly in Muntenia than in Moldavia; on the other hand, it would seem that the Muntenian *vecin* never suffered the iniquities of which his Moldavian fellow was the victim, under the influence of the oppressive customs which were profitably employed by the neighbouring Polish nobles. Most of the time there was some difference between the position of the peasants—both in law and in practice—in the two provinces, but this was mainly in details; so that in such a general sketch as this it is more convenient to treat the two groups as one class, except where differences between them become deeper and more characteristic.

Even without a general decree the reducing of the peasants

to the status of *vecin* was proceeding apace, either by fair means, the peasant consenting to it, or by the use of false witnesses who were made to declare that this or that peasant was descended of *vecini.* In Moldavia the *vecin* was free to go wherever he liked if he could prove that he or his parents had sometime owned property, however small, but this proof was difficult to give and made the peasant's freedom illusory in practice. Formerly the *vecini* had been an exception, but they formed a majority by the beginning of the eighteenth century, and during the first half of that century the bulk of the peasants fell into villeinage. The boiars even attempted to have all the peasants who had lived twelve years in one of their villages declared as *vecini.* But the Prince, Constantin Duca, by no means one of the best, rebuked them severely, for even 'the pagans purchase their slaves and set them free after seven years or sooner; but you are Christian, and pay not for him, and he being Christian like you, yet you would make him a serf for ever'.

The first act in this process of subjection was abruptly brought to an end by Constantin Mavrocordat. The many taxes which the Muntenian peasants had to bear during the Phanariote régime were increased considerably by the abuses of the collectors; there is evidence of this in a decree of Constantin Brăncoveanu who threatens with hanging those collectors who should take more than 'is written'. These abuses caused whole villages to be broken up and flee across the Danube, which reduced the sources of public revenue. Between 1741 and 1746, 77,000 tax-paying families left Muntenia, out of a total of 147,000; for in the democratically organized Ottoman society the rights of landlords over the labour and person of the peasants were more liberally circumscribed. This led Mavrocordat to decree, in 1746, that 'those who return to their lands shall be freed of *rumânie*'. As ruler of Moldavia, Mavrocordat decreed the emancipation of the *vecini* in the northern province also, in 1749. Serfdom, therefore, was abolished earlier here than in the advanced West; and it was abolished at the height of political decadence.

The *urbarial* system inaugurated by Mavrocordat was a mixed régime, reducing serfdom without according complete freedom. It was the first, though as yet vague, enactment towards the emancipation of both land and peasants, mitigating the peasants' obligations but also limiting their rights to land. The system made it incumbent upon the landlords to grant the peasant cultivators certain portions of land—thus recognizing the peasants' ancient rights—and also timber for building and fuel. In their turn the peasants were obliged to pay the tithe and to render certain labour servitudes, from eight to twelve days yearly, the system thus consecrating but also defining and limiting the landlords' feudal rights. Moreover, a formal act, signed by all the boiars and the chief prelates, gave the peasant, a 'brother in Christ', 'subjected by an evil custom', the right to purchase his full freedom by a payment of ten piastres. It was the first step towards transforming the title to land into private ownership, but with the maintenance of servitudes in favour of the peasants. It was also the first attempt of the State to intervene between the two rural classes. Under serfdom the peasants were altogether dependent on the landlords; under the *urbarial* system the State intervenes with precise regulations and with supervising agents. But these agents merely used their powers to carry out the wishes of the boiars, and to enrich themselves. Thus the *urbarial* system was never more than a mild improvement on the full serfdom which had preceded it. Nevertheless, the tendency of these Princes, whose French education made them accessible to the humanitarian views of the eighteenth century philosophers, was to secure the peasants against the abuses of the boiars. Constantin Mavrocordat extracted from the Moldavian boiars a definition of *vecin* which declared it to mean 'a peasant who has his own land, which he has inherited, with no right to abandon the soil'. Among the minor advantages obtained by the peasants in the second half of the eighteenth century was exemption from tithe for their gardens; orchards planted by themselves had never been subjected to tithe.

The increase in servitudes. The efforts of the upper class to tie the peasants to the land were natural enough, for the boiars lived altogether from the tithes of their villages. Later, villeinage became the instrument for extracting, without risk of losing the goose that laid the eggs, ever heavier dues from the villagers.

Not only had the personal needs of the boiars vastly increased, as contact with Turkish pashas and Polish nobles initiated them into the delights of eastern luxury; they also needed a great deal of money to purchase from the corrupt and frequently changing rulers favours or merely freedom. And now they also had a market for any surplus corn. In the second half of the eighteenth century Turkey increased the supplies of corn she had been accustomed to draw from the Rumanian provinces, at prices fixed as it pleased the Porte. This imposition, of course, added another weight to the misery of the peasants, who, in the bargain, were often abused and maltreated by the agents charged with the collection of the grain. But to the landlords it opened a door to easy gain, as what they sold to the Porte had been grown for them with the seed and oxen and labour of their villagers. About the same period Jewish immigrants, coming from Poland, began to settle in Moldavia, and to teach the Moldavian landlords the art of making vegetable alcohol. Stills were set up everywhere for the making of alcohol from corn and potatoes, the landlords engaging to that end increasingly in cultivation on their own account.

Therefore the landlords sought to get from their villagers heavier service, both in its quantity and in its kind. Formerly, when the *județ* needed corn merely for his own consumption, which he obtained from the tithe, and rarely engaged in farming on his own, the servitudes of the peasants, as their name, *clacă* (Serb = *tlaka*)[1] implies, were used for light labours, such as mowing and easy carting. But now the boiars wanted as much corn as they could get; they began to cultivate the spare lands, the villagers doing the ploughing and sowing and harvesting during the days which they had to labour for the landlord. But for this purpose the old due of three days in the year could not have been enough. It is likely that, here and there, the peasants were made to work more, but there is no evidence of it, till we come to the middle of the eighteenth century. In 1742 we find the priors of the monasteries, owners of vast lands, successfully demanding that their villagers should work for them twelve days in the year,

[1] Ordinarily the gathering at which peasants meet to help one among themselves, making a festivity of the occasion; in general, friendly light help they give each other.

with their hands or with their carts, as the priors may wish. There is no corresponding decree for the villages of lay owners; contemporary documents show that in those villages six days were served by the peasants in the year.

From the middle of the eighteenth century the claims of the landlords for more service became frequent and insistent; and from that time conditions differed more widely in the two provinces. The Moldavian boiars had occasion to learn from various contacts how the peasants slaved in Poland and Russia; they demanded and secured from Grigorie Ghika, reputed one of the best princes of the eighteenth century, a decree allowing them to claim twelve days' service from their villagers (Jan. 1, 1766). Moreover, the decree contained an innovation which was to prove the insidious means for the true enslavement of the peasantry. Whereas till then the peasants' service had been measured by the actual time spent in doing it, this decree now fixed the quantity of labour which, according to its nature, each peasant must perform in one day. This *nart* (Turkish, *nark* = tax) was twice or thrice as heavy as that which a normal man could do in a normal day. In a country with only the caricature of government and justice this was an easy way of indenturing the labour of the peasants, though the landlords never ceased to press for an increase in the formal obligations of the villagers. Early in 1775 the great boiars, led by the Primate Gavril Calimach, complained of the losses they had suffered through the abolition of *vecini*, and asked to be compensated by obliging the peasants to give them one-tenth of their working days, just as they had to give one-tenth of their produce. It was only two and a half years later that this demand was to some extent granted, the Prince adding to the decree of 1766 various obligations which amounted to five more fixed labour days, as well as an undetermined number of days for the repair of dams and ditches. In 1805, when the Turkish hold on the principalities was already crumbling away, the Moldavian boiars actually endeavoured to get from the Porte a revision of the peasants' dues. In a petition sent to Constantinople by 'the very humble and obedient Metropolitan, bishops and all great boiars of Moldavia' they admitted that till then the peasants had worked not more than twelve days in the year; yet they begged 'that the Imperial Ottoman Power should take them under its protection', and should authorize them to claim one-tenth of labour days just as they were entitled to one-tenth of the produce. Political circumstances did not leave the Porte an opportunity of dealing with that request, but the boiars obtained locally further privileges which brought the total servitudes due from the peasants to 36–40 days yearly, according to the estimate of the experienced and very conservative boiar Sturdza. Why the peasants should have stood such impositions though no longer bound to the land may be explained only by the state of moral and physical lassitude into which they had fallen; also, all the land was now occupied, if not actually cultivated. But that they were tempted to flee their life of slavery when they had a chance is shown by the reservation made by the boiars themselves in 1805, that the villages adjoining Muntenia and the Turkish districts should be held to only half the labour exacted from the others. Even so, the new arrangements were in many places passively resisted by the villagers and could never be applied.

The subjection of the peasantry never reached such limits in Muntenia, where conditions were less propitious for economic development, and outside influences less corroding. The peasants remained *vecini* till 1746, and it is likely that they had gradually been harnessed to more labour than the traditional three days. But, on the whole, the increase was not severe. We find Alexander Ypsilanti (1774–82) decreeing twelve days' labour, more, apparently, under the stimulus of the Moldavian example than under the pressure of local needs, for the boiars were permitted to transform into payments the servitudes of which they could not make use; and Part VI of the code of Carada (1812–18) had to provide penalties for those landlords who claimed less from their villagers than the prescribed twelve days. Contemporary documents suggest, indeed, that in the last quarter of the eighteenth century and the first of the nineteenth the servitudes of the Muntenian peasants were nearer to six than to twelve days. There is no trace of any claim by the boiars to one-tenth of the labour days; nor, which is more conclusive, that a *nart* was ever fixed for the labour days in the lower Rumanian province.

Limitation of right to land. If the burdens of the peasantry had become very heavy, its existence, at any rate, had been safeguarded by the traditional right to the use of all the available land. That right had been the foundation of ancient custom in all the regions inhabited by Rumanians. Later on it was confirmed by many princely decrees and, in identical terms, by the two oldest written laws, the *pravile* of Vasile Lupu and Mateiu Bassarab. The land law (*urbariu*) of Grigore Calimach reaffirmed it as late as 1768, Art. XV stating clearly that a peasant who paid his dues in kind and labour had a right to all the land he needed. Besides much other evidence there is finally a petition, dated 28th February 1803, from the boiars of the Moldavian divan, which shows that before they could let any surplus of the village 'boundary' the landlords had to ask the villagers whether they did not want it themselves; and further to safeguard them against rash or unfair decisions, the villagers were given a year within which they might claim back land which they had first refused and which in consequence had been let to an outsider. Such old village plans as have been found show that everywhere the arable land was divided into equal strips, without any domain reserve, the lord of the village being evidently only entitled to the use of one or a few of the normal strips. Hence there is no doubt that originally the peasants were entitled to use the whole extent of the cultivable land, and this right remained untouched throughout the worst political decadence. During that period the rulers, mostly foreigners, were appointed by the Porte and frequently changed; they had no need to attach the upper-class to themselves by granting them feudal privileges of lordship over the land.[1] But when Russia assumed the protectorate of the Rumanian principalities and thus prepared the

[1] The Rumanian people had never known the feudalism of Western Europe. The *cnezi* had only had a relative influence; like the later *boiars*, they were merely the Prince's adjutors. That is why the peasants had never risen against their servitudes; for that would have been a revolt not against feudal landlords, themselves opposed to the prince, but against the ruler's own men and establishment. (*Cf.* Leon, *op. cit.*, p. 103.)—Nominally, the *boiars*, divided into great and small, were merely a bureaucratic nobility; but they had gained hold of the land, and on their estates they exercised almost autocratic powers. The clergy were enjoying an autonomous administration, like a medieval guild. The mass of the peasants were servile, under the protection of their *boiars*. The State only had direct control over two restricted sections of the population: the *răzeşi*, who were yeomen

end of Turkish domination (Treaties of Kutshuk Kainardji, 1772 and of Jassy, 1792), and insisted on the appointment of native princes for a fixed period of seven years (Convention of 1802), then only did the ancient land rights of the peasantry suffer their first diminution.

The first measure for limiting the peasants' right to the use of the land was a decree of the Moldavian ruler, Alexander Moruzzi, who in 1803 allowed the landlords to reserve for their own use one-fourth of the meadow land, the peasants' grazing rights being limited for each household to sixteen large animals in Bessarabia, twelve in the Moldavian plains, and six in the mountain regions; 'any hay they may still want the inhabitants shall buy elsewhere' (Art. XIX). This title to a 'domain' was considerably extended after the Convention of Ackermann (1826) which gave the Rumanian principalities internal autonomy, under the joint protectorate of Russia and Turkey, and the right to be ruled by native princes. The first national prince to occupy the Moldavian throne after that dark period was also the first to grant what the Phanariote princes had refused to allow. By a decree of 10th March 1828, Ioniţa Sturdza severely restricted the peasants' ancient rights to the unlimited use of the land. In order, it said, 'that the perpetual landlord shall not remain without the meadow and arable land which he needs, both arable and meadow land, including wood clearings, shall be divided into three parts, and two parts of arable land and meadows shall be given to the inhabitants, but the third part shall be left without fail for the perpetual landlord.' This was the first document to refer to the landlords as 'proprietors'; till then they had been known only as *stăpâni* (*domini*), which indicated a personal relationship between them and the peasants rather than a real relationship between them and the land.

In Muntenia, where conditions remained easier, the rights of the peasants to use all the land suffered no restriction till the coming of the 'Organic Statutes', of which we will speak presently.

Rape of yeoman land. Besides thus securing a privilege on the

farmers, and the merchants and artisans, mostly foreigners, organized into guilds. (EMINESCU, *Works*, 1914, pp. 474–5.)

use of one-third of the available land, the boiars used every means, fair and unfair, to acquire possession of yeoman land. In those villages in which the peasants had been able to retain the old order grazing land was held in common, while arable land, with house and garden, was in the private ownership of each household. Such private land, by old custom, could not be bought or sold except between members of the village community, the *răzeşi*; the main problem, therefore, for a greedy landlord was himself to get possession of a yeoman 'part', by pressure or intrigue, or simulated gift, and thus become a *răzeş* with a right to buy village land. The rest was done for him by the increasingly heavy taxes and other burdens which were depressing the peasantry. In Muntenia, indeed, it happened that whole yeoman villages which were unable to pay their taxes sold themselves to the landlord, together with their land, thus becoming his *vecini*. The rape of yeoman land was considerable in itself, but it was only a small part of the hardships which were bending the back of the Rumanian peasantry. For that the two tendencies to increase the villagers' dues in labour and to reduce their right to the use of the land were mainly responsible; their upward curve can be traced throughout the legislation of the eighteenth century, but they only reach their extreme form after the breaking of the Turkish domination, in the first half of the nineteenth century. The weakening of the central authority, and the growth of Austrian and Russian influence, reduced the Rumanian provinces to a geographical expression, to a political no man's land, in which the boiars did as they pleased. Mr. and Mrs. Hammond wrote in their *Village Labourer* that 'In England the aristocracy had power and no privileges; in France the aristocracy had privileges and no power'. In the Rumanian provinces the boiars had both power and privileges. Nominally high functionaries on behalf of the Prince, they were in fact the keepers and uncontrolled defenders of their own interests as landlords. They had apportioned the land among themselves. According to N. Soutzo's *Statistique de la Principauté de Moldavie* the average size of a Rumanian estate was eighty-five times greater than the average English large estate; small properties were few and insignificant. Realizing that their power depended on their wealth in land, the boiars, in the absence of primogeniture, introduced the custom of deliberate disinheritance; one or two of the children inherited wealth and name, while the others were forced into the monastic state. 'The country', wrote Mihail Eminescu in one of his essays, 'was but a big estate, administered like an estate—a complex of latifundia in which private law is public law, the inheritance of landed wealth the inheritance of power in the State.'

When the first national rising in the Balkans, the Greek *Hetairia* movement, broke out on Rumanian soil in 1821, the Rumanian patriot Tudor Vladimirescu used the opportunity to instigate a popular revolt. He thus explained his purpose in a lapidary proclamation to the people: 'No laws can prevent you from returning evil for evil. If a serpent crosses your path, hit it and kill it, for if it bites you it will probably endanger your life. But these dragons—our ecclesiastical and political chiefs—who have devoured our rights, how long shall we let them suck our blood, how long shall we remain their slaves?' And to Dervis Pasha, commanding at Vidin, he wrote to assure him that 'our rising is directed only against the boiars, who have devoured our rights'. The first revolt of the Rumanian people was consciously aimed, therefore, not against outside political oppression, but against social and economic exploitation by their own upper class.

CHAPTER II

POLITICAL FREEDOM AND SOCIAL DEPENDENCE

The Organic Statutes. Political changes at the end of the eighteenth century were creating new economic conditions on the western shore of the Black Sea; these, in their turn, called forth social revisions—the whole forming a chain of causes and effects which supplied a noteworthy example of how economic prosperity may produce social regress.

In the earlier centuries, according to Carra, not more than one-fortieth of Rumanian soil was under crops. Corn was too bulky to be transported across the rough land routes, and in any case the surrounding countries were generally self-sufficing. But when the Turks lost Crimea, Egypt, and other provinces in southern Europe and northern Africa, they began to draw large corn supplies from the Rumanian principalities, reserving to themselves a priority of purchase. The frequent visits by Russian armies, during the conflicts with Turkey, also raised the demand for corn. Agriculture received a strong impetus, much quickened by the Treaty of Adrianople, which in 1829 put an end to that Turkish corn monopoly and opened the Black Sea to international trade. Between 1831 and 1833 the rent of land doubled and trebled. The great famine which visited Russia in 1833 drew attention to the agricultural richness of the Rumanian provinces. Wheat only began to be exported in appreciable quantities from Muntenia in that year. Jules de Hagemeister, in a book published at Odessa in 1835, stated that one *chila* of wheat which sold at 14 piastres during the Turkish occupation reached 210 piastres in 1833. Pasture and meadows, which not long before had covered 90 per cent. of the arable land, were reduced to 32·64 per cent. by 1860 and to barely 15½ per cent. by 1903.

That change ruined cattle-breeding, which had been the mainstay of the peasants, but it brought great wealth to all who disposed of land and labour for growing corn cheaply, as the landlords were quick to realize, especially during the Crimean War.

The rapid rise in corn and land values, as well as the great loss of population resulting from a plague epidemic in 1812, irresistibly tempted them further to encroach upon the land rights and freedom of labour of the peasantry; and the first step towards national independence was to give them power to satisfy these ambitions.

After some fifty years of efforts to take the place which Turkey held as master in the principalities, Russia settled the issue in her own favour by the Convention of Ackermann (1826) and the Treaty of Adrianople (1829). Her armies occupied the countries for six years and the administration was placed under the control of a Russian Commissary. Discontent was at the time deep and general. The small educated class aspired to gain a share of power; the mass of the people longed to be rid of the plundering taxes of the Phanariote régime. It was in the interest of the newcomer to make herself welcome by a better and, in some degree, autonomous government. The rulers, it was decided, should now be elected for life, from among the leading native families, by the respective *divans*; and these oligarchic assemblies, in which sat the great boiars and the higher clergy, were also to make the laws of their two countries. Meanwhile, Russia appointed as governor of the principalities Count Paul Kisselev, a general equally able, enlightened, and energetic, who at once set to work to endow the countries with fundamental laws. In deference to the terminological susceptibilities of Russia's autocrat the new laws were called 'Organic Statutes'.

In more than one sense these organic laws mark the beginning of modern life in the Rumanian provinces. During the long stretch of Turkish domination the Rumanian Principalities had never been turned into mere pashaliks, but neither had they been allowed any political will of their own. It was a haphazard régime, during which autonomous State life was suspended rather than suppressed. The populations had no political power and the transient rulers had but an indifferent political interest in the country. Legislation, sporadic and ephemeral, in the form of princely decrees, was mainly devoted to securing ever more benefits for the prince and his satellites. Old rules and customs persisted or were changed in random fashion, and basic principles escaped mutilation simply because no one troubled about principles at all.

In some such way had the peasants' right to the use of all the land persisted well into the nineteenth century. With the first measure of independence, however, the boiars found the chance of having their privileges sanctioned by the compelling authority of the law; and no doubt they also thought it timely, on the threshold of a new Rumanian polity, to establish as many of their claims as they could—'to endow the country', as they put it, 'with a modern law.' For this they were in a peculiarly favourable position. Probably they would in any case have been given a monopoly of power, seeing the views which St. Petersburg held on government by the people. But, moreover, they were the only element that counted politically at all, and Russia was anxious to propitiate them, lest they should turn their hopes towards Vienna.

The Moldavian Organic Statute. The new fundamental laws were drafted for each province by a Commission of four great boiars, and after being approved by St. Petersburg, they were submitted to 'Extraordinary General Assemblies of Revision', composed exclusively of great boiars, which gave them final sanction. The Moldavian divan, convened in 1830 to pass the Organic Statute, consisted of forty-six great boiars, i.e. as many as there were of them, representing no one but themselves, and of six leaders of the Moldavian ecclesiastical hierarchy. Yet it is saying a great deal that Europe's most reactionary government should have felt called upon to censure—with little effect—the new agrarian régime which the first autonomous Rumanian assemblies proposed to set up.

Section VII of Ch. III of the two Organic Statutes dealt comprehensively with principles and rules of the new rural order. It has already been shown how the eighteenth century had seen the growth of a tendency to reduce the land rights and increase the labour dues of the peasantry. What the boiars could not obtain even from the dissolute Phanariote princes they now bestowed upon themselves, with the reluctant support of the protecting Power. The Organic Statutes amended the old land rights in two ways, both to the loss of the peasants. A first change, of principle, upset the whole basis of land tenure in the Rumanian provinces: whereas the peasant had been the real owner of the land, and the landlord, as leader of the village, merely entitled to one-tenth of the harvest, now for the first time the lord of the village was established as lord of the land also, and given the title of 'proprietor'. It is true that the peasants' right to the use of the land was confirmed (Art. 118–25). But the landlord was allowed to reserve for himself one-third on the 'narrow estates', where there was not enough ground to satisfy all the inhabitants, after the example of the 'tiers sage' reserved to the landlords when the village communities were broken up in France. And the door was opened to further encroachments by two insidious texts: Art. 118 saying that the new rules shall have force 'until it shall be possible to arrange the leasing of the land by mutual understanding', and Art. 127 that a newly married man should be bound to perform the prescribed labour dues if he claimed his share of land 'and the proprietor is willing to give it'.

The landlord's title was thus exalted into one of full ownership, qualified as yet solely by the obligation to let the peasants till as much as two-thirds of the estate; the peasant's title, however, was reduced to a collective claim to the use of not more than two-thirds of the land, with the warning that even thus restricted it was in the giving of the landlord. The former tithe-owners blossom out into full owners of the land; the former full possessors shrink to little more than privileged tenants.

Besides being restricted to that collective maximum the peasant's share was further cut down individually. Of old he had cultivated as much as he wanted against payment of tithe and labour dues. In 1805 he was limited to certain fixed holdings, the villagers being for this purpose divided into three categories: *fruntaş, mijlocaş, codaş* (leading, middling, and tail-end peasants) —according to the number of cattle they kept. The Organic Statute reduced the holdings provided in 1805 by more than half.[1] Even if together they did not make up two-thirds of the

[1] Kisselev asserted that, according to regions, the Moldavian peasants had in 1805 received 1½ to 3½ times more land than was allowed to them by the first text of the Organic Statutes. In 1805 they received enough to keep twelve big animals in the plain, and six in the mountainous region; now they hardly had enough for five, though cattle-rearing was the peasant's main source of revenue.

estate the landlords were not bound to give any land beyond that. The boiars' aim to lay hold of the land was thus achieved in a large measure.

Labour Dues. They were still more successful in their aim to increase the labour dues of the peasantry. The original three days which the villagers worked for the *județ* had increased to twelve by 1805. This figure was kept by the Organic Statute. But whereas Moruzzi's decree referred to ordinary days, the Statute's twelve days were with *nart*, i. e. with a fixed programme of work. The first, that is, fixed the number of days, the second, in reality, the quantity of work, with the result that the twelve days of the one were far from being equal to the dozen days of the other. Kisselev himself declared that one day, as fixed in Moldavia, was equal to about two days in the Ukraine, and each of these equal to at least two days' real work. The number of actual days which the three categories had to serve in the year reached an average of fifty-six for the villager with four oxen, fifty-eight for the man with two oxen, and sixty for the peasant who had no oxen at all. According to the values of the time, the labour servitudes of the three categories were worth, respectively,

	Lei	129·35	121·05	99·30
the tithe		48·00	39·00	28·00
extra labour for the making of maize barns		8·00	8·00	8·00
Total	Lei	185·35	168·05	135·30
which divided by the holdings to which they were entitled worked out, per *falce*, at	Lei	35·00	43·15	57·60

Yet Kisselev had fixed, for other purposes, and the landlords had accepted, the lease value of a *falce* to be 18 lei. Hence, after being reduced to the state of tenants the peasants were made to pay for their land roughly two, two and a half, and three times more than it was worth at the time; and the poorer they were the more they had to pay for such fields as were left them.

Serfdom disguised and other burdens. Perhaps the landlords were not unaware of how crushing these labour dues were; at any rate, they made sure that the peasants could not run away from them. Art. 135, which had not existed in the first draft, confirmed the abolition of serfdom, but in practice serfdom was revived by the obstacles that were placed in the way of peasants moving from one village to another. They could do so only at the end of administrative periods of seven years each; they had to give to the landlord and to the Treasury one year's notice and to pay tithe and State tax for a whole year in advance (Art. 127). A peasant who wished to move in the course of an administrative period had first to pay the State tax for all the years that had still to run to the end of the period, and to the landlord the value of his labour dues for a whole year. It was only after pressure from Kisselev that some of these conditions were waived for the peasant who had to move because the landlord could not give him the holding which should be his (Art. 123).

Among the secondary servitudes imposed by the Organic Statutes was that of the 'voluntary servants'. Art. 72 allowed the landlords to conscript 10 per cent. of the inhabitants on their estates (and 20 per cent. where their total number was below 200) for unlimited domestic service; they were to be allowed time merely to work their fields. These servants were exempted from the State tax.

The peasants also lost the valuable right to wood for fuel and building which they had enjoyed throughout the worst Turkish times. Later, in 1844, this right was restored to them by the ruling Prince.

About the only alleviation which the Organic Statutes brought to the lot of the peasant was in the degree and manner of taxation. The tithe was maintained, except for garden produce. But the devouring mass of direct and indirect taxes bequeathed by the Phanariote régime was replaced by a single direct tax, the *bir*, amounting to some 30 lei annually, for whose payment the village was jointly responsible—another means of making the villagers police would-be runaways themselves. This improvement in the degree and nature of the State tax, and not least in the manner of its collection, which went a long way towards checking abuses, undoubtedly meant a boon for the peasantry. With this, however, the boiars were little concerned. They themselves remained, as before, altogether exempted from taxes. In addition, they secured compensation for themselves and their widows and their minor children, for renouncing the abusive privilege of possessing *scutelnici*: men, that is, who were exempted from all taxes to the State but laboured continuously for the landlords, being thus bound to full serfdom in practice. According to Al. Golescu there were in 1828 17,000 families of *scutelnici*; they were granted to the boiars in varying numbers, according to rank, as a perpetual right. In Muntenia the compensation was 60 piastres yearly for each *scutelnic*, and half that sum to the boiar's widow and children. As that compensation swallowed about one-fifth of the Moldavian and one-tenth of the Muntenian budgets at the time, what it came to in practice was an indirect tribute paid by the peasants to the boiars. Finally, both Statutes maintained the landlord's monopoly to mill, to sell spirits and meat, to open shops, and so on.

The Organic Statute in Muntenia. The Russian Commissioner found conditions in Muntenia which, in law and in practice, differed a great deal from those in Moldavia. Geographical position and the stimulating intercourse with neighbouring countries had set a much faster pace in the economic development of the northern province. In Muntenia, however, landlords farming for their own account were few and far between; and this economic backwardness translated itself into better social conditions on the land. No measure had come as yet to curtail the peasants' right to the use of the land; contemporary documents, indeed, show that they were being urged to plough as much land as they could. Nor had their labour dues been weighted with unfair reckonings so far. The twelve days were ordinary days; mostly, also, they were commuted into a money payment, generally at the low rate of one leu per day; while in many villages along the Danube, where flight was easier, only six days' work was asked of the villagers in the year. Had the Muntenian Statute, therefore, held to the Moldavian model, it would relatively have caused greater damage in the life of the peasantry. In fact, the two laws differed considerably in their provisions relating to land rights and labour dues; and while the Muntenian Statute was much fairer in the assessment of servitudes, it was much greedier in the curtailment of the peasants' right to land.

Art. 140 of the Muntenian Statute acknowledged the peasants' right to land, according to the number of their cattle; but it also said, like the Moldavian Statute, that these provisions were to stand until the 'leasing' could be arranged by mutual agreement. And Art. 144 opened with the assertion that 'just as the landlord is complete owner of his land. . . .', &c. Here the reservation of two-thirds of the land for the use of the peasants disappeared altogether. Moreover, the holdings allotted to the various categories of villagers were smaller than in Moldavia, though the province was rich in extensive plains.[1] There seems no other way of explaining this than by an excessive greed for wealth, for the Muntenian landlords continued to let to the peasants land in plenty on easy terms, and to demand from them much lighter labour dues than those of Moldavia.

Labour Dues. In Muntenia, as in Moldavia, the peasants had to give twelve days' labour in the year, and here also a *nart* was now fixed for those days. But in Muntenia the various labours were defined so fairly as almost to approach what actually was possible. To give one instance: while the Moldavian rules allowed but one day for the cultivation of twelve *prăjini* of maize, eight days were allowed in Muntenia for the same labour; and even this, being found inadequate, was improved later. Here also they remembered to reduce, proportionately, the labour dues of the villager where there was not enough land to give him the whole extent to which he could lay claim. Again, in Muntenia, the landlord was allowed no more than four men from each hundred families to do him personal service, and here this work was usually divided among the villagers, making some fourteen days' service for each, which were often commuted into money.

For the rest, the freedom of movement of the peasants was

[1]

	Moldavia			Muntenia		
	Land granted		Value (p. falce)	Land granted		Value * (p. falce)
	Hectares	ares	Lei	Hectares	ares	Lei
Peasants with 4 oxen	7	68	35·00	4	42	37·24
Peasants with 2 oxen	4	41	43·15	3	30	38·23
Peasants without oxen	2	20	57·60	2	21	48·23

* (This included the value of the domestic service, which in Muntenia was distributed among all the villagers).

as severely hampered as in Moldavia; somewhat wider latitude being allowed only to the men who could not get their full share of land where they lived, and to those who acquired land by marriage or heritage elsewhere. But Art. 144 decreed that even after fulfilling the prescribed conditions the peasants could leave only two at a time; Bibescu later interpreted 'at a time' as meaning during one year, so that only two peasants could leave each year from one village. Unlike the Moldavian law, that for Muntenia maintained the peasants' right to wood, wherever there were forests on the estate. In other respects the two Statutes were substantially alike.

The Organic Statutes radically changed the whole agrarian system of the two Rumanian provinces. The modern conception of property, as a right in itself, not qualified as before by the professional use of the object, entered Rumanian agrarian law for the first time. An official report, dating from 1834, signed by Stirbey and others, openly declared that 'the purpose of the Organic Statutes has been to raise, at the expense of the Treasury, the rights of property, so as to indemnify the boiars for the sacrifices they made'. And a Commission appointed to consider the position of monastic estates, proposed to Kisselev that the monasteries should 'renounce for the benefit of the State the surplus revenue which has accrued to property from the application of the new law'. The practical effects of the new régime were summed up by M. R. Rosetti as meaning 'for the peasant, on the one hand, a reduction of taxes, but on the other a grievous reduction, almost to nothing, of his old right to use the land on which he was settled; it means heavy crippling labour dues, under a hypocritical appearance of alleviation; obstacles to his leaving the village in which he suffers coming very near to serfdom; denial of any and every civil right; and, finally, a threat that he may lose even the little land that was left him'. After the passing of the Organic Statutes the holdings were so small that for the first time the peasants found it necessary to rent additional land, beyond that to which they had a formal claim. And this although Kisselev had gone to the length of suspending in both provinces the enforcement of the two texts, and had begged and pressed for amendments; insisting, above all, on the peasant's old right to use all the land, on the justice of keeping labour dues proportionate to the holding given to each peasant, and on the need to recognize frankly and finally the peasant's complete freedom of movement. In a letter to Buteniev, the Russian Ambassador in Constantinople, he remarked that the Assembly of boiars 'having constituted itself judge in its own cause, it is only natural that it seeks to extend its own privileges at the expense of the others, who are neither represented nor defended by anyone. That goes so far, that by an insidious clause regarding labour dues they have bound the villagers to the soil, though they are free by right, and every day they tend to make of them slaves, to oppress them the more. . . .' Kisselev succeeded in securing many improvements of detail, but in return he had to allow certain fundamental principles to penetrate into the Statutes which he knew to be wrong, which depressed the peasantry's standing as a class and compromised its whole future; for after having fought them for three years, the boiars appealed to the Tsar, and Kisselev had to give way, for political reasons. But what a trenchant commentary on the indiscriminate exaltation of national government to see Kisselev—an alien Count and general, a representative of Europe's most autocratic ruler—fighting to save some of the birthrights of the Rumanian peasants which, at the first opportunity, the native boiars were rapaciously usurping. Among the boiars themselves, not a single voice was raised to call a halt to that cruel despoiling of their poorest kith and kin.[1]

Revising the Organic Statutes. That the Organic Statutes had lowered most painfully the life of the peasantry was obvious to every observer. More than once in their reports the foreign consuls condemned the hardships of the new régime and raised the warning against the dangers lurking in such a state of legalized misery. The way in which the peasants responded to the call of 1848 proved indeed that discontent was deep and widespread among them; and during the joint Russo-Turkish

[1] Later Kisselev became Russian Minister of Domains (State lands). In that capacity he introduced, in 1845, reforms aiming at regulating the position of the peasants living on such domains, their dues and their rights to land. This measure was an important precedent for the peasant emancipation of 1861, which embodied its provisions.

occupation which followed, the protecting authorities were able to establish the truth of what their consuls had reported. As a consequence, the Treaty which Russia and Turkey concluded at Balta Liman in 1849 charged the native rulers who were about to be appointed for a period of seven years, and who were to govern in the presence of Russian and Turkish commissaries, with the duty of revising the laws governing the relations between landlords and peasants.

In Moldavia. The new Act brought in by Grigore Ghica in 1851, and approved by the protecting Powers, reconfirmed the peasants' right to land, and jointly to two-thirds of the estate where there was not enough to go round; and it allotted to each household twenty *prăjini* more grazing. The number of labour days was maintained, but the work to be done was defined more closely, and a proportionate allowance made to those peasants who could not get their full share of land. The tithe was abolished, except for orchards and vineyards lying outside the peasant's garden. Removal from one village to another was made easier. And the making of contracts on terms worse than these was forbidden. The practical effects, as estimated by Rosetti, was to increase the labour dues of the first two categories of villagers;[1] that increase, however, was worth not more than 15 lei, whereas the tithe, of which the peasant was now released, was worth 72 and 57 lei, and twenty more *prăjini* grazing also represented a rental of 5–10 lei annually. The poorest section of the peasantry benefited from a reduction in labour dues worth 20–8 lei, from the remission of the tithe, worth some 42 lei, and from the increase in grazing land. More than that, the greater clarity with which the rights and duties of the two parties were now defined put a stop to many abuses of which the peasants had been the victims. In this the Muntenian amending law was even more effective.

In Muntenia. Muntenia's new ruler, Barbu Stirbey, had been one of the makers of the Organic Statute. He believed that the troubles of the peasants were due merely to the bad application

[1] Peasants with 4 oxen from 56 to 61 ½ days.
Peasants with 2 oxen from 58 to 63 ½ days.
Peasants without oxen from 60 to 52 ½ days.

of an essentially good law; and in a Memorandum he wrote before the drafting of the new law he asserted that 'complaints have arisen in regard only to those things which the Organic Statute thought best to leave to the free decision of the two parties'. Stirbey was a keen student of agrarian problems, and while himself a big landlord, not without goodwill for the peasants.

His sayings and actions, therefore, make a fair guide to the outlook of the Rumanian boiars at the time. The Preamble to his new law insisted that the peasant settled on another man's estate is the landlord's tenant, paying rent in the shape of dues in money, labour, and kind. The landlords were still obliged to give land to the peasants, but only until a free economic relationship was established between the two parties. Meanwhile, the need for statutory regulations remained; hence the new law. It doubled the extent of grazing to which the peasants were entitled, fixing it at one pogon per head of cattle; it made the conditions easier and simpler on which the villagers could obtain additional land. Labour dues were fixed at twenty-two days all round, which meant that the peasants with oxen had to labour six days more, and the peasants without oxen four days more than under the Organic Statutes; but in return the personal service, established by custom at fourteen days in the year, was abolished. So was the tithe from garden produce.

Art. 144 declared the peasant to be completely free in his person and possessions, allowing greater latitude, though by no means free choice, to the peasant who wanted to move from his village. But more than this Stirbey could not do, for the existing arrangements assured to the peasant the work and to agriculture the labour which was needed. 'This, which it might be difficult or impossible to secure by other means, constitutes the sole energy in this essentially agricultural land; to deprive agriculture of it, would be to kill it.' That mutual need, Stirbey thought, should govern all future legislation, and the best way of satisfying it would be to apply all round the sliding scale provided in Art. 141 of the Organic Statutes. It allowed the peasant to claim a reduction in his labour dues if the landlord were unable to give him all the land to which he was entitled; should not dues be increased in the same proportion if the peasant asked, and the landlord gave, more land? Stirbey seems to have been anxious to encourage peasant cultivation, rightly observing that the Organic Statutes had ensured to the peasant means of existence, but not for expansion. But his logical argument overlooked the fact that the peasant who laboured more for another must cultivate less for himself, that he was not helped by being given more land while left with less time in which to till it. In point of fact, the new law gave a great fillip to peasant agriculture by securing it against abuses. The regulations for applying it supplied printed forms for agreements between landlords and peasants, and it also obliged the landlords to give for each day of labour a printed receipt to be held by the peasant till the yearly settlement. By this and other such means the new régime certainly helped to improve the material standing of the peasantry; and, in general, it introduced into the Organic Statutes all the improvements of detail compatible with the essential injustice of its principles.

The first step to political rights. Stirbey's rural law contained an innovation of great interest as being the first step towards the peasant's political enfranchisement; besides having immediate practical value for the betterment of his life. The law handed over to a village council the administration of the village, the collection of taxes, as well as the examination of disputes between landlords and villagers—all of them matters in regard to which the peasant had had most of the burdens and none of the saying. All official business within the village had to be carried out through the council. The council was to consist of a mayor, a delegate of the landlord, and two or four peasant delegates—two if the number of households were below a hundred, four if it were above. Mayor and peasant deputies were to be elected for one year from among the peasants with four oxen or, if need be, with two oxen, by all the villagers who paid the head tax; they could not sit two years running. In return for their services they were exempted while holding office from performing labour dues. In days when the burdens of the peasants were largely made up from abuses by landlords, their men and the petty local officials, the creation of the village council was a brave attempt of the Muntenian amending law to reform a great evil.

The state of the peasantry on the eve of national independence. Surveying with one glance the whole of these changes, they show that in the 'forties of the nineteenth century, when the Rumanian provinces recovered their national autonomy, their peasant populations lost much or most of their social and economic freedom. Though the new arrangement was clothed in the ordered articles of a modern law, it could not for a moment mislead the peasants, simple folk as they were, as to the real meaning of the change. Rumours concerning the changes that were contemplated sufficed to cause sporadic local risings in Moldavia in 1831, which the authorities could hardly have suppressed without the saving presence of Russian cossacks. Elsewhere, passive resistance delayed or checked the enforcement of certain provisions of the Organic Statutes. Along the Pruth, first individual peasants, then families, later whole villages began to cross into Bessarabia, then in Russian hands. By 1834 that migration was reported from all the frontiers, into Transylvania and across the Danube as well, into Turkish territory, and was affecting even more distant districts. Golesco, writing in 1856, stated that over 100,000 families had crossed into Bulgaria, Serbia, and Transylvania since 1832. There is rejoicing among the peasants, he said, when the Danube freezes, for they can escape across its solid surface from their sufferings at home. A commission inquiring into the causes of that flight reported that they were: insufficient land and abuses in the application of the Statute. No remedies, however, were apparently attempted, until the protecting Powers imposed them some fifteen years later, and by and by the weary peasantry, steeled in misfortune by having borne the brunt of the country's prolonged trials, settled down under the new load placed upon their shoulders in the first hour of national revival.

Under the new régime the peasant for the first time began to know land hunger. With the simple methods of farming in use at the time he needed for himself and his cattle a good slice of land which in many cases he could not now get. The labour dues were a crushing burden, especially as their performance was in no way adjusted to rough soil, bad weather, and other adverse accidents. Being measured by work rather than by time they were elastic enough and could be stretched into endless abuses; in the mixed commission of 1848 a peasant deputy told how after toiling a whole week, with his own cart and oxen, carting maize for the landlord, this was credited to him as one day's labour. The customary tithe had been a share, not excessive, of something which the peasant really had, and related to the conditions in which he found himself each year. But the heavy labour dues had to be performed in full no matter whether his land and his harvest were good or bad. The excessive labour servitudes themselves were bound to retard the peasant's own cultivation and harvesting; the more so as the landlord would naturally claim the best days for himself, and as the labour dues could not be commuted without his consent. So that the new arrangements banged the gate to prosperity in the face of even the most capable and resourceful peasant. It is characteristic that throughout those protracted agrarian debates no one thought of suggesting that some crumb of schooling and training should be provided for the villagers.

And the many wrongs contained in the letter of the new laws were but procreators of a multitude of abuses committed when applying their provisions. The small upper class to whom the Organic Statutes had accorded excessive rights and privileges had in their hands the whole of the primitive and corrupt administration; there were, therefore, no bounds to the nature and number of injustices which could be perpetrated at the expense of the peasants. 'The peasant is the boiar's capital,' a boiar bluntly exclaimed in the national divan. In 1837 the Moldavian ruler, himself a great boiar, denounced in biting words the way in which the peasants were cheated in the measurements with the pole, a shorter pole frequently being used for the land they received and a longer one for that which they tilled for the landlord. The prince had to intervene again, in 1844, on behalf of the inhabitants of the mountain regions, most of them settled on monastic estates, who were made to pay a money compensation for labour dues as there was no arable land which they could be made to till, a difference which the Organic Statutes had overlooked. About the same time steps had to be taken against landlords who were trying the new trick, so popular in later days, of making labour contracts with the peasants during the winter months, when the peasant's needs were at their highest and his resources at their lowest; and likewise against those who exacted heavy fines for straying cattle, even when no damage had been done. The Organic Statutes had permitted the landlords to claim money payment when they could not use all the labour which the villagers were supposed to perform; for such compensation a scale was to be fixed by the Assembly, and it was so fixed as to bear heavily on the peasant.

Similar excesses were committed with the charges for surplus grazing, so that many peasants fled the villages for the towns. Alexander Ghica, the Muntenian ruler, endeavoured to have them mitigated, but after a long fight he had to declare that 'we shall no longer expect from the Assembly the settlement we had demanded, but will limit ourselves to withholding all official interference and aid for the enforcement of any agreements between proprietors and peasants, and we will award proper damages to injured peasants as soon as complaints reach us'. Finally, the charges for surplus grazing had to be fixed officially, in 1844. The ruler also resisted the demands of the landlords for greater powers to prevent the peasants from moving away; though the landlords themselves had been armed with a fearful means of oppression by Art. 126 of the Organic Statutes which allowed them to expel, with the consent of the local authorities, 'troublesome peasants', simply by giving them six months' notice. When he wanted the peasant's labour, however, the landlord could not only fetch him to work with the gendarme, but also generally had him flogged; an indignity which the peasants bitterly resented. 'In the time of the Turks', one of them lamented, 'the sword may have killed, but it did not sting, like the whip.'

The great sufferings of the villagers were voiced in measured but accusing words by the peasant deputies in the mixed commission of 1848. Said Ene, 'the Jerkin-maker': 'the Organic Statute confined us on the landlord's estate as in a walled fortress with iron gates, so that there was no way by which we could get out; and even if we ran away, abandoning house and orchard and vineyard, the work of our parents and our own, and they found our whereabouts, they brought us back chained, just as if we had been slaves, locking us up in their cellars or barns in winter time, with no fire, and even throwing water over us that we should freeze, that our sufferings should frighten the others; so that many have remained crippled and cannot feed themselves to-day.' And old Lipan told his own story: 'Another time my wife was taken to cut corn and I to mow, with the gendarme after me, and the three-months' old child left under the burning sun, with flies sucking his mouth, and wasps and gnats stinging him—was that not slavery perhaps? Slavery and nothing else, brethren! The wife cutting corn from sunrise till dinner-time, and not allowed to go and suckle the child. Why do the gentlemen say it was not slavery, for we know it to have been slavery, that sorrow that we have sorrowed.' And yet, these oppressed peasants never threatened or attempted violence. They were indeed ready to accept the new order, asking no more than a piece of land that would feed them and their cattle, and willing to pay for it, too. Of their old rights they stubbornly defended only one, that of tilling more land against payment of the tithe. Though ground down by centuries of misrule and want, these peasants had yet learnt neither to beg nor to loot; all they claimed was the right to work.

CHAPTER III

NATIONAL INDEPENDENCE AND THE EMANCIPATION OF THE PEASANTS

The New Factors. The Organic Statutes marked the high tide of rural feudalism in the Rumanian provinces. Their feudal web of peasant servitudes and landlords' obligations was shot with the first inklings of *laissez-faire*, presaging individual property in land and free labour contract. Hardly had the Statutes been passed when slowly the reactionary waters began to recede and gradually to open up the broad valley of individual freedom. For laws of their kind, confiscating the land and the labour of the peasantry, were begotten too late—almost two generations after the French Revolution. Even Holy Russia was being stirred by the spirit of the times; the new currents of thought were not to be checked, like Napoleon's grenadiers, by her snow barriers; they swelled and surged until in 1861 the Russian serfs were emancipated.

In the Rumanian provinces those currents found no castle gates to force. No crowned autocrat, no feudal barons guarded the drawbridge. On the contrary, the political impulse of the country was cutting a ready path for them. With the beginning of the century the vision of government by the people had also crossed the Rumanian sky. The ideals set free by the French Revolution, coinciding with the crumbling away of Ottoman power, awakened the subject Balkan peoples to a sense of national freedom. The first rebellious movement, the *Hetairia*, was organized and started on Rumanian soil; and its dreams were whispered all over the land. The handful of great boiars, replete with privileges and wealth, having much to lose and little to gain, saw freedom merely as the ending of Turkish exactions, by the intercession of either Russia or Austria. But the younger and poorer generation had been touched with the magic wand of the spirit of nationality. Under the stimulus of their rediscovered Latin origin they had gone westwards, every year in greater numbers, to be educated, especially in Paris; and from there they returned all afire with the new liberal ideas and ideals. Freedom, to them, meant the end of all alien interference. A protectorate, however 'Christian', appealed not at all to them. But it was clear that they could not hope to check the expansionist leanings of the three neighbouring autocrats, unless with support from the more distant and liberal West. The complaint against the Turkish suzerain and the Russian protector had to be justified, therefore, with a display of zeal for liberal institutions; and by such liberal reforms they would have put a moat between themselves and Russia, to whom, as a result of religious and social affinity, they felt uncomfortably near.

While trusting to ward off external dangers with help from abroad, at home, in the absence of a middle class, they could not hope to break the arrogance of the great boiars without gaining the support of the masses. Tactical needs, therefore, as much as their own convictions, led them to appeal to the people in their new struggle for national self-government. In 1848 the call rang out to the peasants to be up and to fight for their own soil—a call they had not heard since the days of Ștefan the Great. 'Brother Peasants,' the manifesto of the Muntenian revolutionary government called them; it promised them equal rights, the repeal of servitudes, and the free gift of 'a piece of land that would give them sustenance'. But they were also enjoined to go on working for the boiars and to behave nicely, and not a word was said about redressing the wrongs committed by the Organic Statutes. A mixed commission, consisting in equal numbers of landlords and peasants, was appointed to prepare the agrarian reform; but as it could not agree—its sittings becoming more uproarious every day—it was dissolved, and soon afterwards the revolution collapsed. The revolutionaries of 1848 were devoted to the ideal of individual freedom, in the social as in the political field; but they had little understanding for, and, as a class, probably scant sympathy with, the patriarchal rights and minds of the villagers. Their challenge, however, had at least roused the spirit of the peasants, and it had banded their interests together. The 'brother peasants' had heard the exhilarating promise of 'liberty and equality'; the first thing which the Commission had to concede was that 'man is free and his labour sacred'. In that

Commission the peasants had for the first time been given a voice, to speak their plaint and to ask their due; and for the first time, too, their deputies spoke, not divided each for this village or that district, but together for the whole Muntenian peasantry as one body. Beyond that moral satisfaction they got, however, nothing; the revolt came to an end without having brought them even a temporary relief from their hardships. The European Commission which visited the principalities in 1857–8 reported that such protection as the Organic Statutes afforded to the villagers had remained ineffective. Because of corruption among officials 'the action of the Government was paralysed on the land, the peasants were abandoned to the pleasure of the landlords, and a kind of feudal régime was reintroduced in Moldavia'.

Meanwhile, however, political developments which were advancing the cause of Rumanian independence pleaded also for the betterment of the peasant class. The Paris Congresses of 1856 and 1858 restored national autonomy to the Rumanian provinces, as a means of stopping Russia from making use of these provinces as a military highway to Constantinople. The Great Powers took steps to consolidate them, not only to make of them an effective buffer between Russia and Turkey, but also to ensure the safety of the mouths of the Danube and to enable these rich lands to develop the corn-growing needed for the expanding towns of the West. This was bound to open a new market for the manufacture of western industries, and a new source of food supplies for their workers. In 1856 the Treaty of Paris reduced Turkish rule to a nominal suzerainty; the Principalities, placed under the protection of the Great Powers, were to be governed by native princes, popularly elected, and they were to have a Constitution. For these same reasons it was the wish of the western Powers that the populations should at the same time be given a chance to improve their existence by a reform of agrarian conditions. That was the period of the great rural reforms. Serfdom had been abolished in Austria in 1841; in Prussia partly in 1810 and finally in 1850; in Russia the emancipation of 1861 was being prepared. Serbia had rid herself of the Turkish landlords; and in Bulgaria, though she was still a Turkish province, conditions on the land had much improved. So had the position of the Rumanian peasants who were subjects of neighbouring empires. In 1848 the Transylvanian Diet abolished all servitudes and resettled the peasants without compensation; the state of things which prevailed at the beginning of that year was taken as a basis for the reform, with the result that in Transylvania holdings were larger than those which had been given by the Organic Statutes in the Principalities. An imperial patent finally settled the question in 1854 to such good purpose that in the communes inhabited by Rumanians, 75 per cent. of the ground was taken up by holdings up to 50 ha. The villages were at the same time provided with commons and woodlands. In Bucovina, also, imperial patents of 1848–9 did away with all servitudes and granted land to the peasants; as Bucovina had been annexed by Austria in 1774 its inhabitants had escaped the deprivations of 1805 and 1831, so that the reform of 1848 put them in possession of the whole so-called 'rustical' land which they had occupied of old. Even the Bessarabian peasants, fallen under the rule of the Tsar in 1812, were better off than their fellows in the free Principalities. They, too, had escaped the knife of the Organic Statutes, and after their emancipation in 1861 they received everywhere, against a smaller compensation, holdings that were larger than the largest distributed across the Pruth—11–18 ha. as against a maximum of 7¾ ha. in Moldavia and 5½ ha. in Muntenia—besides being provided with common grazing lands; though it is true that this generosity was inspired by political motives and contrasted strongly with the way in which the peasants were treated in the central parts of the Empire.

In the Rumanian provinces alone the landlords were still strong enough to baulk any plan of reform. The temporary rulers, as we have seen, could do nothing more than ease a few of the worst burdens imposed by the Organic Statutes and endeavour to keep closer watch on how those laws were applied, as they were instructed to do by the Convention of Balta-Liman. The delegates of the Powers to the various European Conferences which dealt in those years with the Eastern question repeatedly insisted, therefore, that in the Principalities 'no progress was possible till the rural problem was settled'; and that 'the solution must be imposed from outside, as the only means of making it accepted'. The Paris Treaty of 1856 decided that the constitution and laws ruling in the two Principalities were in need of reform, and to that end it obliged the Porte, by Art. 27, to convene a special assembly in each of the two Principalities, a *divan ad hoc*, 'representing the interests of all the social classes', which were to formulate and submit to the Powers the wishes of the two countries.

The agrarian question roused a passionate debate in the Moldavian divan which met at Jassy in 1857.[1] Inspired by the national ideal of uniting the two Rumanian provinces, the great boiars waived their old ambition to be rulers of the country; and they still found within themselves enough enthusiasm to surrender their political privileges. But neither the clamour from within nor the hints from without could impress them with the injustice of their feudal privileges as landlords. When the agenda for the work of the divan was read, there was not a single item on it referring to the agrarian problem. The divan included, however, in addition to seventy boiars and eight clerics, fifteen peasant deputies—simple village leaders grown wise in misfortune; seeing that what most ailed them did not trouble the others, they tabled a sober but moving address in which they described their sufferings and their rights, and put forth the reasonable things which they wanted. They demanded the abolition of birching; the replacing of all dues by a single tax, to be paid without exception by all the inhabitants; local government for the village, through an elected council; but first and above all, the abolition of the tithe and of all dues towards the landlords. 'We want to buy our freedom', the address said, 'that we may no longer belong to anybody, but only to the soil, so that we, too, should have a fatherland. . . . We do not want to trespass upon any one's rights, but neither do we wish our own rights to be forgotten.' As always when they had a chance of stating their claims, the peasants strained their rights and means to the utmost, offering to compensate the landlords liberally, so anxious were they to be fair. But the landlords angrily denounced these 'communistic tendencies', and in reply to the peasants' demand

[1] The Muntenian divan interpreted the instructions of the Powers as debarring it from discussing internal reforms.

that servitudes should be abolished they suggested that in that case their own obligation to give land to the peasants on their estates should also be abolished—a proposal which would have reduced the mass of the people to landless labourers at one stroke.

The European Commission waiting in Bucarest to receive the conclusions of the two divans had finally to report that nothing had been done to further the solution of the agrarian problem, adding that 'if this reform were to be left in the care of the two interested parties, it will never be dealt with equitably'. And Talleyrand, the French delegate, wrote that 'a solution will not be accepted without conflict unless the principle on which it is based were first to be proclaimed by the Powers'. In the Paris Convention of 1858, which dealt with the future organization of the Principalities, the Powers embodied this advice in Art. 46, which said: 'All the privileges, exemptions and monopolies which certain classes still enjoy shall be abrogated, and the laws which regulate the relations of landlords and peasants shall be revised without delay, with a view to improving the conditions of the peasantry.' In spite of this injunction, contained in a document which constituted the charter of Rumanian national independence, the solution of the agrarian problem was held up for another six years; nor could Talleyrand's policy prevent in the end the conflict which it was meant to avoid.

The Reform of 1864. As soon as the presence and pressure of the Powers was removed, boiars of yesterday and nationalist revolutionaries of to-day found themselves to be of one mind in wishing to be as little troubled with agrarian reforms as they could possibly contrive. The first national assemblies passed on the duty of attending to Art. 46 of the Paris Treaty to the Central Commission—a joint body consisting of eight members from each province and charged with the drafting of bills. After taking a whole year for this preparatory work, the Commission produced a draft based on a view of the peasant as a privileged tenant of the land which he held.[1] As a consequence the draft was

[1] The agrarian legislation of the first half of the century, by trying to circumvent a rural relationship which it dared not or could not boldly abolish, had indiscribably confused the status of the Rumanian peasant. M. B. Boerescu thus described it

resolutely opposed by the first ruler of Rumania[1]—Alexander Cuza, a former officer of the civic guard and a man of the people, elected on the understanding that he would be merely a *locum tenens* until a foreign prince could be secured—and by his intimate adviser, the passionately progressive Mihail Kogălniceanu. That it was possible no longer to justify and maintain the peasant servitudes was tacitly accepted all round. But when Kogălniceanu put forward a bill of his own, which gave the

in his *Mémoire sur les conditions d'existence des Principautés Danubiennes* (Paris, 1856):

'What kind of man is the Rumanian peasant? Is he free or a serf, an emphyteutic holder, a tenant, a usufructuary or a *métayer-colon*? The answer is difficult. He is all that, and yet he is not. He is a strange abstraction created by the Organic Statute.

'The peasant, indeed, is free, for he is master of his own person and of his possessions; he may go where he wants; he hands over to his heirs house and yard, garden and orchard. Yet, at the same time, he is not free, for he is bound to perpetual labour for the landlord; and when there are no heirs it is the landlord, not the State, who inherits the peasant's house and garden.

'He is serf in that sense, that the landlord partly inherits him *ab intestat*, and that he is sold with the estate; the restrictions placed upon his moving almost amount to bondage. But, on the other hand, he is not a serf, for in public matters he stands in direct relation to the State, and he can leave the estate on fulfilling the prescribed conditions.

'He is an emphyteutic holder, for he holds the land in perpetuity and enjoys its produce in return for an annual due. But he is not that altogether, as (*a*) in addition to the annual payment he owes personal labour; (*b*) he can dispose by will *ab intestat* only part of his land, i. e. orchard and other plantations; (*c*) he can in no case estrange his property during his lifetime; (*d*) the landlord may end the peasant's title, even if he continues to pay his dues.

'He is a tenant, as stated in the law of 1851, because of the ease with which he might be ejected by the landlord. But for the rest he has none of the characteristics of the real tenant, seeing that: (*a*) his professional activity is neither provisional nor voluntary, but perpetual and necessary; (*b*) his obligations are rendered in kind, as well as in money; (*c*) the landlord has no obligation to keep the object in a fit state for the purpose for which it was rented; (*d*) the peasant transfers part of his rights to his heirs.

'He is joint-owner in so far as he has a common title to the land of the estate. Yet he is not that, because (*a*) he has no right to dissolve that joint-ownership; and (*b*) because the landlord can deprive him, against his will, of his real title.

'He is a usufructuary, if you like, as he has the right to use and to enjoy the object, and because he has only a life-title to part of the object.

'Finally, one could say that he is a *métayer-colon*, in that he is obliged to give the landlord part of the produce. But his status is simpler than that of a *colon*, as (*a*) his rights to the land have their origin in law and not in a convention between the two parties; (*b*) they are perpetual and belong *ipso jure* to his heirs.'

Till his emancipation, the peasant could not sell his house, nor the vineyards and orchards he had planted, except when he was expelled from the village for 'being troublesome'. On the other hand, the landlord 'could neither take away nor change at his pleasure' the land which the villagers laboured. (AL. GOLESCO, *L'Abolition du Servage*, 1856, pp. 131–2.)

[1] The union of Moldavia and Muntenia in the new State of Rumania was carried through in 1861.

peasants all the land they then occupied, the landlords denounced it as 'communistic' and retorted with a counterdraft in which they offered puny holdings, smaller even than those contemplated by the Central Commission, which the peasants were to hold in usufruct. The Assembly—elected by 3,796 voters, mostly landowners—passed the landlords' bill. Cuza, however, —strengthened in his convictions by a peasant rising in 1862—refused to sanction the bill, basing himself on the clear provisions of the Paris Convention, and entrusted Kogălniceanu with the government and with the task of introducing a new bill. Kogălniceanu pleaded with the Chamber that 'the peasants' past has been anything but happy, and their future must be different'; he warned his opponents 'not to think that we could, with our vote, drown the rights of the peasants for ever'; but all to no purpose. The 'monstrous coalition' of Conservatives and Liberals so amended the bill as to bring the proposed holdings down to the size of the lowest created by the Organic Statutes; while it increased by one-fourth the compensation which the peasants were to pay for the abolition of servitudes. Thereupon Cuza and Kogălniceanu dissolved the Assembly and launched their reform by means of a princely decree, which afterwards was confirmed by a plebiscite.

This had been the second opportunity which the Rumanian landed class had of legislating for themselves. They inaugurated the return to national independence by trying to complete the work of spoliation begun with the Organic Statutes, and to kill two birds with one stone. If they could have shut out the peasant from access to the land, they would have gained a hold commensurately strong on his labour—a scheme used with great effect in the African colonies, in order to compel the natives to work for the white planters. To save the peasants from thus being totally and finally despoiled, the temporary ruler had to resort to a *coup d'état*. All the privileged people, great boiars and smaller fry, looked upon the reform with hate or at least with suspicion. They fiercely denounced, however, any encroachment upon their own political rights granted them by the Paris Convention. Eighteen months after dissolving Parliament a plot drove Cuza out of the country for ever.

The Nature of the Reform. If Cuza and Kogălniceanu were well primed with liberal ideals, they were less well equipped with economic experience. Their reform, promulgated on 14th August 1864, was in its practical effects to bear the mark of that shortcoming deeply. If the peasant problem had become urgent in their eyes it was primarily in its legal aspect; just as seventy years earlier the French revolutionary assemblies, reacting to the clamour of the peasants, thought solely of breaking up the seigneurial system. Nothing could have seemed to these men more abhorrent than the lack of personal freedom: their main ambition was to free the peasant from servitudes.[1] To continue them would not in any case have been compatible with a system of written Constitution. It was for instance largely owing to the introduction of written constitutions that the peasants of southern Germany were emancipated earlier than peasants elsewhere, at the end of the eighteenth and the beginning of the nineteenth century.

Though in Rumania the Constitution was not yet down on paper, it was written large in the minds of the ruling men, the revolutionaries of yesterday. 'Two hundred boiars do not make a nation,' Kogălniceanu had exclaimed. Hence, every measure which hedged in the personal liberty of the peasant was broken through by the rural law.[2] It abolished all restrictions upon his movements; it did away, 'once and for all,' with all the dues in labour and kind, tithe included. For this the landlords were to receive a compensation in State bonds representing the tenfold value of the yearly servitudes owed by each category of peasants, with 5 per cent. interest; of the total, one-third was to be borne by the State, out of payments received from the peasants re-

[1] A beginning had been made with the emancipation of the gipsy slaves who formed the bulk of the servants in all the boiar households. There were about 150,000 of them, divided into three categories: those belonging to the State, to the monasteries, and to private individuals. The first two categories were freed in 1844, the third at the end of 1855. The origin of their status is unknpown, but gipsies never appear in Rumanian history except as slaves. (See Al. Golesco, p. 18.)

[2] The number of servile families was as follows:

	Muntenia	Moldavia	Total
On private estates . . .	210,000	120,000	330,000
On monastic estates . .	70,000	50,000	120,000
	280,000	170,000	450,000

(Al. Golesco, p. 91.

settled on State domains, while the remainder was to be paid off by the peasants within fifteen years, their annuities running from 51 to 133 lei.

Having made the peasant free of his person, the reform next aimed to make him free in his livelihood. The former serfs were settled, as owners, on holdings which varied in the two provinces and which were measured by heads of cattle they possessed.[1] Land under this scheme, however, was allotted only to the *clăcași*, i.e. to the men bound to servitudes; this left outside the scope of the reform the men who were free from servitudes because they received nothing more than house and garden, and who were obviously in need of land. But its main sin was that it aggravated, instead of redressed, the wrong caused to the peasants, from 1805 onwards, by the measures which had gradually whittled down their ancient right to the use of the land. The peasants had accepted the state of things, and in the *divan ad hoc* had claimed no more than two-thirds of the land; their right to this was still good law and practice in 1864, the landlords having to grant holdings to newly-married peasants from that reserve—a custom which had ensured means of existence to each new generation, and thereby had prevented the excessive splitting-up of the paternal holdings. The new measure did in fact take over two-thirds of the land, but on small estates only, where there was not enough to go round. The landlord's part was thus safeguarded where the peasants were too many for the land, but the peasants' part was not protected with equal care where it was larger than the area immediately required for the new holdings. The excess of villagers in the first case, as well as the newly married, were to be settled on State domains,[2] and not, as would

[1]

Categories	In Moldavia	In Muntenia	Number of peasants	Number of hectares
	ha. ar.	ha. ar.		
With four oxen . .	7 87	5 61	71,912	413,201·86
With two oxen . .	5 73	3 72	202,075	882,737·29
With one cow . . .	3 57	2 30	134,132	384,708·20
Land for house and garden	—	—	59,721	85,610·90
Total . . .	—	—	467,840	1,766,258·25

[2] The very popular law of 11th Dec. 1863, had 'secularized' the estates dedicated

have been just, on the excess of peasant land where the quotas set apart for the local villagers did not make up two-thirds of the private estate. The surplus on these estates, which formed the majority, was simply abandoned to the landlords; and whereas the landlords were to be paid for having to forego abusive privileges, that patrimonial land reserve was lost to the peasants without any compensation whatever. This course is the more difficult to account for as official advisers had pointed out that if that surplus land was to be left to the landlords, its value should be deducted from the indemnities they were to receive. This serious error in the reform meant a double loss to the peasantry: they were deprived of private land to which they had a title, and by the settlement of surplus villagers on State domains a great slice of the public reserve was also lost to future generations of peasants. The immediate loss was felt the more as the governments which followed upon Cuza's abdication neglected to carry out the provisions of the law concerning those peasants who had been left without land in 1864. Their settlement on State domains was not begun till 1876, on the eve of the so-called war of independence.

The Background of the Reform. Cuza's great reform had not matured from changed economic conditions and needs. Neither the growth of population, nor that of towns and trade, nor yet the requirements of the new State were pressing for a change to intensive capitalist farming, with relations between landowners and labourers based on a wage bargain. Rumanian agriculture still lingered in a primitive extensive stage. Such manufacture as went on was from the hands of artisans, organized in guilds;

to native and foreign monasteries, which covered about one-fifth of the arable area and made them into State property. The monasteries were invited to put in a claim for compensation, but as they refused to discuss or negotiate, hoping to gain more by foreign support, the State declared the question closed, in 1867. Al. Golesco (pp. 72–8) states that monastic property covered two-thirds of the total area in the two provinces, forming the richest Church possession in Christendom. The gifts had been mainly intended as charitable endowments for the maintenance of churches, of hospitals and schools for the poor, for the upbringing of orphans and the distribution of alms in time of famine, as well as for the provision of hospitality and asylum for travellers in distress. In certain cases, the excess of revenue was to be 'dedicated' as a homage to monasteries in Jerusalem, Mount Athos, &c., for the purchase of candles and oil, and the distribution of alms. But the foreign monks had often succeeded in laying hands on the whole of the revenue, and Russian intervention had prevented the Rumanian princes from putting an end to that abuse.

an intensive production and exchange of goods did not exist. There was no accumulation of liquid capital, and no 'entrepreneurs' urging its use in ways that would bring in quick returns. Class organization, outside the owners of land, was loose; and the budding middle-class had not yet the service and stimulus of a trained element. None of the factors, in short, required for passing from a natural to a capitalistic economy were yet mature in Rumanian agriculture. The way to that transition was indeed further confused by the reform of 1864. The reasons which caused Liberal institutions to spring up in Rumania overnight have already been described. This is not merely a figure of speech, for the draft of the original constitution, copied from the Belgian model, was produced within twenty-four hours, when the Radical leaders learnt that the stern Hohenzollern prince had eschewed the Austrian police and was driving post-haste towards the capital. In the West the system of production defined as capitalism had conquered the economic field before the class connected with it broke through the old political restraints and privileges; Liberal institutions followed in the wake of economic capitalism. In the backward agrarian countries that process was reversed; a Liberal organization of the State was adopted under the stimulus of contact with western Europe, and this opened and cleared a way for the reform of the economic organization. When individual freedom and equality before the law found their way into the political realm, the servitudes which still subjected the peasants to the landlords could not survive in the social sphere. Public relations cannot be segmented into watertight compartments—not legally, that is. In Rumania, therefore, the freeing of the gipsy slaves and the emancipation of the peasants had to be hitched on to the roundabout of political reform. Yet emancipation, though accompanied by a distribution of land, transformed the social structure without altering very much the economic ways of the rural world; and, especially, it left the narrower economic problem of production altogether untouched.

Nothing could better reveal the non-economic springs of the reform than its authors' complete neglect of the future of agriculture. Had it been otherwise they might have followed one or two possible courses. One of them might have been that favoured by the landlords: to emancipate the serfs without giving them land, which would have left a small number of well-to-do peasants as a rural middle class, but the rest with no other choice than to become wage labourers. This would also have forced the large owners to equip themselves with their own live and dead stock. As things were, agriculture was carried on not merely with the peasants' hands, but also with their ploughs and teams; most of the land farmed in Moldavia by its owners and the whole of such land in Muntenia was tilled by this kind of all-inclusive servile labour. Such an attempt at putting agriculture upon the road to capitalistic production, had it been politically possible, would still have had to overcome the lack of liquid capital; there was none worth speaking of in the country, and the little that could be obtained elsewhere was used up in the ambitious modernizing of the urban centres. Neither were men to be found with an interest in such a technical change and the ability to conduct it. The landed class, and, in general, the thin upper layer of the population, was soon absorbed in the machinery of the new State. The political field offered them rich opportunities of satisfying material and other ambitions, and town life invited them with its western refinements. If the landlords were anxious to increase their revenue, they were not at all anxious to change the prevailing system of production, which required no expert supervision, but merely the driving whip of bailiff and gendarme. Nor was there any other group of men capable of acting as entrepreneurs in a new agrarian system; the interpolation over a period of several centuries, of an alien administration between the few landlords and the peasant mass, and the backwardness of economic life, had left no room for the growth of a national middle-class. Alternately, a second line of policy might have chosen to favour a system of small cultivation. It might have handed the land over to the peasants, on the basis of their old rights, according to their powers of working, with a moderate compensation for the landlords. Having to find money for paying that compensation and for maintaining the State, the peasant proprietors could not have failed to develop their cultivation, for which, in such circumstances, they would have had ample scope. The reform of 1864, however, inclined towards neither large nor small cultivation, simply because it was not concerned with production at all. Those who had the peasants' cause at heart went to the length of a *coup d'état* in order to carry through their emancipation, and they also endowed them with a piece of land. But the holdings of the three categories then resettled averaged 6, 4½ and 2½ ha., respectively, which meant that of the former serfs only some of the first category, as A. V. Millo had conclusively shown, had any chance of building up an economic, self-supporting holding at all. As for large-scale cultivation, its permanent interests were not necessarily identical with the great momentary advantages which the landowners undoubtedly got from the reform. That the interests of large property and large-scale cultivation could at times even conflict with each other was to be shown by the later evolution of Rumania's agrarian problem.

The obvious truth was that political circumstances had dictated the change, and the social ideals set free by the French Revolution had spurred it on. In a circular issued during the agrarian debate, Kogălniceanu had urged priests and peasants to pray that 'God may help them to level up society, to lower the highly placed and to raise the humble'. As the springs of the reform so its flow. It made the peasant a free man, but it did not make him an independent producer. If freed him of his disabilities as a citizen, but on the same principle it also deprived him of the legal safeguards which had protected him as a labourer. It set him up as full owner of a piece of land, but not of a piece big enough for him to live by with the prevailing agricultural methods; nor did it give him the training and means by which he might have got from his holding all that his household needed. So that M. Garoflid has been able to say that 'in 1864 only the juridical problem of the peasant, who becomes free in law, was solved; economically only the estates were emancipated'. The latter result was not accidental. Economic development was hampered because most of the land, the country's only wealth, was tied. One-fifth of the whole arable land belonged to the dedicated monasteries; of the rest, two-thirds was either in the possession of the serfs or reserved for their descendants, while the remainder included extensive estates of the princes and of lay endowments, so that only a minimal extent of land was available for commercial transactions. Kogălniceanu himself, in pleading for the reform, 'as a landowner who cares for the patrimony of my children,' emphasized the need of 'emancipating the large property from the servitude of having to reserve two-thirds of its extent. And thus we will establish in Rumania absolute, western property in the place of Slavonic property'.

The landlords, indeed, were released from all the servitudes which had qualified their title to the land, including that of having to give preference to their own villagers when letting the landlord's third of the estate; a release which was to be viciously exploited by them for screwing out of the villagers, bound to their plots, onerous terms for the lease of additional fields. Similarly, the landlords now escaped the elaborate administrative interference which had controlled their relations with the peasants till then. Probably one of the reasons which had guided the authors of the reform was the expectation, based on the most up-to-date economic doctrine, that a free agriculture would prove a progressive agriculture. In the West, indeed, emancipation meant a loss for the peasants, but at least it opened the door to improved farming. But the Rumanian landlords had neither the knowledge nor the means, nor indeed the inclination, for technical development, and they could not, therefore, dispense with compulsory labour. When it was seen that to abolish the present servitudes could no longer be avoided, they concentrated their efforts upon having the peasants endowed with as little land as possible. Their calculations proved painfully right. For when the peasant, economically unprovided and socially unprepared, was turned adrift on the sea of *laissez-faire*, the landlords found it easy to pull him into the backwaters of a servile economy again. Labour dues, abolished as a legal system, continued as an economic practice; but they were no longer compensated with a right to use the land. In addition, the peasant now had to find ready money for paying the indemnity and the State tax. As, moreover, he had no political say whatever in a country administered by corrupt officials and judges, as soon as he was freed from his chains the peasant found himself being sucked down into a whirlpool of ravaging economic and political forces which he could neither grasp nor control; and to which, in the first years, Nature added her parching wrath, as if to try the patient tiller of the soil to the utmost in his new freedom.

Practical Effects of the Reform. The cumulative effect of what was erroneous in the reform of 1864, and of what was made worse by unfair application, was to depress the peasantry to a level of constant misery. Generally, the Rumanian authorities have not troubled to find out the effect of the many rural measures they have enacted. But in this case, a solitary exception, the very able Ion Ionescu, who had acted as neutral chairman of the mixed commission in 1848, was deputed to make a detailed inquiry. He did not come to deal with more than three districts, but these were picked out by him as typical of different regions, and his three full reports form an invaluable guide as to how the reform was applied and what came of it. In more than one way, what was done or not done in 1864 closely resembles the faults of commission and omission of the present reform, and one may, therefore, expect that some of the consequences will bear the same likeness. And, further, but for the shortcomings of the system established in 1864, Rumania's agrarian problem might have evolved in a way which would not have led, as it inevitably did lead, to the need for those revolutionary changes which took place in 1917–21. There is, therefore, more than one good reason for noting briefly what happened on the land after Cuza's reform had become law.

Land tenure, in general, was given a fatal turn in 1864: more than half the arable land was finally given up to a small group of landlords. Moreover, many bad abuses were committed when dividing and transferring the peasant holdings. In more than one place the serfs were evicted before the law came into force, or were transformed, by the sale of a small plot, into 'proprietors' no longer entitled to be resettled.[1] Elsewhere they were duped into refusing to be resettled, partly by false promises, partly by being made to fear the compensation they might have

[1] Ionescu, *Agricultura Română in Judeţul Mehedinţi.* In 1859, there were 25,089 serfs, in 1864 only 21,708, though the newly-married were included in the intervening years—'as if it were possible for those added to the population to exist without working' (pp. 158–63).

to pay. When it came to measuring, it was done *en bloc* for the land due to all the serfs of one village; the landlord's or the State's part, that is, was settled at once, but the parcelling out of the individual peasant plots was in many places not done by 1868, though the peasants were paying their annuities in full already, and this led to much friction in the village and to indifferent cultivation. Very frequently, it would seem, good land which the peasants had been tilling was taken away and bad land substituted for their new holdings; some of the fields given them were so bad that the peasants paid the compensation without taking over their new property, or they left it untilled and rented their old plots on heavier terms than before. Often, too, the peasants were given plots which had no roads leading to them. The landlords also saw to it that of the peasants to be resettled on their estates, as few as possible were classed in the higher categories, and as many as possible in the lowest, so that relatively less land was taken from them and to that extent more from the State domains; which was a despoiling of the public wealth and a further curtailment of an essential land reserve.[1]

The first visible effect of these malpractices was greatly to reduce the extent of land which the peasants had in use; apart from the fact that much of it was now worse in quality.[2] Within one year, so Ionescu asserts, the terms for the leasing of *prisoase* (surplus fields) had doubled and trebled; therefore, the peasants took little or none of them—an important cause of the heavy fall

[1] Had the distribution been fair, the relation between the various categories should have been on private estates similar to that on the State domains. But of the total number resettled, there were in Putna

of the first category	21%	on State domains and	13%	on private estates
with land for house and garden	18%	,, ,,	24%	,, ,,
and in Mehedinţi, first category	10%	,, ,,	8%	,, ,,
last category	4%	,, ,,	18%	,, ,,

(*Ib., Jud. Putna*, pp. 89; *ib., Jud. Mehedinţi*, p. 158.)

[2] In Dorohoiu the peasants had in use

	in 1859—	after the reform
gardens	1,858 falce	4,691 falce
arable and grazing . . .	62,045 falce	45,493 falce

(*Dorohoiu*, p. 377.)

In Putna the serfs had in use in 1859, 38,000 falce, in 1864 received about 22,000 falce. This happened not only on private estates; at Brezinţia, State domain, the peasants had worked before about 2,000 pogons arable land, and were now given 700. (*Mehedinţi*, pp. 610–11.)

in sowings.[1] The shortage of grazing land, especially, was to become severe. Both the otherwise reactionary project of the Central Commission and Kogălniceanu's first draft contemplated the creation of village commons, but instead the final law gave the peasants individual grazing plots. These were ploughed up, soon to disappear altogether, leaving the peasants with the alternative of paying the landlords what they asked for grazing, or of getting rid of their cattle. Usually, they had to do both. Oxen they had to keep for ploughing and carting; but afterwards villages were found with 420 households, and only two cows, or others in which children of ten and twelve years did not know how cow's milk tasted.[2] The result, for a time, was a general decline of agriculture.[3] To this the bad droughts, and the farming landlord's fear that he would not be able to get labour from the new peasant proprietors, both contributed. Ionescu, however, gives instance after instance to show that, where relations between landlord and peasants had been good, cultivation was found 'easier, better, keener' now than when it had been done with servile labour.

After the reform the price of wheat rose rapidly, and landlords and tenants were extending their cultivation. Yet at the same time the price of agricultural labour fell by one-fourth to one-half. Instead of being statutorily fixed as before, the price of labour was now open to fluctuate with supply and demand; the sequence of bad harvests, the insufficiency of their holdings, and the need for cash to pay annuities and taxes, which in general had doubled, forced the peasants to sell their labour—with a growing practice among landlords and tenants to make contracts in winter time—and caused the change to weigh

[1] Sowings in Putna:

	Maize (Falce)	Wheat (Falce)	Total (Falce)
1859 . .	16,615	5,192	30,000
1865 . .	12,586	2,776	16,677
1867–8 . .	14,412	3,189	21,833

(*Putna*, pp. 104–7; *Mehedinţi*, pp. 478, 485–6).

[2] At Corzu they had to pay for grazing ten heads of cattle more than the whole of their old servitudes and part of their hay, usually one-third. (Ionescu, *Mehedinţi*, p. 492.)

[3] In Putna

	Men on the land	Oxen	Horses
1859	24,848	19,055	4,483
1869	21,846	15,580	228

(*Mehedinţi*, p. 94.)

heavily against them. Those of Mehedinţi were able to pay their annuities in full in the first year, but soon they wallowed in difficulties and by September 1867 they were in arrears with almost half of their annuities. Ionescu records as 'noteworthy' the solitary case of a peasant who had been able to pay off, in a lump sum, the whole amount of his indemnity. In Muntenia, especially, besides the increase in the payments in labour and kind, the landowners increasingly indulged in the habit of asking *ruşfeturi*, i. e. 'gifts' of chickens, eggs, and other products from the peasants' household.

Altogether, the peasants were being severely pressed by circumstances; and, for reasons presently to be mentioned, by the landlords also. The rural law confirmed, wherever they had existed, the rights of the villagers to wood for fuel and building, but all accounts agree that in practice it was denied without compensation. Many writers accuse the landlords of having used unfair means for bringing the peasants in their power. When the land was redistributed, for example, they shut off the villagers from their watering-places; or they retained on the edge of the village a field into which the cattle of the villagers were bound to stray, extracting thereafter heavy fines from the owners. 'In one village they even took the well and spring that were within the village and gave drinking-water for men and beasts, and for the watering of the gardens; afterwards demanding twelve days' labour from each man for the water he drinks in the village.' All these burdens, fair and unfair, grew more oppressive as holdings were split up among descendants of their first owners. Formerly, if the lots were restricted, at least the remainder up to two-thirds of the estate was reserved for the newcomers among the peasantry. Now the lots were restricted without there being any reserve available, and newcomers had to be provided from what the peasants had, by dividing the holdings. The former individual limit had been a permanent standard, the new individual limit was a temporary maximum.

A writer who examined in detail budgets of the three categories of peasants resettled in 1864 came to the conclusion that a family belonging to the first category—possessing four oxen and one cow—working all of them, children included, very hard, with the aid of a farm-hand, and living miserably, with meat on not more than fourteen days in the year—such a family, if there were no untoward accidents, could muster a surplus of about 100 lei at the end of the year. The other two categories could in the best circumstances not escape without a deficit of some 200 lei yearly. Among the various reasons for this calamity was the working of the indemnity: as it was paid for the abolished servitudes, not for the amount of land received, its incidence was in inverse ratio to the means of the three categories of peasants. Other writers, recalling boyhood memories, or general observations they had made, confirm that 'since they have become proprietors and free citizens, the peasants have fallen into poverty'.

'I have collected in various places facts', says Ion Ionescu, 'which show that a man cannot pay with such labour as he can perform in a year even the interest on the borrowed capital; for interest, like capital, is paid in labour. . . . Men have died without being able to pay off their debt; but their widows and children are forced to labour and to pay, even though they may have inherited nothing!'[1]

Speaking generally, one can reduce to three the motives which led to the emancipation of the peasants in the West. There was first the technical motive, derived from the teaching of the Physiocrats, which demanded the economic emancipation, by abolishing all servitudes, of both soil and labour. The second motive was humanitarian, inspired by the same philosophical sources, which deprecated all restraints on personal freedom and required the personal emancipation of the peasants. And, thirdly, there was the political motive pleading for the abolition of all political and judicial class privileges, because they were incompatible with the philosophy of the modern State, as based on equal constitutional rights. If one surveys the Rumanian emancipation in the light of those propositions, one finds that the economic motive counted only partially. There was a desire on the part of the landlords for free transactions in land property, but there was no demand at all for a technical change in the system of cultivation, which had been the main incentive in the West. Economically, therefore, the peasants were expected to continue in the same relationship in which they had stood before

[1] *Judeţul Putna*, pp. 99–100.

to the soil and its owners. But to what extent could their persons and their class be safely made free, if their labour was to remain tied? In comparing the plantation systems of Romans and Americans, of the Ancient World and the New, Max Weber says that in both cases it was found that 'slavery was profitable only when handled with the most rigid discipline, associated with ruthless exploitation'. *Mutatis mutandis*, this evidently is true of any system of production which is not based on a free labour contract. As the Rumanian reform emancipated the peasantry politically merely in form—because the electoral laws debarred them in practice from all franchise—it was questionable whether the third wing of the reform—personal emancipation—could remain unclipped in the hands of a landed class which possessed a monopoly of political power and which desired to retain control of the peasants' labour. It could not have been difficult to foresee that either the old agricultural system or the new personal freedom of the peasants would have to give way.

The emancipation of 1864 was an urgent reform, and the handful of men who sponsored it have acquired lasting merit by seeing it through in the face of such wide and violent obstruction. It does not detract from the personal side of that achievement to admit that, practically, it was a failure. The reform cannot be said to have brought economic improvement in its wake. In the West, as Signor Ruggiero has pointed out in his *European Liberalism*, the innovations of the new economic rationalism consisted essentially of simplifications. But in Rumania, the customs which followed the reform, being devised mainly for eluding it, were more mixed and complex than the former traditional relationships on the land. The technique of production, again, was hardly affected.

More palpable still were to prove the social shortcomings of the change. An ideal reform would have made the peasants both economically and politically independent. The reform of 1864 did neither. It did not give them sufficient economic strength to stand up against political inequality; nor did it give them sufficient political power to withstand economic oppression. Subsequent history appears to justify M. Dobrogeanu-Gherea's bitter epigram—that 'the reform of 1864 carried in its womb the terrible year 1907'.

CHAPTER IV

THE NEW STATE AND THE NEW SERFDOM (1864–1917)

In May 1866, when Prince Carol of Hohenzollern mounted the throne, the young Rumanian State acquired real independence. Nominally it remained under Turkish suzerainty until 1877. But from 1866 onwards no outside factor dictated what laws should be made, or watched how laws were applied. At last the new ruling class found itself in undisturbed *tête-à-tête* with the mass of its people; and from that moment the peasantry was shut out from all direct share in Rumania's political life. Its last chances—until 1917—of participating in the country's government had been in the mixed commission of 1848, the divans of 1857 and the plebiscite of 1864. Now the victors in the national struggle had their hands free to proceed to settle the social contest—a contest which they had waged since discovering that corn could not merely be eaten but could also be sold.

In Rumania the new ruling class still consisted in the main of landed people, with a sprinkling of urban traders and members of the liberal professions. The revolutionary ideas of 1848 which preluded the movement for independence, had found an echo not among urban trading and manufacturing class—for that class was small as yet and consisted mainly of strangers—but they hit the ears of members and scions of the landed class who resented the political monoply of the two-score great boiars among themselves. Socially, however, great and small landowners were welded together by property and its interests, by the conditions of their life and its dependence on tithe and labour dues. After a spirited beginning, therefore, the revolutionary government of 1848 hastened to dissolve the joint agrarian commission when 'the sittings began to get troubled'; and thereafter resettlement was not again mentioned by the revolutionary leaders. When, later on, the peasant deputies came forward with their grievances in the Moldavian divan of 1857, they were faced with the displeasure and resistance of small and large owners alike, and the great idea of national union 'almost foundered on the rock of the agrarian question'. Hence the Great Powers had to demand a settlement of that question, so as to give the new State a fair chance of stability; yet even left-wingers of the Liberal group still did not think that rural reform was urgent. They severely deprecated the raising of that question and wanted to begin rather 'with other reforms—with the moral improvement of officials, with guarantees for trade, with credit institutions'.

'All were feudals, and all wanted free trade,' sums up M. Garoflid. The corn trade was bringing rapidly increasing profits, and land values were keeping pace with them. The Black Sea was now open; a European Commission was making the mouths of the Danube safely navigable; in 1860 the first railway was built by an English company across the Dobrogea, to link up the Danube corn ports to Constanţa. With the advent of Prince Carol, an enlightened and capable ruler, road and rail transport were energetically developed; he also inaugurated a determined commercial policy, based on treaties of commerce in which advantages were secured for Rumanian corn. Spurred on by the high profits that could be made by exporting grain, the landowners strained every means to extend its cultivation.[1] They wanted more land and labour; and they wanted freedom to trade. But for the reasons mentioned in the previous chapter, they were not yet ready to admit free production. Hence their bitter opposition to the reform of 1864; and their immediate efforts to mend the breach made by foreign pressure in the feudal wall, as soon as the first effects of the reform appeared to justify their fears.

The absence of all agrarian method from the anatomy of the reform, namely, queered its progress from the very year of its birth. The year's work was to be carried on, in 1864, on the strength of the old arrangements, but in most places the peasants

[1]

	Area cultivated Hectares	Percentage of total area
1860	2,494,220	19·90
1901–5	5,236,332	39·86
	Value of exports Lei	
1866	116,500,363	
1905	457,101,394	

refused to plough the landlords' land any more. Official accounts rendered to Parliament in 1865 admitted the 'grave momentary disturbance of the whole national economy produced by the change in the nature of the former relations between landlords and peasants'. Swayed in turn by resentment for their former oppression and by exultation in their new freedom and property, the peasants were not in a mood willingly to continue serving their old masters, or to hire themselves out as wage earners just when they had become free owners. But the peasants' prepossession was soon cut short by their needs. They needed more land urgently and now they also needed money; as their plots tied them to the village, they were altogether dependent for the satisfaction of both needs on the local landowners. Money, however, they had none; the only thing they had in abundance to proffer in payment for additional land was their labour, and it was inevitable that the terms of such renting should be harder now than under serfdom, when they had been armed with a right to claim land. The other side of the picture was the landlord's doubt whether he would get all the labour he needed, at the proper time. During serfdom he had been able to hold the peasants to it, under threat of their forfeiting all claim to land. Now the peasants were free to give their labour or to refuse it, backed as they felt themselves to be by their own bit of land; and that made the landlords feel that the bottom had been knocked out of the system which had kept them in a life of ease and plenty. They feared the change the more as land values and profits from corn were rising excitingly fast; and town life was making much heavier demands on their time and purse than had the patriarchal existence at their country seats. What they needed was easily stated: more money with less trouble from their estates. Politically that translated itself into an anxiety to prevent more land from getting into the hands of the peasants; and, above all, to make sure that the supply of labour would flow as dependably as it did before the emancipation.

The consequence was almost fated. Cuza's reform had changed land tenure without reforming the country's agrarian economy; agriculture, therefore, found itself unable to stand solidly on these mixed foundations and leaned back upon its old feudal-servile supports. The two tendencies which the landlords had nursed during the period of their political impotence and of national insecurity, now come freely to the surface as guiding aims of their legislative activity in the young State which they dominated; the State itself being the largest landowner and always in need of funds. The first of these two aims was served by the rules they adopted for the sale of the State's extensive domains. And the second was masterfully achieved by means of the laws on agricultural contracts, which revived the old conditions in all but their juridical form. It is from the latter that Rumania's agrarian system derived the original traits that have distinguished it during the first half-century of her independence; for, with mild alleviations, the régime which those laws created—styled by a Socialist critic the 'neo-serfdom'[1]—has prevailed up to the Great War.

Land Policy after 1864. The landlords had not been altogether unsuccessful in their efforts to keep the size of the holdings distributed to the peasants in 1864 as low as possible; and what they failed to get into the law they were able largely to put through in its application. The main sins of that reform in regard to land tenure had been to abandon to the landlords the surplus peasant reserves, leaving many peasants temporarily without land; and, secondly, to have abolished without compensation the right of newly-married peasants to a holding. Articles 5 and 6 provided that land for these two categories of peasants was to be provided on State domains, and the need of the first group was clearly urgent. Yet nothing was done towards carrying out the law till Rumania was faced with the prospect of war. In the first days of October 1876 the Rumanian Premier travelled to Livadia to negotiate the conditions on which his country might co-operate with the Russians in the impending campaign against Turkey. On the 27th of the same month the Government issued an order interpreting Articles 5 and 6 of the law of 1864. In the following May, Rumania proclaimed her independence; the army was mobilized, and as the country-side seemed lukewarm, the Government thought it prudent to let it be known that land would be given to those taking part in the

[1] Dobrogeanu-Gherea, *Neoiobăgia.*

war. Finally, instructions for the actual application of the two articles were given in the spring of 1878, after having left 48,342 newly married couples to exist as landless labourers during thirteen years. The new holdings were, on the average, somewhat larger than those of 1864, but still within the limits of that law. As before, the application of the measure gave occasion for fresh miscarriages, which, in this case, a law of February 1887 tried to correct, ten years later.[1]

The attitude of the ruling class towards the perennial land hunger of the peasantry need, however, not be inferred indirectly from their remissory completing of the law of 1864. Official policy was openly and directly expressed in the series of laws for the sale of the many extensive domains which the 'secularization' of monastical estates had procured to the State, and which were disposed of in ever-growing numbers whenever the Treasury had a gap to fill. The trend of the laws authorizing these sales unmistakably favoured the passing of the land into the hands of the surfeited large owners rather than into those of the land-hungry peasantry.

A first sale of smaller domains was decreed by a law of July 1866. It allowed the land to be divided out into small lots of 3–25 ha., but the sale was to be conducted by public auction, in Bucarest, and the purchase price had to be paid in full within one month. There is no evidence that any land was sold to peasants on the strength of this law; most of the estates, in fact, were sold undivided. Two years later, the law of July 1868 decreed a fresh sale of State domains, and to make bidding easier, the auctions were to be also held in Jassy and Craiova. But on this occasion the mind of the law-makers was allowed to appear without a veil. Purchasers of whole estates were to be asked to pay one-third of the price within one month, while the remainder was to be paid off in twelve annuities calculated at 6 per cent. interest and 6 per cent. capital. But any one buying small lots of 3–25 ha.—the only kind that came into question

[1] M. T. G. Bibicescu, *Chestiunea agrară*, Bucarest, 1907, p. 37, affirms that many of those entitled to land had not received their holdings even then, after more than forty years; but they were made to pay the annuities and even the land-tax for them.

for the peasants—had to pay within one month the whole value of his purchase.

The first sign of a change of mind more favourable to the peasants is to be found in the law of August 1876, passed on the eve of the war of independence. From that date land could be auctioned in the chief district towns as well; it could be divided into lots from 2½ ha. upwards, 'always taking into account the demands of rural inhabitants of Rumanian stock'; and for such lots only one-third of the price was to be paid at once and the remainder within twenty years, in half-yearly annuities at 6 per cent. interest. Again, there is no evidence that any peasants were able to purchase land under this law; and the facilities which it seemed to offer them were in effect nullified by other conditions demanded by the law of April 1881, after the war was over. The promise made by the Government during the campaign of 1877–8 had not been kept, but it had aroused the expectations of the landless peasants. The new law allowed domains that were put up for sale to be sold in small lots, at a fixed price, but only if the offers received covered the whole of the estate; moreover, the law indicated that the smaller domains should be sold in lots from 25 to 130 ha., the larger domains being indicated for peasant holdings, which made it even more difficult to satisfy the other condition. And, in addition, the buyers of such small plots were to be collectively responsible for the whole purchase price. It is interesting to note that this is the only law for the settling of peasants on the land made by the Liberal Party, the exponent of the undiluted nationalist ideal.

If one looks at the law of 1884 together with its amendments of April 1886, one can discern some of the conflicting tendencies in the land policy of the State. Intent, apparently, upon stemming the growth of latifundia that law decided that State domains of more than 2,000 ha. should be sold in individual plots not exceeding 5 ha. each, and only to Rumanian cultivators, village teachers and the personnel of village churches, at twenty times the actual rental value. The amendments of 1886 made available for the same class of purchasers two-thirds of the domains of one to ten thousand ha. and one-fourth of the area on domains beyond that size; but if the reserved part in these categories were not sold off in small plots, it could be sold in lots of 50 to 400 ha. There is here a recognition of the need to make at least some of the State land available for the peasants; but the policy of extreme parcellation was continued with regard to them, no one being allowed to buy more than two plots of 5 ha. each, whereas the large owners were permitted to round off their possessions with as much as 400 ha. at a time. Altogether, as a result of these laws, 641 State domains passed into the hands of large owners and 39 into those of peasants, up to 1886; the first group covered together 153,297 ha. sold to 760 buyers, i.e. an average of 200 ha. per buyer at an average price of £13 a hectare; though some of the sales rose to 2,000 ha., and some of the prices were as low as £5 per hectare.

Land Policy: Second Period. An outbreak of peasant risings, from the spring of 1888 to that of 1889, in various parts of the country, startled the new State out of its day-dream of smug prosperity. The peasants were more violent in their actions than definite in their aims, yet there was much to be learnt from the destructive bitterness with which they attacked the possessions of large tenants, while actually protecting from harm those of the few owners who were still farming their own land. And it was even more instructive to see the peasants rise not in the poor mountainous districts, but in the rich and underpopulated plain, quite close to the capital. Something had changed, after all, in Rumanian agriculture after 1864; but less in the manner of cultivation than in the mode of management. Every year more landowners abandoned the country-side and handed over their estates to tenants, but in Rumania, with rare exceptions, these were not professional farmers, trained for it and equipped with all the stock it needed. Rather were they farming speculators, renting for short periods at exorbitant rents, who had to make sure in the few years at their disposal of such a wide margin of returns as would give them a good profit whatever the natural risks of the trade. Devoting themselves to the higher function of organizing the new State to their views, the Rumanian upper class left it to despised aliens—Greeks, Jews, Armenians—to perform the menial task of coining ever more money for them by wringing the strength out of soil and labourers. The way in which the peasants discriminated in their attacks of 1888–9 showed that the canker of absenteeism was beginning to tell; it has never ceased spreading till the post-war reforms.

Absenteeism found its natural medium in the rich wheat-growing regions of the plain. High rents made it impossible for the peasants to compete or keep pace with the movement there. An inquiry of the Ministry for Home Affairs proved that it was in such districts, where land was especially dear, that most of the peasants were to be found who had let their own holdings; and that in such regions rents for land leased from peasants were lowest and the period of the lease longest, sometimes running to ninety-nine years, which in Rumania was an extreme exception. Often, too, the peasants were tempted or driven to enter into a new lease with a fresh tenant while the old lease was still running; a custom which was known among large owners also. Other figures proved that small holdings were more numerous in the mountainous districts, where corn-growing was limited, than in the fertile plains; their average number being 38·853 in the twelve mountainous districts, but only 27·453 in the twenty lowland districts, of the thirty-two in which the country was divided.[1] These and other facts might have continued to be ignored, had not the risings of 1888–9 given warning that the prosperity of agriculture had not increased the well-being of the peasantry.

The law which was brought in by a Conservative Government in April 1889 was the first to try to satisfy the peasants' need for land. It allowed all the remaining State domains (876 with an area of 1,200,000 ha.) to be sold to Rumanian cultivators, but only in lots of 5, 10, and 25 ha. The two larger categories were to be sold by auction, and the lots could be resold by their purchasers after having been fully paid up. The bulk of each estate was to be reserved for lots of 5 ha., to be sold directly, without any prepayment, the cost to be covered by annuities of 6 per cent. interest and capital; no individual might buy more

[1] In the first districts the average size of large estates, including much woodland, was 661·5 ha., while in the plain the average size rose to 1,011·24 ha. N. R. Capităneanu *Raportul Asupra Recensământului Fiscal din 1905.* (M. Vandervelde has pointed out in *Le Socialisme et l'Agriculture* that it is a general observation that small holdings predominate in the poorer agricultural regions.)

than one of these lots, nor could they be estranged during a period of thirty years (according to Art. 132 of the Constitution). The evident purpose of the law was to satisfy as large a number of peasants as possible, but in many cases the results were unsatisfactory. Exemption from prepayment, indiscriminate granting of lots, together with the absence of all obligation for the purchaser to settle on it and cultivate it, caused much of that land to pass into the wrong hands. Moreover, the peasants had bought no land at all on the strength of those laws up to 1881, and even the law of that year had attracted no more than 4,970 peasant purchasers who bought together 23·069 ha. and 1,717 purchasers who bought larger lots amounting together to 8·228 ha.; but on account of the easy conditions of the law of 1889 the demand for land far exceeded the available area.[1] Its application, therefore, caused much friction, until a Liberal Government so amended the law in May 1896 that State land could be sold no longer except in lots of 5 ha. only, preference being given to peasants who had no land at all, and then to those who besides house and garden did not own more than ½ ha. That these changes were dictated by the pressure of social conditions rather than by any economic policy is shown by the curious provision that would-be purchasers should be entered on a list, from which the actual recipients of holdings were to be chosen by a draw. The peasants made such a rush to buy land under this law that excessive prices were offered for the larger plots, sold by auction, which afterwards many were unable to pay, even in years of good harvests. The law of March 1899, therefore, authorized the Government to reduce, as from the date of the sale, the price of lots of 8–25 ha. sold on the basis of the law of 1889; and the remaining annuities were to be paid at 4 per cent. interest within sixty years. The large purchasers had already benefited from a similar generosity in 1879, when many of them were remitted arrears of payment while being allowed to retain the estates, and even to buy further land from the State a few years later.

[1]

Lots of	5 ha.	were purchased	by	105,165	peasants,	making	526,233 ha.
,,	10 ,,	,,	,,	1,163	,,	,,	11,690 ,,
,,	25 ,,	,,	,,	386	,,	,,	8,670 ,,
				106,714	,,	,,	546,593 ha.

Altogether, including the reform of 1864, 629,583 peasants received or purchased land to the extent of 2,572,579 ha.; of these, 1,378,197 came from State domains. During the same space of time the State sold 164,942 ha. to large owners. These figures do not, however, give a true picture of the progress of land tenure in that period. Lots were sold without discrimination, to stop the clamour, and many got into the hands of people who did not cultivate them but let them to petty local officials or to publicans for the amount of the annuity. At Băiceni, to give an example, in the district of Jassy, an estate sold in 1904, only fifty-five purchasers settled on their lots, while eighty-six were absent; and of the latter fifty-two let their land to others. When the authorities intervened, it was not to cancel such ill-used sales, but merely to hand over to the fiscal authorities the leasing of plots belonging to absentee owners, by auction; with such results as that exemplified by the tenant who, in 1906, held in lease a closed group of over hundred such plots, at about half the local rental value. The prohibition to sell was easily circumvented by long leases. All these measures, therefore, failed to solve the agrarian problem; nor does the evidence suggest that they were meant to achieve that. State land was put on sale whenever the Treasury was in arrears, and with a marked preference for large purchasers. When discontent on the land grew violent enough to give warning that something must be done for the peasants, the ruling powers altogether ignored the evident lessons of the reform of 1864 and went on distributing to as many peasants as possible plots of land from which they could never get sustenance for themselves and their families. In all that span of time only 386 peasants acquired sound holdings of 25 ha. each, amounting in all to 8,670 ha. The average resettlement lot (including a large number of plots sufficient for house and garden alone), had been of 3·87 ha.; the average lot sold to the peasants had risen to only 4·88 ha. As M. Garoflid has shown in detail, in his valuable *Chestia Agrară*, it was impossible for a family to live from a holding of 5 ha. The distributing or sale of land to the peasants had been a social and political rather than an economic measure; it never altogether freed them from the need of selling their labour.

The laws on Agricultural Contracts. Natural conditions—an extreme continental climate—in Rumanian agriculture demand a spurt of intense labour during a short period of time.[1] Could the peasants be expected to devote it to the landlords rather than to their own holdings? It was easy enough to get the peasants to engage their labour; the land reform had left them hungry and a bad drought had left them starving. But the landlords had no security that arrangements, made in effect under duress, would be carried out. A Senate resolution of January 1865 said that to leave them to be enforced, in case of need, by ordinary judicial methods 'would upset agriculture in a very damaging manner'. While the parties were at law, the fields would remain untilled; moreover, it was little good obtaining damages against a resettled peasant, as his house and land and essential stock could not be sold for debts. Above all, 'during the agricultural seasons one could not find other men whom one might engage at the expense of those under contract'. In these words the preamble to the law of 1872 gave the true reason for the legislation that was demanded as soon as the reform of 1864 was passed. Rumanian agriculture had not reached the state when it might work with free labour; as late as 1882 this was admitted in the statement of motives for the law of that year. After declaring that it was yet premature to place agricultural contracts under the jurisdiction of the common law, its authors asserted their belief that 'the country still needs a special law for developing its productive forces; our agriculture, which is yet in a rudimentary stage, with undeveloped means of production, still needs a special protection which shall assist and sustain it in its efforts to become prosperous'. In reply to an inquiry from the Government, the district councils were unanimous in urging that agricultural contracts should be left to be enforced by administrative organs and means. A bill giving effect to this recommendation was introduced in December 1865, and became law in March 1866—one month after the forced abdication of Cuza, its provisions having been stiffened in that short interval.

[1] Even the army is used for agricultural labours. In 1912, e.g., 26,538 soldiers were 'lent' to the large farmers. The practice is said to have continued even after the War.

This law on agricultural contracts was the first of a series which sanctioned relationships on the land differing little in practice and in effect from the régime of the Organic Statutes. They changed the trend of the State's interference from the protection of the peasant's labour while he was tied to the land, to the conscription of his labour, now that he was free to move. The first clause of the reform of 1864 had declared all labour servitudes abolished; two years later they were revived by the first National Assembly, composed in the main of the new nationalist and anti-boiar elements. The law allowed landowners and their tenants to commute into labour obligations any sum owed to them by the villagers for advances in money and in kind, or as rent for land. Debts towards money-lenders, publicans, &c., were also passed over in practice to landowners or tenants and were by them transformed likewise into labour dues. Such obligations formed a privileged claim, second only to rates and taxes. Labour contracts could be concluded for periods of not more than five years; they could be collective, but though the law forbade joint responsibility (Art. 29), it was in fact frequently stipulated and the judiciary enforced it. In many cases the peasants were made jointly responsible for the labour of even those of them who ran away or fell sick. The enforcement of these contracts, as well as of fines and penalties arising from them, were entrusted wholly to the local administrative organs; and there was no appeal from their decisions (Arts. 30 and 31). When a peasant failed to carry out his contract, it was the duty of village mayor and council first to urge him to perform his labour obligations (Art. 13); if that were of no avail, then 'the village mayor, using the executive means at his disposal, will at once bring the defaulter to the labour for which he has contracted', or he may engage other labourers at the expense of the defaulter (Art. 14). The last provision was done away with by the amending Act of 1872, the rapporteur admitting that no spare men could be found during the agricultural seasons. Instead the Chamber added to Art. 13 a clause which instructed the village council, if the debtor was recalcitrant, to 'obtain at once from the district sub-prefecture, at the expense of the defaulters, the military assistance necessary for constraining the villagers who are obdurate or have run away'. Until then the only means by which runaways could be induced to surrender was to persecute their families; now direct action, with the aid of the military, was sanctioned by Parliament. The other change made in 1872 was to allow peasants contracting as a group to be made jointly responsible, this being demanded by the Liberals on *laissez-faire* grounds. The village mayor was obliged to establish the default and to apply the measures of constraint 'on the very day when the complaint against the labourer has been lodged'; if he failed in this, he was liable to a fine of 50–100 lei and also to the payment of damages to the suffering party.

Some mitigating changes were made by the new law of May 1882. In general, its text was clearer and more definite. Contracts, it decreed, were to refer to genuine agricultural work only, to be performed in the place where the contract was made, and they were to be renewed each year. The villagers were forbidden to contract for more labour than they and their families could reasonably perform. Fridays and Saturdays were reserved for the peasants' own needs; they could not be pursued for refusing to work for the employers on those two days. Enforcement with the aid of soldiers was abolished, and joint responsibility was no longer permitted. When a villager failed to carry out his obligations, 'the mayor will urge him to do the work'; if he refuses still, the mayor could engage other men, at local rates, and sell such of the defaulter's possessions as were not protected by law in order to defray the cost. The price from which that forced sale would start was fixed by the creditor. If other labourers could not be found, the employer could claim damages through the Courts. The anxiety caused by the risings of 1888–9 led the people in power to think of modifying the law of 1882, but quiet being restored, the change was allowed to wait till the law of May 1893. Population was rapidly increasing and some of the large farmers were introducing machines; the need for compulsory labour was diminishing. Hence, besides two minor but useful modifications—grazing contracts were to specify the size of the land and not merely the number of animals, and gifts or payments in kind were forbidden if they were not the produce of the rented land—the new law was remarkable especially for its famous final article (Art. 44) which allowed the two parties, if they so wished, to base their labour contracts on the common law. The new text abolished execution of contracts in the person of the peasant, authorizing it in his possessions. In that state the legislation concerning agricultural contracts remained till 1907.

If one surveys the position of the peasants under the laws on agricultural contracts one discovers that it contains all the material elements which characterized their status before emancipation. The first essential aspect of serfdom, bondage to the soil, was in part created by the reform of 1864 itself when it decreed that the new holdings were inalienable; this was confirmed by the Constitution of 1866 for a period of thirty years, which in 1877 was prolonged for another thirty-two years. That restriction also was applied to plots bought at the sale of State domains. Neither could be sold except to a peasant, and no peasant could buy more than 5 ha. of such land, including what he already possessed. No stranger could penetrate into the mass of the resettled peasantry, and the peasants could not expand beyond the limits fixed for them. Small property was thus immobilized, and individual possessions hedged in within appointed limits, as under feudal tenure. But though he was tied to his inadequate plot, the villager or other members of his household might still have wandered away in search of better wages for their spare labour. This was already forestalled in 1866 by Art. 2 of the law on contracts, which decreed that 'only the respective commune may legalize the contracts of inhabitants within its jurisdiction; it may legalize the contract of a strange labourer only if he produces a certificate from the commune in which he resides, showing that he is free to enter into such contracts'. The 'commune' was invariably the local landowner; if he wanted a peasant's labour he merely had to see that he was not certified. And if in spite of that the peasant ran away, the law of 1872 gave the landowner soldiers, for whom the fugitive had to pay, with which to bring him back.

The second burden of serfdom, compulsory labour, was revived in practice by the laws on agricultural contracts. As a concession to the new democratic façade it was made incumbent upon the village mayor to begin by 'urging' the peasants who were disinclined to work for the landlords, but ultimately it came to the peasant labouring under the guard of heavily armed foremen and village officials. The enforcing of labour contracts with the aid of the military was formally abolished in 1882, but there is no doubt that it continued in use for many years after. Finally, the third compound of the servile status—payments in labour and in kind—was the very essence of the methods encouraged by the laws on agricultural contracts. With rare exceptions, the villagers paid for additional land or for loans of money in labour or in produce. The exceptional régime of the agricultural contracts was devised to secure to landowners and their tenants the labour they needed, and its conditions therefore necessarily discouraged progress towards money transactions between agricultural employers and labourers.

That reactionary trilogy was aptly rounded off by the final article of the law of 1866. 'No other jurisdiction', said that text, 'no other authority whatsoever except those named in the present law may interfere in the procedure relating to contracts for agricultural labour.' Before his emancipation the ill-used peasant could make some sort of appeal to the country's ruler. Now he was ruled by a democratic hierarchy, but for him Constitution and Supreme Judiciary was to begin and end with the village mayor. That measure passed, without a single voice being raised in the new national Parliament to protest against such mass outlawing of the peasantry.

Legally, and viewed as a whole, the laws on agricultural contracts could fairly be described in the words of M. Gherea as 'the confirmation of the old servitudes in contractual form'. Even so their full effect can be judged only by placing their texts against the background of Rumania's public life in that period. It is characteristic of the power and temper of the new rulers that their legislation compared badly with that decreed by Barbu Stirbey under serfdom in 1851. Damages for failing to perform contracted labour were already known, and it is likely that obdurate peasants were handled somewhat roughly, but it was left to the new law formally to provide that he might be constrained *manu militari*; and while the older text enjoined fair dealing upon both parties alike, the new laws, so drastic for the peasants, contained no sanctions whatever for any failure of the landlords to keep their part of the bargain. Landlords and their tenants were in truth anointed as absolute potentates of the villages on their lands; all they had to do was to get a man subservient to them elected as mayor. The military commanders charged with quelling the rising of 1888–9 did their duty sternly, but in their Reports they admit that the administration is callous towards the peasant, that bad faith is rampant, that 'the real law is dictated to him by those who have joined hands for the purpose of exploiting him'.[1] Tax collectors also were frequently the nominees of landlords, and these men, by pressing mercilessly for the payment of taxes when the peasants' resources were most strained, left them with no other choice than to borrow from landlord and tenant, pledging in return their labour in advance. The report of an inquiry conducted by the Jassy District Council in 1862 pointed out that onerous labour contracts were often forced upon employers as much as upon labourers by the ruthless collection of taxes; 'for the peasants try to borrow anywhere and at any price, and the cultivators, fearing to see their labourers contracting with two masters, are obliged to assign them again, above their capacity to labour.'

Contracts made in such circumstances were open to every kind of abuse. The evidence that in general they put the value of land and corn high, and that of labour low, is circumstantial. In the last quarter of the nineteenth century 'the tithe rose everywhere, but especially in the corn-growing districts, from one in five to one in one'. 'I have known sober, hard-working peasants', writes M. Radu Rosetti, himself a landowner, 'who laboured fifteen years to pay off a debt they contracted in the winter 1866–7 for maize which they had borrowed to feed their families.' They found it hard enough to carry out such heavy

[1] Mihail Eminescu in an article on 'Rural Hospitals' (reprinted in his *Collected Works*, 1914, p. 581) complained that under national self-government more licence was given to abusive individuals than they had enjoyed before. He cites the case of an Austrian citizen who had exploited the peasants so grossly that the divan of Grigore Ghica had forbidden him 'for ever after' to buy or rent land, or even to be a bailiff on the land. The same individual had now become a great landowner, he had the peasants taken to the fields in crates, like dogs, and at night he had them surrounded with barbed fences, so that they should not run away.

obligations and also cultivate precariously their small holdings. How could they have thought of taking on more land to farm? The system was in itself sufficient to check all efforts of the peasants to improve themselves; in fact, they were depressed ever lower by the abuses which it made possible.

Through the instrument of the laws on agricultural contracts the Constitutional guarantees for personal liberty and for the sanctity of domicile could be set aside for 95 per cent. of the population, at the will and whim of a village mayor; and this not in exceptional circumstances, but in everyday life, in matters arising out of the daily work—the only means of subsistence—of the people. Radishchev said of similar conditions in Russia: 'For our laws the peasant is dead.' Yet, all these measures were approved by the new nationalist Liberals; some of the harshest—like that of 1872 which ordered the military enforcing of contracts—were actually initiated by them and carried in the face of Conservative disapproval. The only possible way to explain that conduct is by their anxiety to prevent the rural masses from gaining economic and political strength and competing for influence with the new and narrow middle-class which ruled in town and State. A congeries of facts, asserts the Conservative M. Garoflid points to the 'constant tendency of that oligarchy to impede the formation of a rural middle-class'. The mistakes of 1864 may have been, and no doubt were, largely due to inexperience, and to the way in which the Radical leaders were engrossed with the political aspect of peasant emancipation. The whole problem reappeared, however, in 1889 as starkly as a generation earlier, yet no better solution was attempted than the old one of putting off the clamouring peasants with a piece of land that could never become a farm. The Conservatives, who favoured the creation of a rural middle-class, had in 1889 provided for holdings of 10 and 25 ha.; but the following Liberal Government cancelled that provision and reduced all holdings to be sold from State domains to 5 ha. each. The general conditions on which the State's domains were sold were strongly coloured with that stepmotherly bias; and all avenues of escape into social and economic independence were barred to the peasants by the laws on agricultural contracts. Nor were any other elements, that might have roused the country-side, suffered to take solid roots on the land. Foreigners and native Jews—who were excluded from all official functions and from many liberal professions—were permitted to rent enormous stretches of land on speculative terms, which inevitably made of them transient exploiters of soil and labour. They were not allowed, however, to buy a farm and settle on it, for this, while making them careful cultivators, might possibly have made them champions of an adolescent agriculture.

With the impecunious peasants bound to the land and possible farmers barred from it, the agrarian system fell into a peculiar compound of serfdom and capitalism; from it landlords and their tenants secured all the advantages of both while the peasants were saddled with all the burdens of both. From serfdom the landlords had all the facilities of servile labour without any of the feudal obligations towards it; while from capitalism they had the freedom to bargain with labour without the restraint of a free labour market. The peasants, however, were subjected to servile labour without its counterpart in land rights; and from capitalism they had all the trials of wage earners without being really free to trade their labours where they willed. One class, says M. Gherea, had achieved for itself 'roses without thorns, while the thorns—and the thorns alone—were left for the peasants'. It was natural, therefore, that emancipation and land reform should have no effect on the organization of Rumanian agriculture. Lack of capital and training for a new technique might have been made good within a short space of time. But what interest had landlords and tenants in changing a system which gave them such a monopoly of control and profit as they had not enjoyed even under serfdom, and which they could never hope to retain in a full-fledged capitalist régime? 'Neo-serfdom' left the landlords free to occupy themselves with politics or pleasures, and also gave them the means thereof. For it attracted tenants, who, without knowledge or equipment, could afford to offer exorbitant rents, the land continuing to be tilled by the peasants. All that these *sui generis* farmers had to do was to drive the peasant as hard as his mortal frame would stand. The laws on agricultural contracts supplied a perfect device to that end. The return to be expected from an estate used formerly to be calculated according to the number of servile labour days which the landlord was entitled to claim; now an estate's rent was dictated by the relation between give and take in the local contracts. Not the fatness of the land but the sweat of the peasant determined the huge profits that were made. And the peasants toiled, and sighed their helpless sorrow in the saying: 'May God never lay upon a man as much as he can bear.'

The Peasant's Burdens. Rumania's agrarian policy followed, during the first half-century of her national independence, two parallel lines—the endowment of the peasants with plots of land and the impressment of their labour—which appear wholly to contradict each other. In reality they were corollaries, and they support Loria's generalization that 'the conditions which caused the binding of labour in favour of the estate were free land and scarcity of labour'. The tendency to reduce the land available for the peasants, if successful to the end, might have resulted in free labour. But in the measure in which it failed the other tendency gained strength, that of tying the peasant to the village, one completing the other, so as together to give to the trading corn-grower a sufficient supply of labour. It was the Rumanian peasant's refusal to live without land that caused him to live without freedom. Consciously or not, the older Russian revolutionaries had summed up soundly the social and economic essentials of the peasants' case in their motto: 'Land and Liberty' (*Zemlja i Volja.*)

Soon the Rumanian, as well as the Russian, peasants were to make a bid for both of them. Those with whose fate we are concerned were nearing the bottom of their cup of earthly misery. It was generally agreed that 'all those parts of the laws on agricultural contracts which were designed in favour of the peasants, are fallen into disuse'. M. Sebastian Moruzi, who wrote these words in 1907, was speaking from intimate knowledge, having himself been several times Liberal prefect of various counties. 'What village mayor, indeed, would dare to inspect closely the boiar's account books? What mayor claims respect for the two days left weekly for the personal needs of the villagers? . . . Or, above all, what mayor insists on the fair measurement of the land given for labour or sold for money?'[1] The minor improvements decreed in 1893 were never applied. When the law of 1882 put an end to the military execution of contracts, the result was 'a lessening of brutality', says M. Radu Rosetti, 'but on the other hand an increase in frauds'. The State itself, which forty years earlier had become owner of one-fourth of the arable area (one-third with the land of endowments and other institutions), had done nothing to better the life of the peasants on its lands. Rather did it welcome every opportunity of raising rents and its revenue, and that example was followed by institutions under its control; so that peasants on these public estates, said M. Spiru Haret—a former Liberal Minister of Education and famous for his agrarian activities—'are at least as badly off as those on the estates of the worst of the large landlords'. From 1862 to 1905 the area under cultivation had risen by 83 per cent., while the population had increased by only 54 per cent.; yet the price of labour had remained stationary, because of the dependent position in which the peasant had been placed as a labourer. In the sixties the price of labour was double what it had been during the régime of the Organic Statutes, because of the rapid expansion in the culture of maize and wheat; that expansion continued rapidly, but a high official affirmed in 1893—and M. Moruzi repeated this in 1907—that an investigation of labour contracts on various estates showed that the rates still paid were those fixed about 1866. And the majority of them never reached even the poverty level of the customary wages. An inquiry conducted by the Ministry for Home Affairs showed that out of 1,265 labour contracts for 1906, chosen at random, only 39·7 per cent. were concluded at the customary wages; others were lower in varying degrees, 13·2 per cent. showing wages upwards of 75 per cent. below the usual rates. They also showed a widespread custom to reckon the 'working' month as of thirty-two days. Most contracts had penalizing clauses intended to press for more labour; some contracts in the district of Teleorman stipulating

[1] According to M. Şerban, of the 1,192 cases of contraventions to the law on agricultural contracts which came before the courts in 1913, 817 or 68·54% (in 1912, 895 or 73·3%) had been committed by owners and tenants, 224 or 18·79% by village officials, and only 151 or 12·67% by peasants; though, of course, the number of peasants parties to such contracts was infinitely greater than that of their employers.

that 'the father of a family who had living with him married sons or daughters who had not contracted for agricultural labours shall be himself excluded from all contract'. Under the old régime the authorities had to enforce upon the peasants the obligation to labour for 6–24 days for the landlord; under the new régime of 'free labour' the peasant had become so dependent on the landowner, that the legislator had to intervene in order to ensure that two days were left each week for the peasant's own fields.

The price of land, on the other hand, had risen fantastically. The severe agrarian crisis between 1875 and 1890 had caused the price of corn to fall some 30 per cent. in Europe; to that corresponded a fall in the value of the land and of its rental. But in Rumania a contrary phenomenon occurred during that period: a continuous and rapid rise in land rents, frequently by fully 100 per cent. in the century's last quarter, without any corresponding rise in productivity. At the same time, however, the Government had to help the peasants with food every few years. In this, as in many other respects, this was a parallel to the Russian situation. After the emancipation of the peasants, in 1861, Russian wheat exports rose so rapidly that they dominated the western markets. But at the same time the Government had to devote increasing sums to the relief of starving peasants; from a yearly average of 800 roubles in the sixties, relief rose to an average of about one and a half million up to 1890, to nineteen millions between 1890–1900, to 118 million roubles in 1901–5. Relief was distributed in the 'sixties to eight departments, in the 'seventies to fifteen, in the 'eighties to twenty-five, in the 'nineties to twenty-nine, after 1900 to thirty-one. The only possible explanation for such discrepancy in the progress of the Rumanian rural classes is that the great and abnormal increase in the revenue of landlords and tenants was obtained by depressing the revenue of the peasants, by means of the laws on agricultural contracts; and that is confirmed by the fact that in sparsely populated districts, where labourers could not be conscripted but had to be attracted, the rent of the land rose not at all or but slightly. Between 1870 and 1906 the rents paid by 67·6 per cent. of the peasant tenants had risen by 100 per cent.;

those of 13·8 per cent. of them by more than 300 per cent.; and those of 3·3 per cent. by over 500 per cent. For the land held in métayage, which they cultivated with their own seed, animals and implements, the peasants were found to be paying one-half to two-thirds of the harvest, and a variety of additional charges in labour and in kind. They found it increasingly difficult to compete for land, and because of their lack of training and means, they could not get out of it a value that would correspond to the compulsory labour they had to give for it, which thus became doubly onerous for them.[1] Under the pressure of such circumstances the peasants were driven to borrow from the landowners and tenants. M. D. Nenițescu declared that an official inquiry, conducted in 1901, established that for such loans 60 per cent. had come to be regarded as a 'friendly' interest; but that sometimes the landlords took from the peasants 125, 250, 365 and even 528 per cent. in the form of labour dues.

To these conditions had to be added the burden of discriminating taxation; the rating census of 1905–7 estimated the revenue of peasant land four times higher than that of large owners. In the debate on a bill introduced in March 1907, 'to equalize the taxation of peasants and large owners,' the Minister of Finance made it clear that it was not intended to raise the land tax paid by the large owners, but merely to reduce to the same level the tax paid by the peasant proprietors; and even then many deputies could only vote for the bill 'by treading on our hearts'. When the peasants did put in a complaint, which was seldom, they were generally less fortunate than those of Fântâna Banului (Dolj county). These had bought some of the larger plots from the State at exorbitant prices and being unable to

[1] According to an official publication, *The Rise in Money and Métayage Rents*, production had in the same period fallen rather than increased, because of bad cultivation, lack of manure, &c.

Average production in hectolitres per hectare:

	Maize	Wheat
1862–6	12·7	12·4
1872–6	12·4	9·1
1891–5	12·8	14·0
1901–5	11.5	15·8

Maize was largely produced by the peasants, wheat by the large cultivators. One must also take into account that the peasants were given the worst land and left with the worst time for their own work.

pay their annuities were threatened by the Ministry of Finance with forcible eviction; a petition sent by them to the Senate remained unanswered for two years, until a chance encounter on the Senate steps with the late Dimitrie Sturdza led to the latter's effective intervention on their behalf. Generally, the peasants, too, trod on their hearts and suffered in silence. Used up in strength and riddled with disease, exploited and derided, they seemed to have fallen too low even for protest or revolt.

Wars and Revolutions. In the spring of 1907 disturbances broke out in northern Moldavia. At first the attack was concentrated against the Jewish inhabitants and looked like one of the endemic anti-Semitic riots which have frequently been engineered in Rumanian political life, as a means of frightening the Government, or, conversely, of cowing the Opposition. How far the peasants had a hand in that beginning one cannot say; there are military and administrative reports, like that on the little market town of Bivolari (Jassy county), which assert that the sacking of the place was the work of imported rowdies, and that far from aiding in this, local and neighbouring peasants gave asylum to the Jewish victims. But whatever the men and the motives which started the agitation, once the country-side was roused the peasants had a terrible account of their own to settle. All the pent-up despair of a breadless and rightless population broke loose upon the land. The spirit of the peasants had been stung not a little by the stories which had reached them about the Russian rising of 1905 (itself a consequence of the Russo-Japanese War). Ever since the appearance of Russian armies as liberators from the Turkish yoke, the Rumanian peasants, like the Russians, believed the Tsar to be devoted to their interests, and they remained sensitive to every happening that affected the peasantry in the neighbouring empire. Their emancipation in 1864 followed the Russian emancipation of 1861; and there is evidence that when they rose in 1889 they had been carried away by a rumour that the Tsar had ordered the big estates to be divided among them. Similar rumours may have helped to rouse the peasants in 1907, but their revolt sprang from issues that lay deep and were vital for themselves. Recently this has been confessed by the latest leader of that Party which has dominated Rumania since her independence. 'None of these [agrarian] reforms,' wrote M. Vintilă Brătianu in 1926, 'had sufficiently eased the needs of the peasant class; and, the satisfaction of those needs having been delayed particularly in the period from 1890 to 1907, we were faced in that last year with a grave internal crisis.' From one end of the country to the other the peasants vented their accumulated anger upon dwellings and belongings, and occasionally also upon the persons, of landlords and tenants, carefully discriminating between the few who had treated them well and the many towards whom they had no reason to be tolerant. The worst excesses, as it happened, occurred in districts which had few Jewish inhabitants or none at all, which belied the supposed origin of the revolt.

The outbreak, apparently, took the ruling class by surprise. It scared them and it enraged them. The rising was unorganized and in many parts was quickly checked by a little reasoning with villagers; yet so great was the panic of the landlords that some among them thought of appealing for the help of two Austrian army corps. But the Rumanian military proved equal to the task by themselves. The Second Army Corps ordered the repression to be carried through to its end—shooting the villagers and burning their houses—even where the rising had subsided. Guns were in some places brought into action against groups of peasants armed with forks and scythes who were completely surrounded. Altogether, about ten thousand peasants were put to death; the real number may never become known, as the official dossiers of these reprisals have disappeared.[1]

The peasant rising, says M. Vintilă Brătianu, 'roused the conscience of the political parties.' There were two memorable sittings of the Chamber, on the 9th and 13th of March, when the politicians vied with each other in beating their breasts and confessing their sins. A Royal manifesto expressed 'the desire of H.M. the King' for reform, and Parliament passed a series of laws, cancelling others but recently enacted, meant to constitute a comprehensive agrarian policy, which should assist and

[1] Debate in the Chamber, 23rd of May 1925. The order of the Second Army Corps was then read by General Averescu, himself Minister of War at the time. It has been suggested, as a possible explanation of the merciless reprisals, that a neighbouring empire had threatened to intervene if the rising was not quelled at once.

encourage the peasant as a farmer and protect him as a labourer.

The law on agricultural contracts of 23rd December 1907, abruptly broke away from its predecessors, which had protected the employers, and devoted itself to protecting the labourers. It laid down a number of principles which were to put an end to that intermixing of leasing and labour arrangements, which confused the peasant and made him an easy prey to every abuse; and, similarly, to provisions for mixed payments in money and labour and produce in one and the same contract. Henceforward every operation was to be settled in a separate contract, for which official forms were provided. When the contract was one of métayage, the harvest was to be divided up within fifteen days; so as to put an end to cases like that reported from Dolj, where the 1906 maize crop was still in the fields in March 1907, the peasants not being allowed to take their only food home because they had some remnant of debts to pay. Now if the landlord delayed longer than allowed by law, he was liable to a fine; if the peasants took the crop away too early, before having carried out all their obligations, they were liable to imprisonment up to one year. District Commissions, consisting of two delegates of the landowners and two of the peasants, under the chairmanship of the district agronom, were to fix a minimum wage based on the average of the last three years, but whereas previous laws had been absolute, this was applicable only to written contracts, but not to verbal agreements or to agreements made within thirty days before the labours to which they referred; and the Commissions were also to fix a maximum rent which was not to exceed by more than one-third the rents paid for similar soil by the large tenants in that region. This put no restraint upon the speculating tenants who could continue to pass on every increase in rent upon their peasant sub-tenants. The food given to labourers was to conform to the standard laid down by the Ministry of Health. School children were to be employed no longer in field work, unless in exceptional cases and under specified conditions.

The same law tried to meet the great need of the peasants for grazing by instituting communal grazing lands. The Liberals wanted to oblige the landowners to supply land for communal grazings, but the Conservatives, led by Take Ionescu, opposed this on the ground that it would be an expropriation in disguise. These commons were to be created for each village, the land being supplied from State domains, or bought from local landowners owning more than 300 ha. at market value plus 10 per cent., and the landowner remaining owner of the subsoil. The cost was to be defrayed by the commune from grazing rates paid by the villagers for each head of cattle, none of them being entitled to keep more than six big animals on the village common.

To assist in a better distribution of seasonal labour the law also set up an agricultural employment bureau. The whole of that medley of State assistance, intervention, and control was to be in the hands of three special organs: (*a*) District Commissions dealing with wages, rents, &c.; (*b*) district agronoms; and (*c*) a Supreme Agricultural Council composed of five members appointed for fifteen years, who were to supervise the general application of the law and to control the district agronoms.

A second set of laws endeavoured to satisfy in some measure the land hunger of the peasantry. Great hopes were placed on the law which created the Rural Office (*Casa Rurală*), half of whose capital was to be supplied by the State, which was given a thirty years monopoly for its operations. Its main purpose was to facilitate the passing of land belonging to large owners into the hands of peasants, the office either buying itself or acting merely as intermediary. Once more the limit of peasant lots was fixed at 5 ha., and in the hills at 3 ha., though one person might buy as much as five such lots once the demand for single lots was fully satisfied. An important provision was that declaring the lots of 5 ha. to be indivisible, the other heirs being compensated in money; the Rural Office was entitled to grant loans for this and other purposes. This clause was justified by excessive division, which had created crazy holdings such as some that were two metres wide and 700 metres long, tracks and dividing marks swallowing up as much as 30 per cent. of their area. The Rural Office was also charged with encouraging and supervising a better cultivation of the land bought from it or through it; and, further, with encouraging the growth of co-operative purchase and marketing among the peasants. Another law decreed that State domains, as well as estates of cultural, religious, philanthropic and other endowments, could be leased to peasant co-operatives only, not by auction, but on the basis of rents fixed by the district commissions and the Supreme Agricultural Council. Finally, an antitrust law forbade any individual to rent, whether himself or through intermediaries, more than 4,000 ha., existing contracts being allowed to run till 1912; in 1905 a single family was renting in Moldavia 159,399 ha. arable land, and altogether 236,863 ha. including woodland.

However sound the theoretical basis of most of these measures, in effect they were fated to disappear without trace in the quicksand of Rumania's public life. Most of them depended to an extreme degree for their success on the initiative and supervision of public authorities, and that left little prospect that the poor and disfranchised peasants would gather in even a small portion of the fruits promised them by the laws of 1907. 'One can predict with certainty', a French writer ventured to say at once, 'that these texts will bring nothing but disappointment; for they tend to replace by artificial and hasty combinations a natural evolution which could result only from a profound change in the habits of the ruling class.'[1] Indeed, the new law on agricultural contracts was 'deliberately ignored', as M. V. Brătianu admitted in 1913. Difficulties were encountered in creating the village commons; the Report of the Supreme Agricultural Council for 1912–13 showed that commons—covering a total area of 182,518 ha. up to October 1913—had been established on all State domains and on 73 per cent. of the estates belonging to various institutions, but only on 26 per cent. of the private estates, either because their owners were holding back, or because the land they offered was too bad or too dear. Four hundred and four of the offers received had to be rejected because of the excessive price demanded, and of these 90 per cent. were those of private landowners; some communes were able to

[1] Léon Poinsard, *La Production, le Travail et le Problème social dans tous les Pays au début du XXe Siècle*', Paris, 1907, ii. 762.

establish grazings at one-fourth or one-fifth the price first asked for them. Where commons had been established, the peasants frequently found it impossible to carry out the obligation of growing fodder on part of them, because the soil was too poor for that purpose. Up to 1913 the Rural Office purchased 115,467 ha., but of these only 34,929 ha. had been parcelled (in 6,881 lots) and only 16,369 ha. had been sold to peasants, the remainder being merely leased to them or farmed for its own account by the Office; and of the 3,318 lots of 3–5 ha. sold only 133 went to peasants who already had more than 4 ha. of their own, so that medium property received as little encouragement as before. Its mortgage loans to peasants amounted in 1912 to 4,070,591 lei, and in 1913 to 4,047,759 lei. The Office itself, at any rate, did so well out of these transactions that its 500 lei shares had risen to 2,000 lei by the middle of 1912. The State itself was slow in applying the clause which ordered that its domains should be leased to peasant co-operatives and continued to farm them for its own account.

Nothing, then, more positive came of the fierce rising of 1907 than a revised and enlarged edition of the existing agrarian laws. The men who were to interpret and apply them remained the same, and, beyond a passing flush, the temper and habits of Rumanian public life were hardly affected by the peasants' show of despair. Occasionally some politician felt moved to confess that—in the words of the mythical Mecklenburghian Constitution—'Everything remains as it was.' But none of the political groups showed signs of having come to regard rural reform as a concern of its own. In a peculiarly agrarian and peasant country the last issue that seemed capable of enlisting political zeal was the problem of the peasants' fate. How strange it is, e.g. that the wise King Carol (1866–1915), who had the welfare of the people truly at heart, should not have used his great influence to bring about a solution of the rural problem; unless it be that he knew that, like Cuza, he would be faced with the ruthless opposition of all political factions. Only some mysterious reason could explain the baffling fact that in the four crowded volumes in which an 'Eyewitness', making use of the king's private correspondence and personal notes, has minutely described the first fifteen years of Carol's reign, the peasant question is remembered in not one single line.[1]

At the beginning of the second decade of the twentieth century Rumanian agriculture was rapidly recovering from the shock of 1907 and from a series of bad harvests. Exports were rising, and landlords and their tenants very likely looked forward to another period of rich profits, with the help of the peasant's capacity to work much and live on little, and of the State's complacency towards doings on the land. In other words, the rural world was returning to normal. Before, however, it could regain to the full its lop-sided stability again, a chain of external events, following each other in quick succession, buffeted it about with such increasing violence as finally to knock Rumania's agrarian system out of the perverse balance in which it had been kept for a century.

The first link in that chain was forged by the Balkan Wars of 1912–13. As Rumania was expecting to be drawn into them at any moment, the 'morale' of the masses was prepared for the effort by the usual promise that land would be distributed after the peace. Rumania's part in the second Balkan War proved a mere excursion across the Danube, with only part of her forces, but that short expedition had an unexpected effect on her soldiers. Having come to hate, so to speak, these peasants and sons of peasants remained to envy. They saw that in Bulgaria, only a short distance from their own places, there was a country which knew not what large property and masterful landlords were; in which every peasant had his holding, and the villages were better built and the fields better tilled than their own. After the return from Bulgaria the late Ionel Brătianu recognized in a letter published in *L'Indépendance Roumaine* the need for land reform; and his then leader, Dimitrie Sturdza, referring to the poor response of the landowners to the demand for land for village commons, declared that their backwardness 'justified the application of the expropriation principle'. The dread word was spoken at last. Till then the ruling class had favoured rather than discouraged the growth of large property, and had

[1] 'Augenzeuge', *Aus dem Leben König Karls von Rumänien*, 1894–1900. The 'eyewitness' was Dr. Schäffer, formerly tutor to Prince Carol.

safeguarded it by Art. 19 of the Constitution, which forbade expropriation except and 'solely for the requirements of public transport, public health, and the defence of the country'. The leader of the all-powerful Liberal Party now admitted that those restrictions might have to go. How far were Sturdza's words a program, and not merely a sentiment?

Before this query could be properly propounded, the existing order of things sustained a fresh and more formidable shock. One year almost to the day after the signing of the Peace of Bucarest, the Great War broke out. It threw Rumania's rulers into a welter of harrowing hopes and fears. During the two years which it took them to make up their minds how to get the best without risking the worst, 'expropriation' and 'resettlement' freely reached the ears of the waiting peasant soldiers. Promises were floating in the air, but no one thought of bringing them to earth, and impatience was met with the usual: 'after the War.' And then—before the rocks of war could be safely rounded—the Russian Revolution burst upon them. King and Government and Parliament were surprised on the edge of it, in their refuge at Jassy—caught in between the German guns and the Russian red flags. The King journeyed to the front and there solemnly announced that the fatherland's soil would thereafter belong to those who tilled it. The wheel of History had swung round—once again the country was in dire need, and the peasant was restored to his old yeoman standing.

PART II

THE NEW LAND REFORM

CHAPTER V

THE LAND LAWS

POLITICAL HISTORY

Though Rumania seemed to develop steadily from year to year, the peasant problem never ceased to rattle ominously at the gates of the young State. In almost every one of its sessions Parliament had to pass some measure of agrarian reform, but these cannot be adjudged to have been more than palliatives. As M. Sturdza pointed out in 1907, all the great changes in the social status of the peasants had been made peacefully, by the instrument of the law. Yet at no time after 1864 did Rumania's rulers take their own legislative efforts seriously enough to try to measure the results. 'We, the leaders', declared one of them, M. Bibicescu, 'had not sufficient curiosity to find out the precise results of that work on which we had based such great hopes.' There were no statistics, no inquiries, no ground book, no social research; nor, in these circumstances, any considered policy of agrarian reconstruction. A sound remedying of the peasants' troubles would have demanded their endowment with land, with training and with credit—above all with land, to which the other two could be only accessories. But except the handful of Socialists—who, of course, were thinking of transferring property to the State rather than to the peasants—neither of the two political parties which shared the power between them admitted the principle of expropriation.

The Conservatives represented by tradition and membership the big landed property. Even those among them who saw the cracks in the agrarian structure merely thought of how to patch it up so as to strengthen the landowners. They advocated fiscal measures which should force the landowners to return to the country and farm their estates, as well as to make them co-operate for credit, purchase, and sale. For the peasants they merely propounded freedom to sell and mortgage their holdings—a freedom which would soon have sifted them into well-to-do peasants and landless labourers: the first to strengthen the class of landed cultivators, the second to provide this class with plentiful hands. These views were elaborated as late as 1914 in a Memorandum of the Union of Agricultural Syndicates, an organization of large owners, which considered that 'by abolishing inalienability a process of selection will be set going among the peasants whereby the land of those who are idle will be concentrated in the hands of those who are industrious.... Thus a middle-sized property will be created in the hands of the more intelligent peasants, and we should have likewise a class of free labourers....' The State should assist this evolution with credits, enabling the more capable peasants to purchase middle-sized holdings from the large estates. Soon after the rising of 1907, the Conservative Government of M. Grigore Cantacuzino played with the idea of expropriating in full all the mortmain estates, in order to forestall a demand for a wider expropriation. The nature of the compensation was not mentioned. But the idea encountered much opposition and was soon abandoned.

Among the Liberals the agrarian element was in a minority, and it had mainly a capitalist interest; in general, that party's tendency was mercantilist. A manifesto of May 1888 even denied the existence of an agrarian problem. That view could no longer be maintained after the disturbances of the following year and the fierce rising of 1907—the year after the triumphant fortieth jubilee of the State's existence. But in 1909 one of the Liberal leaders formally announced his Party to be opposed to all idea of expropriation. Speaking in the Chamber, during the debate on the Speech from the Throne, he declared that expropriation was against their political creed; and that 'the country's agricultural structure could not be changed by revolutionary methods'. Two years later, a Party manifesto insisted that by the reforms they had applied since 1907 the Liberals had been trying to uproot from the peasants' minds precisely 'the false and dangerous belief' that it was the State's function to provide them with land. In 1912 the Party's foremost 'peasantist', Spiru Haret, wrote that if it were in his power to reform Rumania's country-side, he would have the peasants as tenants rather than as owners of their holdings. And M. Vintilă Brătianu confirmed these various statements in a pamphlet published in 1913, when he denounced as 'demagogic' even promises for the expropriation of mortmain estates.

The turning-point came a few months later, after the return from Bulgaria. Ionel Brătianu's public letter in the semi-official *Indépendance Roumaine* conceded that it might be necessary to expropriate the large owners so as to endow the peasants with land.[1] What apparently was contemplated at the time was the expropriation of estates above 5,000 ha. In the following year M. Vintilă Brătianu demanded the expropriation of 1,200,000 ha., to be paid by the peasants at the rate of 1,000 lei per hectare, within fifty to sixty years. The so-called Conservative-Democratic group, led by Take Ionescu, accepted the principle of such a reform, but the Conservative Party remained opposed to it. In 1915, after the outbreak of the Great War, a few Conservatives under M. M. Cantacuzino seceded from the leadership of M. Alexander Marghiloman and likewise accepted the principle of expropriation.

[1] The conversion of M. Ionel Brătianu and his Party to the idea of sweeping land and franchise reforms was too sudden not to cause some surprise and to make people ask themselves how a mere military excursion across the Danube could so greatly disturb a settled outlook. The following is the reported history of the episode.

A Conservative Government, under Titu Maiorescu, had been in power for a few months when Rumania entered the second Balkan war, in 1913. The Liberal Party, and the Brătianu family, had always considered it as one of their privileges, from the time of the formation of the national State, to preside over all the important events which formed the milestones of the country's history. In 1913, therefore, Ionel Brătianu claimed power either for a Liberal government or for a national coalition government, under his premiership. King Carol replied that he could not dismiss, without grounds, a government which was new to office, nor ask M. Maiorescu, the 'grand old man' of Rumania's academic life, to serve under a relatively young politician. Ionel Brătianu apparently took this refusal so deeply to heart—as a rebuff to his family and to his Party—that he departed to the head-quarters of one of the armies in the field, as captain of artillery, fully intending from there to announce his retirement from political life. It was in that state of despondent indignation that he was found by M. Constantin Stere, the foremost 'peasantist' theoretician, and at the time the 'éminence grise' of the Liberal Party. 'Why resign because you cannot gain favour with the King?' suggested M. Stere. 'Stay and become the favourite of the people.' By distributing land and votes a statesman could gain a popularity which no monarch might withstand. The advice was taken, and M. Stere was asked to prepare drafts for the two reforms. What to the one, however, was a great end in itself, to the other was merely a means to an end. The drafts of M. Stere were thought too radical and were severely expurgated. Yet their essence was still sufficiently revolutionary in the eyes of the fine but very conservative monarch. He took fright, and gave M. Brătianu the government, towards the end of the year, presumably on the understanding that the reforms would be shelved; as indeed they were until 1917.—This is the story as privately told by M. C. Stere himself.

In the meanwhile, the Liberal Party having been called to power in December 1913, they proceeded to elect a Constituent Assembly with power to revise the articles of the Constitution restricting expropriation and the franchise. The new Assemblies met in June 1914, and appointed two parliamentary Commissions to work out proposals for the two reforms. Soon afterwards the War broke out and the Constituent Assemblies were not again convened to fulfil their functions till three years later. The Commission for land reform, according to one of its members, met twice during that interval, 'but no work was done at all. In 1914—nothing; in 1915—nothing; that was all its preparatory work.'

Rumania entered the war in August 1916. From that moment many new truths and doubts began to work upon the peasants' minds. As everywhere, they were extolled as heroes in speeches and articles which left a bitter taste when tested by the realities of their existence. The advance into Transylvania had the same effect on the outlook of the Rumanian soldiers as had the advance into Bulgaria in 1913. The physical sufferings which they had to undergo were perhaps unparalleled in any other theatre of war; and as far as these men could see and judge, many of their sufferings were due to the failure of their betters as leaders and administrators. Moreover, these doubts as to the fitness of those who ruled them were fostered for their own convenience by the German, and other, military who occupied two-thirds of the country for nearly two years. Knowing the antagonism which divided the rural classes in Rumania, the Germans came prepared with large quantities of proclamations, printed in bad Rumanian, with which they apparently intended to break, in case of need, the country's power of resistance. 'The Germans', said that manifesto, had come 'to free the peasants from their bondage to the boiars and to distribute the land among them'. It invited them to rise and 'to chase away their landlords and rulers'. Marshal Mackensen did not find it necessary to make use of this disrupting instrument, but the sentiment therein contained no doubt found other channels of expression during the prolonged contact between the foreign troops and the Rumanian peasantry. In Moldavia, where the

Rumanian Government and a large part of the population had taken refuge, the old class distinction continued to filter through even in the emergency legislation destined to ease the stress of the moment. In 1917 the Government found it necessary to decree the compulsory cultivation of land, according to an established plan; landowners and their tenants failing to carry out these obligations were threatened with a fine, whereas peasants refusing to labour were liable to be court-martialled and imprisoned from one to three years.

All these incidents and influences which were fermenting the peasants' discontent were given substance by the sudden collapse of landlordship in Russia. It has already been mentioned how sensitive Rumania's peasantry was to happenings in the neighbouring Empire; how the Russian emancipation of 1861 was followed by the Rumanian in 1864, and the Russian revolution of 1905 by the rising in Rumania in 1907. In 1917–18 that influence made itself more directly felt than it had ever done before. The Rumanian front in Moldavia, namely, was held jointly by Russian and Rumanian forces, the units of the two armies being interspersed. The rank and file of the Russian regiments, like the bulk of the Russian army, at once took matters into its own hands when the news of the Revolution reached them. They began to organize soldiers' councils and the Rumanian soldiers could hear them arguing for the possession of the land. Such discussions touched to the quick what was still the Rumanian peasants' most ingrained belief. None of the changes which had altered the system of land tenure during the previous century could eradicate the peasants' transmitted conviction that they had a right to the use of the land on which they were settled. The persistence of the system of share-cultivation had helped to keep alive that belief, as in appearance it continued, with merely a change in the respective portions, the old tithe system; whereas a money-wage system would have made a clean breach between property and labour. As things were, the peasants took it for granted that when they had not enough land, the land of the State or of the landlords had to be used to meet their needs; this being one of the reasons why many people had opposed the custom of placating the peasants with occasional grants of land. Now the Rumanian soldiers and peasants saw the Russian soldiers abandon their regiments, especially when Communist agitators began to arrive, and trek home to take possession of the land. Nor did the Communists allow the lesson of this example to sink of itself into the hearts of the Rumanian peasants. With unrestrained revolutionary zeal they began to address their incitements to the Rumanian population as well, and to terrorize the Rumanian authorities. Russian regiments gone 'red' paraded the streets of Jassy; the prison was broken into and Rakowsky and other Communists of Rumanian origin were set free; and still more daring acts were being prepared when at the eleventh hour they were forestalled by the Royal Proclamation which initiated the great reforms. The dramatic story of that historic act has been tersely told by Professor N. Iorga in 1925, when he rose in the Chamber to protest against the suspension of an Opposition deputy from Bessarabia, M. Ion Buzdugan. In 1917 M. Buzdugan was serving in the Russian army and after the first revolution became a commissar with one of the regiments on the Moldavian front. At the risk of his life, said M. Iorga:

'he used to spy upon the meetings of the Bolshevik committee and every night he came to report to me, for he had learnt that the Bolsheviks wanted to capture the King in Jassy.

'M. Duca is here now. . . . One Thursday night he said to me: "It is terrible that any one should capture Rumania's King, beloved by his people, in the centre of his own Capital. What shall we do?"

'Indeed, we were all asking ourselves what we should do. And of all these things M. Buzdugan, whom you have suspended, kept me informed. He used to come to me every night, at the risk of his life, to give me an account of what was happening, while I in my turn passed on the information to the American Minister, Mr. Andrews. And on a Saturday, at 11 o'clock, Mr. Andrews told me that: "Thanks to the person who brought you the information, and to my intervention, the danger is removed." In fact, on Sunday the King was no longer in Jassy. He said to me: "I can be of no use here; I am going to the front, among my soldiers." And he went with a proclamation—His Majesty will pardon me for saying it—in the writing of which I am proud to have collaborated: It was written by me, it was written in the street, that proclamation which pledged land to the soldiers and the right to vote.'

At that very moment, at the request of their western Allies,

Russia and Rumania were preparing to take the offensive, in a last attempt at checking the collapse of the eastern front; and some gesture, therefore, was also needed for the purpose of raising the morale of the discouraged Rumanian troops—to inspire them to this new effort just when the insidious Bolshevik cry of 'peace and land' was running like wildfire along the lines. King Ferdinand visited the front early in April and spoke to the troops which were being reorganized in view of the coming offensive, as follows:

> 'Sons of peasants, who, with your own hands, have defended the soil on which you were born, on which your lives have been passed, I, your King, tell you that besides the great recompense of victory which will assure for every one of you the nation's gratitude, you have earned the right of being masters, in a larger measure, of that soil upon which you fought.
>
> 'Land will be given you. I, your King, am the first to set the example; and you will also take a large part in public affairs.'

Then and later, the King insisted in private conversation that the initiative for the land reform had come from him. 'I want it to be known', he said e.g. to the veteran journalist, M. Costaforu, 'that, as regards the granting of land to the peasants, the idea was mine. I leave all modesty aside and say openly that the merit is mine.' Two months later, the principles of the reform had become part of the country's law.

Fixing the Basis of the Reform. In his *Sozialismus und Soziale Frage* Professor Sombart remarks that people sometimes point out how great social reforms were advocated by bourgeois idealists, forgetting that those reforms would never have been carried out without the pressure of the class interested in them. In Rumania the land reform may be said to have been brought about by the potential pressure of a peasantry roused by the War and by the Bolshevik revolution. Though the principle of the reform had by most people been accepted before, its dimensions and character were determined by the events of 1917–18. One might put it paradoxically and say that it was just the extreme violence of the Russian revolution that saved the peace in Rumania, as in other neighbouring countries; that the horror of what was happening next door caused Rumania's rulers and landlords hastily to surrender their shirts, when in less stormy times they might have bargained hard for a button.[1]

Much enlightenment on this point may be gained from the conditions under which the actual bill came into being. The peasant masses, most of them being under arms, spectrally dominated the situation, like the ghost in *Hamlet*; but though they imposed the reform, they had no voice whatever in the making of it. They were not consulted when it was being drafted. In 1856, in the *divan ad hoc* which was to prepare the way for independence, each county had a peasant among its delegates. But the Constituent Assembly of 1917, after half a century of independence, had among its members not a single peasant who

[1] The direct influence of the Russian events on the progress of the reform in Rumania has repeatedly been admitted by Rumanian politicians of all colours. In a discreet way, in keeping with the paper's semi-official status, this was at once confessed by the *Indépendance Roumaine*, in its leader of 22nd May 1917. The leader began by saying that it had been thought wiser not to attempt any extensive reforms during the War. 'But, it having been proved by our eastern neighbour that one can solve, even during the War, infinitely more difficult and complex problems, an understanding came about spontaneously between our Sovereign and the Government, and it has been decided that the agrarian and electoral reforms must not be delayed any longer, at least in so far as regards their insertion into our Constitution.' And the paper added a sentence upon which M. Iorga's story forms a suggestive commentary: 'The reforms', it says, 'will be like the pact of a *new alliance* between the dynasty and the people, between the ruling class and the working classes.' Speaking in the Chamber at Jassy, in July 1917, Dr. N. Lupu, then a leader of the newly formed Labour group, complained that nothing had been done during the three years since the Constituent Assembly was first convened, in 1914; though by passing the reform before entering the fray 'we should have put one more moral conviction into the soldier's knapsack'. And then he added: 'and if on 2nd March 1917, the great Russian revolution had not broken out, which brusquely upset the course of history, I doubt if we would have done anything even now.' In the *exposé de motifs* to the new Constitution, privately circulated to members of Parliament in 1923, M. Chirculescu, the rapporteur, justified the changes by bluntly stating that they had to choose between Constitution or Revolution. When a Unitarian delegation from U.S.A. visited Rumania in 1924, to inquire into the complaints of their co-religionists, M. Alexander Constantinescu, the Liberal Minister of Agriculture, told them that the agrarian reform had been necessary to ward off Bolshevism. Likewise, when defending the reform before the Council of the League of Nations, in March 1927, M. Titulescu, then Minister for Foreign Affairs, insisted that the reform had been 'an instrument of social defence. To maintain social order in Central Europe, Rumania imposed upon her landowners sacrifices without parallel in history.... It had been a case of saving the principle of individual property itself.' A more comprehensive view was expressed in an article which appeared in the Bucarest *Adevĕrul*, on 25th August 1924. It said that the agrarian and political reforms, originally advocated by the Socialists, had been opposed as much by the Liberals as by the Conservatives, and that 'they have been realized in full measure thanks only to the War, thanks to the union [with the new provinces, some of whom had better agrarian and political conditions], and thanks to the great agrarian transformation which took place around us during the War'.

might speak for his fellows in this debate that concerned them most of all. There were some progressive deputies of the third college and about ten of them broke away from the Liberal Party and formed themselves into a 'Labour group'.[1] These deputies considered the Government's proposals inadequate, but none of them was consulted before the preliminary bill was drafted, and they finally voted against it. Nor did the bill issue from the expert labours of the special Parliamentary commissions. Its main principles represented a bargain between the two big parties, worked out in private between Ion Brătianu and Take Ionescu. 'That is as it always has been', bitterly remarked a deputy.

'The dethronement of the national Prince, Cuza, in 1866, the Constitution of 1866—these were the results of a bargain between the parties. The laws on agricultural contracts of 1866 and 1872; that of 1882; that of P. Carp of 1892, and finally M. Brătianu's last law of 1907—all have resulted from a bargain between Liberal and Conservative politicians, and all these bargains were made at the expense of the peasants, and, so far from their having given their consent to them, they were made without their being even listened to, without their being even called in.'

In this case, Parliamentary commissions and Parliament were merely asked to register the new bargain, which was submitted to them with the intimation that 'the agreement between the parties must not be touched'. To simplify the discussion, the two Parliamentary commissions appointed in their turn subcommissions from which, by a useful coincidence, two well-known professors of Bucarest University, members of the majority, were left out—the deputy Basilescu and senator Dissescu—both of whom had laid before the Chamber drafts of their own for the agrarian and political reforms.

This method of work was reflected in the basis on which the discussion proceeded. As the proposed reform was intended to effect a change in land tenure, its practical extent might have been determined with some degree of scientific accuracy by calculating either the area that could be made available, or the area that would be required for providing a given number of

[1] They were Radicals rather than Socialists, but they adopted that name apparently in the hope that their group might act as a bridge between the Russian revolutionaries and the Rumanian 'boiars', towards whom the Russians displayed a contemptuous hatred.

peasants with holdings of a given size. The protagonists, however, followed neither of these two lines in their argument. They simply contended around a figure at which the area to be expropriated was to be mechanically and arbitrarily fixed. The Liberals, who in 1914 wanted to expropriate 1,200,000 ha., in 1917 stood for 2,500,000 ha. The Conservatives considered that 1,800,000 ha. would be as much as they could allow and threatened to cause a popular agitation if the Government went farther. The Labour group demanded the expropriation of all estates beyond a maximum of 100 ha. per owner, a view which, later on, became the standpoint of the Peasant Party formed after the War. According to the calculations of Dr. Lupu, the application of this proposal would have made it possible to endow all peasant cultivators with holdings of 5 ha. each, large property being left with a total area of 417,000 hectares; the Government's proposal left in its possession 1,400,000 ha.

The Government found itself in a quandary. A Constituent Assembly required a quorum of two-thirds, but the Government's supporters were five short of that number, as some of the members of Parliament had remained behind in occupied territory. This and other circumstances were used by the Conservatives as bargaining points whereby they forced a compromise which fixed the area to be expropriated at 2,000,000 ha. Moreover, they forced the Government to abandon its intention of expropriating the subsoil, too; though this was later carried through in the Constitution of 1923. Even so, a Conservative leader, M. Argentoianu (now a leading member of the Liberal Party) denounced the law as being based on a sentiment of class-struggle: 'It is a law of persecution, directed against a whole class.' But the mass of the people accepted it as one of the first signs of grace among the men who ruled the country; and M. Mihalache, the present leader of the Peasant Party, admits that, notwithstanding its shortcomings, 'the reform of 1917 caused a true wave of rejoicing among all of us who were fighting at the front.' [1]

Legislative Evolution of the Proposals. During the preliminary discussions four drafts were prepared by the Government in turn,

[1] The modification of the Constitution was promulgated by the decree No. 721 on 19th July 1917, published in the *Monitorul Oficial* No. 93 of 20th July 1917.

all of them based on the compromise reached with the Conservatives. They varied, however, from each other in form and in details which showed traces of the pressure exercised by various vested interests; so that the final wording, as the Prime Minister admitted, was 'not very elegant'. A few of the changes helped to simplify and to improve the text, but the bulk of them caused much uneasiness among the more progressive elements in Parliament. The original text of the understanding conceded that 'in addition to reasons of public utility . . . the expropriation of arable land is also admitted, in the following measures and conditions, for the purpose of extending the area of peasant property in land'. The meaning of this was clear and frank; there was no pretence that the transfer of land to the peasants could be bracketed with the several measures of public utility mentioned in the Constitution. It was left for all to understand what most people knew, that it was rather a case of political necessity. The final draft, however, stealthily avoided the point and made use of a new political notion by saying that 'for reasons of *national utility*, the area of peasant land property is enlarged by the expropriation of arable land, in the following measures and conditions'. The new text, like the old article of the Constitution which it was meant to replace, again limited the right of expropriation in general to fixed property and then for specific reasons only. The Professor of Law in the Bucarest University, M. Basilescu, found it especially unpardonable that an Assembly elected on the basis of a three-class franchise should thus hedge in the rights of future Parliaments, which would be elected by universal franchise.

Another change concerned the fate of the subsoil. The original draft had said nothing about this, presumably meaning that it should be treated like the surface to which it belonged. But the final text exempted from expropriation 12,000 ha. of land established to bear oil, without indicating how and by whom the oil-bearing character of that land was to be determined. Nor did it grant similar preferential treatment to land containing other minerals. The change was generally regarded as a concession to the Conservative Cantacuzino family, who owned extensive properties in the more renowned oil-fields.

A further point against which much criticism was directed was the decision to take individual estates as the units to which expropriation should be applied; an arrangement which was considered to be bad law and worse practice. Expropriation could juridically affect the rights of ownership of a certain individual, but not the object to which those rights referred and which was not a juridical entity. Moreover, the proposal was deprecated as introducing a haphazard element into the criterion of expropriation; favouring those who happened to own a number of estates as against others who owned a similar or lesser extent in one compact property, and being even more patently unfair to joint owners of an estate who were prevented by legal circumstances from dividing up the property. The Labour group, therefore, and a few other members of Parliament, unsuccessfully urged that expropriation should be applied to each subject in an equal measure, and not to the various and varying objects which formed his possessions.

More disturbing than all these details was the general indefiniteness which permeated the proposed texts. They did not in truth constitute a reform, but merely the enunciation of the principles on which the reform was eventually to be carried out. None of the elements of the reform was finally fixed. The total extent, 2,000,000 ha., to be expropriated from the area privately owned was indeed prescribed, but not the action itself; and this was made worse by the erratic grammar of the formula —evidence of an uneasy birth. After emphatically affirming that 'the peasant property *is* augmented by . . .', the text vacantly changed its mood and tense in the next paragraph, when it came to saying how that increase was to be brought about, and announced that 'there *will be* expropriated . . .'—though the first operation could clearly only follow from and after the second. The complaint then made, that this was a promise of reform rather than the reform itself, appeared justified, for the 'will be expropriated' was made dependent on a special expropriation law to be passed within six months after the end of the War, and—that was the crux—with the two-thirds majority required for amendments to the Constitution. What this amounted to in practice was that the decision made in 1917, to carry through a sweeping land reform, was left to be confirmed by a second decision after the War. The same vagueness permeated other essential provisions of the text. For instance, expropriation was to be applied according to 'a progressive scale', but this scale was not determined and made known. And the fixing of the compensation to be paid to the landowners was to be ultimately left to the Judiciary—an arrangement which threatened to prove as elastic in time as it was in substance. Again, nothing at all was said about the price which the peasants would have to pay for the land. There was merely the King's promise, made at the front, that he would give them land 'like Ștefan the Great', which literally interpreted would have meant without any payment at all. For all these reasons, M. Garoflid charged the work of the Constituent Assembly with having been 'not a piece of scientific reform, but merely an electoral manifesto. . . . It could not be scientific—in the absence of a ground-book, of exact statistics on the distribution of property—to pass as law a measure applicable to a state of things which was unknown to us'.

Memories of how interest in the peasants always slackened after the emergency by which it was aroused had passed, were not calculated to allay the suspicions entertained by those who criticized the Government's formulae. It had been difficult enough to get the reform accepted even in 1917, when Rumania's rulers were besieged on all sides by spirits in revolt. Would it be possible to repeat the feat after the War, with authority again in normal control of events at home and abroad? What would have remained of the Rumanian reform—and of so many others—if the Allied victory had been followed by the success of their intervention in support of counter-Revolution in Russia? Or, alternately, what if the Central Powers had carried the day, in which case no new provinces, impatient of a change, would have been added to Rumania; and power would inevitably have fallen to the Conservative leaders, M. Marghiloman and M. Carp, whose rooted distrust of Russia had caused them to oppose Rumania's entry into the War, and whose attachment to large property was unshakable?

The critics had the shock of seeing their fears come true,

fortunately but for a passing spell, when the latter alternative and its consequences were realized after the second Russian revolution. The eastern front collapsed altogether, and in March 1918 Rumania was forced to accept the separate peace of Bucarest; M. Marghiloman, as the only statesman likely to meet with some condescension from the Central Powers, having in the meantime been entrusted with the government. M. Marghiloman proceeded to dissolve the Constituent Assembly and to hold new elections. The Conservative leader had always opposed the idea of expropriation. At the beginning of the new parliamentary period he therefore took it upon himself to indicate the lines they proposed to follow with regard to agrarian policy. He began by declaring that, notwithstanding the change in Article 19 of the Constitution, the reform was not yet made. 'The whole agrarian problem must be taken up anew from the beginning.' 'The Conservative Party would propose an agrarian reform which would allow the peasants, without brutally despoiling the large owners, to acquire fresh land, and gradually to form a rural middle-class, which is indispensable to the country's social balance.' The country, however, was at the time thrown wholly upon its own badly depleted resources, and these were limited almost altogether to what agriculture could give. Therefore the Government began by re-editing and systematizing the measures for the compulsory cultivation of the soil which had been in force in Moldavia in 1916–17. The destruction caused by the War and the burdens imposed by the separate Treaty of Peace, declared the Minister of Agriculture, were so heavy that 'it will be indispensable for us to produce in the country the largest possible amount of the things we require: hence the necessity of a general plan of cultivation. We must try to produce the utmost possible quantity and to export the utmost possible quantity; hence the need for compulsory labour'. At the same time, the new legislative measures endeavoured to protect the peasants by providing that they should first till their own lands and only afterwards that of the large owners. It also tried to ensure fair measurement and payment for the peasants' labour; and while obliging the peasants to work for the large farmers, it also obliged the latter on their part to lend to the peasants such machines and implements as they had. In September of the same year, the Government followed this up with its own proposals for land reform.

Their author, M. Garoflid, as Minister of Agriculture, was known firmly to believe in the necessity for reducing the latifundia, for social as well as for economic reasons, and of extending the middle-sized peasant property; and also that semi-feudal relations must finally be replaced by money relations on the land. M. Garoflid had acknowledged on an earlier occasion, and he repeated this in introducing his new bill, that one could not reach those ends without the expropriation of the large estates. But he and his Government considered that in the abnormal circumstances in which the State then found itself, such an expropriation would be bound up with technical and financial difficulties 'which risked upsetting the whole economic life of the country profoundly'. The reform they proposed, therefore, was limited to a bill for the compulsory leasing of land to the peasants; or, as M. Garoflid put it, 'for a general redistribution of the allotment of land.' In M. Garoflid's mind this, apparently, was to be the first phase of a sedate agrarian reform. In the second phase, the land was to be expropriated and leased to peasant co-operatives. The third and final phase was to establish the peasants as full owners of individual lots. The whole proposal seemed a balanced scheme for the progressive reform of Rumania's agrarian structure, and may have deserved a fair trial; but to M. Garoflid the chance of propounding it came too late—when the agrarian problem was about to solve itself by bursting the shell of neglect in which it had been allowed to fester. There is some doubt as to the area which M. Marghiloman's Government intended to expropriate. M. Garoflid, it would seem, considered that they ought to carry out the expropriation of 2,000,000 ha., as promised in the amendment to the Constitution. Other members of the Government meant to leave untouched estates up to 1,000 ha.; the two figures being, of course, incompatible with each other. But the proposal was not in any case destined to take effect, and a curious fate reserved to M. Garoflid, three years later, the task of applying a law radically differing from his own war-time projects.

At the end of October 1918 the resistance of the Central Powers broke down and Marshal Mackensen and his troops of occupation hastily left Rumania. M. Marghiloman resigned and was replaced by a transitional Government under General Coandă. The elections held by the Conservative Government during the enemy occupation were declared illegal, Parliament was dissolved and all the laws it had passed became in consequence null and void. Hence the agrarian régime automatically reverted to the position in which it was left in 1917. It only remained to enact the principles then introduced into the Constitution by means of a special law, 'within six months after the end of the War'—without its having been made clear whether that delay was to be counted from the end of hostilities or from the legal termination of the War through the signing and ratifying of a treaty of peace. But once again the will of the statesmen was rushed by the stormy flood of events. The end of war was no longer the self-glorifying parade of victorious authority which it had been formerly. Instead, authority found itself face to face with a population exhausted and restless, growing more impatient with Government as the means for satisfying its needs appeared sadly inadequate. The threads of the 'sacred union' snapped; politicians and military tried to pass on to each other the responsibility for the many failures and excessive sufferings of the War. In Rumania, General Averescu, then worshipped by his soldiers, raised a Cromwellian voice against political inefficiency and corruption, which for a moment threatened to rally all that heaving discontent to an assault upon the old order of things. All around—in Hungary, in Austria, in Bulgaria, not to speak of Russia—the masses were in ferment and the monarchs in flight. In the neighbouring provinces inhabited by Rumanians, revolutionary assemblies were taking power into their own hands, bent as much upon social as upon national reforms. They found themselves at the cross-roads in the progress of their political destiny and were in search of a new allegiance to which to pin their hopes and fair expectations. A syrup of deferred promises would hardly have stilled even for a moment the popular thirst for a better existence. On the day after that on which the general Armistice was signed, therefore, King Ferdinand issued a fresh proclamation in which he solemnly reaffirmed the pledges he had given in 1917:

> 'Circumstances', said the royal message, 'have again given Me the possibility of fulfilling what I promised and what in my heart I have never ceased to desire. My Government will realize the Constitutional reforms which will secure to all citizens universal suffrage, and to the peasants the ownership of 2,000,000 ha. from large private property, as well as the domains of the Crown, of the State and of the charitable endowments. By means of these reforms, we will ensure to all those who labour a social and material existence more just and more plentiful. My Government will without delay take the necessary measures for bringing these reforms into effect in the old Kingdom and in Bessarabia. It is Our will that the given word shall be kept.'

The royal proclamation correctly referred to the old Kingdom and only in addition to Bessarabia, which in the meanwhile had proclaimed its union with Rumania; but not to Transylvania and the Bucovina, whose relations with the motherland were not yet determined. On December 1 a revolutionary Transylvanian Assembly met at Alba-Iulia and proclaimed the independence of the province. At the same time it passed a Charter containing the principles on which it wanted Transylvania to be governed in future. Among these, the wish for a radical land reform took pride of place. That hunger for land among the populations of the new provinces, added to the other circumstances, helped to stimulate the Rumanian Government into prompt action. They knew that they could not do otherwise than concur in the Transylvanian demand, even if such a land reform had not in any case been desirable as a means towards curtailing the power of the alien upper class in the new provinces across the Carpathians. Without waiting, therefore, for the moment when a joint Parliament could be gathered together, the Government brushed aside the stern formalities which it had itself prescribed in the Constitutional amendment of 1917 and decided to enact the measures for the carrying out of the land reform by means of a so-called decree-law—that is, a royal decree having force of law under the reserve of its subsequent ratification by Parliament. The first decree-law, promulgated on 16th December 1918, was based very largely on a draft prepared under the care of M. Fotin Enescu, a former director-general of the Popular Banks, as Minister of Agriculture in the Jassy Cabinet of General Coandă. The decree, signed by M. Duca as Minister of Agriculture, laid down the norms according to which the land was to be expropriated at once and transferred to peasant co-operatives specially established for that purpose. Both these vast operations were carried out during the winter of 1918–19 and the following spring. A series of decrees-law, one for each province—beginning with that for Bessarabia, promulgated on 22nd December 1918—extended the reform to the new provinces, with variations adapted to local conditions.

The Peasantist Interlude. With the issuing of these decrees and their breathless application, large-scale property irretrievably passed away in Rumania. It only remained to classify the legatees and to portion out the heritage among them. The performance of this second act of the land reform fell to the lot of the Coalition Government which M. Vaida-Voevod formed from the new political groupings from Transylvania and Bessarabia and from the young Peasant Party. This was Rumania's first parliamentary Government, the new currents having gained a large majority in the first elections which were held under universal suffrage in October 1919. If the decree-law of December 1918 'represented the only possible formula at the time', as M. Mihalache admitted, 'in view of the composition of Parliament and of the country's situation', the union with Bessarabia and the other provinces, and the coming of universal franchise, had strengthened the current 'in favour of the radical and full resettlement of the peasantry'. By agreement among its members, the Cabinet entrusted indeed the drafting of the law of resettlement to the Peasant Party, which had absorbed some of the keenest members of the ephemeral Jassy Labour group, and whose leader, M. I. Mihalache, was in charge of the Ministry of Agriculture. The new Party eagerly seized this opportunity for fulfilling its program. M. Mihalache's bill bore signs of having been more carefully worked out than the original legislation. It proceeded in many details with greater caution and a better grasp of the working of economic factors; and its whole structure rested not on a bargain between vested interests and political opportunism, but on a clear-cut social conception of a peasantist persuasion.

'Our agrarian legislation', said M. Mihalache, 'is inspired by the ideal that our country is a peasant country . . . that is, a country whose economic life must be based on agriculture, with small property as its typical agent and, at the most, a limited extent of middle-sized property; large property being abolished and reduced merely to certain model farms which shall serve solely for experimental purposes and as schools for the small cultivators. It is true that this ideal is laughed at by the Socialists and treated with contumely by the president of the Agrarian Committee [M. Garoflid, also President of the Union of Agrarian Syndicates]; but it is a natural ideal for any country which is still far from being industrialized. And before coming to that distant Socialist heaven, of which M. Garoflid, too, has a vision, the country must first pass under the sign of the Peasant—a phase upon which, after the War, every agricultural country is now entering through the breaking up of the large estates for the benefit of the peasants.'

M. Garoflid objected that it was not a question of discussing the philosophy of property or of making comparisons with the West, but simply that 'in an adolescent society the role of the large owner as initiator could not be denied'. M. Mihalache agreed that certain elements among the large owners could perform functions which were not as yet within the means of the smallholders. 'But', he went on to say, 'here is the fundamental difference between you and ourselves, that we want to reserve to large property this *role of school, of model farm alone*, this and nothing more—three, four, ten at the most in one county—and consequently *all the excess of large property above what is strictly needed for that purpose must be expropriated in full.*'

M. Mihalache's bill proposed in fact to expropriate each individual owner down to a limit of 100 ha. He and his friends regarded this proposal as the key of their bill, the *exposé de motifs* devoting five of its seven pages to it. This was the basis of the Bessarabian reform, and M. Mihalache contended that as it had been generally agreed that the legislation of the various parts of Greater Rumania must be unified, one should not make an exception with the basis of her new agrarian structure. He also urged that it would be dangerous to allow the feeling to grow up on the land that the Bessarabian peasants were resettled more generously by revolution than the peasants in the Kingdom by Constitution. Above all, M. Mihalache insisted that the arbitrary limits within which the reform had been corseted at Jassy were untenable. Of the historic rights of the peasants no account had been taken at all. The new legislation merely continued the emergency re-settlements of 1881 and 1889, inspired by the same shallow idea that peace could be secured by creating a fresh balance between large property and small. Elsewhere the reform had been framed within social criteria which pursued an end carefully weighed—that of reducing the large estates and, especially, of creating economically viable peasant holdings. Nowhere had the reform been based on a purely mechanical calculation. It was a piece of strange irony that in Rumania, which had neither ground book nor statistics, 'one should have had the peculiar idea of building up a whole legislation upon something which did not exist or which existed only in an unreliable state.' Under such conditions, indeed, it was technically impossible to keep the expropriation exactly to the figure of 2,000,000 ha. What, then, would happen, asked M. Mihalache, if that figure should be exceeded? The possible consequences were realistically suggested by a circular which the Union of Syndicates of Moldavian Landowners sent to its members. 'We, landowners', declared that document, 'are firmly determined to oppose by all means, and especially by impeaching before the Court of Cassation, all those provisions of the decree-law of 1918, of the law for communal grazings and of any future law or decree, which may conflict with the letter of the Constitution regarding the expropriation of land.' And in order 'to re-consecrate the intangible right of property, we demand that a law shall be passed which shall punish with hard labour for life whosoever shall speak of expropriation again'. The Union imposed a levy upon its members in support of the action contemplated in that circular. One notes this outburst merely as illustrating how sharply the temper of the landowners had veered round after the War.

The essential principles of M. Mihalache's bill were the same as those of the Bessarabian law, passed unanimously by Parliament a short while earlier. But as soon as the Peasant bill became public, the new democratic façade collapsed at its touch as quickly as had the idea of national unity in 1857, when the agrarian question was forced to the attention of the Moldavian *divan*. The bill had been adopted by the Cabinet and had for several weeks been awaiting the approval of the King in order to be brought before Parliament. Meanwhile, however, delegations of large owners were denouncing it to the Monarch as a bill that would despoil the proprietors and destroy agriculture; and the Opposition intrigued so influentially that during all those weeks the Minister of Agriculture was unable to see the King. Finally, the Government tried a bold card. On March 12 its followers introduced the measure in the Chamber as a private members' bill. Foreseeing the possible consequences of that step, M. Mihalache wanted it 'to be known that if this Parliament or this Government has to leave, it will be because of a conspiracy of the whole oligarchy from all political parties'. They were not merely disapproving the provisions of the bill; they were also afraid of letting the new Party achieve a popular success so early in its career. 'The certificate of ability for the new Parties would at the same time have been a sentence against those who had ruled the country hitherto.'

M. Vaida Voevod's Government had at that moment been only three months in power. It disposed of a comfortable majority in both Chambers. The Premier himself was in London, working to obtain from the British Government the recognition of Bessarabia's union with Rumania. But in spite of the Premier's absence and of the strength of his following, the Government was in fact dismissed on the day which followed the introduction of the agrarian bill in the Chamber, and replaced by a Government under General Averescu.

The Second Expropriation. The coup against the Vaida Voevod Government was carried to its extreme conclusion when M. Mihalache's place as Minister of Agriculture was taken by M. Garoflid. The task of completing the second act of the reform was dashed from the hands of the Peasant leader to be passed into those of the acknowledged spokesman of the great landowners. Three years earlier he had charged the Jassy reform with having been an 'electoral manifesto' rather than a considered piece of legislation. Now he declared this 'electoral manifesto' to be sacred, and he used it as a shield wherewith to ward off the more radical demands of the peasants. All his life M. Garoflid had condemned as uneconomic the system of small peasant holdings. Now he was called upon to create many more of them, and he frankly recognized that his own bill was 'a compromise between differing social requirements'.

The details of the second act of the reform were contained in M. Garoflid's law, promulgated on July 17, 1921, and in the corresponding laws for each of the new provinces. Their main concern was with the rules and means for the resettlement of the peasants. But they also extended the basis of expropriation, thus admitting by implication that some of the criticisms levelled against the measure of 1917 had been justified by events. The Premier had repeatedly declared that he would not allow land to be expropriated beyond the limits fixed in 1917, but later he had to abandon that standpoint. The first expropriation had failed, namely, to detach from the large private estates the 2,000,000 ha. demanded by the Constitution and had fallen short of that figure by some 450,000 ha. To make good the deficiency, the new law adopted the principle that expropriation should be applied to each owner individually and not to each of his several estates—a principle which had been urged not only by the political Opposition, but, on theoretical grounds, by a Congress of Rumanian agrarian economists as well. For the same purpose, the law expropriated certain additional categories of owners. The new bill further admitted that it would be impracticable to leave the amount of compensation to be paid to the owners to be fixed by the Judiciary, and itself, therefore, laid down the basis on which it was to be calculated. Finally, the bill abandoned the method of handing over the land provisionally to 'associations of resettlement' first, and decreed that individual lots were to be distributed forthwith to the peasants, though the necessary measurements were far from being finished. Many of the evils in the application of the reform may be traced to this decision, which was not demanded by the peasants. There had been complaints against the administration of some of these special associations, but not against the system as such.

Other sensible intentions or provisions—as, for instance, those contemplating a much-needed consolidation of the scattered peasant fields—had likewise to be dropped because of the restlessness on the land or merely because the various political groups were using the reform as a stick with which to beat each other. In 1922 M. Mihalache complained that their propaganda concerning the reform had acquired the character of an auction, each Party trying to outbid the other in criticizing what had been done and in promising what they themselves might do if they came to power; so as later to be able to claim the credit for having given land to the new peasant voters. Demagogic pressure from some of his own partisans thus forced a conservative Minister of Agriculture to go farther in certain directions than even his Peasant predecessor would have gone. Even so, M. Garoflid's bill had an uneasy parliamentary career. Against Constitutional practice, it was first passed through the Senate in March 1921, and by the Chamber only in July, during an all-night sitting, when the left-wingers of the majority had to be threatened or cajoled, and Liberal and Socialist votes called to the rescue.

By that time the attitude of the various parties towards land reform had reached a more or less stable position. The Peasant Party, as we have seen, stood for the expropriation of all estates down to a limit of 100 ha., and that view they reiterated in the programme of 1922; model farms alone being treated as an exception. They also promised a revision of miscarriages in the application of the reform, as well as measures which should prevent the reforming of large estates. The ultimate aim of land reform, as seen by them, was the transfer of the land to those who tilled it, in the form of holdings limited in general by each family's power of working. The other parties formed a group whose agrarian policy was sharply opposed to that of the Peasants, while showing barely any difference as between the members of the group. Its most consistent ingredient was a determined objection to any further transfer of land from the large to the small owners by means of State action. General

Averescu, indeed, when addressing his People's League in 1920, had promised to the peasants holdings of 10 ha. each, without stating how that was to be done; but nothing was heard of this when the General came to power in 1921. The legislation then passed by his Government settled in fact the permanent outline of the agrarian reform, except for certain minor details. In its application, it is true, the law suffered considerable changes at the hands of the Liberal Government which took power early in 1922; and in the 1925 programme of his Party, as well as in the programmatic declaration he made when he came to power for the second time in 1926, General Averescu in his turn promised his partisans 'a revision of the Liberal revisions'. On the latter occasion he further announced 'a completion of the reform in the wooded regions and in the districts liable to flooding'; but none of these intentions were carried out, as General Averescu was never more than a *locum tenens* for the Liberal Party.

During the past few years criticism of the reform has never ceased to provide a daily topic for controversy in Parliament and in the Press. As the laws increased in age, that criticism was ever directed less against their texts and more against alleged abuses in the way in which they had been applied. From this one may deduct with some degree of safety the trend of Rumanian land policy in the immediate future. Of the two political groups which share influence at present, the Liberal Party would no doubt oppose any attempt to tamper with the essentials of a reform whose paternity they claim; though paternal pride would not be the sole motive for their opposition. The National-Peasant Party, on the other hand, contemplates reforming so many of the institutions and customs now prevailing in Rumania, that they would hardly choose to use up their strength in at once tackling an issue which is as complex as it is controversial. To be consistent with their stern and relentless criticism, they may endeavour to correct the grosser faults of omission in the expropriation of the large estates; they would find it less easy to amend miscarriages in the distribution of holdings whose possession has in a way been legitimized by the passage of years. Moreover, the principles of the reform have passed into the

Constitution, and the Constitution cannot be amended until the King comes of age. In their broad essentials, therefore, the new land laws are likely to remain the foundation of Rumania's agrarian structure at least during the lifetime of the next generation or two.

CHAPTER VI

THE NATURE OF THE REFORM

GENERAL LEGISLATIVE PROVISIONS

In the years before the War the idea of expropriation in Rumania was barely discussed practically—as a measure to be considered for its economic and social effects—but on the whole the chief arguments centred round the question of how far a forcible transfer of land would be justified by the letter of the Constitution. That debate was brought to an abrupt end by the events of 1917. Even before the War the old conception of property, based on Roman law, was gradually being displaced everywhere by another more in keeping with the social philosophy of our time. Increasingly under the pressure of philosophical and political criticism, property was coming to be looked upon less as an absolute individual right and more as a social function. This evolution was spurred on by the nature of the last War, which forced every combatant State to call upon all the resources of its nationals. The great ease with which thereafter each State curtailed the property rights of its citizens, whenever its needs were pressing, was in fact a tacit and universal recognition of the new conception of property. In Italy it received more formal sanction in the shape of a decree which authorized the expropriation of land not cultivated by its owner. The Constitution of Republican Germany was the first to proclaim the new view as an established principle in its Art. 153 which said: 'Property carries duties with it. Its use shall at the same time be a service for the general good'; and Art. 155: 'The cultivation and exploitation of the soil is a duty of the landowner towards the community.' The Russian revolution had in the meantime pulled the old conception out by its roots, especially in regard to land property, and this was bound to have a strong repercussion in the neighbouring peasant countries.

In Rumania the new viewpoint was put forward tersely by the peasant leader, M. Mihalache, when introducing his agrarian bill in 1920. Land, he said, could be regarded no longer as a source of rent, but 'as a definite and limited means for employing the labour of a category of citizens whose regular occupation was the tilling of the soil'. And in a speech delivered in 1921 he completed the idea by adding that 'whether landlord or peasant, we believe that if the owner does not fulfil the social and economic duties incumbent upon property, he must be treated as a speculator, and all such people should be expropriated'. It was from a similar point of view that the principles of the Rumanian reform were born. They resemble in a striking degree the program adopted by the Russian Cadets after a heated debate at their eighth Congress, in May 1917. It was then agreed that all excess of land property beyond a 'working norm', which was to be fixed by local committees, should be expropriated for the benefit of peasant cultivators. Private owners were to be compensated according to the normal revenue of their estates. State domains were to be broken up altogether. In Rumania both facets of the new conception of property were to be found in a decree issued as early as August 1917. It imposed upon the peasants the compulsory performance of agricultural labours, and it imposed upon the landlords the duty of cultivating their land. In the Minute with which he submitted the decree for signature to the King, the Minister of Agriculture said that 'property considered as a social function must serve common interests and satisfy the needs of the whole national community'.

These new ideas and the events which helped them along beyond doubt played a large part in preparing the way for reform. In its essence, one must note, the new conception of property for 'use' represents what was but common practice in the Rumanian provinces before the establishment of the national State. Both in the principles it enacted and in the change it caused in the distribution of land property, the Rumanian reform restored therefore in a large measure the agrarian conditions which prevailed before the Organic Statutes undermined them.

Section I

Expropriation

A. *The First Expropriation.* The agrarian reform received its first legislative expression in the shape of successive decrees-law for the Old Kingdom, Bessarabia, Bucovina, and Transylvania.[1]

The decree-law concerning the reform in the Old Kingdom dealt merely with the measures for expropriation, leaving resettlement to be dealt with later by a special law. That dissection of the legislative measure made it possible to elaborate the law of resettlement with more leisure and care and to submit it direct to Parliament. Political circumstances made it necessary to hasten with the first part of the reform at least, so as to forestall the suspicions of the peasants.

Of the four decrees-law by means of which the reform was introduced in the several provinces, that concerning Bessarabia was the simplest and most radical measure. Indeed, the original project worked out by the Sfatul Ţării was even more radical: excepting vineyards, orchards, and other plantations, landowners were to be left with only 50 ha. of arable land each, whereas the final decree raised that limit to 100 ha.

The reform in the Old Kingdom was more moderate and that for Bucovina held an intermediate position. In the Old Kingdom, too, the minimum that could not be expropriated was 100 ha., but expropriation was applied on a progressive scale which left to the large estates 500 ha. arable land in addition to vineyards, plantations, forests, and land unfit for cultivation. The text of the decree-law for the Old Kingdom was categorical and precise and therefore easily applicable.

The purpose of the reform was described variously in the several acts. That for the Old Kingdom paid less attention to the needs of production than, for instance, did the Transylvanian

[1] Decree-law concerning expropriation for reasons of national utility, No. 3697, *Monitorul Oficial* No. 215 of December 16, 1918. Decree-law for the agrarian reform in Bessarabia No. 3791 passed by the Sfatul Ţării, published in the *Monitorul Oficial* No. 220, of December 22, 1918. Decree-law for agrarian reform in Bucovina, No. 3871, published in the *Monitorul Oficial* No. 113 of September 7, 1919. Decree-law for agrarian reform in Transylvania, Banat, and the Hungarian lands No. 3911, adopted by the Great National Council on August 12, 1919, published in the *Monitorul Oficial* No. 117 of September 12, 1919.

decree. The latter sinned by the excessive vagueness of its provisions, most of which, though radical in principle, were facultative in application or studded with numerous exceptions. The Transylvanian decree, however, was based on a wider economic standpoint, whereas the other three decrees were more narrowly agrarian in outlook. Art. 1 of the Transylvanian decree evidenced the legislator's care for production by declaring that the purpose of the reform was to make the small owners economically autonomous, to create a middle-sized property and model farms, as well as to further the interests of industry and of industrial workers. Special attention was paid to the problem of housing and gardens for industrial workers; Art. 9, clause 2, allowing the expropriation for this purpose of even communal plantations and grazings. That wider outlook was natural enough in a province in which industry was much more developed than in the other parts of Rumania. Anxiety for the problem of production was also shown by the provision of the last clause of Art. 39 that land might be taken back from those who should prove incapable of working it. Art. 2, clause 16, expropriated all the landowners who had purchased their properties after July 31, 1914, and who were not themselves cultivators. Art. 36 provided that any land remaining after the needs of the peasants had been satisfied might be used for the establishment of middle-sized holdings and of model farms.

The economic viewpoint appeared likewise in the provision of the Transylvanian decree which exempted a part of the estate from expropriation for the benefit of a son studying agriculture; an exception which later was adopted in the Old Kingdom as well. Further, in the provision which allowed each landowner to choose the part which was to be left him; this was copied in the decree for Bucovina, as well as in that for Bessarabia in which was the proviso that the landowner's choice should not depreciate the remainder of the estate. In the decree for the Old Kingdom it was laid down that expropriation was to take account 'of the economic conditions of the estate' and that 'the expropriated part as well as the part which remained to the landlord should as far as possible be consolidated' (Art. 23).

Production was hardly taken into account in the decree for

the Old Kingdom, unless in the vague statement of Art. 10 which said that the State could reserve part of the expropriated arable land and devote it to some purpose of general interest. Art. 46 of the Bessarabian decree was more definite as it prescribed the establishment of a number of institutions destined to guide and stimulate production and it also decided the area which was to be allotted to them. This useful provision would seem to have been due to the presence in the Bessarabian legislative body of a number of agricultural experts who fought hard to introduce this and similar points in the original drafts. The right of the State to reserve an extent of land for general needs was likewise recognized in the decree for Bucovina.

Perhaps the most convincing proof of the practical and cooler spirit which presided at the elaboration of the Transylvanian reform was its authors' decision to carry out expropriation and re-settlement gradually, according to local needs and demands; though, of course, all the expropriable area was placed from the outset at the disposal of the State. The decree for Bucovina permitted the landowner to continue to use the land until it was actually taken over. In the Old Kingdom and in Bessarabia, where the political atmosphere was more feverish, the whole of the expropriated land was taken over at once.

In the Old Kingdom the decree-law prescribed the total expropriation of (*a*) the arable land on State domains and on the estates of public and private institutions; (*b*) the whole extent of estates belonging to foreigners and to absentees. All those upon whom the double land tax, payable by landowners living abroad, had been imposed during the previous five years were now considered as absentees. From private estates the reform expropriated 2,000,000 ha. arable land, on a progressive scale, as given in the table on p. 125 which exempted properties of not more than 100 ha., while reducing those of 10,000 ha. or more to 500 ha.

All land, including grazing and pasture, which was fit for cultivation was considered as arable land. One of the chief characteristics of this decree was that it expropriated the estates and not their owners, so that a proprietor of several estates could retain from each of them the exempted quota of 100–500 ha., in addition to forests, plantations, and non-arable land. Moreover, a landowner falling under this category was permitted to retain his several quotas in one single estate if the latter were situated in one of the so-called regions of colonization. In the case of a joint estate, each of the participants was entitled to retain a full individual quota; so that in this case the decree expropriated individual owners, whereas in general it was to be applied to each property.

Total area Hectares	Exempted quota Hectares	Total area Hectares	Exempted quota Hectares
100	100·0	1,500	307·8
110	109·0	2,000	324·6
120	117·2	2,500	338·7
130	124·9	3,000	351·4
140	132·0	3,500	365·8
150	138·6	4,000	374·9
160	144·7	4,500	388·2
170	150·5	5,000	396·9
180	155·9	5,500	409·7
190	160·9	6,000	418·1
200	165·7	6,500	426·8
300	201·7	7,000	438·8
400	224·8	7,500	451·2
500	241·2	8,000	459·2
600	253·7	8,500	471·7
700	263·6	9,000	479·7
800	271·8	9,500	491·9
900	278·8	10,000 and above	500·0
1,000	284·9		

In Bucovina as in the other new provinces the basis of the reform was more radical. The decree for Bucovina expropriated in full: (*a*) estates of foreigners, i.e. of those inhabitants who were not Rumanians or who were not citizens of the country on August 1, 1914; (*b*) mortmain estates; (*c*) the estates of absentees (outside the boundaries of Greater Rumania); (*d*) the estates of individuals having lost their civil rights; (*e*) estates farmed out during nine consecutive years before 1919. This latter provision did not exist in the decree for the Old Kingdom and marked a more radical tendency. In the same way, the expropriation scale was more radical than in the Old Kingdom, everything above 250 ha. being expropriated. Moreover, if neither the owner nor his parents had been cultivators, he was not allowed to retain more than 4 ha. Further, expropriation was applied to each individual proprietor, so that even if he owned several estates he could not retain more than 250 ha. altogether. As in the Old Kingdom, the decree expropriated arable land, grazing, and pastures. It exempted farm-yards, country-houses, gardens, vineyards, and industrial establishments together with the area of land they required. Communal property was to be expropriated after reserving the area necessary for village grazings and for other communal needs. Expropriated forests became the property of the State; likewise land unfit for cultivation.

The Bessarabian reform displayed a characteristic levelling tendency. It expropriated in full: (*a*) former State domains; (*b*) former Crown domains; (*c*) all mortmain estates; (*d*) estates belonging to the towns, beyond the area needed for town-planning purposes; (*e*) estates of foreigners—i. e. of all those who on January 1, 1919, eight days after the promulgation of the decree, had not declared for Rumanian citizenship; (*f*) estates which had been farmed out during five consecutive years. Both the latter provisions indicate how much more radical was the Bessarabian decree as compared with that for Bucovina. Monasteries were allowed to retain ½ ha. for each monk, as well as gardens and vineyards. From private property the decree ordered the expropriation of 1,000,000 ha., everything above 100 ha. arable land being taken away without exception. If that measure were not to produce the gross total of 1,000,000 ha., the decree allowed expropriation to go even beyond the limit of 100 ha. Vineyards, gardens, and specialized model farms were exempted. As in Bucovina, the decree expropriated each individual owner down to the limit of 100 ha. without regard to the number of properties he possessed. Joint properties were treated as if they were in the possession of a single owner. Waterways and land unfit for cultivation passed to the State.

In Transylvania the radical character of the reform was sufficiently pronounced, but it was softened by a whole series of exceptions and facultative provisions. The decree expropriated in full: (*a*) the estates of foreigners, i. e. those who on the strength of a subsequent nationality law would opt for foreign citizenship; (*b*) estates of public and private institutions whose residence was outside the country's frontiers; (*c*) estates having a special interest from a scientific point of view. The text did not explain what it regarded as of 'special interest from a scientific point of view'. Other categories of properties might be expropriated in full (Art. 2, clause 2) namely: (*a*) properties which had passed into the hands of their present owners after 1st November 1917, on the strength of the ordinances of the former Hungarian Governments restricting the transfer of real estate; (*b*) properties belonging to private or public institutions even if their residence were within the frontiers of Greater Rumania, except when these properties served directly a scientific, artistic, educational, sanitary, philanthropic, or economic national purpose. This comprehensive exception was improved upon by the subsequent clause which declared that 'such exceptions may even be created', without saying how, why, and by whom. A second exception to this clause exempted from the expropriation of mortmain estates the forests, mountain pastures, and grazings belonging to communes and to the so-called frontier communities, or those which were the joint property of various groups of peasants. A third exception referred to endowments; a fourth to forests which were used for fuel; a fifth to land belonging to religious or educational associations. It is difficult to see what was not exempted. The article further permitted the total expropriation of properties above 20 jugars[1] which since July 31, 1914 had passed, otherwise than by inheritance, into the hands of owners who had not till then occupied themselves with agriculture. Clause 3 of the same article further allowed the total expropriation of arable land: (*a*) from properties of more than 30 jugars in rural communes and more than 10 jugars in urban communes, which had been let on lease during twelve consecutive years—with a number of exceptions; (*b*) from any kind of estates everything beyond 500 cadastral jugars. Below 500 jugars the land was to be expropriated beginning with a quota of 20 per cent. which could be repeated till the exempted minimum of

[1] The German expression *joch* has frequently been rendered into English as *yoke*, which is unknown in English agrarian history as a term of measurement. The Rumanian form *jugar*, closely connected with the original Latin *jugerum*, is more accurate and less likely to cause confusion.

200 jugars was reached. In brief, expropriation definitely applied in full to foreigners only and to institutions whose residence was outside Rumania. On the other hand, Art. 4 of the decree declared that in those communes where the land was not sufficient for the purposes of the reform, any property might be expropriated, i. e. even those with a lesser area than the minimum of 200 jugars. The first clause allowing the expropriation of properties of less than 200 jugars did not indicate how far this might go; the second clause fixed a minimum of 50 jugars but only for urban communes, without indicating why it laid down no clear limit for the rural communes. Art. 4 then had the following severe provision:

> 'If the housing problem cannot be solved with the aid of the area made available through these expropriations, one may pass according to need to an additional expropriation of other areas as well, namely: in rural communes to land situated within a radius of at the most 600 metres from the edge of the commune; and in the urban communes as well as in mining and industrial centres and health resorts within a radius of 1,000 m.'

The text did not indicate whether the 600 metres were to be measured once and for all on the basis of the state of things existing at the time when the decree was promulgated. Its vagueness might have made it possible to encroach upon successive radii of 600 or 1,000 metres. It was only with difficulty that in 1928 the representatives of the national minorities induced the Ministry of Agriculture to adopt the first interpretation. It should be added that the whole of Art. 4 could be applied only by the head of the agricultural department in agreement with the heads of the departments of industry, finance, and social reform.

As in Bucovina and in Bessarabia, the Transylvanian decree applied expropriation not to properties but to proprietors. The decree permitted the expropriation of factories, works and of any establishments and rights connected with the expropriated land. This provision was not introduced in the decree for the Old Kingdom, with the result that in certain cases the owners refused to hand over various installations found on the expropriated area. Water rights, excise rights, and any other royalties and privileges were expropriated for the benefit of the State. The decree exempted from expropriation vineyards, orchards, and communal grazings, except when needed for housing purposes.

B. *The Second Expropriation.* The expropriation of the large private property on the basis of the decree-law fell short by some 450,000 ha. of the required area of 2,000,000 ha. A new measure of expropriation became, therefore, necessary. This was enacted in the Old Kingdom at the same time as the detailed resettlement, by the so-called Garoflid law of 1921. The law consisted of two parts, the first being entitled Expropriation and the second Resettlement.

The main characteristic of M. Garoflid's expropriation measure was the abandonment of the mechanical expropriation scale. As that scale took no account of economic circumstances it threatened to destroy even the few rational agricultural undertakings which Rumania possessed; and because it ignored local conditions it had taken away insufficient land where the demand for it was considerable and in other parts too much of it, so that great extents remained uncultivated. The new law introduced an economic criterion of expropriation. It laid down a maximum regional limit for all estates, but took care not to destroy the more progressive agricultural exploitations and breeding stations. And it likewise took into account the local demand for land and also the fact of a landowner being himself a cultivator or not.

These considerations appeared clearly in Art. 8 of the 1921 law. It left to those landowners who possessed their own dead stock, who occupied themselves with cattle breeding or who carried on agricultural industries, a larger unexpropriated area —100, 200, 300, 500 ha. They were grouped into categories according to the situation of the estates in highland or hilly or plain regions; and, secondly, in relation to the local resettlement demands—whether considerable, middling, or satisfied. Landowners who did not possess a proper equipment in dead or live stock were allowed to retain under similar conditions only 100, 150, 200, or 250 ha. In the second place, the law adopted a social criterion in that it took more land in those districts where the demand for land on the part of the peasants was greater. As the available area was smaller in the mountainous and hill regions, Art. 16 invited landowners in such regions to allow the whole of their estates to be expropriated, offering them in return one-and-a-half times as much land in the colonization regions. Art. 68 of the ordinance of execution classified as highland estates the properties situated in the high regions where the growing of cereal crops was incidental; as hill estates the properties situated in the hilly regions in which orchards and vineyards were the normal plantations, and whose arable area was less than 40 per cent. of the total area of the estate, less forests. Properties entering in neither of these two categories were considered as lowland estates.

The second important characteristic of the 1921 law was the extension of the basis of expropriation. Art. 6 regarded estates and parts of estates situated in the same commune or in neighbouring communes and belonging to the same owner as forming one single property. No landowner therefore could retain more than the maximum limit of 500 ha. arable land, no matter how many properties he possessed. There would seem to be a contradiction between this general provision and the text of Art. 4, which exempted from expropriation properties of less than 100 ha. arable land; so that a landowner possessing let us say ten or more properties of less than 100 ha. each might presumably keep all of them if they were not situated in the same commune or in neighbouring communes. The authorities entrusted with the application of the law apparently interpreted the limit of 500 ha. arable land for one individual landowner as absolute. This limit dominated the text of Art. 10 which decreed the so-called co-ordination of the expropriation measure, i. e. the reduction of each individual's possessions to the same absolute limit which applied to his category, no matter how large the number of parts which constituted them.

The law maintained the full expropriation of foreigners and absentee owners, with one single difference between the two categories. While exempting from expropriation private country-houses and parks, plantations, vineyards, woodlands, and industrial establishments, it obliged foreign owners to sell those objects within three years from the promulgation of the law; no such obligation was imposed upon Rumanian absentee owners. The same Art. 7 expropriated in full estates which had been let on lease continuously between April 23, 1910 and April 24, 1924. Likewise, all land leased in emphyteusis or similar titles were expropriated in full for the benefit of the users.

The law also extended the purpose of expropriation, adding to the original intentions that of creating communal grazings as well as the satisfaction of general economic and cultural needs (Art. 1). It dated the beginning of the reform from December 15, 1918, that being the date on which the State was considered to have acquired the title to the land (Art. 2). All transfer of land after August 15, 1916 was declared invalid, except land sold not later than February 1, 1921 to peasant co-operatives, Popular Banks, or to individual peasant cultivators up to 10 ha. each; as well as land sold for building or factories. The land thus sold was not included in the 2,000,000 ha. to be expropriated (Art. 3). In 1924 this article suffered a modification which recognized the validity of sales made to peasants after February 1, 1921 and until January 1, 1924, up to 5 ha. each, on condition that land sold in that way should be deducted from the part which the landowner was to retain.

Art. 13 considered as arable land for the purpose of the law all land which up to its promulgation had been used for cultivation, for grazing and for pastures, as well as all land liable to flooding but which was used for cultivation or for grazing. Vineyards, orchards, and other plantations laid out up to January 1, 1917, as well as land artificially irrigated, woodlands, &c., did not enter into the calculation of the law. But Art. 14 expropriated in full ponds and river-beds, barren land, &c., without regard to their extent, for the purpose of their being drained or afforested by the communes or by the State.

The 1921 law adopted the idea of gradual expropriation. The part which was to remain the owner's was to be determined at once, but the remainder was to be taken over only when it could be transferred without delay to the peasants. Any surplus of expropriated land could be let on lease to the original owner for a period of three years, if it was not demanded by the peasants themselves (Art. 15).

C. *The Mechanism of Expropriation.* The starting-point of the expropriation process in all the provinces was the landowner's declaration containing all the details referring to his property. The actual execution was then carried out by a number of commissions whose composition and functions varied in keeping with the stronger or milder radical tendencies of the respective measures.

In the Old Kingdom the provisional taking over of the land was entrusted to local commissions consisting of the district judge, of the interested landowner and of a delegate of the peasants. Its labours were revised by county commissions, which consisted of a judge, a delegate of the Central Resettlement Office, two representatives of the landowners, and two of the peasants. These commissions determined the actual area to be taken over, and they estimated the compensation. One could appeal against decisions of the county commissions to regional commissions. These were presided over by a president of the regional Court of Appeal or of the local Tribunal, and included a delegate of the Central Resettlement Office, one of the Superior Agricultural Council, one representative of the landowners, and one of the peasants. These commissions determined finally all the aspects of the issue, except the compensation, with regard to which an ultimate appeal was possible before the Court of Appeal. Where the final measurement established something different from the original indications of the landowner, the whole work of expropriation was revised in relation to the correct area.

In Bessarabia the decree-law entrusted the work of expropriation to commissions consisting of a judge, a delegate of the special institution known as Our Office, an agricultural expert, a delegate of the agricultural department, a delegate of the landowners, and five delegates of the peasants. The chairman of the commission was to be chosen from among its members. The arrangement gave a preponderant influence to the peasant delegates. All the work of expropriation was centralized and executed by Our Office. Appeals were to be heard by a Central Commission and were to be settled in 8–15 days, the interested parties not being legally represented. The whole procedure was drastically simple and expeditious.

In Bucovina the decree established local commissions, as in the Old Kingdom, and a Central Commission as in Bessarabia.

The local commissions comprised two intellectuals, one of whom was chosen as chairman by a drawing of lots, the mayor, a landowner, and two peasants. Final appeals were heard by the Central Agrarian Commission consisting of twenty-two members, among whom were three landowners and six peasants.

In Transylvania local commissions consisted of a judge, the official agricultural expert, the owner of the estate, and two peasant delegates. They had to try to bring about an agreement between the interested parties with regard both to the area to be expropriated and the compensation to be paid for it. The work of the local commissions was revised by county commissions, and final appeals were taken to the Superior Council for the Agrarian Reform, consisting of a president, a vice-president, twelve members, and twenty-four assistant members. Like the Central Commission in Bucovina, the Transylvanian Superior Council worked in the main through sub-committees. It was distinguished, however, from the institutions of appeal in the other provinces in that it consisted altogether of nominated members appointed by the Governing Council of Transylvania. This again shows how the various arrangements reflected the circumstances from which they had issued.

The whole procedure of expropriation was, therefore, in its general lines similar in the various provinces, and the organs entrusted with the execution of the measure were built upon the same principles. In the Old Kingdom and in Bessarabia, however, the central organ for the application of the reform was a State institution. In Bucovina and in Transylvania the work was placed in the hands of two private banks, the Regional Bank of Cernăuţi and the Agrarian Bank of Cluj.[1] In the Old Kingdom the legislator had before him the unsatisfactory results obtained by the Rural Office established in 1907. At the same time, it was thought useful to link up the execution of the reform with a democratic institution known to the peasantry, namely, the Central Office of the Popular Banks and of the Peasant Co-

[1] Decree-law for the creation of the Agrarian Bank destined to apply the agrarian reform in Transylvania, the Banat, and the Hungarian districts (No. 4167) *Monitorul Oficial* No. 125, September 21, 1919. Statute of the Regional Bank of Cernăuţi reorganized for the application of the agrarian reform. *Monitorul Oficial* No. 130, September 27, 1919.

operatives, which functioned at the time as a section of the Ministry of Finance. A decree-law transformed it into an autonomous institution attached to the Ministry of Agriculture. It was to consist of five sections: (1) the Central of the Popular Banks; (2) the Central of Peasant Co-operatives of Production and Consumption; (3) the Central of Co-operatives of Leasing; (4) the Land Mortgage Credit Office; (5) the Survey Office. The first two sections were to continue the work of rural co-operation, on a broader basis. The other three were to apply the agrarian reform.

In Bessarabia the central organ for the execution of the reform was a State institution known as Our Office, to which were entrusted the same functions as those performed by the Central Office in the Old Kingdom. The law declared the Bessarabian institution to be a branch of the Central Re-settlement Office at Bucarest.

In Bucovina the reform was placed in the hands of a privileged private institution, the Regional Bank, which was entitled to carry out all the technical and financial operations connected with the agrarian reform. The board of directors included members appointed by the Minister for Bucovina, and delegates of the co-operatives of the various nationalities in the province. The Central Government had a right of control.

In Transylvania the execution of the reform was entrusted to the newly created Agrarian Bank, in which the State participated with capital and also with representatives on the board of directors. The Agrarian Bank was altogether a private institution having no connexion whatever with the Central Resettlement Office, whereas the Regional Bank in Bucovina was considered to be affiliated to that office, though this was merely a matter of form.

In regard to the institutions entrusted with the technical execution of the reform, the original legislation therefore differed considerably in the four provinces. The difference was due partly to local needs, and even more to local idiosyncrasies and to the desire of the new provinces to keep things in their own hands.

The subsequent laws passed in 1921 modified only slightly the procedure of expropriation and the organs charged with carrying it out in the Old Kingdom. The law for the Old Kingdom maintained the Central Resettlement Office as the main authority for the execution of the reform. But Art. 37 of the law, as subsequently modified by another law published in the *Monitorul Oficial* No. 164 of October 28, 1922, made certain important changes in the composition and activity of the Agrarian Committee, the highest authority in matters of agrarian reform. The membership of the Committee was reduced from eighteen to twelve, six members forming a quorum. Its president was the Minister of Agriculture himself. The Agrarian Committee was to sit *in camera* without calling the parties, working solely on the basis of the dossiers referring to the various cases brought before it. Appeals could be lodged before the Agrarian Committee by either of the interested parties, as well as by the Ministry of Agriculture. The arrangement by which the Agrarian Committee was to work behind closed doors was no doubt due to the great number of cases with which it had to deal. But it was severely criticized, first because it took away from the Committee the appearance of being an impartial Court, so necessary for giving authority to decisions which were bound to be disliked by one side or the other; and, secondly, because the Committee could not in such circumstances handle its work with a full grasp of the details of each case, from an agrarian point of view, especially as, with one exception, all the Committee's members belonged to the judiciary.

More important were the changes made in the procedure which had been originally devised for the new provinces. The main purpose of these changes was to unify the procedure by bringing the whole execution of the reform within the purview of the Central Resettlement Office. The institution named Our Office was abolished by the amendment to the Bessarabian law published in the *Monitorul Oficial* No. 12 of January 18, 1921; and a subsequent amendment published in No. 68 of that official journal on March 25, 1925 extended the authority of the Agrarian Committee to Bessarabia as well. The Transylvanian law, promulgated on July 30, 1921, transferred the functions of the Agrarian Bank to the Central Resettlement Office, and the amendment published on September 20, 1922 completely modified the judicial part of the procedure of expropriation. The same thing was done for Bucovina by the law published in the *Monitorul Oficial* on July 30, 1921, with the amendment published on April 20, 1924.

To these laws were gradually added a number of ordinances which made the actual procedure conform to the above changes in the original legislation.

Section 2. The Resettlement

A. *The Distribution of Holdings.* The resettlement of the peasants was arranged in the new provinces simultaneously with the details of the expropriation. Only in the Old Kingdom was it left to Parliament to work out at a later stage a detailed resettlement law.

The principal question which the re-settlement laws had to decide referred to the size and nature of the holdings which the peasants were to receive. In Bessarabia it was decided to distribute full holdings of 6–8 ha., colonization holdings of 8–10 ha., as well as so-called complementary lots. The size of the latter was not determined, which made their distribution a problem of peculiar difficulty in the absence of any intention to prescribe at the same time the consolidation of existing peasant holdings. A second issue which created considerable discussion in Bessarabia was the establishement of an order of preference among the claimants to land. Ultimately the economic point of view prevailed and the decree-law laid down the following order of preference:

(*a*) Those holdings were to be completed which did not amount to a minimum of 6–8 ha.

(*b*) Full holdings were to be distributed to the peasants living on the estate and who had no land at all.

(*c*) Full holdings were to be distributed to landless peasants living within a radius of 5 versts.

(*d*) Complementary lots to the peasants living within a radius of 5 versts. The peasants in this category could be transferred to the third category, if they ceded to the State their existing properties. This was a beginning towards consolidation and the only one attempted.

The Bessarabian law reserved eight holdings of 25 ha. each for the creation of training colleges for teachers, a peasant holding for each village school, two holdings of 7 ha. each for each regiment residing in the province, to serve both for the instruction of the soldiers and for supplying their food, and 35,350 ha. for various establishments of a general agricultural interest.

The decree-law for Bucovina prescribed the simultaneous carrying out of expropriation and resettlement. Resettlement was to begin on the basis of individual demands, just as expropriation was set going on the basis of the landowner's declaration; the arrangement had the advantage that it established the real demand for land in the various localities. Like the Bessarabian decree, it created three categories of resettlement lots, but not in the same sizes: full lots of 4–8 ha., colonization lots of 5 ha., and complementary lots which could not be of less than $\frac{1}{4}$ ha., and were to be given to the peasants owning less than 4 ha. Full holdings were to be distributed to those peasants who had no land at all or to those who, having only some land, were willing to cede it to the State and to receive a colonization holding instead. Like the similar Bessarabian arrangement, this was a mild attempt at consolidation. Rural schools were to be endowed with one holding, and village priests with two holdings each.

In Transylvania, as in the other two provinces, the decree-law foresaw the distribution of full holdings, of colonization holdings and of complementary holdings, without determining their size. It declared instead that the holdings would be given in accordance with local circumstances and with the capacity of the claimants to work them.

As regards the order of preference in which the land was to be distributed, the Transylvanian decree sought to harmonize the economic with the national point of view. It therefore gave preference to local residents, to those who had suffered through the War, to war invalids capable of working the land themselves. In general, those peasants who had been mobilized received preference. The peasants owning more than 5 cadastral jugars were not entitled to receive land until the demands of all the other categories had been satisfied. Those owning less than

5 jugars were to receive sufficient land to bring their holdings up to that figure. The decree also entitled agricultural labourers and servants to resettlement as well, though these had no means of tilling the land.

Expropriation and resettlement were to go hand in hand in Transylvania, only as much land being taken over as was actually required at the moment. The land was to be transferred to the peasants at once, under a system of so-called compulsory leases, the final measurement and resettlement to follow after the necessary preliminary work had been concluded. As an exception, however, the Transylvanian decree permitted the immediate division and distribution of land where the two interested parties could reach a direct agreement (Art. 18), without any further expert advice and decision. The titles of the resettled peasants were to be inscribed in the ground books which already existed in Hungary before the War.

The decree-law for the Old Kingdom did not concern itself with resettlement, unless one excepts a general indication as to the order of preference to be followed, contained in Art. 38:

> 'The village associations shall include within the limits of the estate the peasant cultivators who do not possess sufficient land of their own, preference being given to those who have taken part in the War and to their successors. Those guilty of desertion or of insubordination during the War shall be excluded from these associations.'

The 1921 law for the Old Kingdom devoted its whole second part to the arrangements for the resettlement of the peasants. Art. 78 determined the following order of preference: (*a*) those mobilized in the War 1916–19; (*b*) those mobilized in the war of 1913; (*c*) war widows, for their children; (*d*) small cultivators without land; (*e*) cultivators owning less than 5 ha.; and (*f*) war orphans. The next article laid down a second order of preference for those falling within one and the same category: (*a*) war invalids; (*b*) peasants who had previously laboured on the estate; (*c*) peasants who had their own stock and a settled farm; (*d*) those having more children and (*e*) those older in years. If the available area were to be insufficient for all those within a category who fulfilled the same conditions, the distribution was to be settled by drawing lots (Art. 80). The law for the Old Kingdom recognized in addition a large number of rural inhabitants as being entitled to receive land, namely: priests, teachers, and all lesser officials residing in rural communes, the holders of agricultural degrees of whatever kind, on condition that all these should reside on the land and cultivate their holdings. Artisans like carpenters, tailors, blacksmiths, fiddlers, &c. who had not previously occupied themselves with agriculture, publicans and merchants, as well as any inhabitants who used to own land but had sold it, were not entitled to be resettled until the claims of all the other categories had been satisfied (Art. 84). Invalid officers received preference for ordinary holdings of 5 ha.; they could receive as much as 25 ha. in colonization regions after the resettlement of the peasants had been finished, on condition that they cultivated the land themselves (Art. 85).

Where there was not enough land to go round, some of the peasants were to be settled in the colonization regions, the choice of those who were to go being made, if possible, by mutual agreement or otherwise by the local committees. On the larger holdings formed on the expropriated area there were to be established model farms for the benefit of holders of agricultural degrees and of invalid officers who undertook to cultivate the land in accordance with conditions imposed by the Central Resettlement Office (Art. 88). From any surplus land remaining after the work of resettlement under the above conditions had been finished a second holding could be sold after the passing of a period of three years to those who had already received one on the strength of the agrarian reform. Preference was to be given to those peasants who possessed stock, who had a larger family, and who paid a greater part of the cost in advance (Art. 89).

The law for the Old Kingdom went farther than any of the others in allowing other categories of rural inhabitants, besides peasants, a claim to land. Whereas the original decree had, indeed, spoken of 'peasant cultivators', the resettlement law widened the expression to 'Rumanian inhabitants who cultivate the land'. A law of March 15, 1927 increased still further the number of non-peasant claimants to land by allowing holdings of 25 jugars in Transylvania and of 25 ha. in the Old Kingdom, Bessarabia, and Bucovina, to officers decorated with the order of Mihaiu the Brave, the highest Rumanian military decoration. Later still, holdings in Bessarabia were attributed to the former members of the Council of the Land, the *ad hoc* Bessarabian Assembly which had proclaimed the union with Rumania; and landowners whose daughters married officers of the Rumanian army were allowed to retain an additional quota beyond the general maximum of 100 ha. These arrangements have been severely criticized from an economic point of view, in addition to the fact that they gave to non-cultivators land for which its previous owners had received merely a nominal compensation.

B. *The Mechanism of Resettlement.* In the Old Kingdom the law fixed the size of the full resettlement holdings at 5 ha. and of the colonization holdings at 7 ha., besides building lots and communal grazings. In addition, of course, there were to be distributed complementary lots, which could not be smaller than ½ ha., to peasants owning less than 5 ha., after taking into account the land which they were likely to inherit in a direct line (Art. 92). Village artisans who occupied themselves with agriculture as well were entitled merely to a building lot and up to 1 ha. arable land. Building lots were to be of 1,000–3,000 sq. m. and were not included in the extent of the holdings.

Buildings found on the expropriated land were preferably to be sold to the communal or county authorities or to the peasant co-operatives (Art. 101).

As an exception to the general trend of the agrarian legislation, Art. 102 allowed the creation of 10 ha. holdings in the mountainous and hilly regions, as well as in those with a dense population, and of farms of 50 ha. in the other regions. A decision of the Agrarian Committee was required to that end. These lots could include buildings found on the expropriated land. Their total extent could not exceed one-eighth of the area expropriated.

In each commune a list of those entitled to receive land was drawn up by a committee consisting of the mayor, the priest, the head master, and four peasant delegates. Appeals were judged by a district committee which included the local magistrate and an agricultural expert as delegate of the Central Resettlement Office. In both cases, decisions were taken by a majority vote (Art. 104). The whole procedure was made as expeditious as possible, the period of appeal from decisions of the local committees being of fifteen days, and decisions of the district committees having to be given within twenty days. Appeals were allowed from the district committees to the Agrarian Committee.

The actual division and distribution of the holdings was carried out by the organs of the Central Re-settlement Office. Before this parcellation the Agrarian Committee had to reserve from the expropriated area: (*a*) the land necessary for general public and educational interests; (*b*) land for the creation or completion of village grazings; (*c*) the land necessary for the establishment of new villages or for the widening of existing villages, for roads, &c., as well as for afforestation in the colonization regions or on land unfit for cultivation or grazing (Art. 111). If the creation of those reserves reduced the area available for resettlement, the number of the holdings to be distributed was to be restricted in the same proportion.

C. *Communal Grazings and Woodlands.* (1) *Grazings.* The conception which in the past had generally animated the Rumanian legislator, under the impression created by the profitable development of corn-growing, was that any land taken away from the cultivation of cereals was an economic retrogression. In the discussions of the Jassy reform one could discern the persistence of that outlook, and the text of the reform itself contemplated the re-settlement of the peasants only on arable land, so as to maintain the agricultural standard of the country. Likewise, the decree-law of December 1918 provided for the distribution of arable land, but made no mention of grazings. In the mountainous regions, where corn-growing was not possible, it was allowed by Art. 17 of the decree to expropriate land for grazing and pasture; but that obviously was not looked upon as economically necessary, but merely as a way of compensating the peasants of those regions for not receiving proper agricultural land. And in consequence a circular, issued on April 25, 1919, by the Minister of Agriculture, interpreted Art. 17 as meaning that expropriation could in no case be applied to properties of less than 100 ha., which was the general minimum laid down by the law.

In the following year M. Garoflid had to give way to the peasants' insistent demand for grazing and to enact a special law, published in the *Monitorul Oficial* No. 138 of September 27, 1920. Those who had insisted on the need for such a measure pointed out that it was more useful to create grazing for a whole village than individual holdings for 7–8 peasants. The whole progress of cattle-breeding depended on them. If the breeding-stations which the reforms wanted to encourage were to stimulate by their example, and not merely to cause envy, they had to exist side by side with village grazings. Without such common lands the development of fodder crops was not possible: they could not be grown on the scattered peasant fields, which were not fenced in to protect the crops from the inroads of animals. Just in the higher regions, where the keeping of cattle played a more important part than cultivation in the peasant economy, most of the communes did not possess village grazings. The 1920 law allowed in the mountainous districts the expropriation of land even below the limit of 100 ha. for the creation of communal grazings and extended that provision to the hill region as well. Its purpose was said to be the completion of the peasant economy by enabling peasant cultivators to keep two oxen and a cow in the lowlands and in the hills; and in the mountains, where rearing cattle was the peasants' main occupation, five large animals, in addition to what they could graze on the Alpine pastures. This special law was regarded in part as merely re-editing the provisions, concerning communal grazings, of the law on agricultural contracts of 1908, as it affected in the first place those landowners who had at that time failed to give from their estates the area required for communal grazings. Art. 12 allowed the expropriation even of forests if the purpose of the law could not be achieved otherwise.

The critics of this law complained of the many exceptions which confused its provisions, and that this was made worse by the unsympathetic interpretation given to its texts by the authorities concerned. Art. 13, for instance, indicated that the area to be expropriated was to be calculated either by taking the number of the heads of families in each commune, and allowing three large animals in the hills and six large animals in the mountains for each household; or, alternatively, by taking the actual number of animals and allowing for three heads of cattle 1 ha. in the hills and 2 ha. in the mountains. Subsequent instructions of the Ministry of Agriculture decreed that only the second criterion was to be used, though the effect of the War had evidently been to reduce the existing number of animals below the pre-war level. A second circular instructed the officials concerned with the application of the law to take into account not the total number of the inhabitants of a commune, but only those who had been inscribed in the resettlement lists. This was clearly in direct conflict with the text of the law. It gave insufficient grazing even for the existing inhabitants of a commune, as M. Mihalache complained (speech on June 29, 1921), instead of being far-sighted enough to leave a margin for newcomers among the peasant cultivators.

The 1921 law for the Old Kingdom further extended the provision for communal grazings by its Art. 23. It allowed the expropriation in the mountains of land fit only for grazing or hay down to a limit of 25 ha. In the mountains and in the hills the expropriation was further permitted of clearings of less than 20 and of 10 ha., as well as of the intermediary surface if there should be two or more such clearings at a maximum distance of 200 metres from each other. Finally, the said article suggested that in the absence of any expropriable property, the peasants might create from their own holdings a communal grazing, if the majority of the inhabitants agreed to do so.

In Bessarabia the decree-law provided for the completion of communal grazings at the same time as the distribution of individual holdings to the peasants, the Bessarabian legislator showing much understanding of the important role which grazings play in a country of peasant cultivators. Transylvania and Bucovina were on the whole already provided with communal grazings before the present reform. Art. 24 of the Transylvanian law permitted the expropriation of existing communal grazings in so far as they were in excess of the normal needs of the villagers. Grazings were to be created or completed on the basis of the number of households; Art. 26 allowing for each household up to 10 jugars in the mountains, 5 in the highland districts, and up to 2 in the lowlands. Where cattle-breeding was the main occupation of the villagers the law allowed up to 22 jugars in the mountains and up to 10 in the highland districts.

(2) *Forests.* In the Rumanian provinces, as elsewhere, forests had originally been communal property from which the inhabitants of the neighbouring villages had satisfied their needs for fuel and building material. The first measure to restrict these communal rights to the equal use of the available timber was a decree of Moruzi, of November 28, 1792, confirmed two years later by Mihaiu Sutzu. On the pretext that the peasants were destroying the forests, the boiars demanded, and the Prince granted, that they should have 'the right to guard and to protect' young forests and plantations, and they were also allowed the sole right of keeping sheep and pigs in them, a right which they were entitled to sell. The peasants were allowed to take freely only: (*a*) dead wood from any forest; (*b*) wood for fuel from anywhere except from plantations and young timber; (*c*) any building timber for the needs of their own household; (*d*) and, finally, timber of any kind and from any forest for sale, as long as they gave the landlord a tithe of one in ten, as they gave the landlords in the lowlands from corn and hay. It will be seen that the landlords were only granted the title of protectors over the forests on behalf of the State, with certain privileges in return, and that the peasant rights, though restricted, were still sufficient to satisfy all their needs in wood for building and fuel.

These rights were reduced almost to nothing by the Organic Statutes of 1829, which transformed the landlords from guardians into proprietors of the forests. Art. 129 of the Moldavian Statute declared that 'the landowner alone . . . has the right to the use of mills, ponds, forests and such things'. It was the only mention of forests in the Moldavian Statute, and it summarily swept away all the ancient peasant rights to the use of timber. An almost identical text was contained in Art. 146 of the Muntenian Statute. Some of the peasant rights were, however, maintained in Art. 140, which said that on those estates which had forests the landowners should allow the serfs 'to take wood for fuel from the woods and copses which he himself will indicate and solely for their own indispensable needs, in the way in which this has been done hitherto'.

During the short wave of reform which preceded the Independence, the peasant rights to wood were restored by the Moldavian law of Mihaiu Sturdza, in 1844, and by the Muntenian law of Barbu Stirbey, in 1851. Both allowed the peasants to take dead wood for fuel, and any kind of timber for building and working from the places indicated by the landlords, against a yearly payment of 6 lei per household. These renewed privileges did not survive the great reform of 1864. Art. 9 of the rural law declared that 'the right possessed by the Moldavian peasants to the use of timber, according to Art. 44 of the Mountain Law . . . and by those on the Muntenian estates in virtue of Art. 140 of the law of April 23, 1851, shall be secured to them undiminished for the future. After fifteen years the landowners shall be entitled to demand the freeing of the forests from this servitude by friendly agreement or by decision of the Courts.' It is generally admitted that after the reform the peasants were denied all rights to timber, nor was any measure passed during the prescribed period of fifteen years to regulate the peasant rights to wood. On the contrary, a new Forestry Code went beyond even the Organic Statutes and completely stopped the peasants' access to wood, thus finally abrogating a right which had persisted through centuries of foreign domination.

Neither the decree-law of 1919 nor the law of 1921 made any attempt to deal with this old wrong and with the needs of the peasants for timber, especially under the circumstances created by the agrarian reform. The Peasant Party alone pressed this question and pledged itself to create communal woodlands in addition to communal grazings when it should have an opportunity of doing so.

The more radical Bessarabian reform expropriated all forests and woodlands, whenever the State should find it financially possible to take them over (Art. 11). The Transylvanian decree likewise allowed the expropriation of forests and woodlands wherever this should be deemed to be in the economic interest of the population. It permitted even the cutting down of forests when this should be demanded by a general economic interest.

From the laws passed in 1921, the Transylvanian law was the only one to concern itself with the expropriation of forests, in its Chapter III, entitled 'Communal Woodlands'. Articles 32–7 of the law dealt with the means for the creation or completion of communal woodlands, taking for this purpose first the woodlands available on those estates which had been expropriated in full. Where this did not suffice, the law permitted the expropriation of forests belonging to institutions and to private individuals down to 100 cadastral jugars in the lowlands and to 200 in the mountains. Finally, the law allowed the expropriation, in case of need, of communal forests or of forest properties with a communal character, when their area exceeded the limit laid down in the law (3–5 cadastral jugars per household, according to the greater or lesser agricultural character of the region).

The neglect of this problem by the agrarian laws for the other three parts of the country was made good by a provision introduced in Art. 132 of the new Constitution promulgated on March 29, 1923, which decreed the expropriation of forests in the rest of the country. For the carrying out of this constitutional provision a 'Law for the satisfaction of the normal requirements in timber for fuel and building of the rural population in the Old Kingdom, Bessarabia, and Bucovina' was passed in 1924, as published in the *Monitorul Oficial* No. 140 of July 1, 1924. The law consisted of seven chapters and nineteen articles. It provided for the creation of communal woodlands on the basis of 3,000 sq. m. per household in the lowlands, 5,000 sq. m. in the hills and 1 ha. in the mountains. The State was obliged to place at the disposal of the population the timber which it possessed within a radius of 20 km. from the centre of the commune. Where such State forests did not exist, or were required for some general public interest, the law allowed the expropriation of forests belonging to public or private institutions and situated likewise within a radius of 20 km. from the centre of the communes. And when that, too, did not suffice, private forests situated within that radius could be expropriated as well. Institutions and private owners were allowed to retain at least 100 ha. The expropriation was made in favour of the State.

The exploitation of such communal woodlands was to be made under the control of the State, preferably through peasant co-operatives. Wood for fuel and for building was to be sold by the State to the peasants at prices fixed in a yearly tariff, which was to be the same for State forests and for expropriated forests.

Chapter VI of this special law dealt with the administration of communal forests in Transylvania, as well as with that of communal and co-operative forests in Bucovina.

D. *Colonization.* All the four decrees-law contemplated the settlement in undeveloped regions of peasants who had no land at all and of those who had insufficient land and were willing to exchange it for a full holding elsewhere.

In the Old Kingdom colonization was dealt with in detail in Chapter XV, Articles 114–19, of the law of 1921. It entrusted the whole work of colonization to the Central Resettlement Office, which was to take the necessary steps for the creation of new villages or for the enlargement of existing villages. The law foresaw the granting of colonization holdings 'up to 7 ha.'. This meant that they could be of only 5 or 6 ha., and the arrangement was severely criticized, the Peasant Party insisting that the more capable peasants would either not agree to be settled in a new region on 5 ha. or would be unable to make good on such a holding. These undeveloped regions generally had remained so because of their less fertile soil. Hence the Peasant Party advocated the granting of colonization holdings of 10–25 ha., according to the size of the household and to the equipment it possessed, and this was to be exclusive of communal grazings.

For the assistance of the new settlers, the Central Resettlement Office was to establish depots of building materials, implements, and seed; it was to grant credits and to execute the technical works necessary. Likewise it was to supply plans for rural buildings and estimates of their cost, eventually giving assistance for their execution. The sums required for the establishment of such smallholdings were to be advanced by the State, half of the expenses remaining to be paid by the colonists within a period of forty years, which was to begin to run five years after their settlement; the other half was to remain a charge of the State. For the rest, the general provisions of the law, in so far as they were not in conflict with the articles of this special chapter, applied to colonization as well.

The Bessarabian decree-law also made provision for colonization. After a discussion as to the best way of establishing such settlements, preference was given to the creation of whole villages, rather than of isolated farms, as being more in keeping with the customs of the Rumanian people, as well as with the cultural and administrative needs of the moment.

Section 3. The Problem of Compensation

In the discussion of the agrarian reform, the problem of compensation played a prominent role. Apart from political conditions, the economic position of landowners towards the end of the War—when the shortage of labour, of live stock, and of implements became acute—caused most of them to resign themselves willingly to the necessity of giving up their land. If they showed any resistance, it was merely in an attempt to obtain as large a price as possible for it. The Constitution said that the price was to be fair and paid in advance, and that it was to be fixed by the judiciary. But in this respect opinion varied greatly, and perhaps nothing shows the change in the political background so clearly as a comparison between the ideas which the rulers of Rumania had in their minds with regard to compensation before the War and the actual arrangements they made in 1917.

When in 1914 M. Vintilă Brătianu advocated the expropriation of 1,200,000 ha., he meant the peasants to pay for the land at the rate of 1,000 lei per ha. (i.e. about £40 in gold values). Similarly, M. Take Ionescu only accepted the idea of expropriation in 1914 on condition that the landowners should be paid in cash, i.e. in gold. As the total cost would have amounted to some three milliard gold lei, that condition made the reform impracticable. After the outbreak of the War M. Take Ionescu apparently agreed to the arrangement by which the expropriated owners were to be paid in State bonds.

During that short interval the whole attitude of public opinion towards property altered rapidly. Everywhere the State had felt itself entitled to take what it needed from the possessions of its citizens for the pursuit of the War; and in the degree in which that need caused it to make greater demands on the mass of its subjects, the mass acquired a proportionate claim on the goodwill of the State. If the right of the State to create taxes for the levelling of private wealth was tacitly accepted all round, how much more was it in its right to take over part of the wealth in land from a few of its citizens, for the purpose of solving thereby social and economic problems on which the very existence of the State perhaps depended? The form of that imposition was dictated by the nature of the problem. The mere reduction of latifundiary land property could have been achieved by a progressive land tax. In that case, however, there would have been no assurance that the land would be acquired by the peasants. As this was the very aim of the reform, for social even more than for economic reasons, expropriation asserted itself as the only means of attaining that end.

If the reform was in the first place a measure for the assistance of a vanquished social class, then the main point which required to be kept in mind in fixing the compensation was that it should be just for the peasants.[1] What was socially just in this case appeared economically wise. The nature of the compensation was bound to have a determining effect on the success of the new peasant proprietors to be settled on the expropriated land. For a number of reasons the price of land at the time of the reform was exceedingly high, in relation to the exceptionally high price of agricultural produce at the end of the War. If compensation were to have been paid in accordance with that price, the land would have been acquired by the peasants on conditions that must have become oppressive when the price of agricultural produce began to fall. The new smallholders would in such circumstances have become bonded to the State, and the annuities would have swallowed the whole net profit of their cultivation. It was essential, therefore, so to fix the price that its payment should fall lightly on the shoulders of the new peasant proprietors. The legislator had two alternatives before him: either to fix the price very low, or to make the State take

[1] See a characteristic discussion of this issue by Dr. G. N. Leon in the *Neamul Românesc*, June 23 and 24, 1917.

part of it upon itself. The arrangements adopted in the Rumanian reform were in fact based upon both those possibilities.

The imposition of a tax on capital, which in the case of land property would inevitably have taken the form of a transfer of part of the property to the State, was discussed in the War period in most of the belligerent countries. In Rumania the amendment to the Constitution of 1917 and the subsequent decrees-law which realized the reform contemplated no connexion whatever between the agrarian problem and the financial problem. The agrarian reform imposed itself quite apart from any financial need of the State; in fact, it was clear from the beginning that the State would have to bear some of the cost of the reform. Hence the needs of the peasants coincided on that point with the difficulties of the State. The problem resolved itself, therefore, into finding a method which should take account of those impecuniary circumstances of both State and peasants, while basing compensation on a criterion that would be as fair as possible and generally applicable. The market value of land did not offer reliable guidance because of the speculative and social elements which entered in its formation; nor did the rental value of land, especially in Rumania, where, at any rate before 1907, it rose above the value of the revenue that could be obtained under normal conditions from the respective estates. Those cases in which landowmers or tenants invested part of the realized profits in improvements, which would at least have maintained if not increased the real value of the land, were very rare indeed; so that the annual profits were generally higher than would normally be achieved by the use of the instrument of production and the activity of the entrepreneur. These circumstances likewise made it difficult to base the estimate on a third possible formula, on the German theory of 'utilization value', meaning the value that could be obtained by using the object properly. The only way to approach as near as possible to reality in Rumania was to base the size of the compensation on the net value of production.

All theory, however, went by the board in the summer of 1917. Circumstances were such that the State could not demand much from the peasants and could offer very little itself to the landowners. Speaking in the Chamber, M. Take Ionescu addressed an appeal to the landowners to bear well in mind the nature of the vote that was being asked of them:

> 'That is why, since August 1914, I have been saying that it will no longer be possible to pay in cash, and therefore that our expropriation will have the character also of a spoliation, of a confiscation—if you like —of a reduction of wealth. That is the truth about the expropriation which we are proposing to you. It is as well that when voting upon the measure which we are submitting to you, the landowners should know that they are voting not merely for a transformation of values, but at the same time for a reduction of values.'

A. *The Expropriation Price.* Even so, the fixing of the compensation proved a thorny problem, especially in the Old Kingdom, where the price of land, as well as its revenue, varied greatly from region to region and from year to year. Scientific inquiries on the various factors determining the value of land did not exist. There was no map of the country's rural economy in relation to its geology, or of the soil's fertility. In consequence, the authors of the decree-law were forced to give an elastic construction to the text of Art. 18, which indicated the ways and means for estimating the value of the land, allowing almost any factor to be taken into account for that purpose—sale price, regional rental, estimates made by credit institutions, the net revenue per ha., expert estimates of the quality of the land, the land tax, &c., &c. The only settled indication was that the price to be fixed was in no case to be higher than twenty times the regional rental fixed in 1916 for arable land, and fifteen times for grazing (the original draft having taken the rental fixed in 1907, which, of course, would have been lower). But the division of the country into regions and the fixing of the regional rental by the Superior Agricultural Council had not followed a unitary and scientific method. Local circumstances, especially the relations between landowners and peasants, frequently caused estates of a similar nature and quality to be classified in different categories. A Memorandum of the Moldavian Landowners, published in 1920, asserted that no landowner had ever let land to the peasants at the fixed regional rent, except in return for labour, the price of which was likewise fixed regionally. Before 1907 the land was given to the peasants in return for labour. The reform law of that year demanded two separate contracts for labour and for land, with values calculated in money. When the contracts were signed in the autumn, the rates for labour were 35–40 per cent. lower than when the labour was actually performed, and the rental was fixed accordingly low. The Commissions which fixed those rates did not concern themselves with the actual value of the land, but merely with the relation between price of land and cost of labour, on the basis of the exchange that had been customary in the region. When the land was let for money, the rent was 70–80 per cent. higher, or even double the 'regional' rent. On the other hand, many spokesmen of the peasants asserted that the regional rent was seldom obtained in practice, and that e.g. none of the peasant co-operatives of leasing, which generally paid high prices, had leases at the regional rents, but always below them. As a result, the policy laid down in Art. 18 created as much dissatisfaction among the peasants as among the landowners.[1]

The multiplying of the regional rent by twenty corresponded to a capitalization at 5 per cent. Payment was to ensue as soon as the final measurement of the expropriated area had been determined. The original agreement arrived at in 1917 between Liberals and Conservatives apparently contemplated the expropriation of 1,800,000 ha., while the remaining 200,000 ha. were to be purchased by the State at 10 per cent. above the expropriation price, that difference remaining to be paid in cash.[2] The revised draft of the constitutional amendment abandoned that idea and the decree-law provided that payment was to follow in State bonds bearing 5 per cent. interest and payable in fifty years. This interest was apparently to be intangible, even if the State might later be able to convert on better terms. Institutions were to be paid in perpetual bonds. The State was to take upon itself up to 35 per cent. of the expropriation price 'in order to lessen the burden of the resettled peasants'. This was considered good policy, not only as a means for giving the peasants a chance to make good, but also because it dis-

[1] The bill prepared by the Peasant Party in 1920 proposed to base the price on the average between the estimates of the taxing authorities and the regional rent. This would have simplified the whole problem.

[2] See Basilescu, *op. cit.*, p. 85.

tributed among all the social classes the cost of a reform considered essential for the welfare of the country. And, thirdly, it was thought just that the peasants should not pay the whole price as they received only the surface, while the subsoil passed into the possession of the State.

If the estate concerned was mortgaged, the creditor was to receive that part of the debt which attached to the expropriated area in expropriation bonds, retained from the compensation due to the proprietor.

Until the final transfer of the land, the landowners were to receive for the expropriated area a rental in accordance with the regional tariff, or the rent indicated in leases, where such leases existed and the agreed rent was less than the regional rent.

In Bessarabia, the price of the land was calculated on the basis of the rent obtained by the owners in the period 1910–14. That rent was capitalized at 5½ per cent., the difference of ½ per cent. corresponding to the higher returns on money in that province. For those estates which had not been let out, the Bessarabian decree took the net profit reduced by half—one half of the profit representing, according to the Bessarabian legislator, the value of the personal work and the remuneration as entrepreneur of the former owner. Both factors were to be corroborated by means of various criteria of evaluation, which were named in the decree and were similar to those laid down for the Old Kingdom. In general, the price of the land was fixed by the Courts at 800 lei per ha., not without many protests from the peasants, who insisted that the land had been given them by the Revolution. The State was to take upon itself 25 per cent. of the expropriation price. Here the State's share was smaller, but it was definitely fixed, whereas the share in the Old Kingdom was laid down as a maximum limit. Payment was to follow, as in the Old Kingdom, in bonds bearing 5 per cent. and payable in fifty years. Mortgages were to be paid off from the bonds to which the landowner was entitled and in the order of their inscription. If such debts exceeded the value of the bonds which the landowner was to receive, the remaining debt was to become a charge upon that part of the property which the landowner retained.

In Bucovina, the price was to be established after a number

of criteria, taken at their pre-war standing. Rent was to be one of them and was to be multiplied as in the Old Kingdom by twenty. The decree for Bucovina had two interesting differences in that it laid down the possibility of the landowners being paid in cash as well; it did not indicate when and how this was to be done and it is more than probable that in fact no cash was paid at all. In the second place, the landowners were to receive not State bonds, but special land bonds, apparently guaranteed with the expropriated land. The bonds were to bear interest at 5 per cent.

In Transylvania, the price was to be estimated with the help of various factors similar to those indicated in the other decrees. But the commissions were not bound by a rigid limit, determined by the real or regional rent, but were left a greater latitude than elsewhere in fixing the price. As a safeguard the decree merely said that if in consequence the price were to be fixed excessively high or exceedingly low, it could be modified 'in virtue of strong reasons'. Appeals regarding the expropriation price were settled ultimately by the Court of Appeal. Payment was to follow in Transylvania more promptly, as the landowner was entitled to it as soon as his land had been declared expropriated. If the price was not finally fixed, the landowner was to receive a corresponding advance payment from the Agrarian Bank. The peasants could therefore be convinced that all connexion between the former owner and the expropriated land had ceased, whereas the payment in the Old Kingdom of rent during the intermediary period vaguely maintained the title of the landowner. The rent due for the land which had not been finally transferred to the peasants was to be paid by them to the Agrarian Bank.

The 1921 law introduced certain changes in the matter of compensation. The rapid fall of the Rumanian exchange had caused the compensation previously fixed to lose the greater part of its real value. When General Averescu came to power, therefore, he agreed to raise the compensation to sixty times the regional rent fixed in 1916. Opposition from within and from outside his party forced the Government to compromise at forty times the rent fixed by the regional commissions in 1916 for the period 1917–22, this being the maximum limit. The concession was criticized as raising excessively the burdens that would fall upon the State; and because it granted a definite compensation for a loss caused by temporary circumstances. At that time the Liberal Party still hoped that the exchange could be revalorized and they preferred an arrangement by which the landowners would have been given in return for the fall in the exchange a higher interest, for a certain period, or a part payment in cash. Land fit only for grazing and expropriated for the purpose of communal grazings, was to be paid at best with twenty times the regional rents for that period. The latter provision applied to vineyards, orchards and other plantations as well (Art. 36). Articles 69–76 of the 1921 law determined the methods of payment which, with insignificant exceptions, remained the same as those established by the decree-law. The second change of importance made by the 1921 law was that it increased the share that was to be borne by the State at 50 per cent. of the expropriation price, in view of the higher level at which it was to be calculated. The third change referred to matters of procedure.

The price was to be fixed by the District Commissions, both parties having the right to apply to the Courts of Appeal. A confidential circular of the Ministry of Justice had given the Courts of Appeal freedom to go beyond the regional tariffs in fixing the price of land. Some members of the Courts took advantage of this permission, others stuck to the letter of the decree-law of 1918; with the result that prices varied greatly from one district to another. To stop this confusion the united Opposition submitted in December 1920 a bill to suspend all appeals in matters of compensation, till the passing of the final law. Notwithstanding a protest to the King from the Union of Agricultural Syndicates, the 1921 law did in fact remove the matter of compensation from the jurisdiction of the Courts. The Judiciary had not the means of inquiring into the details of each case; and the procedure originally fixed by the Constitution for occasional expropriations threatened, when applied to a national expropriation, to swamp the Courts for many years to come under a flood of such appeals. Various landowners took the matter to the

Court of Cassation (the Rumanian supreme Court), asking it to condemn the change as a breach of the Constitution. But in a test case introduced by the Ministry of Agriculture a full bench of this Court decided on April 1, 1922, that it had no jurisdiction, and sent all the cases to the Agrarian Committee, as the highest authority in matters of expropriation.

The decrees-law further contained a number of provisions referring to the position of tenants, and to the payment of crops found on the expropriated surface. The decree for the Old Kingdom cancelled the lease of a tenant in so far as it referred to the expropriated land. If that area exceeded a quarter of the whole estate, the tenant was entitled to give up the whole lease. The same provision was included in the decree for Bucovina. The Transylvanian decree only mentioned the cancelling of the lease in so far as it concerned the expropriated area. According to the decrees for the Old Kingdom, Bucovina and Transylvania, the crops found on the expropriated part were to be taken over by the peasants. In Bessarabia such a provision would have been superfluous, as the land was in fact already in the hands of the peasants when the decree was promulgated. In Bucovina the landowner was entitled to the refund of expenses made with the expropriated crops. A similar provision gave much trouble in the Old Kingdom, either because the parties could not agree or because the peasants had not the means with which to pay for the crop. A subsequent modification allowed a sharing of the crop between landowners and peasants, as in the customary métayage system. Likewise, the Transylvanian owners were entitled to compensation for expenses incurred. The same arrangement was to be followed for the payment of stock taken over by the new smallholders.

The price of the woodlands expropriated in virtue of the special law for the Old Kingdom, Bessarabia and Bucovina, was to be calculated and paid in a way similar to that for the arable land. The respective provisions were contained in Chapter III of the special law of 1924.

B. *The Resettlement Price.* In Bessarabia the resettled peasants gave in return for the document recognizing their title to the land another by which they undertook to pay the annuities.

These were calculated at 5 per cent. interest and 1 per cent. amortisation and expenses. Failure to pay within a given period caused the holding to pass for temporary administration to 'Our Office'. In case of continued failure to pay the holdings were to be sold at public auction to cultivators only. Until the full payment of the price the land was unsaleable.

In Bucovina the expenses for parcellation and administration were added to the original price. On the other hand the State took upon itself 50 per cent. of it, as in the Old Kingdom. The peasants could pay in whole or in part, the remaining amount becoming a mortgage payable in fifty years, at 5 per cent. Even if paid in full the holding could not be sold or mortgaged during a period of ten years.

In Transylvania the holding could in any case not be sold till 1930.

The 1921 law for the Old Kingdom allowed the peasants a period of twenty years for the payment of the price, on condition that they paid at the outset at least 20 per cent. of the sum. The Agrarian Committee was empowered to waive this condition in the case of poor peasants (Art. 143). Payment for the communal grazings created by the Law of 1920 was to be made under the same conditions by the communes concerned. On taking power General Averescu issued a decree-law in March 1920, which provided that the peasants were to pay only the original cost, while the interest was to be borne by the State. Afterwards the Government discovered that the burden would be excessive and the promise was rescinded.

Everywhere, therefore, the peasants paid for the land less than the price received by the former owners. This was accepted by all Parties as inevitable. The only change which the Conservatives proposed through M. Marghiloman, in 1920, was one of proportion, not of division. They wanted the peasants to pay the whole of the price fixed on the basis of the decree-law of 1918, and the owners to receive a 'fair' compensation, fixed by the Courts; the difference remaining to be borne by the State. It is clear, however, as the peasants form the bulk of the taxpayers, that the major portion of that part of the expropriation price which in form has been remitted them by the State will in fact still be paid by the peasants themselves. Most of what they gain on the annuities they will lose on annual taxes.

Section 4. Measures for the Protection of Production

A. *The Associations of Resettlement* (*Obştii*). To safeguard the continuation of the agricultural work, notwithstanding the disturbance which was bound to be caused by the application of the reform and by the general shortage of live and dead stock, the land though expropriated at once was not handed over, in the Old Kingdom, direct to individual peasants. Instead, using as a model the generally successful peasant co-operatives of leasing, it was decided to transfer the land in each commune in the first place to specially created *obştii*, i. e. associations of those peasants who came within the purview of the provision for the granting of land. It is interesting to note that a similar proposal, though meant in practice to go farther, had been made by the landowners themselves. Fearing probably that they might be unable after the expropriation to secure enough labour, or only on heavy conditions, the Moldavian landowners proposed in 1919 that there should be formed on each estate an association of production. The landowner was to bring the remainder of his estate and such stock as he had, and the peasants theirs, so that nothing should remain untilled.

The actual legislation contemplated merely the establishment of village associations modelled on the co-operatives of leasing, each peasant cultivating an individual part, for a duration of some three to five years, so as to gain a respite for the gradual carrying out of the resettlement, with due regard to economic considerations. The experience of the Rural Office created in 1907 had shown that the settlement of new cultivators could not be made hurriedly. Some 2,300 of such village associations were in fact created on the strength of the 1919 decree-law by the Central Resettlement Office. They were governed by the law for Popular Banks and Village Co-operatives. They took over the expropriated land as soon as it was detached from the respective estate. Each association was in the charge of an official expert, who divided the fields among the peasants, and who had to see to it that the necessary seed, animals and machines were forthcoming and properly used. The associations thus acted as a channel for the distribution of the means of production that were lacking after the War, as well as an instrument for the economic and technical education of the new smallholders. The somewhat chaotic state of things which followed the expropriation in Bessarabia brought out the advantages of the system used in the Old Kingdom. In the province across the Pruth, the resettlement authorities found it beyond their powers to keep in touch with the individual peasants, especially during the early period of the reform.

In view of these obvious and considerable advantages the arrangement by which the land was handed over to such associations, until the completion of the measurement and of the formalities of sale, was confirmed by Articles 66–8 of the Law of 1921, with certain important modifications. The texts concerned contemplated giving to these bodies the character of 'associations for agricultural exploitation'. They were to be directed by an administrator appointed by the Central Resettlement Office and by two delegates elected by the peasants. These changes were made because in many of the associations formed on the strength of the decree-law the directing official experts had abused their powers and the peasants had had no means of controlling them. That increased the suspicion, entertained by some of the peasants, that they had been freed of the landlords only to be brought under the tutelage of the State. There is general agreement that if the peasants complained of the manner in which some of the associations were administered, they did not object to the system itself. This was also the opinion of the majority of the agricultural councillors when they were consulted by M. Garoflid in 1920, and it was then agreed to change the name and the working of the associations. But the politicians had discovered a fresh opportunity for posing as the protectors of the peasants, and they magnified the sporadic complaints into a general grievance. Under pressure from his own chief, General Averescu, M. Garoflid had finally to drop the whole system; the Liberals afterwards concurring. After 1921 the land was handed over to individual peasants as soon as it was taken from the landowners.

B. *The Cultivation of the New Holdings.* Some wise, and in part drastic, provisions for the advancement of cultivation were laid down by Articles 137–41 of the law of 1921. The Central Resettlement Office was empowered to devise and impose means for the 'better cultivation of smallholdings, for the best possible use of the communal grazings', as well as for the marketing of the smallholders' produce. All recipients of holdings situated at a distance greater than 15 km. from their place of residence were obliged to move, within three years, to the commune to which the new holding belonged.

Failure to observe these obligations might involve the loss of the holding, at the demand of the Resettlement Office and upon a decision of the local tribunal; the holder receiving back whatever part of the price he had paid already.

Section 5. Measures for the Safeguarding of Peasant Property

A. *Consolidation.* The piecemeal distribution of peasant property made of consolidation a measure destined, in the words of M. Mihalache, 'to revolutionize our whole system of dwarf property.' He brought examples from a trip through Transylvania to show that the peasants were not against it on principle. They disliked it where it had been abused in order to deprive them of their good land, but wherever it had been carried out fairly the peasants were asking for a second consolidation. M. Mihalache insisted that such a complex operation should be carried out on the basis of a proper survey and ground-books, neither of which existed in Rumania. On the other hand, the reform offered a unique opportunity for initiating a process of consolidation. The measurement and distribution of fields had in any case to be undertaken on a large scale; and when the State came with a gift in its hands, it was in a better position to impose a rearrangement of fields upon the peasants without rousing their suspicions.

For these reasons the bill prepared by M. Mihalache provided in Art. 71 the consolidation of holdings, but hesitated to make it compulsory. Existing holdings and the expropriated area were to be pooled together and fresh compact holdings given from this to all those entitled; but for this the consent of one half of the peasant owners was necessary in each village. M. Garoflid likewise introduced the principle in Art. 136 of his law, remarking that consolidation, which elsewhere had been an improvement for cultivation, in Rumania would mean a new agriculture altogether. But he admitted that the problem was as complex as it was important, and he therefore agreed to deal with it in a special law. The bill was prepared and the draft is apparently to be found in the archives of the Central Resettlement Office. But politicians intervened once again and barred a measure which they feared might make them unpopular with the peasants. The question has remained in abeyance to this day.

In Bessarabia the problem was discussed in the agrarian commission appointed by the Council of the Land to prepare the reform, but the deputies could not be induced to adopt it. In its comments upon the first reform bill the Society of Agronoms insisted on the enormous importance of the problem and on the unique opportunity which they now had of solving it. Failure to legislate such a measure must be attributed solely to the anxiety of politicians to do nothing that might make them unpopular on the land; though it would be difficult to prove that the peasants were really against it.

B. *The Maintenance of Peasant Holdings. Inheritance.* The excessive splitting-up of peasant holdings had been largely caused by the complete equality among the several heirs to a land property, in Rumanian civil law. Hence, the recognition of the need to consolidate the scattered peasant fields gave rise to the complementary demand for a measure which should check in the future that disintegrating process. The former agrarian laws had made the peasant holdings inalienable, but had done nothing to prevent their being broken up. The absence of an industrial outlet for the population and the fanatic attachment of the peasants to the land had in consequence led to the continuous division of the peasant holdings.

The 1921 law for the Old Kingdom decreed by Art. 126 that arable land could not be divided by inheritance below 2 ha. in the lowlands and 1 ha. in the mountains and hills. The same provision was laid down by Art. 95 of the law for Bucovina. In

Transylvania the indivisible minimum was fixed at 2 jugars; the Bessarabian law said nothing about this. No restriction was placed upon the division of farm-yards, kitchen gardens, vineyards, plantations, &c. Experts considered that 3 ha. was the smallest area on which machines and implements could be rationally used; and that was the minimum which the bill of M. Mihalache had proposed to declare indivisible.

In the second place, Art. 127 gave each landowner the right, as an exception from the Civil Code, to leave the property to one of his heirs alone, the others being compensated by the new owner in money. Provision was made for the fixing of the compensation in such a way as to prevent excessive obligations being incurred by the new owner. If the testator failed to indicate an heir in this manner, and the division of the holding would conflict with the provisions of the previous article, the law empowered the judicial authorities to designate among the heirs one who should take over the holding on the conditions fixed by this Article. Thirdly, smallholders were empowered by Art. 134, as an exception to the Civil Code, to leave the indivisible minimum to a single heir even if its value exceeded that individual's proportional share, and without obligation for testator or heir to contribute the difference in money.

Art. 133 allowed any landowner, in the Old Kingdom and in Bucovina, to declare indivisible an area up to 50 ha. of his property, which thus would be inherited under the conditions laid down in this eighteenth chapter of the law. An obligatory minimum of 2 ha. and a facultative maximum of 50 ha. were thus brought by the law into entail, in the Old Kingdom and in Bucovina. The Transylvanian figures were the same, but referred to cadastral jugars.

Sale and Mortgage of Holdings. All the previous agrarian laws had endeavoured to protect the peasant property by making it inalienable. The peasants' consequent inability to sell or buy chained them to their holdings, which were continuously diminished in size by inheritance; as a result their labour lost value with the loss of mobility, and all selection among the peasant cultivators was checked.

The new agrarian laws abandoned a system which ensured

nothing but the peasants' poverty. Holdings, as we have seen, could not be broken up below a certain minimum, while they could be left as a whole to one only of the heirs. They could henceforward be bought and sold—after a certain period and on certain conditions.

While abolishing inalienability, so as to give free play to a natural selection among cultivators, the legislator wanted to ensure that smallholdings would remain in the possession of the class for which the reform had been enacted. The safeguard of the original law being deemed insufficient, a new measure was promulgated in March 1925, which gave the State a right of pre-emption on holdings secured under the latest reform, as well as on all those purchased in earlier years from the sale of State domains or through the Rural Office. If the State did not exercise its right, then the peasants were free to dispose of their lots on the following conditions, laid down in Art. II:

(*a*) The purchasers have to be Rumanian citizens, cultivators or graduates of an agricultural school, residing and carrying on agriculture in the commune in which the holding is situated;

(*b*) holdings could be sold and bought only to the full extent in which they were originally obtained;

(*c*) they could only be sold five years after their owners had obtained the final title-deeds which followed the payment of the price in full;

(*d*) the purchaser should not own more than 25 ha. arable property constituted on the basis of the present and previous land laws. (Art. 122 of the 1921 law had fixed this limit at 25 ha. in the highland and hill regions, and at 100 ha. in the plain.) The State's right of pre-emption was to be exercised through the Central Resettlement Office. The land thus obtained was to be given to peasants who had not yet been resettled, in the order in which they were inscribed in the resettlement lists.

The same law provided that house, garden, and farm-yard—up to a maximum of 1 ha.—could not be touched. They could not be mortgaged, and they could not on any ground be seized and sold by a third party. The rest of the holding, up to 25 ha., could be mortgaged only with the Central Resettlement Office, the Popular Banks, or some other credit institution authorized by the State. Beyond 25 ha. the peasant owners could dispose of their properties as they liked.

In Bessarabia, resettlement holdings could be sold to cultivators only. None of them were allowed to own more than 20 ha. arable land.

The Restriction of Large Property. The essential aim of the reform had been to transform land tenure from a system of large property into one of smallholdings. To prevent the revival of large property, the State reserved to itself a right of pre-emption on all land sales involving more than 50 ha. This provision was interpreted by its author, M. Garoflid, as referring to transactions which concerned an area of at least 50 ha. Many direct sales of lesser lots passed therefore unchallenged. Subsequently, this interpretation was changed by M. Al. Constantinescu to mean that the State's right came into play whenever any land was to be sold from an estate of more than 50 ha., no matter how much or how little of it was up for sale. That would appear to be the current interpretation of the State's right. It will be noted that the measure only means a limitation of sale, not a limitation of possession, as was applied to peasant properties formed from resettlement holdings.

The State's right of pre-emption on land for sale revives in fact the old peasant right of *protimis*, which gave the villagers the first claim to any land from the village *hotar* which the landlord wanted to let or sell.

Within the limits and conditions laid down in these laws property is guaranteed by Art. 17 of the new Rumanian Constitution. Expropriation is permitted for reasons of public utility alone, after fair and preliminary compensation. Expropriation is specifically permitted in the interests of public transport, public health, military and cultural works, and of works required by the direct general interests of State and public authorities. Other cases of 'public utility' will have to be determined by special laws, voted with a majority of two-thirds.

Section 6. Variations in the Several Agrarian Laws

The enactment of the reform through the instrument of four different laws—one for the Old Kingdom and one for each of the three new provinces—was in the first place the result of political circumstances which did not synchronize. Though passed in every case by peaceful means, the reform sprang, in two of the provinces at least, from a revolutionary state. The character and limits of the reform were, in such circumstances, dictated by the momentary situation and were passed into law by special revolutionary assemblies. All that could be done later by normally elected Parliaments was to elaborate the details keeping as close as they could, or as they had to keep, to the original principles. The Bessarabian reform was decided upon by the *ad hoc* revolutionary assembly known as the Council of the Land, organized in October 1917. A few months earlier the principles of the proposed reform had been adopted for the Old Kingdom by the Parliament at Jassy. The Transylvanian reform did not materialize till December 1918, when a revolutionary assembly assumed power in Transylvania, after the collapse of the Austro-Hungarian Empire. The reform for Bucovina alone was adopted during normal conditions, after the end of the War. The variation in time and political circumstances therefore explains why the reform was set going by different laws for the various parts of new Rumania.

In the second place, these variations were dictated by the considerable difference in the agrarian conditions of the several Provinces. Transylvania had never been linked up politically with the Rumanian provinces and had followed a rural evolution of its own. Bucovina passed to the Austrian Empire in 1774 and Bessarabia to the Russian Empire in 1812, before the emancipation of the serfs, which being carried out variously in these several countries, destroyed the former uniformity in the rural structure of the Rumanian lands. The upshot was a considerable difference in the distribution of property, as well as in the laws governing land tenure in the several provinces. When the peasant leader, M. Mihalache, as Minister of Agriculture, first brought his bill before the Cabinet, in 1920, he intended applying one and the same measure to the whole of the country. But it became clear during the Cabinet discussion that by enacting one principle for the whole country its application would have to be varied by innumerable instructions, so as to adapt it to the needs of the several provinces. It was therefore concluded that it would be simpler and more practical to prepare at the outset a bill for each of the new provinces, in continuation of the decrees-law promulgated for each of them as soon as the War was over.

The Transylvanian reform was initiated by the resolution which the National Assembly of all the Rumanians of Hungary adopted at Alba Julia, on December 1, 1918. Point 5 of Article III demanded 'a radical agrarian reform. The expropriation of all properties, expecially of the large estates. By abolishing the entailed estates, on the basis of that expropriation, and by reducing the latifundia according to needs, the peasants shall be enabled to acquire at least as much land (arable, grazing, woodland) as they can cultivate with the help of their families. The guiding principle of this agrarian policy is on the one hand the promotion of social levelling and on the other hand the furthering of production'. On the basis of this revolutionary decision, a Commission appointed by the Transylvanian Directorate drafted the text of the decree-law which introduced the reform into Transylvania. The subsequent modifications of the Transylvanian measure were likewise drafted by a Commission composed of Transylvanian deputies. Although these drafts went farther than the reform for the Old Kingdom, they did not satisfy a more radical group of Transylvanian deputies, who formed, in 1920, a short-lived National Radical Party. The programme of the Old National Party was itself revised in 1920 with regard to land reform on the principle that 'land must belong to those who cultivate it, to the total exclusion of rent without labour'. The Party pledged itself not to allow any changes to be made in the reform to the detriment of the peasantry, either with regard to the extent or to the price of the expropriated area, when the reforms for Transylvania and Bucovina should come before the legislative assemblies for revision. Moreover, the Party demanded for the Old Kingdom a new law more in keeping with the needs of the peasants and with the democratic views which had inspired the reform in the new provinces. The first draft prepared for Transylvania by M. Garoflid under the influence of these radical tendencies evoked a protest from the national minorities against its drastic provisions. A second draft, as presented to the Senate, proved more tolerable. But when it came before the Chamber, the Minister in charge, under pressure from the Transylvanian deputies, accepted several new articles which again widened the powers of the Law.

In Bessarabia, the reform began as part and parcel of the Russian Revolution. From July 1917 onwards, the peasants began to lay hands on the large estates. These sporadic actions became general in October, and by the end of the year two-thirds of the large property had been seized, as well as much of the live and dead stock. Of the two currents which then traversed the revolutionary movement, one wanted all land to be given to the peasants, without any payment at all. An article proclaiming its views declared that 'henceforward this gift of God should neither be sold nor bought, but free for any man to use, like the water and the air'. This point of view was raised into a formal demand by the first Peasant Congress, held at the beginning of May 1917. The Peasant Group, inclining socially Leftwards, was on the whole averse to union with Rumania, as it mistrusted the social outlook of the politicians who ruled the Old Kingdom. The second current was represented by the National Moldavian Party, whose program had a nationalist rather than a social bias. By an inversion of the usual run of prejudices, the large landowners had become nationalist and stood for union with Rumania, hoping thereby to save their properties. But in the third month of the Revolution the National Party was already forced to adopt a radical agrarian program, as the peasants did not understand 'autonomy' without land.

In March 1918 the various groups voted firmly for union with Rumania, but added the condition that the gains through the Revolution should be respected. This condition was waived at the end of November, after the Council of the Land had finally passed the agrarian law for Bessarabia. The law was ratified by the decree-law of December 22, 1918. Of the seventy-three Articles composing the measure passed by the Council of the Land, fourty-seven Articles were modified by the law for Bessarabia adopted by the Rumanian Parliament in 1920. In keeping with these circumstances, the Bessarabian reform passed through three stages. At first the large properties were completely swept away by the Revolution, without any compensation. Subsequently, the leaders of the Council of the Land succeeded in getting 50 ha. returned to each owner. Finally, under pressure from the Old Kingdom, whose rulers were anxious to establish some equilibrium between the various reforms, the exempted minimum was with difficulty raised to 100 ha. for each owner. The general compensation was fixed by the Courts at 800 lei per ha. A 'Memorandum of the Bessarabian Cultivators Owning More than 25 ha.', issued in 1921, demanded for its authors equality of treatment with the landowners in the Old Kingdom. That demand was reiterated by a Congress of Bessarabian owners held at Chişinău, in February 1923; by that time it had become altogether impracticable.

In Bucovina the change to the new régime happened quietly, after the end of the War. A general Congress of the province met on November 28, 1918, at Cernăuţi and voted the union of the province with Rumania. This decision was taken unconditionally, and the Congress made no attempt to proclaim at the same time the social principles on which the province was to be governed in the future. The reform was initiated by a decree-law, of 1919, when the province stood under the influence of the Democratic-Unionist Party led by M. I. Nistor. The final Law of July 1921, passed by the Government of General Averescu, modified the provisions of the decree in favour of the landowners; it reduced the extent to be expropriated and hardened the procedure of expropriation and resettlement.

The Main Differences between the Several Laws. A comparison of the law of the Old Kingdom with the laws for the new Provinces shows the following main differences between them.

1. The agrarian law for the Old Kingdom expropriated in full through Art. 7 only the arable land of crown domains, of the Rural Office and of all public and private institutions. The law for Transylvania made a distinction between public and private institutions. Art. 6 expropriated in full the whole of the properties belonging to institutions which pursue a public interest—such as corporations, endowments, churches and monasteries, universities and schools, hospitals, local authorities, &c.; whereas Art. 7 expropriated in full only the arable part of land

properties belonging to institutions with a private interest—such as banks, limited companies, professional syndicates, joint holdings, &c.

It so happened that in Transylvania the property belonging to institutions was an important factor and covered about 40 per cent. of the whole area. The expropriation of land belonging to churches and schools did not affect merely the great estates of the Catholic bishoprics, but also the lesser properties of the various religious communities. The first clause of Art. 6 exempted from expropriation a maximum of 32 jugars from the arable land of the parishes; 8 for the parish priests; 16 for the schools. In addition the law allowed 10 jugars for the general needs of each church, and 5 for the training farms in forestry attached to the public schools. The law further exempted from expropriation 200 jugars for each archbishopric, 100 for each bishopric, and 30 for each monastery. It is not yet possible to know exactly how much land has been expropriated from the possessions of the Transylvanian churches, but the total extent must be considerable, as the fourteen Lutheran parishes of the Braşov district alone have lost 4,000 jugars of their land. Apart from the cultural consequences of this loss, the measure was criticized because much of the church estate had been constituted partly by free contributions in land from the members, as was often the case with the Lutheran communities, for the purpose of creating an estate from the revenue of which the work of churches and schools could be supported, but which remained in the use of the peasant members themselves.

The law also affected the communal properties of the various national communities in Transylvania. These properties consisted of considerable pastures and forests for general use, the revenue forming the budget of communal organizations and activities. The law allowed the expropriation of such communal pastures in so far as they extended beyond the average laid down by the law for each household.

In addition to the properties of national communities, there were in Transylvania a number of other forms of joint properties, formed either during the process of consolidation, or by co-operative purchase, &c. Generally, the part of each member of these associations was registered separately, but the object was used jointly. In addition, there were the properties of the former frontier regiments. All these kinds of joint holdings came under the provisions which related to communal properties. That is to say they could be expropriated if their area were larger than the needs of the members, as estimated in the provisions of the law.

2. The law for the Old Kingdom expropriated in full the estates which had been leased out and cultivated by tenants from April 23, 1910, till April 24, 1920, without interruption. Art. 8 of the Transylvanian law expropriated only the arable land, whether rural or urban, of private owners whose properties had been leased out during a period of ten agricultural years between 1904 and 1918; exempting from expropriation 30 cadastral jugars from properties situated in rural communes and 10 from properties in urban communes.

The law for the Old Kingdom apparently went farther than the Transylvanian law. But whereas the latter exempted from the effect of this provision only the properties of minors, the first exempted the properties of women as well, of public officials and of officers on active service. The period inscribed in the law for the Old Kingdom, moreover, though shorter, had to be continuous and ran up to the time of the reform; whereas the Transylvanian arrangement left room for such illogical possibilities as, for instance, the expropriation of an estate which had been let to a tenant up to 1914, but was afterwards cultivated by its owner till the expropriation. It would seem that the landowner who had bought a property in 1914 and had since farmed it himself, could nevertheless be expropriated because the property may have been let on lease by its previous owner, between 1904 and 1914. The efforts made by representatives of the national minorities during the discussion of the law to have this clause amended remained unsuccessful.

Of great importance in this connexion is the difference between the meaning attributed by the law for the Old Kingdom to the term 'let on lease', and the interpretation placed upon it in Transylvania. Art. 17 of the ordinance for the application of the reform in the Old Kingdom explained that the provision for the expropriation of estates which had been let on lease did not apply to those estates which had been let by their owners direct to the peasants. The Transylvanian law did not make this very natural concession.

In Bucovina the agrarian law expropriated in full the arable land of rural estates which had been let on lease during more than nine years, that is, over ten harvests at least, between Jan. 1, 1905 and Jan. 1, 1919.

3. The law for the Old Kingdom by its Art. 8 expropriated the arable land beyond 100 ha. of private estates which were let on lease and cultivated by tenants on April 23, 1920. A similar provision applied to the Transylvanian estates let on lease on May 1, 1921, namely, to whatever exceeded 50 cadastral jugars in the highlands and in the hills, and 100 in the plains.

4. The law for the Old Kingdom expropriated the arable land of private estates cultivated by their owners beyond the following areas:

100 ha. in the highlands and in the hills.
150 ha. in the plains, where the demand for land was great.
200 ha. in the plains, where the demand for land was moderate: and
250 ha. in the plains, where the demand for land was satisfied.

Those owners who on February 1, 1921, had considerable investments in live and dead stock, in agricultural buildings or in installations for agricultural industries, were allowed to retain instead of the above quotas 100, 200, 300, and 500 ha. respectively.

The Transylvanian law made no distinction between the owners who cultivated their own estates and those who in the same conditions possessed considerable investments in live and dead stock. It only took account of the regions in which the estates were situated and it expropriated them beyond the following areas:

50 cadastral jugars in the mountains;
100 cadastral jugars in the hills;
200 cadastral jugars in the plains, where the demand for land was moderate; and
500 cadastral jugars in the plains, where the demand for land was satisfied.

The Agrarian Committee was entitled, however, to keep the exempted quota down to 200 jugars, for purposes of colonization. It would seem that in fact 200 jugars was the area normally exempted in Transylvania. It is true that the Transylvanian law (Art. 22) likewise permitted the exemption of 500 jugars from model farms, but only in the regions where the demand for land was moderate or satisfied; and this was to be an exception requiring in each case the consent of the Agrarian Committee. Moreover, this concession was rescinded for all practical purposes by a decree of the Minister of Agriculture of September 20, 1922. It instructed the authorities concerned to apply in Transylvania the general norms of the law; the owner of a model farm could receive what exceeded those norms, and up to 500 jugars, merely on lease.

The effect of these differences between the two laws was the more weighty as most of the Transylvanian land was classified as belonging to mountain or highland regions, whereas in the Old Kingdom most of the area was lowland.

The law for Bucovina distinguished between landowners who at its promulgation or on August 1, 1914, had important agricultural establishements, breeding farms or agricultural industries, and those landowners who did not possess such investments. The latter were to be expropriated of their arable land beyond 100 ha.; the former were to be expropriated according to a progressive scale which reduced e.g. to 100 ha. all the properties between 100 and 105 ha., those between 200 and 210 ha. to 165 ha., those between 300 and 320 to 201 ha., those between 400 and 420 ha. to 224 ha., those between 500 and 525 ha. to 241 ha. and those above 600 ha. to 250 ha.

The law for Bessarabia expropriated from private estates all the arable land above 100 ha.

5. The area of the estates to be expropriated was assumed by the law for the Old Kingdom to be that which they legally had on August 15, 1916 (the date when Rumania entered the War); all transfers of land made after that date were considered invalid for the purposes of the agrarian reform. The Transylvanian law was based on the legal position of the estates on December 1, 1918; and the law for Bucovina on their legal position on September 6, 1919.

6. The size of the typical resettlement holding was fixed at a maximum of 5 ha. in the Old Kingdom, with 7 ha. for colonization holdings; in Bessarabia they were of 6 and 8 ha.; in Bucovina of 4 and 6 ha.; while in Transylvania the maximum was of 7 jugars, and of 16 for colonization purposes.

7. There was a certain difference between the authorities charged with the application of the several laws. The law for the Old Kingdom (Art. 104) entrusted the work of execution to (*a*) local committees, and (*b*) district committees; with powers for the Central Resettlement Office to provoke a fresh decision of the district committees if the re-settlement lists appeared to be inaccurate.

Article 101 of the Transylvanian Law created three authorities for the carrying out of the resettlement: (*a*) local committees, (*b*) district committees, and (*c*) county committees.

Like the Transylvanian law, that of Bucovina established three resettlement authorities, with, however, a Regional Commission instead of the county committees.

8. The original laws for the Old Kingdom, Bucovina and Bessarabia, left the forests untouched. They were expropriated, however, by the special law of July 1, 1924.

The expropriation of forests for the purpose of creating or completing communal forests was in the Transylvanian case decreed by Art. 32 of the law, down to an exempted quota of 100 jugars in the lowlands and highlands, and 200 in the mountains. In the Old Kingdom, the expropriated owners were to retain from each forest a minimum of 100 ha., so that apparently the same owner could retain as many quotas as the several woodlands he possessed. Moreover, young plantations or land which was under process of being afforested were exempted from expropriation in the Old Kingdom.

9. In regard to compensation, the law for the Old Kingdom fixed as maximum the regional rental for the period 1917–22, multiplied by forty. The Transylvanian law took as guiding criteria the rental, the land tax, &c., for the five years before 1913, capitalized at 5 per cent. The compensation could in no case exceed the price of land in 1913; or, in the case of forests, the average price in the quinquennial period before 1913. For the

purposes of the law one Hungarian crown was taken as equal to one leu (their value at par being about the same). As an exception, two crowns were taken as equal to one leu for the payment of land expropriated for the creation or completion of communal grazings.

The method of payment was in Transylvania the same as in the Old Kingdom. An exception was made only for land expropriated on the strength of Art. 9 of the Transylvanian law—down to 50 jugars, or in the case of non-cultivators even to 10 jugars, for the satisfaction of special categories of claimants—and of Art. 14—for the solving of the housing problem. The price in these two cases was to be paid in cash.

In general, therefore, the Transylvanian law offered in payment twenty times the pre-war rental; the law for the Old Kingdom granted forty times the rental officially fixed in 1916, which no doubt was lower than the market value. No such arrangement for the fixing of rent had existed in Transylvania, and this was one reason why Parliament rejected the proposal, made by the Transylvanian Minorities, that the expropriation price should be calculated in the same manner as in the Old Kingdom.

10. The Transylvanian law had some special provisions concerning the leasing of land. Art. 45 decided that land which had not been expropriated could not be let on lease for less than seven years, preference having to be given on equal conditions to local cultivators and to co-operatives of leasing. That applied even to properties of no more than ten jugars.

The State reserved for itself the same right of pre-emption as in the Old Kingdom for all sales of land involving more than 50 jugars; except when the transaction took place among close relations. Likewise, the State had a right of pre-emption on all holdings acquired through the agrarian reform (law of March 11, 1925). If the Central Office made no use of that right on behalf of the State within sixty days, the holding could be sold privately on the following conditions: (*a*) the purchaser had to be a Rumanian citizen and to cultivate the soil himself; (*b*) or he must hold an agricultural degree and reside in the commune in which the holding was situated; (*c*) the sale could not take place until five years after the payment of the full resettlement price; (*d*) the purchaser must in no case own more than 25 ha. arable land, including the area to be sold. These provisions resembled those laid down in the Old Kingdom.

Nationalist Tendencies in the Reform. Some provisions of the reform appear to have been devised against the interests of non-Rumanian landowners. The following are the main instances:

11. Art. 7, clause (*c*) of the Transylvanian law expropriated the whole arable land of private estates purchased between August 1, 1914, and July 30, 1921, the day on which the law was promulgated. Areas up to 100 cadastral jugars were exempted if purchased by priests, teachers, or their dependants; if purchased for the sake of more intensive cultivation from landowners who sold their estates in that same period—such purchases being expropriated in accordance with the provisions of Art. 8. Apparently the measure contained in clause (*c*) of Art. 7 was directed against war profiteers, but it was applied in Transylvania only.

12. Clause (*d*) of Art. 7 allowed the total expropriation of land acquired by their new owners after November 1, 1917, on the strength of the war measure decreed on that date by the then Hungarian Government. Each sale of land had to be sanctioned by a special commission; the decree gave the Ministry of Agriculture a right of pre-emption, within the conditions of the intended contract, on all land for which permission to sell had been refused. The measure was applied to thirty-five counties, of which all but one were largely Slovak or Rumanian. It was excused as being aimed at war profiteers (just as the Rumanians afterwards excused the measure described in the previous paragraph), and as a measure which later was applied to all Hungary; but the Hungarian figures show that it was applied with greater severity in Transylvania than elsewhere in Hungary. The effect of the above provision was to rescind the action of the former Hungarian Government, wherever it may have pursued nationalist ends; the application of the Rumanian text was accordingly made facultative, being left to the discretion of the Agrarian Committee.

13. Art. 10 of the Transylvanian law expropriated the land

of colonists, settled after 1885, up to the limit of the holdings allotted by the reform to claimants in the respective districts. This measure was directed against the Hungarian peasants who since the 'eighties had purchased from the Hungarian State holdings of about 16 jugars each, and settled on them; it was to be applied even to settlers who had been natives of the place. The purpose was to acquire land for distribution to Rumanian peasants, and thus to bring the average holdings of the district to about the same level. The settlers would seem to have been left with 4–7 jugars each. A number, 2,285 of them, appealed to the League of Nations, which approved a compromise ultimately offered by the Rumanian Government, on the strength of which the settlers were to receive 700,000 gold francs instead of the expropriation price of 300,000 gold francs for the 24,015 jugars of land they had lost. Such a measure, applied to smallholders who owned much less than the minimum generally exempted from expropriation, could have had only a nationalist aim.

14. The reform laws dealt very severely with absentee owners. Art. 7, clause (*b*), of the law for the Old Kingdom expropriated their whole property; though clause (*g*) allowed the State to restore to them forests, vineyards, country houses and parks, if the owners demanded it and the Agrarian Committee approved it; provided that such properties were not required for some public interest of an economic, sanitary or cultural nature. Here the law made a clear distinction between foreign absentee owners, who were obliged to sell such exempted hereditaments within a period of three years, and Rumanian absentee owners, upon whom no such obligation was imposed.

The law for Bucovina, Art. 5, expropriated in full the absentees who owned more than 25 ha. land. Art. 6 of the Transylvanian law likewise decreed the total expropriation of estates belonging to absentee owners who possessed more than 50 jugars. This exemption was introduced in the laws for Bucovina and Transylvania in favour of peasants who had temporarily migrated to America.

The chief difference between the several laws lay in the definition of absentees. The law for the Old Kingdom treated as absentees those landowners who had had to pay, during the five years which preceded the promulgation of the reform, the double land tax, in virtue of the special fiscal law, for having lived uninterruptedly abroad during that period. The Bucovinian law regarded as absentees those landowners who, in the period from August 1, 1909 to August 1, 1919, had during five consecutive years spent more than six months yearly outside the boundaries of Greater Rumania or of Austria, without imperative reasons; and those landowners who without being absent on some official mission did not reside within the frontiers of Greater Rumania from November 28, 1918, and till the promulgation of the law. The latter arrangement formed likewise the basis of the Transylvanian definition, the respective period running from December 1, 1918, when Transylvania proclaimed its union with Rumania, to March 23, 1921, on which day the law was submitted to Parliament.

15. A special group of landowners affected by the provision of the Transylvanian reform relating to absentees were the so-called optants, i.e. Hungarian inhabitants of Transylvania who after the war opted for Hungarian citizenship. Art 63 of the Treaty of Trianon gave them one year within which they might do so. If they made use of that right, they had to transfer their residence together with their allegiance, but were entitled 'to retain their immovable property in Rumanian territory'.

The Transylvanian leaders who had devised the first tenòur of the reform had been careful to leave the would-be optants outside its scope, their property remaining to be treated as the Treaty of peace, which was then under discussion, might decide. The Trianon Treaty having permitted them to retain their immovable property, this right was acknowledged by the author of the 1921 law, M. Garoflid, by means of an official interpretation issued on November 4, 1921. It explained that the clause relating to the expropriation of absentee owners did not apply to those who had been abroad on official duty 'and to foreigners'. This respected the letter and spirit of the Treaty; nor was it any more than fair, as Hungarian nationals had been refused visas for entering Rumania from the time of the dissolution of the Hapsburg Monarchy till the spring of 1921—a fact established by the Collection of documents relating to the case of the optants issued by the Hungarian Foreign Office. In July 1922, however, M. Al. Constantinescu, as Minister of Agriculture in the following Liberal Government, issued a new ordinance which declared that 'an absentee is one who was absent from December 1, 1918, till March 3, 1921, whatever his nationality or present domicile'. This completely reversed the former interpretation and caused the optants to be expropriated of all their property. Their case was thereupon taken by the Hungarian Government to the League of Nations, in whose annals it is likely to remain famous for the protracted and eloquent arguments to which it gave rise. The case has been before the Council of the League since the spring season of 1923, but has not yet been solved.

16. Foreign owners who were not absentees were expropriated of all their property in the Old Kingdom and Bessarabia, whether they were aliens by birth, by marriage, or from any other reason. As Art. 7 of the old Rumanian Constitution did not permit aliens to own rural property, such cases could have arisen only through a tacit disregard of the Constitution. On the other hand, Art. 11 of the Constitution placed foreign citizens on an equal footing with Rumanian citizens in the eyes of the law, and any discriminating legislation conflicted in general with that principle. Moreover, it was argued that the amendment to Art. 19 of the Constitution, passed at Jassy, referred merely to arable land, and that in consequence foreign landowners could not legally be expropriated of all their rural possessions. A concesssion in that sense was made later, apparently at the instance of M. Take Ionescu, as was mentioned under point 14.

The laws for Transylvania and Bucovina treated foreigners in the same way as Rumanians, expropriating them partially, if they did not fall under the category of absentees in general and of optants in special.

In practice some foreign owners had the benefit of a favoured treatment. It would appear that while M. Take Ionescu was trying to obtain the consent of the Western Powers to the union of Bessarabia with Rumania, he was prevailed upon by the French and British Governments to promise full payment to a few of their subjects who had acquired, through marriage, estates in Bessarabia. The arrangement was kept secret, but it exists in writing. It is certain that under it a number of French and British citizens have received the full value of their expropriated estates,[1] not through the open and successful affirmation of international principles, but thanks to those governments' ability to use as a lever Rumania's momentary need of diplomatic support.

17. The case of southern Dobrogea, i. e. of the strip of territory known as the Quadrilateral which Bulgaria ceded to Rumania after the second Balkan War of 1913, is in a category of its own. Most of the land, according to Ottoman law and custom, was formally the property of the State. It was held by the peasants in a kind of emphyteutic tenure, known as *mirie*, which had often left a holding in the hands of the same family for centuries, against an annual payment in kind. Only an inconsiderable part of the land was held in freehold, and was known as *mulk*. After the annexation of the district, M. Al. Constantinescu, as Minister of Agriculture in the then Liberal Government, passed a law on April 1, 1914, demanding all landowners to prove their titles, and then to surrender to the State one-third of the land to which they had thus established a claim, or to pay its value in cash. It will be seen that the measure was modelled on the arrangement made in Rumania when the serfs were emancipated, when two-thirds of the estate was reserved for the peasants while one-third became the property of the landlord. In this case the Rumanian State considered itself as having acquired the title of ownership formerly enjoyed by the Ottoman, and later by the Bulgarian, State; though Bulgaria had confirmed before the annexation the title of the holders. The application of that measure was interrupted by the outbreak of the Great War.

In 1921 the government of General Averescu, on the suggestion of M. I. Camarăşescu, who was Prefect of the district, passed a fresh law which upset the measure of 1914. The rights of the inhabitants were recognized in full, provided that they could prove their title; a demand which was complicated by the fact that many title-deeds deposited with the Rumanian authorities

[1] See the paper read by Miss Lucy Textor before the Anglo-American Historical Congress, Richmond, Virginia, January, 1925.

and conveyed by them, together with many other things, to Moscow for safekeeping when the Central Empires invaded Rumania, could not now be recovered.

The Liberal Party having returned to power, M. Constantinescu came in 1922 with a new law which abrogated that of 1921 and reverted to the measure he had passed in 1914. The arrangements for the proving of titles were somewhat simplified, but those holders who passed the test successfully had to surrender one-third of their holding; payment in cash was no longer allowed, as the land was wanted for colonization. Holders who could not prove their title to land previously considered as the property of the Ottoman and Bulgarian States, risked losing their entire holdings.

Protests against this measure were numerous, even from the ranks of the Liberal Party. It was pointed out that it would be difficult to apply it: in the Durostor county most of the holdings were small; in the county of Caliacra they were medium-sized, but were generally held on a family basis. The partition would be a complicated affair, and the creation of 10 ha. holdings, as contemplated by the law, would require an elaborate process of consolidation. Moreover, there was no local demand for resettlement. 'We need a Rumanian guard there,' was the explanation of Dr. N. Hasnaş during a debate in the Senate on March 4, 1926. Of the population 45 per cent. were Bulgarians, 35 per cent. Turks, and the remainder Rumanians and others. The measure had a purely nationalist purpose.

On coming into power, in the autumn of 1928, the National-Peasant Party set about redeeming a promise made in opposition. M. I. Mihalache, as Minister of Agriculture, issued a statement to the effect that the whole question of land tenure and expropriation in southern Dobrogea would be revised, so as to meet the just complaints of the Bulgarian and other smallholders. In December 1928 M. Mihalache appointed a commission to deal with this problem.

The Rumanian land laws having been applied extensively in provinces like Transylvania and Bessarabia, in which a large part of the population was non-Rumanian, it has not unnaturally been widely affirmed or assumed that one of the mainsprings of the reform had been a desire to use it for nationalist ends. And it would be easy to support such a view by culling from speeches and articles a whole list of statements of the kind made by M. Octavian Goga in 1920: 'We regard the agrarian reform as the most potent instrument in the Rumanization of Transylvania.' It is not a mere chance that M. Goga, a Transylvanian himself, has since turned in enmity from his Transylvanian friends and joined hands with the politicians of the Old Kingdom. Among these rather than among Transylvanian and Bessarabian leaders were to be found the men who saw eye to eye with M. Goga on that point. But put in this way the argument is both exaggerated and out of focus. It is a generally ascertainable fact that ardent nationalists make indifferent social reformers. In Rumania those few politicians who spoke in the same temper as M. Goga were out of tune with the great purpose from which the reform sprang. To say that they wished for such a sweeping reform because they would use it to a narrow nationalist end, is clearly paradoxical; the truth being rather that because the reform was accomplished, they strove to make the most of it to that end. And being men with influence in the country, they provoked some of the discriminating texts of which mention has been made before.

There are in this connexion a few points which it is useful to clear up. In the first place, there is nothing to warrant the view that the reform could have stopped short at Transylvania. It is another question—which will be discussed in the next chapter—whether its application there was justified by the existing distribution of land property or not; but it was certainly unavoidable in the political conditions which prevailed at the end of the War. Just as the Russian Revolution, of which the Bessarabian reform was part and parcel, imposed the reform upon the Old Kingdom, so it would have been out of the question to distribute land to the peasants in two-thirds of the country and deny it to them in the remainder. And, further, it is a strange fancy to suppose that the Rumanian landowners voted the reform at Jassy, in 1917, for the purpose of driving a nationalist wedge into the neighbouring foreign populations. If they were so sanguine as to look forward to a day when with that reform they might cut off the heads of the Hungarian landlords, the only thing of which they could be certain at the time was that they must begin by decapitating themselves.

Indeed, it is safe to say that the Bessarabian and Transylvanian leaders would have carried through a broad land reform even if no one had thought of it in the Old Kingdom. That also suggests why the laws of those two provinces were different and more drastic than the one passed in Old Rumania. If the latter went as far as it did, though granted by King and Parliament themselves, is it to be wondered at that matters went much farther in Transylvania and Bessarabia, where Dynasty and Government had foundered and where the reform was carried on the crest of a revolutionary wave? It is probable enough that the reform would nevertheless have turned out more mildly in the two provinces if the large owners had been of the same stock as the mass of the peasants there. Yet it must be noted that the outlook of the Transylvanian leaders is more radical in every other respect, and not merely in matters of land reform, than is that of the politicians in the Old Kingdom—largely no doubt as the result of the state of suppressed opposition in which they had been kept by the former Hungarian régime; just as the Bessarabian leaders, who were born and bred under autocracy, struck still more to the Left in their first act of freedom, passing a reform which was much more stringent than that of Transylvania, just as the latter exceeded in stringency the reform of the Old Kingdom.

The manner in which the reform was executed will be discussed in subsequent chapters; and one hopes that it may soon be possible to express its nationalist effects in precise figures. Until then, it is but true to say that the real carriers of the reform, the mass of peasantry, were concerned merely with getting the land and cared little whence it came. A small minority among the ruling class, especially in the Old Kingdom, were animated by nationalist prejudices and instilled them into the reform as far as they could. Yet the differences between the several laws being so much fainter and fewer than their similarities, the worst that could safely be affirmed is that one edge only of a very big piece of social legislation has been tainted with a nationalist bias.

PART III

THE APPLICATION OF THE REFORM AND ITS RESULTS

CHAPTER VII

THE EFFECTS OF THE REFORM ON THE DISTRIBUTION OF LAND PROPERTY

SECTION 1

THE CHANGE IN THE OLD KINGDOM

THE discussion and the legislative evolution of the reform—as well as the lack of all systematic inquiry, and the arithmetical basis of execution—all produce the unmitigated impression that the agrarian problem was never considered in its many economic complexities. Only the social angle of the problem was taken account of, i. e. the peasant's hunger for land, and in consequence the solution contemplated never went beyond a wish to meet that demand by transferring a certain extent of land from the big owners to the peasants.

Restricted within that simpler frame, the reform has had a revolutionary effect. M. Basilescu, in pressing for a generous solution, urged his fellow deputies in 1917 to 'be quite clear that what we are doing to-day is a real revolution—the upsetting of one state of things, which we replace with a totally different state of things'. M. Garoflid, the author of the law of 1921, though not altogether in sympathy with it, could not refrain from exclaiming with a measure of pride in introducing his bill that 'this is the mightiest social revolution ever recorded in history. It leaves in the shadow even the historic agrarian revolution carried through in France after 1793; for notwithstanding the Great Revolution, large property still retained in France 30 per cent. of the land.' In Rumania large property above 100 ha. had covered 48·69 per cent. of the arable area; after the reform its share fell to 7·78 per cent., according to official figures.

It is essential whenever using agrarian statistics in this study, to warn the reader that they are merely approximate. Especially is this necessary in the case of figures relating to land property, as Rumania has no ground book and as very few estates have surveys and plans of their own. The distribution of land property was in fact merely guessed at until, in 1905, M. R. Căpităneanu, of the Ministry of Finance, extracted from the available fiscal data an estimate of the extent of land which was owned by large owners and by peasants. Then only was the excessive spreading of large property realized and the intense debate, which prepared the ground for reform, begun. In 1906, Dr. Creangă compiled in his turn statistical tables on the distribution of land property in Rumania which have since been used by almost every writer and speaker on the subject. But M. Garoflid and others maintained that neither of the two sets of figures was quite reliable, those of the Ministry of Finance attributing to the peasants the possession of 200,000 ha. more than the figures of Dr. Creangă. Nor were these serious differences composed on the occasion of the reform. The Central Resettlement Office, e.g., gave in its statistics 400,000 ha. less to large property than the figures of the Ministry of Finance.

Using, therefore, the various figures with a certain approximation, one finds that land property was distributed in 1905, according to the figures of the Ministry of Finance, as follows:

Categories	Extent	% of total	Class of property	
Hectares	Hectares			
up to 10	3,153,645	40·29	small	40·29%
10–50	695,953	8·89	medium	11·02%
50–100	166,847	2·13		
100–500	816,355	10·43	large	10·43%
500–1,000	803,084	10·26	latifundiary	38·26%
1,000–3,000	1,236,420	15·80		
3,000–5,000	434,367	5·55		
5,000	520,095	6·65		
Total	7,826,766	100·00		100·00%

This table makes an attempt to systematize the classification of the various properties. In common usage, however, properties below 1,000 ha. were always regarded as small estates, which suggests a somewhat medieval notion of size. The above total refers only to the arable surface. To this would have to be added 930,366 ha. vineyards and orchards, belonging in a considerable degree to small owners; and 4,378,587 ha. forests and wastes, which, with insignificant exceptions, were in the possession of large owners. Up to 1907, therefore, 920,080 peasant families representing 92 per cent. of the population, owned 40·29 per cent. of the arable area; medium sized owners, representing 4·01 per cent. of the population, detained 11 per cent.; and large owners with properties above 100 ha. had in their hands 48·68 per cent., though they themselves only formed 0·56 per cent. of the country's population.

The moderate reforms adopted after the rising of 1907 somewhat altered the proportion of the land held by the two extreme classes of owners—those with properties of less than 10 ha. and those owning more than 100 ha. The total effect of these changes was as follows:

	Hectares	Hectares
To the property up to 10 ha., which in 1907 amounted to		3,319,695
There were added:		
Sales through Rural Office	100,000	
Communal grazings established by the law on agricultural contracts of 1908	150,000	
Sales from State domains between 1907–18 . . .	12,500	
Direct purchases from large owners between 1907–18 .	150,000	
		412,500
So that property of 0–10 ha. covered at the beginning of the land reform a total area of		3,732,195
From large property above 100 ha., which in 1907 covered a total arable area of		3,810,351
There were detached, as shown above, during the period 1907–1918		412,500
So that its total area was at the moment when the land reform began		3,397,851

To sum up, arable land at the beginning of the agrarian reform was distributed as follows:

	Hectares	Per cent.
Property up to 10 ha. . .	3,732,195	46·7
Property from 10–100 ha. . .	860,953	10·8
Property above 100 ha. . .	3,397,851	42·5
Total	7,990,999	100·0

A. *The Expropriation.* The first expropriation, carried out on the strength of the decree-law of December 1918, produced the following results:

	Hectares
From State domains	143,895·95
From Crown domains and mortmain estates	437,030·28
From foreign owners	90,115·00
From absentee owners	40,879·23
From private owners	1,512,668·04
Total	2,224,588·50

As the amendment to the Constitution demanded the expropriation of 2,000,000 ha. from private owners alone, the first result fell short by about 450,000 ha. of the required area. This deficiency was made good through the second expropriation, decreed by the law of 1921.

On the strength of the two legislative measures for the expropriation of land the area given in the table below was detached from the arable land owned by large proprietors in the Old Kingdom:

	No. of estates expropriated on basis of decree-law 3697/918	No. of estates expropriated on basis of agrarian law of 1921	Total area expropriated
			Hectares
1. From private owners	3,697	1,930	1,759,084·86
2. From State domains	334	502	509,374·00
3. From mortmain estates	396	178	388,432·26
4. From foreign owners	20	16	51,847·93
5. From absentee owners	20	88	67,662·38
Total	4,467	2,714	2,776,401·43

After setting aside the various extents required for communal forests, communal grazings, &c., the use to which the expropriated land was put showed the following distribution, on September 1, 1927:

	Hectares
1. Distributed to individual owners	2,037,293·04
2. Communal grazings	524,720·87
3. Forests administered by the State and about to be allotted	21,027·90
4. Land unsuitable for distribution	17,677·44
5. Reserves of general interest	175,682·18
Total expropriated	2,776,401·43

The Bessarabian law specifically provided (Art. 44) that a certain area should be set aside from the expropriated land for various public requirements. In the law for the Old Kingdom there was a vague reference to 'State needs', but by an oversight no definite provision was made for their satisfaction. The omission had to be made good by administrative measures, and a total area of 175,682·18 ha., as shown in the above table, was reserved for various public needs—such as the extension of towns and villages, the building of roads and railways, the establishment of military shooting ranges, of aeroplane factories, &c. For these reserves the State paid the same compensation as for land distributed to the peasants.

The transfer of the expropriated area produced the following changes in the extent of land held by the various categories of owners:

	Hectares	Hectares	Per cent.
To property up to 10 ha. covering .	3,732,195		
there were added through the reform .	2,776,401		
		6,508,596	81·43
Property of 10–100 ha. remained unchanged with an area of		860,953	10·80
From property above 100 ha. the reform detached 2,776,401 ha. leaving it with .		621,450	7·77

These official figures are not accepted as correct by everybody. During the discussion of the law, M. Mihalache quoted M. Garoflid as maintaining that the two expropriations would leave in the hands of the large owners 577,000 ha. arable land; Mihalache himself cited figures which showed that, on a moderate calculation, the large owners would retain at least 946,742 ha., representing 13·6 per cent. of the arable surface.

Whatever the precise figures, neither party was fully satisfied with the extent of the expropriation. M. Garoflid, speaking in the Chamber, declared that the large owners did not object to the principle of the expropriation, but condemned its extent. The reform had gone too far, and that was the more unfortunate as large property had to play a more important role than before after a reform which made extensive cultivation no longer possible. To be economically profitable, an agricultural exploitation should retain at least 200 ha. in the densely populated regions, 300 ha. in regions less densely populated, and 500 ha.

in regions sparsely populated. Otherwise 'the result would be disastrous economically, as estates too reduced in extent could be used only for intensive agriculture. But in our social and economic circumstances, the regions of intensive agriculture are very restricted. All those small estates will gravitate towards small property and will disappear in less than a generation'. The Memorandum of the Moldavian Landowners pointed out that all the calculations necessitated by the law had been based on the state of things existing in 1916, and they therefore claimed that land bought by the peasants between 1916 and 1920 should be included in the area to be expropriated.

The spokesmen of the peasantry maintained, on the contrary, that the reform was wanting in fairness towards the peasants. M. Basilescu complained that no account had been taken of the peasants' historic rights. From 1864 to the end of the Great War, an area of 2,572,045 ha. had been distributed on various occasions to 616,280 peasants. Assuming that the arable surface was on that date 7,998,000 ha., the peasants would be entitled, according to old custom, to two-thirds of that area, i. e. to about 5,000,000 ha. Instead of which the reform only gave them little more than half of that extent.[1]

The Peasant Party, as we have seen in an earlier chapter, wanted to apply in the Old Kingdom the same norm as the Bessarabian law and to expropriate all the arable land up to a limit of 100 ha. per owner; they estimated that they would obtain thereby 6–700,000 ha. more than was secured through the Garoflid law, not including forests and land liable to flooding. Altogether, their proposals would have expropriated, according to their own estimate, not more than 3,400,000 ha., together with common grazings. Here again there is much disagreement as to the figures concerned; but in his Memorandum to the King M. Garoflid maintained that the Peasant proposal would leave

[1] The argument is mentioned here as showing how uncompromisingly even a learned spokesman of the peasants felt in 1917. But M. Basilescu's figures refer only to what the peasants had received since 1864; they do not include what the peasants purchased directly since that date, nor that peasant property whose owners, for various reasons, were not found to be entitled to receive land in 1864 and on subsequent occasions. Even accepting M. Mihalache's figure, it has been seen that the total arable area retained by the large owners did not exceed after the reform 950,000 ha., which is very far from M. Basilescu's implied surplus of about 2,500,000 ha.

merely 350,000 ha., i. e. 4·6 per cent. of the arable land in the hands of the large owners, and M. Mihalache concurred in that estimate. The peasants' spokesman, moreover, maintained that the reform was bound to err on the side of the large owners, as it was based on their own declarations; in his speech in the Chamber, in 1921, M. Mihalache mentioned the admission of M. Enăşescu, a large owner himself, that many declarations were incorrect and that as a result 'there were estates left after the expropriation of 1,000 and 2,000 ha., whereas according to the law they could not have been larger than 500 ha'.

B. *The Resettlement.* In the Old Kingdom resettlement began on the strength of a decree-law, first through the channel of the associations of resettlement, the whole operation being afterwards revised as soon as the agrarian law was passed. The law established a preferential claim to resettlement, in the order of the categories mentioned below:

1. War invalids.
2. Minor children of soldiers killed in the war, and war orphans born not later than 1903, possessing agricultural equipment.
3. Village priests and teachers.
4. Those having taken part in the campaign 1916–18.
5. Those having taken part in the campaign 1913.
6. Children of soldiers killed in the war who were not of age at the time of the reform, and who do not own agricultural equipment.
7. Small cultivators without land of their own.
8. Cultivators owning less than 5 ha. land.
9. War orphans who were not of age on August 15, 1916.

On the basis of this preferential order tables of those entitled to receive land were drawn up in each commune by a local commission. These tables were revised by district commissions, and appeals provoked by that revision were finally adjudged by the Agrarian Committee.

As a result of that preliminary operation, 1,053,628 individuals were registered as being entitled to receive land. Of that number 630,113 individuals received, up to September 1, 1927, holdings covering a total area of 2,037,293·04 ha.; in addition there were allotted to them 524,720·87 ha. as communal grazings.

The detailed application of the law's general principles was even more sternly criticized. When so many peasant cultivators had to be left without land, it was felt to be wrong in principle and economically unprofitable that land should be given to artisans also, 'to bind them to the soil', as well as to petty officials, as, e. g., to those on the State Railways. Land used in that way was clearly lost for the new conception of property of production. Moreover, these holdings were often several kilometres distant from the station where their owners were at work; and whereas in western Europe the allotments of workers took a subordinate place in their activity, the Rumanian railway worker, &c., still had the peasant mentality in him. He thought first of his land and therefore lived in the village, spending his limited leisure in journeying to and fro and in tilling his field. To that doubling of his activity and the resulting fatigue, an important official has attributed many of the frequent accidents on the Rumanian railways; he pointed out that the arrangement also made it impossible to move the railway workers and lesser officials about according to their ability and to the needs of the service. Much criticism has been likewise directed against the granting of land to gipsies who had served in the War, because they very rarely engaged in agriculture and merely became absentee owners on a small scale. Finally, M. Garoflid criticized the bringing of mountaineers into the lowlands, who never became good cultivators, thus restricting the area available for the real farmers. 'It is the same policy of settlement which has impoverished the large villages of the plain, situated on the State's domains, when these estates were broken up to be divided into lots of 5 ha.'

Most critics, whatever side they represent, agree in declaring that the holdings distributed were too small. In 1864 some account was taken at least of the means of production owned by the various peasants who received land, who were divided into three categories according to the number of draught animals they possessed. Subsequent re-settlements on State domains granted stereotyped holdings of 5 ha. each, with the exception of the measure of 1889 which also provided lots of 10 and 25 ha. Altogether, after 1864, 777,630 ha. were distributed in lots of less than 10 ha. each and only 28,588 ha. in larger holdings. As the peasant holdings were divisible without limit by succession or sale, they were split up in time to a degree which made of most of them mere allotments. The following table indicates the character and distribution of peasant holdings shortly before the War:

Categories	Number of holdings	Per cent. of total properties	Extent in Hectares	Per cent. of total area
Hectares				
½	62,832	6·60	26,426	0·34
½–1	81,039	8·50	72,757	0·93
1–2	147,900	15·20	237,029	3·01
2–3	131,630	13·60	337,000	4·30
3–4	172,446	17·90	631,964	8·08
4–5	148,717	15·40	711,033	9·08
5–7	131,145	13·50	743,486	9·50
7–10	45,230	4·70	393,950	5·05
Total	920,939	95·40	3,153,645	40·29 [1]

The agrarian law for the Old Kingdom endeavoured to check that pulverization of property by fixing the minimum holding to be distributed to those without any land at all at 2 ha. But so anxious were the authorities to satisfy as many claimants as possible, that in practice that provision was disregarded and many lots of a lesser size were distributed. Rumanian agricultural economists generally agree that 5 ha. are not sufficient for a peasant family. M. Garoflid sets the lowest limit at 7 ha.; and while that would ensure the existence of a peasant family, it would not exhaust its labour power. That, according to M. Garoflid, would require for a family of four persons a holding of 15·5–16 ha.—which one assumes to refer to the conditions which prevailed about 1907, when his book was written, and which imposed upon the peasants a primitive extensive cultivation. The economic size of a peasant holding varies indeed with the quality of the land, its situation, and the kind of farming for which it is used. In Germany, Roscher placed the minimum at

[1] M. Şerban, *op. cit.*, p. 20.

6–7 morgen land of good quality and 24 morgen land of mediocre quality; the great Italian agricultural inquiry of 1879–82 concluded that holdings should vary between 7–15 ha.; in France, Souchon, in *La Propriété Paysanne*, suggested a minimum of 5·5 ha., even for prolific cultures, while Caziot, in *La Terre à la Famille Paysanne*, suggested 8–20 ha., according to region, fertility, and character of the crops. In Rumania the Peasant Party project had in view indivisible minimum holdings of 3 ha. in the highlands, 5 ha. in the lowlands, and 10 ha. in the regions to be newly settled.

The reform did not of course distribute only complete holdings, but also many so-called complementary lots to peasants already owning less than 5 ha. land. According to local needs and possibilities, the size of the lots thus distributed was fixed by the Central Resettlement Office between 0·5 and 5 ha. So far no figures exist to show how many lots of the various kinds were given and to what categories of peasants, so that it is not yet possible to know how the peasant properties are classified at present.[1]

Criticism of Application. Writing in 1919 M. Ionescu-Sisești urged the need of applying with implacable fairness 'this decisive measure'. 'The real kernel of the problem lies in this, much more than in the principles and details of the law. Our agrarian problem for half a century has been one long story of good intentions and mediocre execution.' If some of the principles adopted by the legislator for the transfer of land were not of the best for social selection and economic development, it is generally admitted that their application was more deficient than the principles themselves. That was partly due to the political circumstances of the period; for the rest, to the lack of reliable statistical material and

[1] An attempt to secure some indication of the new state of things from the contents of local publications merely ended in the capture of a strange example of local statistics. In 1928 the Chamber of Commerce of Botoșani, in Moldavia, published a year-book with elaborate economic figures referring to the four counties of Botoșani, Dorohoi, Fălticeni, and Hotin. The year-book contained three tables on the distribution of land, neither of which was compatible with the other two. The third table gave the number of owners in the various categories of property and the percentage they represented from the total number of landowners in the four counties, but even by attributing the least possible area to the number of owners in each category the minimum total far exceeded the total area of the four counties, as given in the first table.

of a personnel prepared technically and morally for the proper execution of such an extensive reform. The task imposed upon the country's judicial and agricultural authorities was truly immense. In most cases all the three expropriation authorities were called upon to give a decision concerning the 15,000 properties which were expropriated; about 5,000 of which came up for a second time before the expropriation bodies.

The first difficulty was created by the exceeding haste of the reform. Court and politicians who had been in refuge at Jassy returned to Bucarest in November 1918. Within ten days a new Government was formed—on the 1st of December—and eighteen days later the decree for the application of the reform was issued. In the first days of January the expropriation commissions were at work. The first post-war Cabinet of M. Ionel Brătianu remained in power ten months; during that period five-sixths of the area to be expropriated had been taken over. That feverish haste has been severely criticized and made responsible for much that is faulty in the reform. Even the Peasant Party proposed that expropriation should proceed only gradually, and step by step with resettlement; expropriation should, in their opinion, have been proclaimed at once in principle, but a period of five or even ten years should have been set for the actual taking over of the land. The Government defended the line they adopted as calculated to prevent the peasants suspecting that the provision of the Constitution might remain a dead letter. Some agricultural experts indeed believe that the rapid application of the reform has had a salutary effect by enlisting at once all the means and energy of the peasants in the service of agricultural reconstruction. The large owners needed credit and the labour of the peasants, neither of which they could have got easily as long as the fate of the reform remained uncertain; so that at a given moment the interests of State, landowners, and peasants coincided, this alone making possible the quick execution of such a radical measure.

A second and considerable difficulty was caused by the inadequate means available for the measurement of the land. There was no groundbook and hardly any private estate plans. The number of trained surveyors was very limited and a surveying school had to be improvised for the occasion. Similarly, the available surveying instruments were altogether inadequate for the needs of the reform, and as they were costly and the State was short of money, an effort had to be made to manufacture them on the spot. In the Old Kingdom the detailed measurement for resettlement purposes was almost everywhere made with the chain, and only occasionally by means of analytical parcellation. The results were bound to show errors.

The means for choosing the land to be expropriated were as rudimentary as those for its measurement. The local commissions, who were the chief factor in making that choice, did not have at their disposal any detailed and precise material concerning the nature and quality of the land in their districts. Being in a great hurry, they had inevitably to rely in many cases on the statements of the parties interested, and so it came about that much arable land was exempted as grazing or as being liable to flooding. As a consequence the first expropriation did not secure the 2,000,000 ha. demanded by the Constitution, while some of the land expropriated was of little use. The general figures given above show that of the area taken over 17,000 ha. were altogether unfit for cultivation, and that gives some point to the complaint of Dr. Lupu that certain landowners have given barren and stony wastes, while retaining the fertile soil for themselves.

The need of carrying out in great haste a technical work of a lasting character naturally strained to the utmost the country's resources in personnel and material; and the great deficiency of material placed a correspondingly greater burden on the personnel. The merits of their unusual performance, therefore, are the more enhanced. The results of their work form a valuable foundation for the eventual establishment of a survey. The work was begun in 1919 with twenty surveying teams; their number reached 300, engaged in field work, by 1925, assisted by an office establishment of about 200 calculators and draughtsmen. The land was divided up in a provisional manner on the strength of the decisions of the lower authorities, and was handed over to the peasants to be used first through the associations of resettlement and then in individual holdings, on a preliminary resettlement. The final resettlement was carried out in the measure in which the technical work advanced. The development of this work appears from the following figures:

1919 area measured	100,487 ha.	and parcelled out	ha.
1920 ,,	528,021 ,,	,,	47,969 ,,
1921 ,,	653,670 ,,	,,	13,201 ha
1922 ,,	765,634 ,,	,,	22,938 ,,
1923 ,,	1,169,259 ,,	,,	135,317 ,,
1924 ,,	1,073,193 ,,	,,	417,728 ,,
1925 ,,	908,046 ,,	,,	415,239 ,,
Total	5,198,310 ha.		1,052,392 ha.

The total expropriated area was,	5,889,709 ha.
with the area to be expropriated in Southern Dobrogea	110,291 ,,
Total	6,000,000 ,,
Measured till 1925	5,198,310 ,,
Remaining to be measured	801,690 ha.
To be parcelled out 6,000,000 ha.	
Parcelled out till 1925 1,052,392 ,,	
	4,947,608 ,,
Remaining to be measured and parcelled out	5,749,298 ha.

The latest figures of the Survey Directorate concerning the work executed by its organs from 1919 till December 31, 1928, are contained in the following table:

Region	Measured area Hectares	Area surveyed for parcellation Hectares	Area to which parcellation applied Hectares	Area for village housing Hectares
Old Kingdom	4,539,291	1,888,947	1,594,441	50,074
Transylvania	1,812,001	942,160	658,965	14,573
Bessarabia	1,360,683	429,644	381,043	7,057
Bucovina	187,518	75,786	75,366	—
Total	7,899,493	3,336,537	2,709,815	71,704

Because of the hasty application, again, the organs entrusted with it had to be greatly decentralized. The execution of the measure was primarily in the hands of special local bodies whose composition and functions have been described in the previous chapter. The procedure under the Duca decree was simple and expeditious. The local commissions had the character of bodies whose task it was to bring about an understanding rather than a judgement—the judge being more in the position of a friendly arbiter. These commissions in most cases in fact reached a friendly agreement; as the extent to be expropriated was rigidly fixed, the respective provisions were easy to apply and discussion centred mainly round the choice of the spot where the land was to be taken. Appeals from these decisions were as a result few and most of them were settled in favour of the peasants. The Garoflid law, however, by introducing more complex economic criteria of expropriation, required appropriately larger and more elaborate organs of application, and likewise more elaborate courts of appeal. However well intentioned the new commissions may have been, their decisions were rather in the nature of a judgement, after pleadings by both sides, each of course putting forth an extreme case; and where in the first case the decision had generally meant an agreement, in the second it frequently displeased both parties to the case. The second system produced a large crop of appeals and most of them were settled in favour of the landowners. That disclosed a significant change of psychology: in the former period public sympathy was on the side of the peasants, who had suffered so long; after the first expropriation sympathy veered to the side of the dispossessed owners, especially after the collapse of the exchange.

While expropriation on the whole proceeded smoothly, the second part of the reform has given rise to many abuses and to consequent ill-feeling. The lists of those entitled to receive land were prepared by local committees who would seem to have been to an unfortunate degree imposed upon by the more greedy and vociferous villagers. The real conditions were difficult to establish because the taxation registers were not up to date. Expropriation appeals, moreover, were dealt with by the county tribunals, but in the case of resettlement appeals the local judge considered the issue on the spot, assisted by experts and delegates of the parties, so that the proceedings were no longer conducted in the juridical atmosphere which surrounded the courts. The judicial part was not sufficiently separated from the administrative part in the second case. Moreover, the final authority, the Agrarian

Committee, worked behind closed doors, the parties not being present at the proceedings; and inquiries on behalf of the Committee were conducted in a similar executive manner by officials of the Ministry of Agriculture—all of which was apt to leave in the minds of those dissatisfied with the decision a suspicion that there may have been something wrong in the procedure.

Whether deliberate fraud or unfortunate error, the list of complaints against the detailed application of both parts of the reform is very formidable. Any one may collect from the Parliamentary debates a whole volume of well-documented cases, and high officials do not deny that a good many of these complaints must be justified.[1] The only excuse which one of them attempted was merely to insist that such frauds 'represented only a percentage of the whole measure—what elsewhere may have been 1 per cent. is perhaps 10 per cent. here'. The Peasant Party, especially, has been unsparing in its denunciation of such abuses and has pledged itself to redress them when coming into power. Against such an intention all Rumanian agricultural experts seem to be united. They admit that mistakes have been made, but contend that they were inevitable in such a hasty application of an immense measure of reform. To try to amend them would be useless, as errors and frauds would be as possible now as they were a few years ago; and a revision of the reform would at the same time be economically ruinous as it would mean a prolonged state of insecurity for the whole agricultural industry. Yet security and stability are essential if agriculture is to advance technically. Agriculture has suffered during the past ten years just because it found itself in an unstable period of transition. To create another such period through an attempt to revise the application of the reform would be disastrous. The reform was carried out on a social-political basis, which politicians continued to keep to the fore. But the technical experts being interested merely in the economic aspect, they are all of them against any attempt at revision.

[1] The comprehensive table on p. 227 shows, e.g., the curious fact that many properties above 250 ha. still exist in Bessarabia, though the Bessarabian law was supposed to expropriate everything above 100 ha.

While doing full justice to the standpoint of the agricultural experts—summarized above from statements heard from many of them in almost identical words—one must record the fact that the application of the reform has left behind in almost every village one or more disputes which in certain cases have developed into conflicts. Whether a revision is practically possible, except in a restricted number of flagrant cases, must remain questionable. M. Negură declared in the Chamber, in May 1924, that the Agrarian Committee had to deal within a period of twelve months with 500 appeals against expropriation decisions and 71,000 against resettlement decisions—which worked out at a rate of some 200 cases each day. A decision to revise the application of the reform would beyond doubt call forth an avalanche of complaints. At best, therefore, the revision would be a very long affair; and the experience already made with procedure suggests that it would be humanly impossible to dispense pondered justice, especially as many—if not most—cases, if they were to be handled with care, would necessitate an expert inquiry on the spot.

Section II

The Effects of the Reform in the New Provinces

A. *Bessarabia.* The distribution of land property in Bessarabia was affected, as in the whole of the Russian Empire, by the reforms of 1861–6, when the peasants were emancipated and provided with land in the collective form of the *mir*, and afterwards by the so-called Stolypin law of 1906 which was intended to further the establishment of individual peasant holdings. The latter purpose was financed by the Peasant Bank created in 1882; to it were transferred in 1906 considerable Crown domains which were to be sold to the peasants. To check the speculations of intermediaries, an ukaze of November 1908 created agrarian commissions whose function was to facilitate the transfer of estates from the large owners to the Peasant Bank, and thereafter from the Peasant Bank to the peasants. The total area of Bessarabia, according to the figures of the Russian Central Statistical Committee, was of 3,834,824 dessiatines. Of these,

1,658,109 dess., i. e. 43 per cent., belonged to private owners; 1,864,023 dess., i. e. 48·6 per cent., were *nadyel* land, allotted to the peasants when they were emancipated or through the Peasant Bank, under similar safeguards regarding mortgages, debts, &c.; while State, Church, and other institutions possessed 314,692 dess., i.e. 8·2 per cent.

The general distribution of land property before the agrarian reform, according to official Rumanian statistics, may be seen in the table below: [1]

Categories		Number of Properties	Total area Hectares	Average area Hectares	Per cent. of total area
1. Private individual properties	Nobles .	1,474	818,744	555·4	20·1
	Priests .	126	4,308	34·2	0·1
	Merchants &c.	1,944	253,867	—	6·7
	Peasants	7,718	135,384	17·5	3·2
	Foreigners	13	30,390	2,337·6	0·8
	Various .	10,102	94,157	9·0	2·3
	Large owners	275	169,416	616·3	4·2
		21,652	1,506,266	69·4	36·8
2. Joint holdings .		540	201,591	373·3	4·9
3. Nadyel land .		2,209	2,111,940	956·0	50·7
4. State domains .		—	53,648	—	1·3
5. Churches and Monasteries		—	204,190	—	5·0
6. Town properties .		—	23,600	—	0·6
7. Private institutions .		—	30,362	—	0·7
		24,401	4,131,597	166·3	100·0

It will be seen that Bessarabia had a greater variety of categories of properties, according to the social standing of their owners, than the other parts of new Rumania. If one excepted the so-called *nadyel* land, the largest category was that of noble estates with a total area of 818,744 ha., prevalent mainly in the centre and in the north of the province. Their total number was 1,474, which gave an average of 555·4 ha. per estate; though in the northern region they reached, an average of 2,099 ha. in the Tighina county. The other categories of private properties were much smaller, large property not belonging to the nobility being represented by 275 estates with a total of 169,416 and an average of 616·3 ha. A special category, which played an important part in the discussions on the compensation to be paid

[1] E. Giurgea in *Buletinul Statistic*, 1919, No. 2, pp. 324–7.

for the expropriated land, was that of the foreign owners; they held thirteen estates covering a total area of 30,390 ha., with an average extent of 2,337·6 ha.

A noteworthy fact was the small number of peasants who owned individual holdings. They were merely 7,718, possessing a total area of 135,384 ha., which gave an average of 17·5 per owner. This peasant property was to be found especially in the southern districts, which corresponded to the lesser area held by estates of nobles, and also with the numerous colonies of foreign cultivators established in the southern region at various periods. The average of 134 ha. per peasant owner in the county of Cetatea Albă shows how prosperous some of these colonies were. The *nadyel* land was the largest kind of property, with 2,111,940 ha. divided into 2,209 properties, the largest average being again found in the county of Cetatea Albă with 4,791·5 ha. per property; the average per peasant family was 7·08 ha. Taken as a whole, land properties were distributed according to their size, as follows:

Category Hectares	Total area Hectares	Per cent. of total area	Number of owners	Per cent. of owners
up to 10	2,156,827	51·6 }	285,663	98·4
10–100	180,984	4·3 }		
over 100	1,844,539	44·1	4,480	1·6
Total	4,182,350	100%	290,143	100%

Expropriation was applied in Bessarabia to 4,271 landowners, from whom a total area of 1,491,920.24 ha. was taken. This area was put to the following use:

	Hectares
1. For resettlement	1,098,045·50
2. Forests	198,404·60
3. Land unfit for distribution	82,888·44
4. Reserves for general needs	112,581·70
Total	1,491,920·24

The Bessarabian law determined the following categories of cultivators as being entitled to receive land:

(1) cultivators who lived on the estate and owned less than the area fixed for a resettlement holding in that region;

(2) cultivators who lived on the estate and owned no land at all;
(3) cultivators who lived within a radius of 5 km. from the estate and had less land than the fixed re-settlement holding;
(4) cultivators who lived within a radius of 5 km. from the estate and owned no land at all.

The list of those entitled to receive land was to be established by local commissions, on the basis of the above norms, and to be finally settled by the Central Commission, which also determined the size of the resettlement holdings. In reality no resettlement lists were drawn up at all, but the authorities concerned simply confirmed in their possession those who had seized the land. In Bessarabia one may say that all the peasants were resettled, the size of the lots being empirically determined by local reserves and needs. They varied with these local circumstances between 1–6 ha. As a result were resettled:

262,536 Rumanians
94,480 other nationalities

357,016 individuals, who received altogether 1,098,045 ha.

Up to the revolution of 1917 the peasants owned

	Hectares
Arable land	2,041,040
Forests	67,692
	2,108,732
Purchased and distributed by the Peasant Bank	48,095
Total	2,156,827

The reform procured the following increase in the land held by peasants:

	Hectares
From private owners, arable land	608,568
Land purchased by the Peasant Bank, but not yet distributed by 1917	176,388
Land of former German and Bulgarian colonists	58,884
Land of monasteries and foreign owners	143,729
Area under communal roadways	7,060
Building land and gardens, formerly belonging to private owners	39,990
Ponds and waterways	63,426
Total	1,098,045 [1]

[1] P. V. Synadino, *Insemnătatea Reformei Agrare*, p. 9.

The land reform caused the following changes in the distribution of rural properties:

Hectares		Hectares	Hectares	Per cent. of total area
To property of up to 10 ha. .	2,156,827·00			
there were added . .	1,491,920·24	3,648,747·24	Reserves and forests . 393,875·24	9·42
			Small holdings . 3,254,872·00	77·52
Property of 10–100 ha. left unchanged		180,984·00		4·33
From property above 100 ha. of . . .	1,844,539·00			
there was expropriated	1,491,920·24	352,618·76		8·73
Total		4,182,350·00		100·00

B. *Bucovina.* According to the ground-book, the total area of the province covered 1,044,458 ha. Before the reform the land was divided among 199,185 properties, of which 2,540 belonged to the State, to Churches, and to other institutions. The distribution of these properties according to size was as follows:

Categories	Number of properties	Per cent.	Area in Hectares	Per cent.
Hectares				
up to 10	191,737	96·27	270,730	25·92
10–100	6,606	3·32	134,115	12·84
100–500	585	0·29	130,939	12·53
Above 500	257	0·12	508,674	48·71
Total	199,185	100·00	1,044,458	100·00 [1]

Among the 257 large estates there were 63 with an extent of more than 2,000 ha. each, covering together 30·21 per cent. of the province. An area of 75,967·35 ha. was expropriated from 561 landowners and used for the following purposes:

[1] After Livius Lazar, *La Mise en Œuvre de la Réforme Agraire*, p. 64.

	Hectares
1. For resettlement	42,832·25
2. Communal grazings	5,831·85
3. Communal forests	4,377·72
4. Forests remaining to be distributed . .	8,523·84
5. Land unfit for distribution . . .	605·69
6. Reserves for general needs . . .	13,796·00
Total	75,967·63

The law for Bucovina established the following order of preference among those claiming land:

(1) peasant cultivators, war invalids (or their families), who had less land than the size of the lot fixed for their commune;

(2) peasant cultivators who served or had served in the army and who had less land than the typical lot;

(3) peasant cultivators with less land than the typical lot and whose possessions had been damaged in the war;

(4) the Orthodox parishes;

(5) rural schools;

(6) peasant cultivators, war invalids (or their families), who had no land at all;

(7) peasant cultivators who served or had served in the army and had no land;

(8) peasant cultivators whose possessions had been damaged in the war and who owned no land at all.

The list of those entitled to receive land was established by various commissions on the strength of the above indications, and the Regional Commission fixed typical lots varying between 0.25–2.5 ha. As a result there were inscribed on the list

	47,866 Rumanians and
	30,045 other nationalities,[1]
altogether	77,911 individuals or families.

Of these 71,266 were resettled up to September 1, 1927,

	Hectares
on a total area of	42,832·25
in addition to which were granted as communal grazings	5,831·85
and as communal forests	4,377·70
so that the total area allotted to them was . . .	53,041·80

[1] An article in *Economia Naţională*, August 1927, gave the following numbers of non-Rumanians as having received land: Ruthenians, 21,140; Germans, 5,683; Hungarians, 868; Jews, 493; Gipsies, 406; Russians, 98; other nationalities, 1,357.

The effect of the land reform was to change the distribution of property in Bucovina as follows:

		Hectares	Per cent. of total area
To small property of 405,000 ha. there were added 75,967 ha.	=	480,967	92·49
From large property of 115,000 ha. there was expropriated 75,967 ha. leaving		39,033	7·51
Total		520,000	100·00

C. *Transylvania.* The modern agrarian structure of Transylvania had its origin in the reform which followed the revolution of 1848.

Serfdom had been general in the Hungarian lands. A certain mitigation of its hardships began with the passing of Transylvania under the rule of the Hapsburgs in 1691. Early in 1714 the Diet which met at Sibiu adopted regulations, under pressure from Vienna, which forbade the landlords to 'force the serfs to labour more than 208 days yearly' for them. A further step was made by the letters-patent issued by Maria Theresa in 1769, forbidding the imposition of unfair taxes and fines, and limiting corporal punishment to twenty-four birch strokes for men and twenty-four strokes of the whip for women, at the most. Most of these regulations seem to have been disregarded by the landlords. Their attitude provoked the anger of Joseph II, who in 1765 wrote in a Memorandum that 'politics can have one foundation only, and that is the people—the masses—for they supply the soldiers and pay the taxes. Hence it is the mission of the State, and of the ruler especially, to protect the people against the privileged classes. One should not skin 200 peasants for the sake of a lazy landlord. . . .' This was followed by a decree given at Sibiu, in 1773—Joseph II travelled a great deal—which allowed the serfs to marry without payment of a tax, to learn handicrafts, and to move about freely. Life could not have improved much, at any rate for the Rumanian serfs, as they attempted a desperately futile rising in 1784. Their leaders—Horia, Cloşca, and Crişan—who have remained legendary in popular annals, were broken on the wheel. As late as 1847 a law was passed in Transylvania regulating the dues in kind and labour which the serfs had to pay: tithe from field and garden, from flax and wool, &c., and a yearly service of fifty-two days with oxen and 104 days with their hands from those who had the use of a normal holding; the *zileri* who had but a house and garden, were to give eighteen days' labour, and the peasants who had not even a house, six days. Many historians regard that law as the main local cause of the revolt which followed. The national assembly which gathered at Alba Julia in 1848 declared that 'the Rumanian nation, conscious at last of the individual rights of man, demands the immediate abolition of serfdom, without any payment from the servile peasants. . .'

The reform which followed abolished serfdom—compensation being paid by the State—and gave the servile peasants the ownership of some of the land which they had been cultivating. Considerable confusion and friction arose out of the variety of titles to the land, and a whole series of laws, beginning with that of 1880, endeavoured to regulate land tenure and to reorganize it on a more economic basis, by segregration on the one hand and by consolidation on the other. Pastures and forests were involved in that regrouping. But while the central idea of the measure was sound enough, its application was badly vitiated by abuses at the expense of the former serfs, who were given bad land in exchange for good. On all these occasions the Rumanian peasants appear to have suffered additional losses through national discrimination. The famous Memorandum addressed to the Emperor in 1892 detailed some of their complaints, showing how they had been deprived of ancient rights, especially in regard to grazing and wood, and how many lawsuits between landlords and former serfs, arising out of the reform of 1848, were still before the Courts, after the passing of forty-four years. At the wish of the Hungarian Government the Memorandum was returned from Vienna unopened, but its authors were tried and sent to prison. It is undoubtedly true that until lately the Hungarian, Saxon, and other villages were better provided with grazing and forests than most of the Rumanian villages; the grazing and wood rights of the former serfs having been transformed into communal rights on the occasion of these reforms.

The distribution of property after these changes was established for the first time in 1895. The statistics gathered in that year offered evidence of a considerable disproportion between the area occupied by small cultivators and that in the hands of large owners:

Categories	Per cent. of total number of properties		Per cent. of total area		Average size of properties
jugars					jugars
up to 5	99·01	52·02	52·34	5·84	1·69
5–100		46·99		46·50	15·75
100–500	0·99	0·80	47·66	15·37	285·11
Above 500		0·19		32·29	3,158·00 [1]

Holdings of less than 5 cadastral jugars made up therefore 52·02 per cent. of the total rural properties, but covered merely 5·84 per cent. of the total area; while properties above 500 jugars made up only 0·19 per cent. of the total, but covered 32·29 per cent. of the land. The considerable difference between the average extent of the two middle categories showed that the ascent was not graduated through a chain of well-balanced medium-sized farms.

The statistics of 1915 indicated a slow improvement in the distribution of property. Small property had gained in that period of twenty years 2·26 of the total area, as indicated in the table below:

Categories	1895	1915	Increase or Decrease
Jugars			
up to 5	5·84	6·2	+0·36
5–100	46·50	48·4	+1·90
100–500	15·37	14·2	−1·17
Above 500	32·29	31·2	−1·09 [2]

The latest figures referring to the situation before the reform were those collected by the Secretariat of the so-called Governing Council (the Provisional Government of Transylvania), in 1919. The table below is based on them and gives the number of properties in the various categories and the total area occupied by each category:

[1] After Livius Lazar, *op. cit.*, pp. 48–9. [2] *Ib.*, p. 50.

Categories	Total number of properties	Total area occupied	Per cent. of total area			
			1895	1915	1919	+or−
Jugars		Jugars				
up to 1	132,084	40,847				
1–5	304,638	853,231				
5–10	223,874	1,622,320	52·34	54·60	58·75	+6·41
10–20	182,852	1,557,078				
20–100	110,049	3,782,267				
100–200	3,838	523,967				
200–500	2,368	742,997	47·66	45·40	41·25	−6·41[1]
500–1000	1,035	732,967				
above 1000	1,198	4,026,951				

There was a striking increase, by 4·15 per cent., of the total extent of land in the possession of smallholders, between 1915 and 1919. In so far as these figures were correct, that increase must have been due to the relative well-being of the peasants during the War and to the land purchases they made in consequence. Possibly landowners in Hungary, as in other countries, exploited the high price of land and forced their tenants to purchase or to quit. Large owners in Hungary occupied a priviledged political and social position, yet one notes that here, as elsewhere, they were constantly losing ground.

The reform applied in Transylvania after the union of the province with Rumania expropriated a total area of 1,663,809·03 ha. from 8,963 estates situated in 3,583 communes. The expropriated land was put to the following use:

	Hectares
1. For resettlement	451,653·96
2. Communal grazings	418,361·43
3. Communal forests	484,805·24
4. Forests in possession of the State and remaining to be distributed	179,162·00
5. Land unfit for resettlement	36,442·78
6. Reserves for general needs	93,383·62
Total	1,663,809·03

The Transylvanian reform law fixed the following order of preference for the distribution of land:

(1) war invalids, heads of families; or the widows and families of those killed in the War;

[1] After Livius Lazar, *op. cit.*, pp. 54–5.

(2) demobilized soldiers who were heads of families;
(3) war invalids without family;
(4) demobilized soldiers without family;
(5) those mobilized by order of the Governing Council;
(6) heads of families who had not been mobilized;
(7) men without family and who had not been mobilized;
(8) returned emigrants.

On the basis of these indications, tables of those entitled to receive land were drawn up by local committees; they were revised if complaints had been lodged by district commissions and finally by county commissions. The latter also determined the communal needs in grazing and woodland.

The size of the lots to be distributed in each region was fixed by the county commissions and varied between 1–7 jugars. Altogether 490,528 individuals, of whom 363,664 were Rumanians and 126,864 of other nationalities, were found to be entitled to land. Of these were resettled till September 1, 1927, 227,943 Rumanians and 82,640 other nationalities. Total 310,583. They received arable land covering 451,653·96 ha., to which were added communal grazings 418,361·43 ha., and communal woodlands 484,805·24 ha., making a total allocation of 1,354,820·63 ha.

Arable land was distributed here in a much lower average than in the Old Kingdom, but grazing was given more generously; that was in keeping with the highland character of the province and the predominant position of cattle-breeding in its economic life.

The effect of the reform on the distribution of land among the various categories of owners has been as follows, according to the official figures of the Central Resettlement Office:

(*a*) *Before the reform*

Categories	Hectares	Per cent. of total area	Number of owners	Per cent. of total number
Hectares up to 10	2,536,738	34	843,448	87·6
10–100	2,153,117	29	113,887	11·8
Above 100	2,751,457	37	4,601	0·6
Total	7,441,312		961,936	

(*b*) *After the reform*

	Hectares	Hectares	Per cent. of total area
Property up to 10 ha.	2,536,738		
Expropriated area	1,663,809		
		4,200,547	56·45
Property of 10–100 ha. remained unchanged with		2,153,117	28·94
Property above 100 ha.	2,751,457		
Expropriated	1,663,809		
		1,087,648	14·61
Total		7,441,312	100·00 [1]

D. *The Change in the Distribution of Land among the various Nationalities.* The Rumanian land reform has proved a thorny branch of the minorities problem in the new State. Its rulers first laid themselves open to a suspicion of having pursued nationalist ends through having enacted laws differing from each other for the several parts of the country. The reproach, as we have seen, is valid only in part. The situation in the several provinces at the end of the War made that differentiation unavoidable, and to some extent, at any rate, it came about independently of the will of the rulers. Nor is it easy to accept as justified the suggestion made by some critics that the vagueness of the Transylvanian law was deliberate and was meant to leave the officials with a free hand to do the kind of thing for which the legislators could not openly assume responsibility. Whether deliberate or not, however, the lack of precision in many clauses of the Transylvanian law no doubt gave undesirable

[1] The above figures, like all those given in this chapter to illustrate the results of the reform, were supplied by the Central Resettlement Office and have therefore an official character. It must be remarked, however, that there have undoubtedly been quite a number of cases in which land was expropriated in Transylvania from properties of less than 200 jugars—which corresponds roughly to 100 ha. There is no means of establishing yet how many such properties were touched by the reform and how much land they lost through it. But the figure in the table above, which gives the total area covered by properties of 10–100 ha. as having remained unchanged, cannot be more than approximately correct, and must be a general estimate rather than a real calculation of the actual state of things. Moreover, the table credits small property with all the land expropriated; whereas, in fact, as shown higher up, over 300,000 ha. were still in the hands of the authorities at the end of 1927, and some 130,000 ha. were reserved or unfit for distribution.

latitude to those who had to apply a sweeping reform in great haste, in a province whose mixed populations had not been on the best of terms with each other.

The reproach levelled against the authors of the reform that they have allowed their national bias to appear more starkly in the execution of the measure must remain unanswered until the day when the authorities can supply figures detailed and definite enough for an impartial observer to check the standpoint of the reformers and of their critics. As far as Bessarabia is concerned, one might take it for granted that national discrimination could have occurred merely in a very limited degree, as the land was in fact taken over and divided by the peasants themselves. Some fifty thousand acres of land were lost to German and Bulgarian colonists; that probably happened before the reform, as a result of some Russian War measure, because those colonists belonged to enemy nations. Discrimination must have occurred as a matter of fact in southern Dobrogea, where the reform was mainly concerned with colonizing the area expropriated for the benefit of the Rumanian State. Of the 150,000 ha. which were available, some 40,000 have been distributed to 4,500 families up to the end of 1927. The law prescribed that the State's third should be taken over on the basis of special rules concerning the measurement and the division of the land. But the deputy, M. Pencov, asserted in the Chamber, on July 28, 1929, that those rules had not yet been drafted, the verification of the deeds not yet finished and the land not yet surveyed, but that nevertheless the State's third was being forcibly taken over by the local authorities. From this a large number of lawsuits, expensive and dilatory, has resulted between owners and State. Some of the colonists came from the Old Kingdom, but another part were Vlachs from Macedonia, brought over in pursuance of some extravagant nationalist idea. One suspects that the temper of these Macedonian half-nomads was not calculated to reconcile the established population to the loss of some of their land; and the quartering of the newcomers upon the local Bulgarian peasants, often for several years, because of the absence of credits for building and farming, has caused serious friction and even bloodshed. Worse still, many of these people came without being asked, attracted by the rumour of free land, and squatted in groups on land that no one had assigned to them. The director of the Central Resettlement Office found during a tour of inquiry, in the autumn of 1927, that about one-half of the colonists had come without the sanction of the Office, and that most of these had settled in districts where there was no surplus land available. In southern Dobrogea, therefore, a somewhat irrational scheme of colonization has been made worse by its erratic application; and psychological factors have deepened the feeling among the minorities that they were being deprived of some of their land because of national prejudice.

The main field to which the contention refers, however, was Transylvania. There the extent of the reform has been considerable, and in that province the number and inextricable admixture of nationalities gives the minorities problem international importance. To clear the issue, in so far as this is possible before the detailed accounts of the reform are closed, one must begin by looking into the distribution of land among the various nationalities before the reform. One of the circumstances which offered a serious difficulty to the redistribution of land, so as to meet the needs of small cultivators, was the large area which in one form or another was for practical purposes taken out of the real estate market, as shown by the following figures:

	Per cent.
1. State domains	7·65
2. Communal properties	15·45
3. Endowments, churches, schools	4·53
4. Societies and corporations	1·55
5. Joint properties of the inhabitants of certain communes	9·05
6. Entail	0·82
7. Private property	60·95

Well over one-third, therefore, of the total area was in the hands of local bodies, institutions, and various organizations; and given the political and social subjection in which the Rumanian population was held in Hungary before the War, it is not improbable that the bulk of such possessions must have belonged to non-Rumanian bodies.

The figures compiled in 1919 by the provisional Transylvanian

Government gave the total area of the transferred districts as of 14,882,625 jugars, 7,613,555 of which were arable soil. The total was divided between Rumanians and the other nationalities inhabiting Transylvania as follows:

Nationality	Total number of population	Property below 100 jugars	Property above 100 jugars	Total property jugars
Rumanians . .	3,316,345	3,448,602	150,067	3,598,669
Other nationalities .	1,891,942	5,407,141	5,869,815	11,276,956
	5,208,287	8,855,743	6,019,882	14,875,625

Therefore the Rumanian population owned little more than 1 jugar per head, whereas the other national groups possessed nearly 6 jugars for each of their members. More important was the fact that among the 8,435 proprietors owning more than 100 jugars each there were only 209 Rumanians, with altogether 150,067 jugars; and among the 1,190 landowners with more than 1,000 jugars, 27 only were Rumanians.

In other words, class division in Transylvania had largely coincided with national division. Any measure altering the social structure of the country was bound to affect one nationality more than another, and no land reform could have avoided having also a nationalist effect. Even if a similar reform had been applied by a Hungarian Government, they could not have prevented it from following nationally the same trend—i. e. of taking the land mainly from Magyar, Saxon, and other such owners and of transferring it largely to Rumanian peasants. That state of things could hardly have resulted altogether from the working of normal economic factors. Without going farther into the history of the abuses committed during the process of consolidation and on other occasions, one can find traces of the causes which gave national colour to the distribution of land in the settlement policy pursued by Hungary before the War. In 1894 a settlement fund of 3,000,000 florins was created for the purpose of settling peasants on estates in those districts of Transylvania which had a considerable non-Magyar population. It was estimated that in 1911 the fund had properties valued at more than 7,000,000 florins in various Transylvanian districts. Because of the political friction which that policy caused, the Hungarian Government transferred the possessions and the activities of the settlement fund to the so-called Altruist Bank. It is affirmed that the offers of Rumanian peasants inhabiting neighbouring communes were disregarded on principle when State domains were parcelled out or when estates were sold through the Altruist Bank, and Hungarian settlers were brought from elsewhere. The Rumanians defended themselves as best they could by following a similar policy of acquisition through special banks. In May 1913, in a speech delivered at Cluj, Count Stephen Bethlen spoke anxiously of the many large Hungarian estates which were being bought by Rumanians. He estimated their purchases at some 7,000,000 florins yearly. He promised that if they should come into power the Coalition which he represented would give 5,000,000 florins yearly for purposes of settlement.

It was to be expected that the memory of that competition, in which the Rumanians were bound to be at a disadvantage, might influence those officials and private individuals in whose hands lay the execution of the new reform. It was of course foolish to punish the present citizens of Rumania for the misdeeds of their former rulers; and even more so to punish Hungarian small-holders for the policy of magnates who had shown little kindness to the mass of their own kinsmen. But the wisdom of forgetting the past has seemingly not been able altogether to restrain those who applied the new reform. This is evident, e. g., from the way in which they dealt with the joint properties which formed a valuable feature in Transylvania's agrarian economy.

There were several kinds of joint holdings in Transylvania. One group included those established in certain frontier regions during the reign of Maria Theresa (1740–80), when those regions were militarized and the inhabitants of the respective communes were bound to certain military services. The Transylvanian militarized territory, established in 1764, ran from the Iron Gates to the boundary of Bucovina and was divided into five regimental districts. Two of the regiments were Rumanian and three were regiments of Szeklers. In return for their military services, the inhabitants of these frontier regions received land and forests for joint possession and use, the title to them being rather different in the case of nobles from that allowed to former serfs. In addition, these militarized populations had the use of the communal woodlands and pastures. When, for various reasons, they did not use these joint possessions themselves, the land was let and the rent paid into the communal treasury, or into a joint fund in the case of possessions of the several military districts. Land property was gradually individualized through a number of legal measures, even in these frontier regions, but a certain proportion of joint holdings remained in being and were administered by special organs; e.g. the joint property of the communities of the former second Rumanian frontier regiment at Nasaud, the Szekler frontier communities, as well as the Caraseverin community, in the Banat. A second kind of joint property was that created during the process of consolidation, within the last fifty years. Peasants inhabiting a certain commune and belonging to the same national group, in many cases surrendered some part of their individual holdings and acquired instead a share in a common grazing or pasture. This was merely a community of use, as the title of each member was separately registered.

It has been estimated that about 30 per cent. of the Transylvanian population had a share in one or the other of these joint holdings. Nevertheless, it was first intended to treat these joint properties as large estates belonging to private institutions and in consequence to expropriate them completely. Only the fierce resistance of those concerned prevented that intention from being carried out. In the end, however, some of the joint holdings, even those belonging to Rumanian communities, had to suffer. There have been suggestions that the Transylvanian intellectuals were especially incensed against the Saxon University, and that in order to be able to deprive it of its considerable possessions, they did not hesitate to pass decisions which damaged the interests of Rumanian joint properties as well. Though M. Garoflid had formally recognized in Parliament that the possessions of the Saxon University represented a communal property, which should have exempted its forests from expropriation, the University lost 8,000 jugars woodland and grazing in the Talmesch district.[1] The agrarian law for Transylvania expressly exempted from expropriation, by Art. 24, clause 2 (*c*), the communal pastures, and by Art. 32 (*c*), the woodlands belonging to the communities of the former second Rumanian frontier regiment at Nasaud. During the application of the reform, the joint properties of the Banat frontier regiments were likewise altogether exempted from expropriation. On the other hand, the Agrarian Committee decided to expropriate in full, and without any compensation whatever, the joint properties of the Szekler frontier communities, on the plea that the land was State property which had been given to the Szekler communities for use in return for certain services. As those services were no longer performed, the State was entitled to claim back its possessions. 'One might perhaps admit the validity of that argument, but then it applied with equal force to the Rumanian frontier communities.'[2]

The spokesmen of the national minorities in Transylvania were loud in asserting that the nationalist bias was shown especially in the manner in which the reform was applied. They declared that in many cases even smallholders were expropriated for the building of schools or churches, i. e. of Rumanian schools and churches, and that they were given in exchange either the expropriation price or a piece of bad land. The Saxon People's Council for Transylvania—at its plenary meeting, November 18, 1922—passed a resolution which declared that after criticizing certain provisions of the original bill, they had loyally accepted the law in its final form. But they demanded that the law should be respected by the authorities as well.[3] As to the alleged corruption of officials, one can only say that it cut both ways. A corrupt official was if anything more amenable to the wishes of a rich landowner than to those of a poor peasant. Among the cases brought before Parliament there were quite

[1] Fritz Connert, article in *Siebenbürgisch Deutsches Tageblatt*, August 24, 1922.

[2] Fritz Connert, 'Zur Frage der Agrarreform in Siebenbürgen', *Nation und Staat*, Vienna, December 1927, p. 262.

[3] See *Kronstädter Zeitung*, November 21, 1922. A number of flagrant miscarriages in the application of the reform are described on pp. 256–64 of the important article published by Herr Fritz Connert in the review *Nation und Staat*.

a number showing that rich Hungarian landowners had been able to evade the provisions of the land laws. The Transylvanian deputy, M. Ioan Iacob, a specialist on the subject, wrote in the book he published in 1924 that 'the Rumanian landowners were subjected to a rigorous expropriation, whereas the Transylvanian counts and barons, those great magnates of the land (all of them Magyars), have already reconstructed their domains by other means. The time will come when I shall publish some statistics of that sad state of affairs'.

If the figures, or rather estimates, published so far can be accepted, it would seem that the minorities were better treated when it came to resettlement. To the American Unitarian delegation which visited Rumania in 1924, M. Al. Constantinescu, as Minister of Agriculture, declared that 134,000 non-Rumanian peasants had received land in Transylvania. There may be a misunderstanding at the basis of the statement which appeared to this end in the Commission's Report. A detailed list of those entitled to receive land in Transylvania, published by the Bucarest *Argus*, on December 26, 1923, gave the following figures:

			Per cent.
Rumanians	396,342	(roughly)	75
Hungarians	87,426		16
Saxons	31,195		6
Serbs, Croats	6,124		3
Jews	330		
Other nationalities	9,277		
Total	530,694		100

It will be seen that the figure of 134,352 represented the total number of individuals belonging to national minorities who had been found to be entitled to receive land. No authoritative data are available so far to show how many of them have actually received land and how much of it. A Saxon writer, Herr M. Englisch, affirmed that the Saxon communities in the Nösnergau lost through expropriation 20,282 jugars, of which 16,054 were communal land, 1,920 were church property and 1,362 were private property. At the resettlement, 134 claimants belonging to these communities received together 177·5 jugars. M. de Szász quotes from a Hungarian publication the following numbers as having received land, in fifteen counties of Transylvania:

			Per cent.
Rumanians	179,940	(roughly)	72·9
Hungarians	36,481		14·7
Saxons	20,643		8·3
Others	9,944		4·1
Total	247,008		100·0

This would show that of those entitled to be resettled, a larger percentage actually received land, up to 1925, among the minorities than among Rumanians. It does not clear up the question as to whether the resettlement lists were drawn up fairly; nor does it indicate how much land was given to the various national groups. Only detailed figures will make it possible to check the assertions of one side and the denials of the other that the agrarian reform was permeated by a nationalist bias. It is as well to remember at the outset that the whole reform sprang from a revolutionary temper; that it was applied in a region which had been the very centre of the national friction which provoked the Great War; and that it was carried out during a period when European nationalism was celebrating one of its supreme orgies, even in the more mature and sedate West. Whatever the exact figures, they will not affect the conclusion that, because of the way in which land was previously distributed, the reform has reduced the extent of land in the hands of the minorities; but that for the same reason it has increased among the minorities the number of peasants who own some land of their own. M. Ioan Iacob, who had been rapporteur of the law for Transylvania in 1921, complained in his book that 'the agrarian reform not only did no harm to the minorities element, but, on the contrary, it strengthened it. For if its practical effect has been to reduce the estates of a few hundred Hungarian magnates, it has given land to thousands of Hungarian peasants.' From a general standpoint the truth certainly is that the land reform has in a large measure dissolved the provokingly artificial national barrier which formerly separated rich and poor among the rural populations of Transylvania and Bessarabia.

Section 3.

Summary of the Effect of the Land Reform on the Distribution of Property

The following table gives the total area expropriated in Greater Rumania on the strength of the four separate agrarian laws:

PROVINCE

	Old Kingdom	Transylvania	Bucovina	Bessarabia	Total
	Hectares	Hectares	Hectares	Hectares	Hectares
Arable .	2,269,192·27	470,389·56	48,544·64	1,210,627·20	3,998,753·67
Grazing .	27,386·94	80,745·60	9,742·63	—	117,875·17
Pasture .	442,988·68	398,257·50	8,296·69	—	849,542·87
Forests .	19,156·10	663,967·24	8,420·54	198,404·60	889,948·48
Orchards, vine-yards, &c.	2,466·18	12,114·07	—	—	14,580·25
Building land, farm-yards	360·11	14,006·35	357·16	—	14,723·62
Barren .	14,851·15	24,328·71	605·69	82,888·44	122,673·99
Total .	2,776,401·43	1,663,809·03	75,967·35	1,491,920·24	6,008,098·05

Not all of that area has been handed over to individual smallholders. A considerable portion of it has been used for the establishment of communal grazings and woodlands, as well as for the creation of a land reserve from which public needs—such as the building of roads, town extensions, model farms, &c. —might be satisfied. The following table shows the extent of land devoted to these various purposes:

	In the Old Kingdom	In Transylvania	In Bucovina	In Bessarabia
	Hectares	Hectares	Hectares	Hectares
1. Distributed to the peasants	2,037,293·04	451,653·96	42,832·25	1,098,045·50
2. Communal grazings .	524,720·87	418,361·43	5,831·85	—
3. Communal woodlands .	—	484,805·24	4,377·72	—
4. Forests administered by State and remaining to be distributed . .	21,027·90	179,162·00	8,523·84	198,404·60
5. Land unfit for resettlement . . .	17,677·44	36,442·78	605·69	82,888·44
6. Reserves for general needs	175,682·18	93,383·62	13,796·00	112,581·70
Total expropriated .	2,776,401·43	1,663,809·03	75,967·35	1,491,920·24

The work of resettlement showed the following summary results on September 1, 1927:

	No. of peasants entitled to land	No. of peasants re-settled	Area distributed to peasants	Area given as communal grazing	Area given as forests
			Hectares	Hectares	Hectares
In the Old Kingdom	1,053,628	630,113	2,037,293·04	524,720·87	—
In Transylvania .	490,528	310,583	451,653·96	418,361·43	484,805·24
In Bucovina .	77,911	71,266	42,832·25	5,831·85	4,377·72
In Bessarabia .	357,016	357,016	1,098,045·50	—	—
Total .	1,979,083	1,368,978	3,629,824·75	948,914·15	489,182·96

The change in the distribution of land among the various categories of owners is summarily described in the two tables below:

(*a*) *Before the Reform*

	Property up to 100 hectares		Property above 100 hectares	
	Hectares	Per cent.	Hectares	Per cent.
In the Old Kingdom . .	4,593,148	57·5	3,397,851	42·5
In Bessarabia . . .	2,337,811	55·9	1,844,539	44·1
In Transylvania . . .	4,689,855	63·0	2,751,457	37·0
In Bucovina	405,000	78·0	115,000	22·0
Total	12,025,814		8,108,847	

The arable area of the whole country was therefore divided as follows:

	Hectares	Per cent.
Small property . . .	12,025,814	59·77
Large property . . .	8,108,847	40·23
Total	20,134,661	100·00

(*b*) *After the Reform*

	Property up to 100 hectares		Property above 100 hectares	
	Hectares	Per cent.	Hectares	Per cent.
In the Old Kingdom . .	7,369,549	92·22	621,450	7·78
In Bessarabia . . .	3,829,731	91·57	352,619	8.43
In Transylvania . . .	6,353,664	85·38	1,087,648	14·62
In Bucovina	480,967	92·49	39,033	7·51
Total	18,033,911		2,100,750	

The arable land of the whole country is therefore divided as follows:

	Hectares	Per cent.
Small property	18,033,911	89·56
Large property	2,100,750	10·44
Total	20,134,661	100·00

(*a*) One of the first points which arise out of these figures is that not all the peasants who were found to be entitled to receive land were actually provided with it. Notwithstanding the radical extent of the expropriation, of the 1,979,083 individuals whose names were placed on the resettlement lists, over 600,000 were altogether left out of this wholesale distribution of land. It should of course be remembered that preference was given to those peasants who had no land at all. Hence it is probable that many or most of those who remained outside the reform owned some land of their own, though not enough for independent farming.

(*b*) The more fortunate two-thirds of the claimants, who received some measure of land, did not get as much as was originally intended to give them. It is probable that in a majority of cases the minimum lot fixed by the various executive organs had in practice to be reduced. Dr. Aurel Vlad asserts that none of the holdings distributed in Transylvania was larger than 3 jugars. The official summaries cited above show that an average of 2·65 ha. arable land was distributed per head; to which would have to be added an individual average of 0·61 ha. in the shape of communal grazings and 0·35 ha. as communal woodlands.

(*c*) Rumania's land problem had been a compound of two opposite evils—on the one side excessively large estates and on the other side excessively small peasant holdings. The reform has remedied only one of these two aspects of the problem: large property has been abolished. But small property has not been raised to a level where it might become economically autonomous. No figures have been collected as yet to show the new distribution of peasant property. It is clear that the reform has increased the number of families owning land, but not in the same proportion the number of those who could derive an existence from their holdings alone. Many landless labourers have been made poor owners, but not independent cultivators. Even less has the reform created that medium-sized property which, by general agreement, could initiate an advance in technique during a period of transition.[1]

(*d*) In a certain measure the reform has aggravated the uneconomic organization of small property in Rumania. The reform has not been complemented by a simultaneous process of consolidation; hence the reform has multiplied, in those cases in which the claimants were entitled merely to additional land and not to a full new holding, the number of small fields which generally form a peasant property.

(*e*) The general land problem has, therefore, not been finally solved. Nor could the demands of those who have remained altogether landless, or of those with insufficient holdings, be met through a new expropriation. In answer to a question put to him in the Chamber, on December 12, 1927, M. Argentoianu, the Minister for Agriculture, replied that the State had no further reserve for those who had remained without land. On the other hand, in the autumn of 1928 a circular from the Ministry of Agriculture instructed its subordinate authorities to hasten the execution of those expropriation cases which had remained pending, so that the claimants might be able to take over the holdings to which they were entitled. That means that a certain extent of land still remained to be transferred from the large owners to the peasants. Moreover, it would seem that by circumventing the provision of the reform laws, a number of large estates have been reconstituted—either by purchase under a fictitious name, or by putting together parts of an estate belonging to members of one family, or by successive sales to the same individual of lots of less than 50 ha. each from the same estate.[2] Officials of the Ministry of Agriculture confirmed that in the steppe region of the Bărăgan, newly formed estates can

[1] The new National-Peasant Government is endeavouring to make good that deficiency by means of an Act, passed in 1929, which cancels the restriction to sell holdings obtained under the land reform. Such holdings may now be sold freely, to peasant cultivators, up to an area of 25 hectares per individual.

[2] The State has a right of pre-emption on all sales of 50 ha. and more; here there seems to be a gap in the law which only legal proceedings could fill, but officials of the Central Office are doubtful whether such proceedings could succeed.

be found running up to 3,000 ha. Yet, taking together these two categories of estates which may still be available, the total area cannot be otherwise than insignificant in relation to what would be needed to satisfy the remaining peasant demand for land. One might add to the two sources of land named above a third —namely, voluntary sales from the remaining large estates. Either because they may be afraid of a further expropriation, on account of the growing political strength of the Peasant current, or because they find large scale cultivation under the new conditions unprofitable, many large owners have been selling the land which the reform left them; especially as the peasants seem willing to pay almost any price for such land as still comes into the market, being afraid perhaps to lose the last chance of getting land in their own localities. With the assistance of dummy bidders the landowners force up the price and make it difficult for the Central Resettlement Office to compete. The *Aurora* stated on February 4, 1926, that in the Ilfov county alone nineteen estates had so far been sold to the peasants, at fabulous prices ranging from 60,000 lei per ha. upwards. The Central Office would seem to have purchased until the end of 1928, on the strength of the State's right of pre-emption, about 6,000 ha. at an average price of 20,000 lei per ha., and about 4,100 jugars at an average price of 8,000 lei per jugar; in addition to buildings, mills, &c. The bulk of that land has been handed over to peasant co-operatives of leasing and purchase.

(*f*) Altogether, the extent of large property still available for eventual transfer to the peasants is very reduced. Nor could it for the time being be subjected to further expropriation. The expropriation law has a constitutional character, and provisions of the Constitution cannot be changed until the King comes of age. As King Mihaiu is seven years old, eleven years must pass before the Constitution could be touched. None of the present political parties would think of breaking that rule; the Peasant leaders, who are pledged to abolish large scale property altogether, consider the issue as shelved for the time being. What they might attempt is to revise the grosser abuses in the application of the law. For the rest, the Peasant spokesmen hope to solve the agrarian problem rather by the intensification of agriculture, a full-fledged system of co-operation, and the development of local industries using agricultural and other raw materials available on the spot and the surplus of peasant labour.

(*g*) Until such a comprehensive program can be applied, and begins to bear fruit, the land problem will remain acute. For the mass of the peasants still regard it as a duty of the State to provide them with land whenever they need it. The last expropriation, in their opinion, has merely recognized and applied that ancient right and, in addition, has freed them of certain obligations towards the landlords. One of the village elders of Ruşet gave it as his view that 'the land would have been his even without the reform, for if he gave half of the crop he could always take from the estate as much land as he wanted to till. And it was fairer in that way, as everyone took according to the number of souls in his family, according to needs and to power, and not as is being done now, five hectares to each and everyone alike.'[1] The new generation 'has not the least doubt that in their turn they will receive land. They therefore look longingly at every piece of land still available.' Nor are those peasants who already have been given something, and who form the mass of the villagers, altogether satisfied. 'They have the owner's egoism towards their own piece of land, but look with the coveting eyes of landless serfs upon the land of the boiar.'

This attitude has not been assuaged by the granting of land to officials, to gipsies and to others who are not cultivators. For the peasants will not admit that land has been given as a recompense for bravery in the War, but rather in recognition of the ancient peasant right to the land on which they live. The War has been merely the occasion, but not the cause for the distribution of land. M. Stahl justly remarks that 'where the use of labour in a capitalist form on a money-wage basis, is not widespread, the tiller of the soil always looks upon himself in a confused way as having a title to the land'. The peasant considers, that is, that he has a right to a piece of land which should provide

[1] This and the following quotations are from manuscript notes made by M. Henry Stahl in 1927, during a sociological inquiry under the leadership of Professor Dimitrie Gusti, and kindly communicated by the latter.

him with the means of existence. And the fact that economically the old servile tithe system passed into the present-day *métayage* system—which appeared to the peasant to involve merely a change in the size of the respective shares but no change of form—has helped to perpetuate that proprietary outlook of the peasants towards the land. As long as any land remains in the hands of large owners—and, especially, in the hands of owners, large and small, who do not cultivate it themselves—the peasants' relentless claim to the land will not be subdued.

The Distribution of Land Property (Arable) in 1927

Provinces	–5 hectares		5–10 hectares		10–50 hectares		50–250 hectares		Above 250 hectares		Total	
	No. of owners	Total hectares	No. of owners	Total hectares	No. of owners	Total hectares	No. of owners	Total hectares	No. of owners	Total hectares	No. of owners	Total hectares
1. Old Kingdom	1,683,591	3,489,754	189,184	1,269,600	45,454	820,608	10,861	882,996	1,652	851,615	1,930,742	7,314,572
2. Transylvania .	937,554	1,567,681	106,144	713,313	55,376	797,720	4,540	362,812	709	294,964	1,104,323	3,736,490
3. Bessarabia .	412,343	1,004,541	136,012	908,161	46,655	748,454	2,378	218,944	194	173,380	597,582	3,023,481
4. Bucovina .	197,975	219,018	4,375	28,779	1,375	25,909	343	40,507	42	15,713	204,110	329,926
Rumania .	3,231,463	6,280,994	435,715	2,919,853	148,860	2,392,691	18,122	1,505,259	2,597	1,305,672	3,836,757	14,404,469
Percentages												
1.	87·20	47·59	9·80	17·22	2·35	11·07	0·56	12·62	0·09	11·50	100	100
2.	84·87	41·96	9·62	19·09	5·02	21·36	0·42	9·71	0·07	7·88	100	100
3.	69·06	33·23	22·73	30·04	7·78	24·75	0·40	7·24	0·03	4·74	100	100
4.	97·04	66·37	2·11	8·72	0·67	7·86	0·16	12·28	0·02	4·77	100	100
Rumania .	87·54	47·29	10·16	18·77	3·96	16·26	0·39	10·46	0·05	7·22	100	100

(From the figures of the fiscal census of the Ministry of Finance).

1927 { Large property . . 12·09% of the arable land
Small property . . 87·91 „ „

1926 { Large property . . 13·30 „ „
Small property . . 86·70 „ „

1925 { Large property . . 12·44% of the arable land
Small property . . 87·56 „ „

1924 { Large property . . 12·42 „ „
Small property . . 87·58 „ „

(Large property = above 100 ha.) (Small property = up to 100 ha.)

(Figures of the Ministry of Agriculture)

CHAPTER VIII

THE EFFECTS OF THE REFORM ON THE ORGANIZATION OF FARMING

WHEN passing to an estimate of the effect the reform produced on the economics of Rumanian agriculture, one must begin by pressing the point that the recent legislative changes were not conceived—either in their origin or in their texts or in their application—as a reform of agriculture. They were discussed and decreed as a reform of land tenure, and such consequences as followed from them in Rumanian agriculture are indirect and, on the whole, still rudimentary. In such circumstances any attempt to prepare a scientific survey of the economic effects of the reform would be premature. Yet it is desirable to try at least to discern their beginnings, in order to round off the picture presented so far. But it will be clear from the nature of the case that any views put forward in the chapters which follow must be taken as tentative—as indicating tendencies rather than settled currents; and as continuing the illustration of some of the conflicting social tendencies and policies out of which the Rumanian agrarian problem has issued.

Even later, when time has crystallized the new character of Rumanian agriculture, it will still remain difficult, for a number of reasons, to provide an accurate study of the economic effects of the great land reform. There is, in the first place, the obstacle caused by the absence of comprehensive and reliable statistical material. As Levasseur said in a speech before the Société Nationale d'Agriculture, 'as regards reliability, agricultural statistics are the most frequently complained of.' In Rumania agricultural statistics have been neglected in a manner which may well drive the student of economics to despair. Rumania has no ground book; the figures concerning the distribution of property are collected by local officials, village mayors, &c., from the statements of the parties concerned, from taxation returns, or from leases and acts of sale. The same method is used for measuring the distribution of the various crops and their annual production. The data is taken from questionnaires completed by village officials, in the case of peasant cultivators, and by the farmers themselves in the case of farms above 100 ha. For these reasons, one finds frequent discrepancies between the figures given by various authorities—or even by several Government departments—on a particular point. When the difference is serious, special attention will be drawn in this study to the fact. But, in general, we shall have to assume that while the figures are approximate, they are relatively true enough to allow valid conclusions to be drawn concerning tendencies in Rumanian agriculture before and after the reform.

A second and more disconcerting difficulty for such a study of effects is presented by the sequence, or even concurrence in time, of the War and of the reform. In the case of Rumania, especially—where two-thirds of the country had during a prolonged period been under enemy occupation, and where much of the official archives were lost during the Rumanian retreat and then again during the hasty departure of the enemy—it is an altogether impossible task to try to disentangle with any precision the effects of the reform from the effects of the War. One finds a good illustration of this difficulty in Yugoslavia, where one self-contained administrative district, the old Serbian Kingdom, did not come within the scope of the agrarian reform at all, as the land was in the hands of the peasants already. Yet in that district production suffered a serious decline, as may be seen from the following figures:

	1909		1923	
	Area cultivated in hectares	Production in quintals	Area cultivated in hectares	Production in quintals
Wheat	378,048	4,388,875	369,326	3,793,274
Maize	585,144	8,751,659	443,356	3,895,511
Barley	113,907	1,374,709	61,600	493,358
Oats	108,412	843,299	75,364	445,844
Rye	49,738	445,591	32,380	222,204

The total decrease in the area cultivated with these five crops amounted therefore to 253,223 ha., and the fall in production to 6,952,942 quintals. On the other hand, the figures for Croatia and Slavonia show an unexpected rise:

	Yearly average 1911–14		1923	
	Hectares	Quintals	Hectares	Quintals
Wheat	332,177	3,745,045	367,051	4,051,213
Maize	421,138	6,533,848	460,893	6,556,598
Rye	90,850	834,470	43,208	378,839
Barley	62,736	548,791	65,607	601,479
Oats	99,595	767,324	97,681	970,165

Yet, in Croatia and Slavonia the land reform was not only applied on a very wide scale, but it was also misapplied, with the result that many progressive large farms were completely disorganized, as I had occasion to see during my journeys; while many of the newly resettled peasants neglected their holdings, being in doubt whether they themselves would be finally confirmed as owners. Only the havoc caused by the War among the manhood and possessions of the Serbian peasantry could explain this paradox.

Finally, a third difficulty springs from the economic policy adopted by the Governments which ruled in Rumania from the end of the War till November 1928. For reasons and by means which will be described in a later chapter, those Governments devoted the resources and favours which the State could command to the creation of a national industry. Agriculture was starved of all support just in that period in which it had to try both to overcome the destruction caused by the War and the inevitable temporary disturbance caused by the reform. In other words, some of the phenomena which will be noted in the following chapters—as far as possible statistically—are not the effects of the reform. On the contrary, they may be said to be the product of circumstances which prevented those effects from working themselves out in a normal economic evolution. The upshot of the reform has been vitiated by Rumania's attempt to carry through simultaneously two different, and, in part, conflicting transformations of her economic life: in agriculture, a change from large to small ownership and production; and, at the same time, a substantial transfer of the nation's capital and energies from agriculture to industry. To which must be added, as a supplementary complicating factor, the advent of several new provinces differing from the Old Kingdom and from each other in economic equipment and needs.

One other word may be allowed by way of introduction, in order to define the standpoint of this study. The Rumanian reform has not remained unnoticed by western economists. If its principles and progress have been somewhat severely treated by them, that is no doubt because in general they have discussed it merely from the angle of production; and that in a period when the problem of agricultural suppplies was greatly worrying some of the Continental States. But for this circumstance, it should hardly have been necessary to point out that in Rumania—and in eastern Europe generally—the reform imposed itself as a pressing social need. Its whole structure and progress must be studied, therefore, as a many-sided social reform; and economic criticism must be tempered by taking into account the wider factors which caused the reform, and which the reform, in its turn, is meant to correct.

Nor is the economic aspect of the question simply a matter of gross production and of a surplus for export. In agriculture, small-scale and large-scale cultivation do not differ solely in the technique of production, but even more profoundly in the ends of production. Large farms and smallholdings each have crops and products which are peculiar to them, which are best produced, i. e. in their respective type of agricultural undertaking. A change from one dimension of farm to another generally involves a change of the whole character of farming. So that before one attempts to calculate the effect of the reform on production as such, one must endeavour to clear the way by a process of elimination. It is necessary to find out in what measure and in what manner the system of production has altered: first, as regards the size of agricultural undertakings; secondly, as regards equipment and technique; and, thirdly, as regards the nature of the products to which they are devoted. Only by a preliminary discussion of these deliberate changes in the organization of production will the final estimate of the effect of the reform on volume and quality of the produce conduce to a true conclusion.

Section 1

Changes in the Scale of Agricultural Exploitations

If one views the Rumanian land reform from the angle of agricultural economics, its significance would seem to be above all a vast experiment in transition from large-scale to small-scale cultivation. As such, the experiment may contain valuable lessons for rural economics in general. But if these lessons are not to be misinterpreted, one must first attempt to establish the extent to which the revolutionary change in the distribution of property has also meant a change in the units of farming—and, therefore, in the methods of farming.

One of the dominant characteristics of Rumania's agrarian structure before the War was the wide discrepancy between the area owned by large proprietors and that covered by large-scale agricultural exploitations. The respective distribution of properties and farms according to categories may be seen from the following table: [1]

Categories	Distribution of property in 1896, 1902, 1905 (arable, grazing and barren land)		Distribution of land among farms in 1913 (without pastures, plantations and barren land)	
	Area in hectares	Per cent. of total	Area in hectares	Per cent. of total
Up to 2 ha.	335,212	4·3	572,167	9·7
2–5 ,,	1,679,997	21·5	1,546,132	26·5
5–10 ,,	1,137,436	14·6	1,118,592	19·2
10–50 ,,	695,953	8·9	815,395	14·0
50–100 ,,	166,847	2·1	107,088	1·8
100–500 ,,	816,385	10·4	588,070	10·1
Above 500 ,,	2,993,966	38·2	1,092,177	18·7
Total	7,825,796	100·0	5,839,621	100·0

The difference of nearly 2,000,000 ha. between the two columns is due to the absence from the agricultural statistics of about 500,000 ha. of barren land and of 1,200,000–1,500,000 ha. of permanent grazings. The latter, especially, belonged almost completely to large owners; in 1910 only 27,721 ha, i. e. 2·4 per

[1] Ministry of Agriculture, *Agricultorii şi Repartizarea Pământului Cultivat in 1913*, Bucarest, 1915.

cent. of the permanent grazings belonged to small owners. Hence, 61·7 per cent. of the arable area was occupied by properties below 500 ha. and 38·3 per cent. by properties above 500 ha. Agricultural exploitations of less than 500 ha. covered, however, 81·3 per cent. of the total arable land, and those of more than 500 ha. only 18·7 per cent. of that area.

That difference was not caused by the division of large estates into a number of farms, in keeping with the needs of sound agricultural practice. On the contrary, Rumanian farming displayed an excessive concentration of agricultural units. Properties and farms above 100 ha. were distributed in the following proportions according to number:

Categories	Number of properties	Number of farms [2]
Hectares	Per cent.	Per cent.
100– 500	61·5	71·4
500–1,000	20·8	18·2
Above 1,000	17·7	10·4

With the exception of the first category, therefore, in which the number of farms was higher than the number of properties, the inverse relation was found throughout Rumanian agriculture, in a measure which rose with the extent of the exploitation. This anomaly could only arise from the fact that some of the larger farms were formed by putting together a number of neighbouring properties, instead of the large estates being divided up into a number of smaller farms. This peculiarity of Rumanian farming is even more striking when compared with conditions in Great Britain and the United States, both of them reputedly countries of large scale agriculture. Whereas one found in England 5,207 estates above 1,000 acres, with an average of 3,500 acres, there were only 603 farms above 1,000 acres, with an average of 1,300 acres—or 520 ha. Rumanian properties above 500 ha. numbered 2,171, with an average extent of 1,470 ha., which nearly equalled the average of the English properties; but Rumanian farms above 500 ha. were 1,180, with an average of 1,208 ha., that is more than double the English

[1] C. Garoflid, *Chestia Agrară*, pp. 164–5.

average. (And the English figures probably included pastures, whereas the Rumanian figures did not.) One might extend the comparison to the United States, where farms above 100 ha. had an average extent of only 208 ha. M. Garoflid mentioned the case of a 10,000 ha. estate in the county of Ialomiţa; it was divided into three farms, but the three were worked jointly.

The explanation of that state of things must be sought for in the considerable proportion of large property which was let to tenants. In 1917 the deputy M. M. Carp affirmed in the Jassy Parliament, that 64 per cent. of all arable land above 100 ha., was let out—a percentage which in certain counties rose to 72–75 per cent., and in one county to over 92 per cent. More definite and detailed figures concerning the area let out from the various categories of properties were supplied by the agrarian inquiry which was instituted after the rising of 1907. At that time the area held in tenancy in the four Rumanian provinces was as follows:

Categories of farms	Area held by tenants				
	Moldavia	Muntenia	Oltenia	Dobrogea	Rumania
Hectares	%	%	%	%	%
50–100	32·09	33·07	23·18	10·23	24·09
100–500	55·08	54·70	40·26	29·62	50·17
500–1,000	55·57	57·99	58·46	57·10	58·23
1,000–3,000	59·19	61·27	52·92	26·92	58·33
3,000–5,000	61·84	76·28	63·45	100·00	73·36
Above 5,000	100.00	68·82	71·30	—	72·43 [1]

Statistics collected in 1913 showed that of the total area cultivated in that year, 60 per cent. was farmed by owners and 40 per cent. by tenants:

	Hectares	%
Freehold land	3,504,921	60
Leasehold and *métayage* . .	2,335,700	40 [2]

These proportions vary with the various geographical regions. Cultivation by owners preponderated in the highlands, where small holdings were more numerous, whereas tenancy was

[1] M. Şerban, *op. cit.*, p. 29. [2] Ministry of Agriculture, *op. cit.*, p. 28.

more widespread in the plains, where large estates under cereal crops predominated, as may be seen from the following table:

Region	Percentage of cultivated area	
	Freehold	Leasehold or *Métayage*
Moldavian highlands	76·0	24·0
Dobrogea	69·8	30·2
Muntenian highlands	69·2	30·8
Lowlands of Siret and Pruth	56·8	43·2
Lowlands of the Danube	52·5	47·5 [1]

The tendency of Rumanian landowners to let out their land is confirmed from a different angle by the relative growth of the various classes forming the rural population. There are, unfortunately, no figures which might enable us to follow that evolution gradually. The only reliable statistics, before those of 1913, are those collected by D. Marţian in the *Analele Statistice* for 1860. The latter referred only to Muntenia and Oltenia and the comparison therefore has to be limited to those two provinces. The confronting of the two sets of figures reveals a decline in the proportion of owner-cultivators, notwithstanding the several distributions of land from 1864 onwards, and a great increase in the proportion of tenant-cultivators:

Categories	1860		1913 [2]	
		%		%
Owner cultivators	398,958	87·5	638,792	74·7
Tenants	2,980	0·6	100,064	11·7
Total number heads of family (rural)	457,270 [3]		856,246	

(Percentage = that of total number of rural heads of families.)

These data do not sufficiently justify the conclusion that there had been a concentration in land property during that period—especially as general conditions and the methods of classification

[1] Ministry of Agriculture, *op. cit.*, p. 29.

[2] *Ibid.*, *op. cit.*, p. 12.

[3] The first figure in this column is erroneously given in the original as 408,958, and the third as 307,270. The figures here used, which correspond to the percentages, have been corrected from the details of the original table.

had changed. Certain figures available for 1905 make possible the following comparison:

Categories	1905		1913[1]	
		%		%
Number of cultivators .	1,018,079		1,112,578	
Owners	876,570	86·1	967,897	87·0
Tenants (money or *métayage*)	141,509	13·9	144,681	13·0

During that short period, therefore, the number of owner-cultivators increased relatively more rapidly than the number of tenants. In general, however, Rumanian agricultural statistics relating to individual years are not reliable material for comparison. The greatly varying climatic and economic conditions frequently induce startling changes from year to year in the area cultivated and, consequently, in the amount of land rented by the peasants. In this particular instance the 1913 figure relating to owner-cultivators was undoubtedly influenced by the transfer of land to the peasants after the rising of 1907.

As interesting as the rapid rise in the number of tenants from 1860 to 1913 was the contrary tendency in the number of agricultural labourers. The total area brought under the plough, and the cultivation of wheat, increased enormously during that space of time. Yet the proportion of agricultural labourers among the total number of heads of rural households only rose from 11·9 per cent. to 13·6 per cent.—a much slower growth than that of the rural population as a whole:

Number of Labourers (and other Professions)[2]

		%
1860	55,332	11·9
1913	117,570	13·6

This curious stagnation was by no means due—as the uninitiated perhaps might believe—to the extensive use of elaborate machines; quite the contrary. Placing the rapid increase in the number of tenants side by side with the slow increase in the number of labourers, one discovers a further characteristic of the Rumanian agrarian system. Not only was a considerable

[1] Ministry of Agriculture, *op. cit.*, p. 13.
[2] *Ibid.*, *op. cit.*, p. 12.

proportion of land let out from the large estates, but, moreover, much of it was let to small cultivators. In the agricultural statistics for 1907, published by M. L. Colescu, the area let to small cultivators from properties above 100 ha. was given as 1,037,000 ha., to which had to be added grazings, about 800,000 ha., making a total 1,837,000 ha.—that meant about 48 per cent. of all the arable land owned by large proprietors. M. Șerban cites the following table from the *Report on the Application of the Law on Agricultural Contracts*:

	1910	1911	1912	1913
	Hectares	Hectares	Hectares	Hectares
Land let to peasants on a money rent	352,409	258,971	304,774	410,212
Land let to peasants in *métayage*	522,742	510,692	479,704	424,788
Grazing let to peasants by the ha.	73,651	63,879	61,706	65,095
Grazing let to peasants by heads of cattle (no. of animals) . .	497,852	410,500	441,313	405,663
	1,446,654	1,244,042	1,287,497	1,305,758[1]

These figures only referred to land let to the peasants on the basis of written and duly registered contracts, but not to such land as was let merely on a verbal agreement—a widespread custom which enabled large owners or tenants to impose upon the peasants higher prices than those officially fixed. Arable land and grazing let to the peasants by verbal agreements rose in certain years to 600,000 ha.; and to this would have to be added permanent and artificial pastures. In 1910 alone, when the number of written contracts was 1,767, the authorities were able to track 864 verbal agreements; but a large number remained undiscovered, frequently with the connivance of the peasants themselves, who were afraid of being refused land in the future if they disclosed the hard bargains which large owners or tenants had extracted from them.

The state of things thus revealed in the official reports was confirmed by a private inquiry conducted by M. Șerban, mainly in 1911. He investigated in detail the situation on twenty estates, from various parts of the country and representing various kinds of agricultural undertakings, a Crown domain

[1] *Op. cit.*, p. 35.

being included among them. He found that of these estates only three did not let land to the peasants—and two of them were farmed on progressive lines quite exceptional in Rumanian agriculture. Of the total area of 79,323 ha. covered by these estates, 58,078 ha. was arable land. From this the peasants rented 20,001 ha. in *métayage* and 7,947 ha. on the basis of agricultural contracts; i.e. altogether 27,948 ha. or 47·31 per cent. of the total arable land. The first two tables given in this section had shown that 19·6 per cent. out of a total area of 38·2 per cent. covered by estates above 500 ha. were worked in farms of a lesser size; and as the categories of farms between 10 and 500 ha. were, on the whole, insignificant in number and extent, it followed that half of the large property above 500 ha. was before the war farmed in small holdings up to 10 ha. each. This general conclusion coincides with that reached by M. Șerban's special investigation. M. G. Mantu has stated that the extent of the cultivated area rose but little between 1906 and 1915, but that the area in the hands of the small cultivators increased from 65 per cent. to 73·88 per cent. of the arable land. These figures are not easy to check; but it is probable that peasant cultivation increased substantially after the rising of 1907, partly through the application of the legislative measures described in Chapter IV, and partly through the voluntary action of the large owners and tenants. Some of them must have found that farming no longer attracted them when the new laws inaugurated a stricter supervision of the wages which the peasants received for their labour and of the prices they paid for the land they rented. This forced many landowners either to sell out or to cultivate more intensively themselves, by investing more capital, which very few of them were in a position to do. In Rumania the movement was not so accentuated as in Russia,[1] especially as the legislation of 1907 remained on the whole a dead letter. But one can trace a similar effect of the 1907 rising in the figures given by M. Mantu, as well as in the growing change from *métayage* to money rent for land taken over by peasants from large owners.

[1] In Russia, after the rising of 1905–6, some 27,000,000 acres passed by purchase into the hands of the peasants, not only because of the Stolypin legislation, but also because of the rise in wages and the fall in the rent of land.

This appears from the table M. Șerban quotes from the *Report on the Application of the Law on Agricultural Contracts*, and even more clearly from the table below which refers to farms of less than 100 ha.:

Year	Area cultivated by owners	Area held on money rent	Area cultivated in *métayage*
	%	%	%
1909	66·47	15·29	18·24
1910	67·00	16·41	16·59
1911	66·86	16·39	16·75
1912	66·63	17·94	15·43
1913	65·65	20·03	14·32 [1]

In a study of Rumania's agrarian problem recently published in the second issue of the *Agrarna Probleme*—the Bulletin of the International Agricultural Institute in Moscow—the author, M. Timow, characteristically draws from the above table the conclusion that capitalist farming was on the increase in Rumania. That, of course, is in keeping with the strict Communist view that any peasant who owns land is a capitalist. A more catholic reading of rural economics will see in these facts rather the trace of a contrary development—the growth of peasant farming at the expense of the capitalist farmer. That applies with special force to Rumania, where the large-scale tenant seldom had either knowledge or equipment for farming, and literally brought no other contribution to the process of production than the cash or credit he could command. M. Timow's view might have had some foundation if the change had been solely one from rent in kind to rent in money. But that was only one side of the evolution. The second and, evidently, the more significant side was a change from large-scale to small-scale renting—a change which increasingly excluded the capitalist. Moreover, M. Timow has drawn his conclusion from the figures without taking sufficiently into account some of the peculiar circumstances which qualified those figures and gave them their true meaning. The money rent mentioned in the contracts was generally merely a nominal factor, used as a means of measurement, and transformed in practice into labour obligations which, in their turn,

[1] Ministry of Agriculture, *op. cit.*, p. 35.

were reckoned on a money basis. That arrangement was made necessary by the legislation of 1907, which demanded separate contracts, on a money basis, for labour and for renting; but the mutation was allowed by the law on Agricultural Contracts, and the Supreme Agricultural Council itself stated that it was adopted by many large cultivators as an arrangement more profitable to them.[1]

To carry on the argument as to the relative part played by large and small cultivators in Rumanian agriculture, one may note the figures given below which show the distribution of cultivators farming for their own account, without regard to the form of their tenure:

Categories of farmers	Number	Percentage from total
With less than 2 ha.	476,493	42·0
„ 2–5 ha.	441,479	39·0
„ 5–10 ha.	161,563	14·3
„ 10–25 ha.	42,996	3·8
„ 25–50 ha.	5,698	0·5
„ 50–100 ha.	1,553	0·1
„ 100–500 ha.	2,376	0·2
„ more than 500 ha.	1,044	0·1 [2]

Both the total and the relative number of small cultivators was, therefore, overwhelming and far in excess of the similar relation in other European countries. According to figures cited in Rumanian statistical publications, that relation was elsewhere as follows:

	%	
In Rumania, farmers below 5 ha.	81·00	(1913)
In Germany farmers below 5 ha.	76·40	(1907)
In Hungary (up to 5·7 ha.)	72·70	(1895)
In France (up to 5 ha.)	71·29	(1892)
In Belgium (up to 5 ha.)	68·91	(1904)
In Denmark (up to 5 ha.)	53·50	(1903)
In England (up to 8 ha.)	51·48	(1895)
In Holland (up to 5 ha.)	46·70	(1895)

The validity of this comparison is only relative, as all these countries, Hungary excepted, were countries of intensive cultivation, in which a large number of small cultivators was in keeping

[1] See M. Şerban, *Problemele noastre Sociale şi Agrare*, 1914, p. 35.
[2] Ministry of Agriculture, *op. cit.*, p. 15.

with the nature of their agriculture itself; whereas Rumanian agriculture was primitively extensive. A broader classification of Rumanian farms may be seen in the following table:

Categories of farms	Number	Percentage
Dwarf holdings, below 2 ha.	476,649	42·0
Small holdings, 2–10 ha.	602,886	53·3
Medium holdings, 10–100 ha.	50,247	4·4
Large farms above 100 ha.	3,420	0·3
	1,133,202	100·0[1]

Four-fifths of the Rumanian cultivators farming for their own account, therefore, held less than 5 ha.; 4·7 per cent. alone farmed more than 10 ha. and only 0·3 per cent. farmed over 100 ha. We find in this case the usual geographical variation again, smallholders predominating in the highlands and the larger farmers in the lowlands, where cereal crops were grown.

Another aspect of this point, namely the extent of arable land which was in the hands of these various categories of farmers, can be followed in the table below:

Categories of farms	Number	Percentage of total number	Total area covered	Percentage of total area	Average of the category
			Hectares		Hectares
Less than 2 ha.	*476,649*	*42·0*	*572,169*	*9·7*	*1·20*
2–5 ha.	441,336	39·0	1,546,311	26·5	3.50
5–10 ha.	161,550	14·3	1,118,409	19·2	7·34
From 2–10 ha.	*602,886*	*53·3*	*2,664,720*	*45·7*	*4·42*
10–25 ha.	42,996	3·8	622,174	10·7	14·47
25–50 ha.	5,697	0·5	193,128	3·3	33·91
50–100 ha.	1,554	0·1	107,182	1·8	68·96
From 10–100 ha.	*50,247*	*4·4*	*922,484*	*15·8*	*18·36*
100–500 ha.	2,377	0·2	587,549	10·1	247.50
Above 500 ha.	1,043	0·1	1,093,699	18·7	1,047·10
Above 100 ha.	*3,420*	*0·3*	*1,681,248*	*28·8*	*491·59*
Total	*1,133,202*	*100·0*	*5,840,621*	*100·0*	*5·15*[2]

Farms up to 10 ha., that is, covered 2,664,720 ha. or 45·7 per cent. of all the land cultivated, which gave an average of 4·42 ha. per farm in this category—the most numerous being near the lower limit. Altogether, therefore, farms with less than 10 ha. repre-

[1] Ministry of Agriculture, *op. cit.*, p. 19.

[2] *Ibid.*, *op. cit.*, p. 23.

sented no less than 95·3 per cent. of the total number of farms and covered 55·4 of the total arable land.

Finally, one might go a step further and note from the table below the proportion of freehold and leasehold land in each category of farms:

Categories of farms	Cultivated Area			
	Freehold	%	Leasehold	%
	Hectares		Hectares	
Less than 2 ha.	*421,712*	*73·7*	*150,457*	*26·3*
2–5 ha.	1,045,525	67·6	500,786	32·4
5–10 ha.	647,518	57·9	470,891	42·1
From 2–10 ha.	*1,693,043*	*63·5*	*971,677*	*36·5*
10–25 ha.	396,764	63·8	225,410	36·2
25–50 ha.	141,263	73·1	51,865	26·9
50–100 ha.	74,408	69·4	32,774	30·6
From 10–100 ha.	*612,435*	*66·4*	*310,049*	*33·6*
100–500 ha.	281,871	48·0	305,678	52·0
Above 500 ha.	495,860	45·3	597,839	54·7
Above 100 ha.	*777,731*	*46·3*	*903,517*	*53·7*
Total	*3,504,921*	*60·0*	*2,335,700*	*40·0* [1]

The two modes of tenure were not equally distributed in the various categories of farms. The bulk of the small exploitations consisted of freehold land, whereas the larger farms were to a considerable extent composed of rented land. Above 100 ha., the rented area exceeded that held in freehold:

	%
Above 100 ha.	53·7
100–500 ha.	52·0
Above 500 ha.	54·7

The same phenomenon is illustrated perhaps more clearly by the next table, which shows what percentage of all the freehold (arable) land and what of all the leasehold (arable) land was included in the main categories of agricultural undertakings:

Form of Tenure	2 ha.	2–5 ha.	5–10 ha.	10–25 ha.	25–50 ha.	50–100 ha.	100–500 ha.	Above 500 ha.
Freehold	12·0	29·8	18·5	11·3	4·0	2·1	8·1	14·2
Leasehold	6·4	21·4	20·2	9·7	2·2	1·4	13·1	25·6 [2]

[1] Ministry of Agriculture, *op. cit.*, p. 30.
[2] *Ibid.*, *op. cit.*, p. 34.

Here, again the distribution varied according to regions. In the Moldavian lowlands 60·6 per cent. of the area occupied by farms above 500 ha. was rented, and in the Muntenian lowlands 52·7 per cent. These regional peculiarities may be followed in the table below, which gives the regional percentages of the area cultivated by its owners:

Categories of farms	Rumania	Lowlands of the Siret and Pruth	Lowlands of the Danube	Moldavian highlands	Muntenian highlands	Dobrogea
Below 2 ha.	12·0	11·2	6·7	19·5	25·4	0·7
2–5 ha.	29·8	31·5	32·4	38·7	34·8	4·8
5–10 ha.	18·5	19·6	20·7	15·0	15·8	17·1
10–25 ha.	11·3	7·2	8·8	4·6	7·2	37·6
25–50 ha.	4·0	1·5	1·9	2·2	2·5	18·6
50–100 ha.	2·1	0·9	1·2	1·7	2·0	7·3
100–500 ha.	8·1	8·9	8·8	7·8	6·7	6·9
Above 500 ha.	14·2	19·2	19·5	10·5	5·6	7·0[1]

There was, therefore, a great similarity in the conditions which prevailed in the plains of the several provinces, as well as in their mountainous regions. The difference between highlands and lowlands arose from that competition for land in the corn-growing districts to which attention has been drawn before.

There is no material to show from which categories of property the tenanted land was obtained. But a comparison with the distribution of property makes it clear that the bulk of the land rented by the small and medium-sized farmers came from the large owners, and not from the mutual letting of land among small and medium-sized owners. 'So that the present organization of Rumanian agriculture does not reflect the present distribution of property, but, on the contrary, there is a considerable transfer of land from the large owners to the medium-sized and especially to the small cultivators.' [1]

The Rumanian agrarian problem consequently resolved itself largely into the need for establishing a more direct connexion between ownership and farming on the land. In spite of this, no agricultural census has been undertaken so far to ascer-

[1] Ministry of Agriculture, *op. cit.*, p. 30.

tain the precise effects of the reform in that respect. A great deal may, nevertheless, be learnt from the figures prepared for the purpose of this study from the fiscal census for 1927 of the Ministry of Finance.[1] They show that in the lower categories of property only a very small proportion of the land is now let out, the bulk being therefore used by the owners. In the higher categories, the proportion of land let out is still considerable. It is unlikely, however, that such land is still rented by speculating capitalist tenants, of the kind who flourished before the reform, but rather by peasants who work the land in *métayage*. Its persistance was established by the inquiry which the Ministry of Agriculture conducted in 1922 into the extent of the various systems of cultivation in use:

Province	Total area cultivated	Area cultivated in *Métayage*	
	Hectares	Hectares	%
Moldavia	1,293,164	76,742	5·9
Muntenia	2,322,207	478,932	20·6
Oltenia	906,561	54,439	6·0
Dobrogea	783,891	37,070	4·7
Bessarabia	2,466,785	88,580	3·5
Bucovina	233,093	28,697	12·3
Transylvania	2,332,588	382,707	16·4
Total	10,338,289	1,147,167	11·0 [2]

In support of the point made before it should be noted that share-cultivation was previously unknown in Moldavia, whereas renting on a large scale was widespread; after the reform, of the land retained by the large owners much was let instead to peasants, in *métayage*.

The 1,147,167 ha. which the inquiry of 1922 found to have been worked in *métayage* far exceeded the area let out in 1927; according to the fiscal census the latter reached a total of only 869,647 ha. The explanation of this apparent discrepancy is no doubt that in 1922 a part of the expropriated land had not yet been transferred to the peasants in a final form, but was used by them temporarily in *métayage* (an arrangement to which reference has been made in Chapter VI, in the section on compensa-

[1] See Tables on pp. 246–7.

[2] G. Ionescu-Siseşti, *Structure Agraire et Production Agricole*, p. 24.

tion). Share cultivation was still extensive in Muntenia, where it predominated before the reform as well. It had decreased greatly in Oltenia, where large owners and peasants—always reputed to be more enterprising than other sections of the country—had quickly organized their farming on an autonomous footing, as was most practical after the reform. In Transylvania and in Bucovina the application of the reform was still in progress in 1922, and share cultivation was correspondingly prevalent. The lowest percentage was found in Bessarabia; there the reform had been quick and drastic, leaving the former landowners with merely 100 ha. each.

Share cultivation—which, of course, meant cultivation by the peasants—was used, therefore, on 1,147,167 ha. in 1922, or on 11 per cent. of all the cultivated land. An area about half as large was rented by the peasants on the basis of annual labour contracts, calculated in money. The rest of the 10,338,289 ha. cultivated in 1922, consisting mainly of the small and medium-sized peasant property, was farmed by its owners.

The figures obtained from the fiscal census for 1927, showing the distribution of arable land among the various categories of properties, also indicate how much of that land was let out in each category (see Tables, pp. 246–7):

Of the smallholdings up to 5 ha., which covered almost half of the arable land in the Old Kingdom, hardly anything was let at all; of the small properties up to 10 ha., which among them included almost two-thirds of all the arable land, only 1·44 per cent. was let out, and of the medium-sized property only a moderate proportion. But of the larger properties above 50 ha. as much as 28·7 per cent. of their arable land was let out. In every category the percentage of the land let out was greater than the percentage of its owners, which suggests that it was the larger properties which in each case were not farmed by their owners themselves.

The position after the reform is made clearer by the table on p. 248, which summarizes the proportion of arable let out in the four main provinces.

Everywhere, therefore, almost the whole of the arable land held by small owners was worked directly by them; whereas the

(1) *The Old Kingdom*

	Let				Unlet				Total	
	Number of owners	%	Hectares	%	Number of owners	%	Hectares	%	Number of owners	Area in hectares
up to 5 ha.	14,849	0·88	32,440	0·92	1,668,742	99·12	3,457,314	99·08	1,683,591	3,489,754
5–10 ha.	4,943	2·60	35,908	2·80	184,241	97·40	1,233,692	97·20	189,184	1,269,600
10–50 ha.	2,606	5·7	54,900	6·6	42,848	94·3	765,708	93·4	45,454	820,608
50–250 ha.	1,651	15·2	183,762	20·8	9,210	84·8	699,233	79·2	10,861	882,995
Above 250 ha.	538	32·6	314,493	37·7	1,114	67·4	537,122	62·3	1,652	851,615
Total	24,587	1·2	621,503	8·4	1,906,155	98·8	6,693,069	91·6	1,930,742	7,314,572

(2) *Transylvania*

	Let				Unlet				Total	
	Number of owners	%	Hectares	%	Number of owners	%	Hectares	%	Number of owners	Area in hectares
up to 5 ha.	8,982	0·96	15,697	1·0	928,572	99·04	1,551,984	99·0	937,554	1,567,681
5–10 ha.	1,201	1·1	8,722	1·2	104,943	98·9	704,591	98·8	106,144	713,313
10–50 ha.	1,036	1·8	20,102	2·5	54,340	98·2	777,618	97·5	55,376	797,720
50–250 ha.	419	9·2	37,229	10·3	4,121	90·8	325,583	89·7	4,540	362,812
Above 250 ha.	75	10·6	35,551	12.0	634	89·4	259,414	88·0	709	294,965
Total	11,713	1·6	117,301	3·1	1,092,610	98·4	3,619,190	96·9	1,104,323	3,736,491

(3) *Bessarabia*

	Let				Unlet				Total	
	Number of owners	%	Hectares	%	Number of owners	%	Hectares	%	Number of owners	Area in hectares
up to 5 ha.	2,873	0·69	9,247	0·92	409,470	99·31	995,295	99·08	412,343	1,004,542
5–10 ha.	936	0·6	6,538	0·7	135,076	99·4	901,623	99·3	136,012	908,161
10–50 ha.	560	1·2	11,589	1·5	46,095	98·8	736,865	98·5	46,655	748,454
50–250 ha.	413	17·2	42,702	19·5	1,965	82·8	176,242	80·5	2,378	218,944
Above 250 ha.	45	28·3	34,788	31·2	149	71·7	108,592	68·8	194	143,380
Total	4,827	0·8	104,864	3·4	592,755	99·2	2,918,617	96·6	597,582	3,023,481

(4) *Bucovina*

	Let				Unlet				Total	
	Number of owners	%	Hectares	%	Number of owners	%	Hectares	%	Number of owners	Area in hectares
up to 5 ha.	6,027	3·04	6,831	3·1	191,948	96·96	212,188	96·9	197,975	219,018
5–10 ha.	44	1·0	336	1·16	4,331	99·0	28,443	98·84	4,373	28,779
10–50 ha.	107	7·8	2,670	10·3	1,268	92·2	23,238	89·7	1,375	25,909
50–250 ha.	94	27·4	11,240	27·7	249	72·6	29,267	72·3	343	40,507
Above 250 ha.	9	21·9	4,900	31·1	33	78·1	10,812	68·9	42	15,713
Total	6,281	3·0	25,977	7·8	197,829	97·0	303,948	91·2	204,108	329,926

larger proprietors still let out a considerable proportion of their estates, especially in the Old Kingdom and in Bucovina.

PERCENTAGES OF ARABLE LAND LET OUT FROM THE VARIOUS CATEGORIES OF PROPERTIES

Provinces	up to 5 ha.	5–10 ha.	10–50 ha.	50–250 ha.	Above 250 ha.	Total area let out
1. Old Kingdom .	0·92	2·80	6·6	20·8	37·7	8·4
2. Transylvania .	1·00	1·20	2·5	10·3	12·0	3·1
3. Bessarabia .	0·92	0·70	1·5	19·5	14·0	3·4
4. Bucovina .	3·10	1·16	10·3	27·7	31·1	7·8
Whole country .	1·02	1·70	3·7	18·3	29·0	6·0

The comparison with the pre-war situation suffers to some extent from a difference in the nature of our material. The tables which had been given on earlier pages to show the distribution of leasehold before the reform referred to the extent of land rented by various categories of cultivators; whereas the figures of the Ministry of Finance for 1927 refer to land let out by the various categories of landowners. But by taking for the period before the reform the portion of the cultivated area which was rented, and for the period after the reform that portion of the cultivated area which was let out, one is enabled to establish a sufficiently close comparison of the area which in the two periods was farmed not by its owners themselves, but by tenants.[1]

While speaking on the land problem in the Senate, on March 6, 1907, M. D. A. Sturdza, a former leader of the Liberal Party, declared that the large owners held 3,810,361 ha., i. e. 48·69 per cent. of all arable land, and that they were letting out from these 2,293,961 ha., i.e. 60·2 per cent. M. Colescu's figures for the same year gave 1,837,000 ha. as being rented by peasants

[1] One of the incidental effects of the reform has been the disappearance of large estates whose owners lived abroad. In 1926 the number of owners living abroad, and the area of the land they held, was as follows:

Old Kingdom . . .	569	owners holding	14,965 ha.
Transylvania . . .	2,253	,, ,,	44,550 ,,
Bessarabia . . .	407	,, ,,	3,857 ,,
Bucovina . . .	148	,, ,,	5,512 ,,
Total . . .	3,377	,, ,,	68,884 ,,

The average size of the holdings having been merely 20 ha., it is clear that many owners were peasants, who had probably migrated but meant to return.

from estates of more than 100 ha., which meant that only some 460,000 ha. were let to large tenants. As not a few of these tenants re-let some of the land to the peasants, it is reasonable to assume that on the whole the peasants had been renting about two million ha. arable land from the large estates. That is just about the area which the reform has taken from estates above 100 ha. in the Old Kingdom; hence, to that extent, the change in ownership has not meant any change in the class of the users. In the Old Kingdom only the other 562,024 ha. arable distributed to the peasants have really passed from large-scale to small-scale cultivators; with such additional land as the large owners may now be letting to them.

The extent of arable land farmed out after the reform can be derived from the figures of the fiscal census for 1927. The census employed a different scale of categories, so that large property will be taken to include all properties above 50 ha. From these, 498,255 ha. were let out of a total of 1,734,611 ha., which meant 28·7 per cent. The drop from 60·2 to 28·7 per cent. (which would be greater still if the census figures, too, had referred to properties from 100 ha. upwards) in the area not farmed by the owners of large and medium properties themselves represents one of the direct effects of the reform. Further, M. Sturdza put the arable area covered by properties up to 10 ha. at 3,153,645 ha., and that covered by properties of 10–100 ha. at 862,800 ha; he did not state how much of them was let out, but by applying to them the percentages established for their categories in the 1927 census, 108,300 ha. appear to have been let out from the first and 56,945 ha. from the second of the two categories named above. Together these make up a total of 2,459,206 ha., i.e. 31·42 per cent. of all the arable land as having been farmed out before the reform; in 1927 the total farmed out was 625,504 ha. or 8·4 per cent. Therefore the drop from 31·42 to 8·4 per cent. of all arable land let out in the Old Kingdom constitutes one of the definite and significant effects of the reform.

The proportion is more favourable still when the new provinces are included. Information on this point is not available for all of them. M. P. V. Synadino, who dislikes the reform, has given figures for Bessarabia to show that before the reform the peasants used to rent about 40 per cent. of the 608,568 ha. expropriated from the large owners, and half of the 143,729 ha. taken from monasteries and foreign owners, which gives a total of some 315,290 ha. or almost 42 per cent. of the land taken from the said categories; to which if one adds the 176,388 ha. found at the time of the reform in the hands of the Peasant Bank and 58,884 ha. expropriated from colonists belonging to ex-enemy nations, it follows that of the 1,098,045 ha. now distributed to the Bessarabian peasants, 550,562 ha. or fully 50 per cent. had been farmed by them even before the reform. The figures of the Ministry of Finance give 77,490 ha. as having been let out in 1927 from the 462,324 ha. arable covered by Bessarabian properties above 50 ha., or 16·8 per cent.; and the total area then let out was 104,865 ha. or merely 3·4 per cent. of all the arable land of the province.

In the whole of Greater Rumania 664,666 ha. were let out in 1927 or 23·64 per cent. of the 2,810,930 ha. arable land included in properties of more than 50 ha. From all properties, large and small, 869,647 ha. were let out; which in other words meant that of all the arable land merely 6 per cent. was not farmed by the owners themselves.

From what has been said in the foregoing pages it is clear that the great extension of large property before the reform had not meant that farming on a large scale was widespread. The greater part of the area covered by large estates was let out, and the bulk of it was let to peasants. Together with such land as they themselves owned the peasants farmed for their own account at least two-thirds, and possibly three-quarters of all the arable land in the Old Kingdom. The comparisons attempted above show that the transfer of land is much reduced in the new state of things. The reform has led to what one may call a symbiosis of ownership and cultivation; and that is a process which is always adopted when encouraging a more intensive standard of agriculture.

While the reform, therefore, as described in the previous chapter, has caused an enormous change in the distribution of property, it has not, on the strength of the facts and figures discussed in this section, caused any substantial change in the organization of farming—as was readily and rashly assumed by most western critics. That is the first proposition which it is necessary to keep in mind in trying to determine what effects the reform has had on production. The reform's main result in that respect has been to produce an approximation between the distribution of property and the distribution of agricultural undertakings, in the sphere of large-scale cultivation as well as in that of peasant farming—a consequence which marks a great advance on the earlier conditions. The peasant who owns a tolerable holding has a better chance and a greater stimulus to cultivate efficiently than has the peasant who rents that land from year to year and has to barter his best labour for it. This aspect of the change has been welcomed even by Professor Max Sering, who in other respects has frowned severely on the post-war reforms. The placing of the eastern European agrarian structure upon a system of peasant ownership, he considers, is not merely an advance on what existed before, but also 'in itself an improvement on conditions in western Europe, where the agrarian system is based upon peasant tenancies supporting a mostly unproductive class of large landowners. The predominance of peasant ownership, though it may have drawbacks for the present, permits us to hope that in the measure in which education, farming and communications gradually improve during the next few decades, the peasants' hard work will raise agriculture to a flourishing state.'[1]

These considerations apply not less strongly to such large-scale property as is left in Rumania. The figures we have cited show that the bulk of the large owners are at present farming their land themselves, which means a radical break with the past. Through the new conditions of ownership and of labour which it has brought about, the reform has opened the way towards greater economic autonomy among the various classes of farming. Hitherto a cramping interdependence of large and small cultivators has been one of the dominant traits of Rumania's agrarian system. The peasants were to an oppressive degree dependent on the large owners or tenants for land; and the large cultivators almost altogether depended on the peasants for

[1] Introduction to *The Agrarian Revolution in Europe*, ed. Prof. Max Sering, p. 20.

means of production. Now the large farmers have little or no surplus land to give, and the small cultivators, as they progress in technique, will have no stock and labour to spare. The two kinds of farming are being thrown back upon their own resources, and that must gradually lead them to organize their production after the manner which has been found best for their kind in the regions where agriculture stands on a high level of achievement. For hitherto the working arrangements in Rumanian agriculture have been so peculiar that the reform, notwithstanding its vastness, has caused a still lesser change in the means of cultivation than it has in the scale of cultivation.

Section 2

The Change in the Means of Production

The circumstances discussed in the previous section established the fact that a large proportion of the big estates had always been in the hands of small cultivators, and that in consequence the land reform has caused a much lesser change in the organization of farming than in the division of property. To complete that picture, one must go further and consider, besides the change in the size of farms, the change in the methods of farming; the transforming effect of the land reform will then appear still more reduced.

During the discussion of the reform, its critics frequently argued that if large property was useless, large-scale cultivation was still necessary; especially in a country with so backward an agriculture as Rumania's. The argument touches on the perennial and seemingly inexhaustible question as to the respective merits of large and small-scale cultivation. Though raised already in the eighteenth century, by the Physiocrats—when the Academy of Arras offered a prize for the ablest reply to the question 'Which is the best repartition of the soil?'—the problem is far from having been solved. The Physiocrats, of course, were protagonists of large-scale production; though there were exceptions even in their own camp. The old Liberal school took the same standpoint; and so did a number of Conservative writers, largely for political reasons. If Adam Smith, Jean Baptiste Say,

Rau and other economists drew attention to the importance of smallholders it was to point out that they gave a larger total output than the big farmers, though they admitted that the net output of the latter might be higher; a circumstance from which, in their opinion, society derived considerable economic and social advantages.

'The small owner', says Philippovich, 'who to a large extent consumes his own produce, must consider above all his total output. That, however, may be said to be a point which concerns the nation's economy as well: it is to its interest that as large an output as possible should be obtained from the soil, rather than that the difference between cost of production and market price—which constitutes the net return—should be reduced. That is why the old Populationists, like Süssmilch and Sonnenfelds, praised small property as favouring the increase of population, whereas large property contributes to depopulation, just because a wide extent of land is distributed among a small number of owners.'[1]

In general, economists looked at the issue from the angle of industry and of the rapidly growing industrial centres which required a generous supply of cheap corn; and there is no doubt that their attitude was largely influenced by what then seemed an unalterable analogy between industry and agriculture. They took it for granted that the concentration and mechanization which were giving such marvellous results in one field of production would achieve the same wonders when adopted in the other. For both reasons, Marxists, and Socialists in general, not only agreed with them but went even further and plunged for a program of national ownership and large-scale cultivation by 'armies of labourers' (Communist Manifesto), on the assumption that the peasant produced mainly for himself but the large estate mainly for the market.

Two developments have in the main helped to invalidate these assumptions in the field of European agriculture. The first has been the fabulous expansion of corn-growing in the virgin lands across the seas, which has put the European corn-grower out of competition. That was just the field in which large-scale farmers could show certain advantages over the small farmers, and the consequence has been that since the severe agricultural

[1] Philippovich, *Agrarpolitik*, Part I, Ch. 2.

crisis in the seventies of last century many large estates and farms have passed into the hands of small owners and tenants. Nor has that change meant a loss to the national economy of the various countries. For the second development, the progress in the science of agriculture, has shown that the laws of industrial production do not also hold good for the production of food-stuffs. In agriculture production follows a natural process which does not allow an indefinite division of labour, but only an accumulation of labour; and this form of intensifying production has been proved to bring in returns which, for a number of reasons, diminish in the proportion in which the size of the agricultural undertaking increases, as illustrated by the so-called circles of Thünen. More recent inquiries have shown that this is true not only of the total output, which was often conceded, but also of net production. It might be useful to quote here one inquiry, because of its clear results and of the great competence of its author. The director of the Swiss Peasant Secretariat, Professor Ernest Laur, who is a member of the League of Nations' Committee on Agricultural Questions, having worked out returns on capital for various categories of Swiss farms over a period of twenty years (1901–21), has obtained the following averages, in Swiss francs:

Size of farm	Value of total production per hectare	Value of sold produce per hectare
3–5 ha.	1,180	795
5–10 ha.	1,005	740
10–15 ha.	900	700
15–30 ha.	825	660
above 30 ha.	710	695

This is not the place to join issues with the protagonists of one or the other school, especially as the theoretical basis of the discussion is not yet sufficiently strengthened with experiment. It is undeniable, however, that capitalist society has regularly followed the dictates of its economic interests—in the mechanized nineteenth century more than ever—and that it would inevitably have steered towards a concentration of property and production if it had found that form as profitable in agriculture as it did in industry. Statistics show instead that everywhere the trend has been away from the large unit of production—a unique instance of the deviation of a whole industry from the despotic progress of the large economic unit in the capitalist era.

But in Rumania that evolution has been altogether different: very large agricultural undertakings persisted and prospered. On the face of things it was therefore plausible to argue that they persisted because they still had a function to fulfil; even if the argument in favour of large owners and cultivators was qualified by a reference to the backward state of the country's agriculture. It was implied that, whatever economic progress the peasants might achieve in a near or more distant future, they had still to be guided towards it by the large cultivators. And those who made use of this argument for the purpose of opposing expropriation or of pleading for a limited reform, based it on the part which the large cultivators had played in opening up Rumania's unploughed soil. All the economic prosperity and the civilization of Rumania, affirmed M. Garoflid, was due to the large cultivators; it was they who after the Treaty of Adrianople and, later, after the denunciation of the tariff convention with Austria, forced the peasants to change from pastoral to agricultural pursuits. Another economist, who takes a sympathetic interest in the peasants, privately admitted that the much-abused tenants had been an important agent in such progress as Rumanian agriculture had made, though the contrary view had generally been expressed for social or nationalist propaganda.[1] The tenants, he pointed out, were *entrepreneurs*, whose only interest was to obtain the greatest net return from the land they rented. Hence their capital investments went into productive channels—dead and live stock, selected seed, &c.—and they supervised cultivation themselves, often living throughout the period of agricultural labours in a peasant hut. The large owners, on the contrary, led a patriarchal life; their main investments were in roomy country-houses and other amenities for their private life, and most of their time was spent away from their estates, not infrequently abroad. Tenants who after years of

[1] Of the two million ha. arable land which the large owners used to let out, over 800,000 ha. were let to Jews and to foreigners.

hard work had become owners themselves, not seldom fell into the same wasteful ways.

A Conservative writer, M. R. Mandrea, put the argument for the large cultivators on a wider basis and maintained that the whole system they represented could not yet be dispensed with for reasons of State. 'We believe that at present large property is still the only productive factor in our country.' Extensive farming was still needed for the payment of indispensable imports; and though two-thirds of the land was in the hands of small cultivators, 'yet the only produce that is exported, the only one which can compete in the foreign markets, is still the produce of the large estates alone'. The peasant lands were miserable, their cultivation primitive, and their produce, when they had a surplus, unsaleable, unless mixed with corn from the large farmers. M. Garoflid used a similar argument in the 'Memorandum of the Large Cultivators' presented to the King, in 1920. It began by pointing out that the organism of the State was maintained with resources obtained from exports, and that in Rumania such resources could be supplied by agriculture alone. In the West it had been possible for the land to pass without danger to the peasants: economic expansion created towns which offered a market for agricultural produce, while exports were kept up by industry. In Rumania, however, conditions limited export for a long time to agricultural raw materials, but their export was profitable only as long as they were produced extensively. So far, the professional training and the economic impulse of the peasants had not reached a point where small cultivation might take the place of the large extensive cultivation in supplying a surplus for export.

Even the defenders of large cultivation in Rumania, therefore, did not claim for it an enduring superiority. They supported it with the historical argument of its useful past and with the political argument of its immediate services to the State. But when it came to the technical argument, they merely demanded a stay of execution until small cultivation should have finished its apprenticeship. This discussion on the relative merits of large and small cultivation in Rumania has more than a mere theoretical value for the purpose of our study. Only by elucidating the characteristic aspects of Rumania's former agrarian structure, which are in themselves of great interest for the rural sociologist, can one help the reader to a proper understanding of the new reform. Only a study of those earlier conditions could explain how such a revolutionary redistribution of property—revolutionary in extent and in speed—could possibly have come to pass without upsetting for a prolonged period the economic activities of the countryside.

The former organization of Rumanian agriculture displayed features which had not existed at all or had long disappeared elsewhere. For that reason the weighing of the respective merits of large and small cultivation could not in this case be measured by the usual standards. Neither form of farming was comparable to its western counterpart. Both of them were still tethered to a primitive technique, small cultivation being extensive and large cultivation latifundiary. An illuminating discussion of the latter form of exploitation, which gave Rumanian agriculture its distinctiveness, is to be found in M. Garoflid's *Chestia Agrară* (1920 edition, pp. 163–203).

M. Garoflid started by establishing that whereas elsewhere land properties were being broken up, Rumania displayed a contrary phenomenon. 'Concentration of agricultural exploitation is growing and large-scale farming is overcoming small cultivation. We have here a peculiar economic organization which enables excessively large exploitations to be profitable.'[1] The advantages of latifundiary exploitation were rooted in the belated unfolding of Rumania's agrarian problem, and not, as might seem at a distance, in the superior outfit of the large-scale producer. For, in the first place, latifundiary exploitation was not justified by a better organization for production. The few farmers who, about the turn of the century, attempted to carry on such an exploitation with the usual capitalist means, especially in the steppes of the Ialomiţa and Brăila counties, failed in fact badly. Nor was the success of that system in Rumania explained by a greater fertility of the soil; the much higher fertility of the

[1] It should be noted that M. Garoflid's book was originally published in 1908. We have pointed out that after the rising of 1907 the general tendency towards the break-up of large estates began to make itself felt in Rumania as well, and in a very pronounced degree in the neighbouring Russian empire.

virgin American soil has not produced anything similar. Nor, finally, was it explained by the cheapness of rent. The average revenues given in M. R. Capităneanu's *Recensământul Fiscal* (Fiscal Census), 1905, were:

Lei 30·41 per ha. for properties of 100–500 ha. and
„ 26·16 „ „ „ „ above 500 ha.

Rent, therefore, decreased with the greater extent of the property, but the difference was not sufficiently great to explain the advantages enjoyed by latifundiary cultivation. None of these factors offered a clue to its existence in Rumania; nor did they make clear the reason why 'the tendency of all Rumanian large-scale cultivation was to increase its revenue by concentrating exploitation and not by raising production'.

To get to the bottom of that economic paradox one must begin by noting that the essential feature of farming on such a large scale was the performance of agricultural labours at a great distance. The possibility of such labours explains the existence of latifundiary farming. Now, this phenomenon was peculiar to Rumanian agriculture; it applied not only to the area exploited on a large scale, but likewise to the land which the peasants rented from the large property, and even to a considerable portion of their own small property, frequently situated at a distance of several kilometres from the village. As that form of cultivation involved an enormous wastage of time this had to be compensated by an excessive number of teams, the various labours having to be performed within limited periods. In 1900, the number of draught animals employed on the land was 1,148,312 oxen and 710,929 horses, together 1,859,251 animals, which gave an average of one animal to 2·79 ha. of cultivated area; and reckoning four animals to each team, the proportion was one team or plough to 11·16 ha. That coincided with the number of ploughs —517,463 working 6,129,000 ha., in 1905. In the opinion of Krafft, the economic margin in a three-field system would be three ploughs to 100 ha. It is true that in a system of smallholdings the number of cultivators determines the number of ploughs and teams, but in Rumania their excess was due no doubt as much to the great distance at which the labours had to be performed. In France, e.g., where small culture predominated and where the average size of a holding was about the same as in Rumania, the average worked out in 1892 at one animal to 6·5 ha.; and that was generally considered excessive. It is also true that one must take into account the quality of the animals; but the poor quality of the Rumanian stock was offset by the light nature of the ploughing, 10–12 centimetres deep as against 20–25 centimetres in France. Moreover, the relatively large number of teams in France was due to her intensive agriculture, which required repeated labours; though the elementary character of Rumanian farming was in its turn partly offset by the widespread cultivation of corn, which caused a great rush of carting, demanding many teams, in July. Taking all in all, there is no doubt that the excessive number of teams in Rumania was due to latifundiary cultivation, to make up for the distance at which the various labours were carried out. And as all agricultural economists agree that labours performed even at a moderate distance are apt to swallow up the whole profit, it is undeniable that 'latifundiary cultivation is uneconomic and that it forms an impassable obstacle to the intensification of farming'. For in the measure in which farming becomes more intensive the various agricultural labours become heavier and more frequent, the continuous care of the crop a necessity, and the masses to be transported more bulky. Hence, such intensification was barred as long as farming was carried on at a great distance, i.e. as long as that form of 'nomad cultivation', as M. Garoflid called it, which kept men and beasts during weeks on end in the fields, continued.

The obnoxious effects of that form of cultivation were indirectly established by certain hypothetical calculations undertaken by M. Garoflid. If that form of cultivation had disappeared and the average number of animals had become the same as in France, 958,300 animals would have been sufficient for the needs of Rumanian agriculture. That would have meant an economy of 900,000 animals, which were simply used to overcome the distance, and would at once have set free, either for crops or for cattle breeding, the whole of the area used for the feeding of that excess of teams. The whole system was made worse by the long shape of most Rumanian estates, a heritage from the old pastoral farm. On an average estate of 2,500 ha., the distance from the home farm to its extremities would have been about 12·5 kilometres. Assuming that no more than 8 kilometres had to be covered for each kind of labour, that the normal speed of an ox-team is 2–2·5 kilometres per hour, and that the summer-day lasts about fourteen hours—it would appear that, with moderate halts for resting and feeding, there could have been left only four to five working hours each day. One must add to that the very bad state of the roads, so that the animals—as an earlier economist, M. Maior, had pointed out—'arrive half tired at the place where they have to begin their labours'.

These conclusions were evidently applicable to all agricultural labours performed under such conditions. How, then, could the system persist? 'It is profitable only because the whole of such labour is performed with the animals, the implements, and the hands of the peasants. *This* is the *technical means of production* of large-scale cultivation; and the whole loss resulting from the uneconomic organization of labour falls entirely upon the peasants.' The peasants were in need of land, and the large cultivators entered into agricultural contracts only with those peasants who had teams of their own. The inquiry conducted by the Ministry of Agriculture in 1899 established that 92 per cent. of all the draught animals belonged to the peasants and 8 per cent. to the large cultivators; while 93·7 per cent. belonged to the peasants and 6·3 per cent. to the large cultivators according to the statistics of 1903. Likewise, the inquiry of 1907 found that the peasants owned 92·67 of all the ploughs and 95·8 per cent. of all the carts while the large farmers owned merely 7·4 per cent. of the ploughs and 4·2 per cent. of the carts. As, therefore, most teams and implements belonged to the peasants, the loss resulting from their uneconomic use fell in the same proportion upon the peasants, too.

On any kind of calculation it was evident that large-scale cultivation would have disappeared long ago if it had been carried on with its own teams. As the large cultivators exploited 2,083,000 ha., they would have required 189,000 teams of four oxen each, on the average of one animal to 2·7 ha. Reckoning the cost of upkeep for a team, together with interest on capital, at 851[1] lei yearly, the total yearly cost would have been 160,000,000 lei. On the other hand, the cost of the various labours for the cultivation of the above area, including the carting of the crops to the railway station, amounted at the rates current when M. Garoflid's book was written to 41 lei per ha., so that the total outlay of large cultivation for labour reached 85,403,000 lei. The difference between the two gross sums would have been nearly equal to all the rental value of the area covered by large-scale cultivation. No doubt, if the large farmers had kept their own teams they would have reduced the total number. M. Garoflid pointed out that the carting of the crop, which required 'the maximum of labour in the minimum of time', determines the number of teams and the area they work. And he demonstrated by careful calculations that if the large farmers had kept teams of their own merely for the ordinary labours, carting being done with additional hired teams, the cost of the carting alone would have been about 24 lei per ha.; whereas under existing conditions the latifundiary form of cultivation could only afford about 6 lei for that purpose. The difference would have sufficed to swallow up its whole profits. Indeed, 'carting with their own teams would alone suffice to destroy the excessively large undertakings, as the remaining profit of 9·34 lei per ha., reckoning the other labours at contract rates, is not sufficient' to cover interest on capital, profit and risks. And if that were true of wheat, it was truer still of barley and oats, as the cost of production was about the same, whereas the gross value per ha. of the produce was much lower. 'If nevertheless our soil produced millions of hectolitres of corn in such uneconomic conditions of labour, it is because the whole loss resulting from their cultivation fell upon the peasant.'

The origin of latifundiary cultivation resided in the relative prosperity of the Rumanian peasants till almost the end of the eighteenth century, a prosperity which found expression above all in the abundance of live stock. Had the introduction of corn growing not found the peasants thus equipped with teams and ploughs, latifundiary cultivation could not have come into being.

[1] All these figures refer to the pre-war rate of currency.

Later, in their foolishness and greed, the large owners and tenants overreached themselves and almost destroyed the teams which supported their exploitation by constantly raising the price of grazing. But the measures for the establishment of communal grazings, which were meant to help the peasants, really saved the extensive cultivators; they came in the nick of time to enable the peasants' animals to exist and, in their turn, to keep latifundiary cultivation going. The system of cultivation by contracting with peasants who owned their own teams was so old and widespread that the problem of distance never troubled the large farmers. They left all the labours in the care of the peasants, just as they left the soil in the care of nature. That state of mind was illustrated by the case of a cultivator who farmed an estate 20 kilometres long. He had contracts with villagers living at both ends of the estate, and in order to guard his crops against pilfering, which was easier nearer the home of the labourers, he gave to each of the two villages land at the other extremity of the estate; so that the villagers had to travel about 40 kilometres to and from their labours.

It will be seen that these large agricultural undertakings were comparable only in part to big industry. No big industry could raise its profits by debasing its equipment as a matter of continuous policy. But in the latifundiary farms the instrument of production was not owned by capital, and that dissociation enabled the one to make a good living from the depreciation of the other. The latifundiary farm had the form of a capitalist undertaking of production, but not its ends. The expansion of a large industry was limited by the possibilities of the market; that of a latifundiary farm was limited solely by the degree to which the instrument of production could be depreciated. Hence these undertakings could adopt the purest system of extensive cultivation, as their profits increased arithmetically with the area of the exploitation. In contrast to the evolution of farming everywhere else, Rumanian farming was actually showing a tendency towards concentration. The number of the large capitalist farms decreased from 1904 to 1906, but the total area they exploited increased from 1,934,317 to 2,206,263 ha., and the average extent from 357 to 470 ha. The system created a class of capitalist entrepreneurs, as one may see from the fact that tenant farming rose in direct relation to the size of the undertaking. This is shown by the figures for 1905 relating to farms above 100 ha.

	Tenants	Owner-farmers
Number of farms	51·8%	48·2%
Total area	61·6%	38·4%
Average area	526 ha.	384 ha.

The creation of enormous trusts proved that the system was profitable.[1] It multiplied its profits not by raising the productivity of the soil or by creating accessory agricultural undertakings, either of which would have required much capital and effort, but solely by the stringent restriction of outlay and the constant depression of the value of labour. In years of crisis no attempt was made to meet the situation by improving production, but simply by reducing expenses all round, especially for labour. In 1894 and the following years, when the fall in corn prices caused a serious fall in profits, the conditions of agricultural contracts became suddenly and materially harder. Since 1850 the rent of land had risen tenfold and more, but the price of labour had remained stationary. As production did not keep pace with the increase in rent, that increase could have taken place only at the expense of labour. With better production rent might have been as high without such a severe depreciation of labour's reward. But in Rumanian agriculture, as M. Garoflid said, rent did not represent merely a compensation for the use of the land. The soil, like the labourer, was being 'sweated'. 'Together with corn we also export soil—therein lies the whole secret of our agricultural system, which enables us to compete with the corn produced by western agriculture with the help of manures.'

This sketch of a system whose curious structure and workings would deserve a more detailed description, is fully supported by the general figures of Rumania's agricultural statistics; as well as by the more instructive, though painfully few, special inquiries. One of the best was the inquiry made by M. Șerban

[1] That of the brothers Fischer, in Moldavia, rented 138,424 ha. in 1903 and 159,399 ha. in 1905. It paid a total rental of 3,411,343 lei =£136,544 yearly.

into the detailed organization of twenty estates, immediately before the War; incomplete data referring to eighty other estates being used in checking the conclusions. The estates were chosen by M. Şerban so as to represent a variety of geographical regions and of types of farming; they included one big Crown domain and two model estates, and altogether they were of a better class than the majority. The area farmed by the owners themselves was above the general average, and so was the quantity and quality of the live and dead stock. Whereas these estates, e.g., included merely 2 per cent. of the total area covered by large property, they possessed 10 per cent. of the number of motor ploughs in the country.

One cannot give more than a summary of the results arrived at by M. Şerban's inquiry here. It established that the total capital invested in these undertakings averaged 94·01 lei per ha.; this fell much below the minimum of a number of German inquiries, which found averages running from 206·25 lei to 1032·50 lei per ha. The Rumanian figure was divided as follows: 26·15 lei for live stock, 27·41 lei for dead stock (taken at its original cost), and 46·44 lei working capital (cash only), representing respectively 2·62 per cent., 2·75 per cent., and 4·65 per cent. of the total capital value of the undertaking. These proportions were inverse to the norms prevailing elsewhere, the value of live stock in the more developed countries being generally twice as high as that of the dead stock, and the working capital about one-half of the real estate.[1] Again, the buildings on the above estates represented 8·6 per cent. of the total real estate (land and buildings), which compared badly with the German figures of 25–50 per cent. The value of improvements on the Rumanian estates only amounted to 0·3 per cent. of the real estate. It must be repeated that M. Şerban had picked out exceptional estates, as his figures are higher than those of other inquiries.

[1] On the Moldavian estates direct cultivation was more general and they displayed in consequence a better equipment. In Moldavia the value of the draught animals was 21·76 lei per ha. and in Muntenia only 8·72 lei. This was confirmed by the higher proportion of working capital in Muntenia; it did not prove a more active exploitation, but, on the contrary, a greater reliance on hired teams and implements, whence resulted a greater need for cash.

As regards labour, M. Șerban established an average of 10·5 permanent labourers for each hundred ha. of cultures. For purposes of comparison one must take into account the quality of that labour; the execution of the various labours for the growing of one ha. of wheat required 6·737 days (of ten hours) in Germany and 10·2 days in Rumania. On the moderate assumption, therefore, that the qualitative relation of Rumanian to German labour was as 8 : 10, the Rumanian average fell to 8·4 permanent labourers for 100 ha., as against a German average of 16·57. The Rumanian figures remained even below those calculated by Thaer for Germany more than a hundred years ago. And, in fact, certain large estates were found by M. Șerban to be worked with 4·22 permanent labourers for 100 ha. of cultivated surface. In latifundiary cultivation the farmer had no interest in employing more labour if this raised production merely by an amount equal to the cost of that additional labour; nor had he any interest in repeating labours solely for the purpose of distributing his fixed yearly outlay over several of them. That would have brought him no special profit, though it certainly would have been a gain to national economy.

Finally, on none of these estates was any use being made of the farmyard manure. Artificial manures were quite unknown.

These facts and figures indubitably proved that the large farms were poorly equipped with dead stock and worse still with live stock; and that in general they were being worked on a low margin of capital, the average level falling much below a rational minimum. To make the picture more accurate one would have to go beyond figures and describe how grossly inadequate the farm buildings e.g. generally were. As much of the harvest was sold at once no effort was made to build proper and sufficient barns; and animals were, as a rule, poorly housed, notwithstanding the severity of the Rumanian climate. M. Șerban's practical inquiry produced, therefore, the same conviction as M. Garoflid's theoretical discussion: that if the large farmers 'can nevertheless maintain themselves, this can be explained only by the great disproportion in economic strength, by the aid of which they can advance their interests and exploit, in the worst sense of the word, the means of production—the working powers of the peasant cultivators'. The mass of those large farms 'do not represent organized economic values; and for that reason they are an obstacle to effective progress'.

M. Şerban attempted also the more difficult task of studying in figures the organization of the smallholders. Calculations were hampered by the fact that the peasants did not use their stock solely for themselves, that the area which they cultivated varied from year to year, and that often they possessed draught animals and implements without owning any land at all. In their case, therefore, any inquiry into individual households would have been too vague. M. Şerban chose instead six groups—three from Muntenia and three from Moldavia—which together included seven large and five smaller villages; as well as one example from the mountains. The date of this inquiry is somewhat earlier; hence, the figures mentioned below should be regarded as minimum values, whereas the figures relating to the large farms represented rather maximum values.

M. Şerban worked out averages for these groups and obtained the following results:

		Lei	Lei
Real estate—Value of land	per ha.	535	
Value of buildings	,,	120	
		—	655
Investments—Live stock	,,	177	
Dead stock	,,	53	
		—	230
Working capital	,,		105

He concluded, therefore, that the average capital investments of the smallholders were much higher than those of the large farmers. This was especially so for live stock, the value of which was about nine times higher (including the value of the draught animals which, of course, were also used for labour on the large farms). The value of dead stock and of buildings was likewise higher, if not in the same proportion. The actual cash which in the case of the large farmers had amounted to 44 lei per ha., was only 8·50 lei per ha., with the peasants; they had no outlay for wages and for the hire of teams. Moreover, the value of live and dead stock was above the minimum indicated in the German inquiries and nearer their middle averages; and in the case of the peasants the relative higher value of live stock than of dead stock was in line with the normal conditions ruling generally in the more developed countries.

These several studies showed that before the reform the peasants owned the invested capital of Rumanian agriculture, while the large owners and tenants held the liquid capital. Hence the two categories complemented each other in many respects, and indeed constituted only jointly a full agricultural unit. This proposition is supported by the figures obtained in the two inquiries described above. The averages resulting for each of the two categories of cultivators differed considerably from the averages for the country as a whole, secured by a quite different method. But if one took the two categories together and worked out joint averages for both of them, the approximation between the figures of the special inquiries and the national averages was remarkably close.

Conservative and Radical economists were, therefore, generally agreed that it was economically unsound to continue the system of latifundiary cultivation. In 1908 M. Garoflid had written that the very nature of the system excluded any improvement in production or the establishment of accessory agricultural undertakings; and that it was compatible only with the cultivation of two or three cereal crops, easily stored and easily sold. It prevented specialization and it demanded the growing of the same plant in all the regions and in all the soils. And in 1920, in the Memorandum already quoted, he again admitted that with the bringing of the available land under the plough, the function of extensive cultivation had been consummated. 'Extensive large-scale agriculture had borne all the fruits of which it was capable. Beginning with the twentieth century its part in the evolution of our national economy had come to an end, as the shepherds had become cultivators. A change in the agricultural system was indispensable, but for that end the latifundia had to be reduced.' Under the existing system, the peasants had no chance and the large cultivators no interest in raising production.

M. Garoflid's final conclusion, therefore, was that the crisis in Rumanian agriculture expressed a problem of production rather than one of the distribution of property; and on the strength of that conclusion he pleaded against the extermination of the large owner. But just as his earlier conclusions had been ignored by Rumania's old rulers, so his plea was fatally destined to meet with little response from the class which had become politically influential after the War. To the argument that large-scale cultivation was still necessary M. Mihalache, the Peasant leader, retorted that they had their own experience of several centuries to look back upon, and that they could not overlook it and pay attention rather to what was happening elsewhere. They could not ignore what had been in the past and take into account what might be in the future. The past history of large cultivation in Rumania 'is for us one more reason why we should abolish it altogether, with the exception of the strictly essential model farms'. 'We cannot accept the liabilities of large property as assets when we are about to build up a new agrarian régime.'

The better equipment of the peasantry had been mentioned by Ion Ionescu, as early as 1869. He found, e. g., during his inquiry into the Putna county that the peasants had ten times as many draught animals and ploughs as the landlords. Since that date, cultivation in general, and large-scale cultivation in particular, has expanded enormously, but the supply of the technical equipment has remained as much in charge of the peasants as before. The estimate of 1890 attributed to the peasants 92 per cent. of the big animals, 92·6 per cent. of the ploughs, and 95·7 per cent. of the carts in the whole country. The census of agricultural machines and implements taken by the Ministry of Agriculture in 1905 established a similar or worse proportion:

	Total number	Large owners	Peasants	Average no. of hectares to one machine
Ploughs	519,463	44,720	474,743	10·58
Harrows	448,260	44,728	403,532	12·22
Carts	614,272	24,964	589,308	9·71
Cutting machines	18,451	7,521	10,930	
Seed-cleaning machines	29,461	8,801	20,660	165·87

There were in the country fifty-five motor ploughs and fifty-seven machines for the spreading of artificial manures, all belonging to the large owners.[1] M. Şerban estimated the total value of the dead stock, in 1913, at about 311,000,000 lei, or 38·88 lei per ha. of cultivated land.

The census taken in 1900 established a similar relation with regard to animals. It found them to be distributed among the various categories as follows:

Categories	Horses	Oxen	Sheep	Pigs
	%	%	%	%
Large owners . . .	4·8	7·8	10·9	8·3
Small owners . . .	76·7	80·3	70·8	76·9
Peasants without land .	8·1	6·7	5·2	6·5

Statistics of agricultural buildings have never been collected. The 1912 census merely stated that there were in the 2,620 rural communes 1,303,828 buildings, of which 1,216,411 were inhabited. Their average value was estimated by M. Şerban at 100 lei per ha. of arable land, not including public elevators, steam mills, sugar distilleries, &c. Another inquiry M. Şerban conducted into the equipment of fifty estates varying in size and kind, and including together 3 per cent. of the large property, found that the total outlay for buildings had amounted to 6,570,000 lei or 63·62 lei per ha. We have seen from the inquiries to which we referred earlier in this section that the superiority of the small owners held good in regard to buildings and to agricultural capital as well. In every respect, therefore, the means of production in Rumanian agriculture were, absolutely and relatively, to an overwhelming degree in the hands of the peasants and not of the large cultivators.

All these circumstances rendered the question of productivity in Rumanian agriculture extremely perplexing. In general, peasant farming gave lower average returns than large-scale cultivation. Seeing that the bulk of the large farms were worked

[1] M. E. Giurgea points out in his studies on Bessarabia (*Buletinul Statistic*, 1919, No. 2) that large property prevailed in the northern counties, Hotin and Bălţi, whereas agricultural machines were most numerous in the southern counties, Tighina and Cetatea Albă. To some extent this may have been due to the more level lie of the ground, but, 'above all, to the presence of extensive French and German peasant colonies' in the southern counties.

by the peasants, with their own teams and implements, how was it that the same men using the same methods obtained worse results for themselves than when they worked for employers? The question is evidently still more important for the future of Rumanian agriculture than it has been for its past; it is therefore worth our while to consider briefly the reasons which caused that difference in the returns of large and small cultivation, in order to see whether they were of a permanent character or were merely the accidents of temporary and remediable conditions.

Once more it is necessary to make some comment first on the validity of the figures which will be given below. A distinction between the returns of large and small cultivators was not made in Rumanian agricultural statistics before 1904. From that date the figures relating to farms of less than 100 ha. were registered separately from those concerning farms of 100 ha. and more. No accurate estimate of the productivity of smallholdings as such, i.e. of holdings up to 100 ha., is therefore possible. Nor are the figures, even with that limitation, altogether reliable. Those which refer to peasant cultivation were collected by village officials, inevitably in a somewhat rough and ready fashion. The figures for large-scale farming were based on returns supplied by the farmers themselves, and it has been alleged that they frequently declared a higher production for the purpose of obtaining large credits, and a smaller area for the purpose of paying less land tax. But such as they are, these are the best statistics available.

In the West the production of smallholders has been everywhere showing higher proportional returns than that of the large farmers. But in Rumania, during the decade which preceded the Great War, large farms gave a yearly average per ha. which was higher than that of smallholdings by 13·1 per cent. in the case of wheat, 18·65 per cent. of rye, 15·9 per cent. of barley, 19·4 per cent. of oats, and 19·5 per cent. in the case of maize. The opposite tables give in detail the average production of large and small cultivators per ha., in hectolitres, for the years 1906 to 1915.

Small cultivation produced 25·3 hectolitres more buckwheat in the second period; 74·3 metric quintals more flax in the first period; 8·7 hundred more cabbage heads in the second period; 16·7 metric quintals more of other vegetables in the second period.

One point which strikes one at once on looking through this table is the considerable variation in returns from year to year, which happened to both categories of farming alike. Remember-

Crops		1906	1907	1908	1909	1910	1911	1912	1913	1914	1915
1. Wheat	L.	21·2	9·2	10·9	13·0	21·6	17·6	16·1	19·2	9·2	18·5
	S.	18·3	8·1	10·6	10·5	18·3	16·6	14·1	17·1	7·4	14·9
2. Rye	L.	18·7	7·4	8·3	10·5	12·3	16·0	14·4	17·1	9·1	16·2
	S.	16·8	6·1	6·1	7·5	15·3	13·1	11·5	14·2	8·1	13·3
3. Barley	L.	23·0	16·6	10·1	14·7	21·3	19·7	18·0	19·6	17·4	19·6
	S.	20·5	13·0	6·3	12·1	18·0	17·7	14·1	16·3	15·4	17·8
4. Oats	L.	26·8	21·0	13·9	22·4	25·7	26·5	23·2	27·2	23·3	26·9
	S.	22·4	16·0	11·5	17·0	22·2	21·3	17·4	21·7	19·4	22·1
5. Maize	L.	26·3	13·2	15·4	14·1	22·6	22·5	21·3	21·4	20·8	16·8
	S.	20·7	9·9	13·5	11·2	17·7	18·1	17·1	18·5	17·1	14·2

		Yearly averages for the periods		Minus difference of small cultivation (per cent.)	
		1906–10	1911–15	1906–10	1911–15
1. Wheat	L.	15·2	16·1	13·2	13·0
	S.	13·2	14·0		
2. Rye	L.	12·8	14·6	19·5	17·8
	S.	10·3	12·0		
3. Barley	L.	17·1	18·9	18·1	13·8
	S.	14·0	16·3		
4. Oats	L.	22.0	25·4	19·1	19·7
	S.	17·8	20·4		
5. Maize	L.	18·6	20·6	21·5	17·5[1]
	S.	14·6	17·0		

ing the methods of cultivation described in the preceding pages, when it was said that the crops were left to the care of nature, it is evident that such wide discrepancies were due to climatic conditions which took no account of the size of farms. But given this primitiveness in the nurture of the crops, and the fact that the bulk was raised in the case of large cultivation as in the case of small by the same men with the same animals and implements, how—once more—is the difference in results to be explained?

[1] Ministry of Agriculture, *Statistica Agricolă pe Anii* 1911–15, Bucarest, 1918. pp. 82–3.

The only possible explanation is that the peasants were economically and socially labouring under such oppressive disadvantages that they could use only a residue of the factors of production in their own interests.

Their first and most serious disadvantage lay in the quality of the land which they tilled for themselves. When describing the reform of 1864 we quoted the evidence of Ion Ionescu as showing how, on becoming full proprietors, the landlords took away the good fields which the peasants had been cultivating till then and gave them instead the poorest land on the estate. There is not a writer on agricultural economics who does not produce some fresh example of that form of abuse. The truth of the complaint was proved by the kind of land which the landowners offered in pursuance of the law of 1907 for the creation of village commons. The tide of public sentiment was beginning to turn in the peasants' favour, yet the land proffered was so bad, that much of it had to be refused outright, while most of it proved too poor to be planted with fodder crops, as the law demanded. The land let to the peasants was always the worst, and as they never got it for more than one year, there was no inducement to improve it. It was the custom of large owners and tenants to give land in *métayage* for maize and other crops which required hoeing and ridging, and on the land thus cleared of weeds to plant wheat for their own account the year after. The bad quality of the land was made worse by the disjointed nature of the peasant holdings. Most of them consisted of a number of strips scattered in different directions; and the peasants showed a determined bias for life in compact villages, especially in the plain, while their fields usually lay several kilometres away from the house and farm-yard. Finally, a majority of the peasant holdings were too small to form sound economic units.

It is true, however, that peasant cultivation was as a rule even more careless than that of the large farmers. Because their holdings were generally insufficient, the peasants depended for their existence on getting more land from tenants or landowners, in return for a contract to labour for them. In Muntenia 30·1 per cent. of the peasants were left with less than forty-seven days for their own labours, and in Rumania 8·5 per cent. with only twenty-seven days and 25 per cent. with forty-two days in the year. The average peasant holding of 3·42 ha., planted with the customary peasant crops, required a minimum of 78·89 days' labour alone for the cereal crops, without taking into account the care of animals, &c. Now the cereal system in such an uncertain climate as Rumania's gives to agricultural labours a concentrated character, demanding i. e. the greatest possible amount of labour in the shortest possible space of time. Large owners and tenants were able to mobilize the whole village when the moment was propitious for the performance of this or that labour for which the villagers had contracted. By the time the land of the large farmers was tilled, the weather had as likely as not changed for the worse. And even if heaven was kind, the peasants came to their own fields with tired hands and tired beasts, so that their work was done in a hurry, with such drooping strength as was left in them. Moreover, in such a hot climate a few days generally suffice to ripen the corn, and a few more to over-ripen it; late cutting, therefore, always meant a considerable wastage of seed, estimated by a recent inquiry to reach almost 30 per cent. of the total crop. One must remember also the perpetual state of underfeeding and of chronic ill-health in which most of the peasants lived, in order to weigh rightly what effect the placing of their own work at the tail-end of each season's labours had upon their farming. The number of individuals capable of work between fifteen and sixty years, was as follows, per 1,000 rural inhabitants:

Rumania	534
Germany	569
England	574
Austria	579
France	613

According to ability and power of work, M. Şerban had put the effective working days of a Rumanian peasant at 115 per year, which multiplied with the above proportion gave a total of 707,500 effective working days in the year. This compared with the German peasant's working contribution of 200 days yearly, or a total of 1,707,000 days which for France even reached 1,840,000 days.

All these circumstances warrant the conclusion that the

inferior production of the peasants was attributable mainly to reasons which had little to do with the type of the agricultural unit. One finds the first proof of this in the more careful performance of agricultural labours before the reform of 1864, which was supposed to have advanced Rumanian agriculture, but which has certainly completed the ruin of the peasants. In his three careful and detailed inquiries Ion Ionescu repeatedly mentioned with regret the deterioration of agricultural methods. Up to that time, of course, there was little else beyond peasant cultivation; yet Ion Ionescu affirmed that, generally, the corn was very clean, because of the habit of alternating wheat, oats, and other corn crops with maize, which cleaned the ground of weeds, as maize required careful hoeing and ridging. He referred to the peasants' traditional conviction 'that only wheat which had well ripened should be used for seed. It was specially selected and left to stand longer before being cut. And the wheat thus selected was treated with special care when it was cut and tied and threshed and carted.' But extensive cultivation for immediate profit, by tenants holding the land on short leases, corrupted those time-honoured habits. Wheat was sown over enormous extents, and year after year on the same surface, without any manuring at all. It is important to note, therefore, that the difference between the output of large and small cultivation at present is much narrower than the difference between the erstwhile productivity of the Rumanian soil, when it was almost wholly in the hands of the peasants, and the results obtained after a period of extensive cultivation on a large scale. Dr. Maior wrote in his *Manual de Agricultură Naţională* (1895) that the production of wheat had fallen from an average of 20 hectolitres per ha. to 15, and later to 12·13 hectolitres. There was, moreover, a similar decline in the relation between the quantity of seed used and the total crop. Once upon a time the crop had given 24 times the quantity of seed, but the return had fallen to 15 times and finally to 4·04 times when Dr. Maior was writing. Further, the loss in quantity was accompanied by a loss in the quality of the produce. With few exceptions, the latifundiary large-scale cultivation was the purest 'Raubwirtschaft'. To complete the picture of its influence on rural life one should also mention the decline or decay of many by-products of agriculture. Not only had nothing been done by way of improving life on the land since 1864, but once widespread and flourishing domestic industries—like the keeping of bees and the growing of silk-worms—have totally disappeared. The quality of animals declined to such a point that four, six, or even more oxen were needed to pull a light plough.

Another set of facts which tend to establish the incidental character of the deficiencies observed in peasant farming consists of significant exceptions in Rumania itself. In the neighbouring Rumanian provinces, now united with the mother-country, where the peasants were economically more independent, their farming was correspondingly better. Especially in Transylvania and Bucovina the peasants were generally better off even on smaller holdings than the average in the Old Kingdom. The flourishing colonies of Saxon and Suabian peasants in Transylvania, and the German, French, Bulgarian, and other colonies in southern Bessarabia and Dobrogea, offer convincing illustrations of what peasant farming might achieve even in the eastern regions, their production being fully equal to that of the large cultivators. Nor were such exceptions lacking in the Old Kingdom itself. The figures from which the above table was compiled show that in certain districts of the Muntenian highlands the difference in the average production of wheat was negligible, falling to 0·7 hectolitres in the Prahova county and to 0·1 in Muscel. In 1907 peasant farming gave a higher average production; but 1907 was an abnormal year, in which a considerable part of the large estates remained untilled. The same phenomenon repeated itself in 1913–15 and in 1919, all of them years of abnormal conditions on the land, because of wars and of the reform. These periodical exceptions, therefore, are not valid evidence for estimating the productivity of the two forms of cultivation. But they are all the more striking as an illustration of the helpless state in which large-scale cultivation found itself when circumstances deprived it of the peasants' assistance.

Less hypothetical were the examples described by Dr. N. Lupu, during the discussion of the reform at Jassy. They showed that the peasants could easily outstrip the achievements of the large cultivators when they joined hands for co-operative farming. The co-operative which took over the Bordeiu Verde estate was a case in point. The former large tenant had paid a rental of 120,000 lei; the peasants offered 214,000 lei yearly. In spite of that high rent they did extremely well. Buildings, machines, implements, animals were more numerous and of better quality than before. The co-operative was able to invest 100,000 in War Loan and had reserves amounting to 200,000 lei.

'That cannot be explained except by an enormous increase in production to the hectare. Indeed, according to figures which have been checked, the highest average obtained by the larger cultivator in the best years was 24 hl. per ha., whereas in the best years the co-operative raised on the same estate 40 hl. per ha. The co-operatives increased the number and improved the breed of animals. Neighbouring landowners were bringing their mares and heifers to be served in the stable of the peasant co-operatives.[1]

In brief, it will be seen that the peasants had had none of the qualifications required for good farming. They had neither general education nor special training; they were underfed and overworked; they were burdened excessively with taxes and impositions, but aided scantily or not at all with credits, &c.; above all, they were left with the worst land, and with the worst time for tilling it. And when all that is said, it still remains to be pointed out that a comparison of large-scale production is bound to be misleading if it is limited to cereal crops. These form a speciality of the large farmers, especially in a system of extensive cultivation. But on an equal area the quantity of animal and dairy products which the small cultivators put on the market more than makes up for any deficiency in their corn crops. In Rumania, indeed, the growth of large-scale cultivation has not meant the development of agriculture as a whole, but merely the one-sided expansion of wheat at the expense of rural activities, traditional and beneficent. The change depressed the economic position of the peasant, and social and political oppression totally ruined him. But it was an inferiority of the peasant's class rather than of his type of holding. There is no reason why the latest social and political reforms, when they get fairly under

[1] *Monitorul Oficial*, July 27, 1917, p. 374.

way, should not have the contrary effect. The peasant does not know what it means to stop working. He is all the while seeking to cultivate the greatest possible extent of land because he only knows extensive cultivation. It is merely the lack of knowledge that prevents him from putting the same amount of labour into a smaller extent of land, so as to obtain from it larger returns; for that is the problem which concerns him above all—that he should be able to extract from his land the food which he needs for himself, his family, and his animals. Even the exponents of nationalist economics agree that in this case the past should not be allowed to prejudge the future. The Liberals opposed the partition of the land precisely on the ground that small cultivation was less productive; but in the Senate, in 1923, M. Vintilă Brătianu expressed his conviction that after a few years, when properly settled and equipped, the smallholders would produce more than had the large cultivators in the past.

It is indeed doubtful apart altogether from the effects of the land reform, whether latifundiary cultivation could have lasted any longer. We have explained the circumstances which had enabled it to command the peasants' labour, and why this was essential for its existence, as nearly all the instruments of production were in the peasants' hands. This situation, which had never been different throughout the history of Rumanian agriculture, was accentuated by the War, whose effects made the large cultivators still more dependent on the peasants' means of production. The loss in labour and implements and animals, caused by the War, was bound to affect in the first place those cultivators who had no equipment of their own, unless they were in a position to conscript for their own use such means of production as were still available. And we have seen that, in fact, the first thing which the Marghiloman Government did after the Peace of Bucarest was to oblige the peasants to work for the large farmers. Such an imposition was no longer possible at the end of the War.

In addition to the loss of 800,000 men, there was a general loss in working power. The remaining labourers were no longer capable of the same effort as before. As everywhere else, the nation's manhood was physically and nervously tired, and the masses were in a mood which suffered no further obligations without adequate reward. The loss in working animals was even more severe; they were greatly reduced in numbers, especially in the regions which had been under enemy occupation, and they were badly weakened through poor feeding during the War. The number of horses was reduced by half, that of the large horned animals and of pigs to two-thirds, and the number of sheep to one-half. The new provinces had suffered less, as they had remained outside the battlefields. The War had played a similar havoc with machines and implements, which had been used with little care and without being renewed during the period of hostilities. All over the country, old implements long discarded were summarily repaired and pressed into service again; as had been the case with the primitive all-wooden plough from Bucovina which was on view, during the past winter, in the modest but excellent exhibition arranged by the Sociological Seminar of Bucarest University. For while after the War the need for new machinery was great, even the most essential implements were not easy to replace. Their price had increased considerably, and, on the other hand, the fall in the exchange and the general shortage of money among large and small cultivators alike, not to speak of a short-sighted tariff policy, hampered the purchase of these foreign products. The large owners, especially, who had lost much land but got little money in return, found the problem of equipment difficult to solve. This was even more true of agricultural machines. To the difficulties already mentioned there was added that of personnel, mostly foreigners, who had been employed to drive and repair such machines but who had been scattered by the War. Even when credits could be obtained, the farmers feared to incur foreign debts because of the continuous depreciation of the currency. To some extent the deficiency was made good by production from the factories which were already in Transylvania. But the tables below—the first of which gives comparative figures of agricultural machines imported before and after the War and the second the size of that production—show that the replacement of the dead stock advanced slowly; especially if one takes into account that the post-war figures refer to a territory and to a population twice as large—which included Transylvania and Bucovina where the use of machines and implements was more developed—and to an agricultural situation in which the introduction of intensive farming had become imperative after the agrarian reform:

Import of Agricultural Implements and Machines

Year	Quantity in Kgs.	Value in gold Lei[1]
1885	507,694	652,029
1890	1,136,486	1,136,486
1895	721,552	721,552
1900	967,660	967,660
1905	1,283,733	749,637
1910	12,272,707	11,753,074
1919	3,076,344	687,812
1920	2,610,780	861,562
1921	3,325,691	1,173,500
1922	3,742,251	1,598,000
1923	3,119,292	3,376,938
1924	2,140,506	2,464,687
1925	3,841,211	4,631,312
1926	5,681,453	10,306,437

Internal Production of Agricultural Machines and Implements

Year	Number of Factories	H.P.	Value of the Factories[2]	Value of Production	Number of workers and employees
			Gold Lei	Gold Lei	
1922	17	1,247		1,360,600	1,909
1923	21	1,256	1,771,373	1,878,000	2,514
1924	19	1,467	1,263,222	2,420,469	1,976
1925	26	1,525	2,357,408	3,449,469	1,903
1926	28	1,309	3,653,001	3,407,687	1,874
1927	37	1,640	4,121,875	3,645,900	2,073

Generally speaking, therefore, because of the loss in labour, in animals, and in implements, latifundiary cultivation could no longer have disposed of the cheap peasant labour after the War.

[1] The post-war values are calculated at the average rate of 800 lei to the pound.

[2] The factories dealt with in this table produce other goods besides agricultural machines and implements. The general figures refer to these undertakings as a whole, except those giving the value of production, which refer to agricultural machines and implements alone.

On the other hand, it would have found it extremely difficult to acquire an inventory of its own, with the prevailing high prices and shortage of money. Even without the agrarian reform that type of cultivation would hardly have been practicable again. At the end of the War large-scale cultivation found itself almost wholly divested of live and dead stock and was in danger of being unable to plough the land it possessed. More than one Rumanian economist considers that the immediate transfer of land to the peasants has saved Rumanian agriculture from disaster. In the face of a host of natural and artificial obstacles, which will be discussed later on, the small cultivators have been able to revive agricultural production in a relatively short space of time. Moreover, they have succeeded in these few years in restoring the country's live stock almost to its pre-war size—an achievement which would have been altogether beyond the powers of the large-scale cultivators. Hence, one is led to the conclusion that the increase in smallholdings was as justified for economic, as it was necessary for social reasons.

Unfortunately, no information whatever has so far been collected from which one might draw a scientific conclusion concerning the effects of the agrarian reform on the technique of the various kinds of farming. One must be satisfied with such indications as personal observation and conversations have been able to give. The reform found the large cultivators altogether unprepared for the task with which it presented them. But just as the peasant emancipation in the West provoked an all-round improvement in agricultural methods, so in this case such large property as remains, and which is farmed to a larger extent than before by the owners themselves, displays a marked tendency to intensify its production. The higher cost of labour, &c., as well as the landowners' natural anxiety to recover at least some of the revenue they used to derive from the lost area by higher returns from such land as is left them, has provoked in most cases a more careful tilling of the soil. The depreciation in the value of agricultural products, as compared with that of manufactures, likewise made it necessary to produce relatively more of the former. There is also the fear lest the new spirit which is abroad and the new currents which dominate public life should no longer tolerate the retention of land property in incapable or indifferent hands. A suggestive article published in the *Frankfurter Zeitung* on November 9, 1927, described how a similar fear has brought back the noble landowners in Baden not only to residence on the land, but to farming with their own hands. All these circumstances combined to confront the landowners with a totally new problem, a problem which seems to admit of no other solution beyond either improving cultivation or selling out.

While the peasants were provided, even if somewhat inadequately, with means to cultivate the land which was given them, they had not the means and even less the training for the kind of intensive farming which their new state and the country's needs demanded. How the policy of the governing class took away from them all incentive to fresh effort, will be discussed later on. For the moment, one may note the statement of M. Ionescu-Sisești, now director of the new Institute of Agronomic Research, that the peasants work as well as they did before the reform, or rather that they do not work worse. That would seem to be generally true of Muntenia and Oltenia, as well as of the new provinces, but less true of Moldavia. There the cultural level has been lower and alcoholism higher on the land, and with the disappearance of the driving power of the large tenants, the peasants are apt to fall into slack ways. Everywhere, M. Ionescu-Sisești affirms, one can establish a parallel between the grade of culture of a group of villagers and the quality of their farming; a circumstance which suggests that the partial decline in the quality of peasant cultivation may be due not to the reform but rather to the failure of Rumania's former rulers to educate their masters. Another informant put it that many peasants farm better now than before, while some of them farm worse; this being especially true of what one may call the marginal beneficiaries of the reform, artisans and others, many of whom have land of their own now but no oxen.

The facts discussed in this and the previous section help us to put the economic sense of the reform in its proper perspective. To sum up, 83·3 per cent. of the rural householders were cultivating on their own account; and small cultivation produced about three times as much as large-scale cultivation. On both counts, therefore, the peasants formed the pre-dominant factor in Rumania's agricultural system. They dominated the organization of production still more. The custom of cultivating with animals and implements and seed belonging to the peasants, which had been universal during serfdom, had remained almost intact after their emancipation as well, the tithe having merely changed into métayage, i.e. a sharing of the crops. The peasants received a piece of land and in return cultivated another piece of land for the tenant or owner, or they gave him part of the produce, as they had done when they were his serfs. Landowners and large tenants supplied neither animals nor implements nor adequate capital—'the tenant takes to farming only with a walking stick', the peasants used to say—and only a disciplinarian sort of guidance. In no economic sense, therefore, were the landowners essential factors in the agrarian system, and that explains how it was possible to expropriate them so drastically without wrecking the wheels of production.

This was merely repeating the experience of the French Revolution. Max Weber remarks in his *Wirtschaftsgeschichte* that the Revolution found it possible to expropriate the landlords 'because the French landlord was not a farmer, but a courtier, seeking a career in military and civil functions, upon which he had, in a way, a monopolistic claim. Therefore no organization of production was destroyed by that act, but merely a rent relationship'. If one leaves out 'courtier', every word of that remark applies with equal force to the Rumanian land reform. Because the estates of Rumanian landowners, unlike those of most Czech and Polish owners, were not organized as units of production—because, in other words, the capitalist division between labour and the instrument of production had hardly begun in Rumanian agriculture—the disturbing effects which the reform might otherwise have had remained relatively insignificant. When the peasant was made an independent farmer he was not quite helpless, although the State made no attempt whatever to see him started on his new venture with a proper equipment. The agrarian reform, in brief, has meant an enormous legal change, but only a very moderate economic change. Production is, on the whole, carried on by the same men with the same means as before. Broadly speaking, it has been not so much a change from large-scale to small-scale farming, as a change from farming by small tenants to farming by small owners. 'The size of the holdings now corresponds almost exactly to the methods of farming', wrote a contributor to the Russian volume edited by Prof. Max Sering. 'While in 1916 large land-ownership went together with small-scale farming', now system of tenure and system of cultivation are in harmony with each other.

There is in such an evolution small matter for the gloomy forebodings which many western critics of the reform entertained; and equally scant reason for expecting that a mere change in the form of tenure would produce a sudden flowering of fine agricultural methods and harvests. The reform was bound to have a good and immediate psychological effect, calculated to overcome the nervous exhaustion left by the War, and we shall see how this has indeed contributed to the recovery of production; and security of tenure has always been found to induce a more careful tilling of the soil. In the measure in which any fresh tendencies are discernible at all they serve to confirm that experience. There is evidence that the large farmers are adopting more intensive methods of cultivation, in the endeavour to balance higher costs with higher returns; and that more intensive methods are employed by the peasant farmers as well. The advance is slow and somewhat erratic, which is in the nature of the case, being the result not of a systematic policy but rather of the energy, ability, and means of individual peasants, and of local variations in leadership and opportunity. It is, therefore, probably true to say that formerly there were more differences between various regions, whereas now the differences are rather between villages and individual peasants, the greater freedom of movement encouraging a natural process of selection; and, as a general observation, that more intensive methods are spreading all-round among the peasants, automatically, so to speak, because of the smallholders' well-known inclination towards raising crops which require more effort but yield stronger harvests.

CHAPTER IX

THE EFFECTS OF THE REFORM ON PRODUCTION

Section 1

The Change in the Distribution of Crops

Any attempt to estimate the effect of the land reform on production is further complicated, as we have pointed out, by the change which has taken place in the distribution of crops. The peasant emancipation has everywhere had the effect of increasing the cultivation of industrial crops and grasses, and of other crops which are profitably grown on a small scale and are better adapted to a rural economy based largely on the keeping of animals.

In Rumania, after the abortive emancipation of 1864, which freed the land but left the peasants tied to it, agriculture showed two characteristic and concordant tendencies. Wheat cultivation expanded rapidly, and that was accompanied by a further decline in cattle breeding. The total area cultivated, including grassland, rose from 2,963,940 ha. in 1862 to 5,850,950 ha. in 1900, and wheat cultivation from 697,220 ha. in 1862 to 1,600,000 ha. in 1900; i.e. an increase on 128 per cent. in wheat cultivation as against an increase on 83 per cent. in the total arable area. At the same time the number of oxen fell from 70·2 to each hundred inhabitants in 1860, to 37·7 in 1911, and the number of pigs from 27·8 to 12·8.

It was to be expected that when the emancipation of the peasants was completed, after the Great War, and 90 per cent. of the arable land passed into their hands, those two tendencies would be reversed: that a reduction in the area under wheat would be accompanied by an increased interest in cattle breeding and of the crops which the peasants preferred—maize, barley, industrial crops, &c.—as being more suitable for cultivation on a small scale and because they yield larger gross returns.

Leaving aside for the moment the question as to how far such a change served the interests of Rumanian agriculture as a whole, it is clear that it was indispensable for the peasants themselves. As the peasants keep proportionately a much larger number of animals, they could not, like the large farmers, put one-half or even two-thirds of the land under wheat; quite apart from the obstacles which the narrow and scattered fields, frequently making up the peasant holdings, present to such a course. The peasants require for their kind of farming a large proportion of fodder crops, or of crops like maize, &c., which leave a residue for the feeding of animals. The personal needs of their households likewise require a greater variety of crops, as the Rumanian peasant is largely a vegetarian. To this also tends the evolution of the local market. If the growing of maize and wheat was so widespread hitherto, that was due to the country's social structure as much as to the prevailing system of cultivation. With fifty-four inhabitants to the square kilometre, and 80 per cent. of them living on or below the poverty line, the market for the more refined products of agriculture was extremely limited. But with the betterment of the peasants' standard of living after the War, and with the relative growth in the urban population through the addition of the new provinces, a greater demand for garden and dairy produce has been created. To such considerations of consumption may be added others equally weighty relating to the technique of production. The peasants find a more rational employment for their live and dead stock, and especially for the surplus labour of their families, in the raising of more profitable industrial crops, such as sugar-beet, tobacco, &c. It is indeed a dominant impulse with smallholders to find out ways and means for employing to the full the working powers of their household. They find therein one of their main advantages in competition with capitalist farming. That peculiar advantage of the smallholders becomes accentuated in the measure in which the cost of labour rises and the increase in the size of farms makes supervision more difficult. Small cultivation, therefore, stands to gain over large with all crops requiring more intensive labour; and a corollary of this proposition is the certain expansion of such crops wherever large property is broken up and passes into the hands of smallholders. Careful calculations made by Thaer,

Krafft, and others show that the change from an extensive three-field system to a more intensive crop-rotation necessitates an increase of about 70 per cent. in manual labour. Similarly, the transition to industrial crops and to market-gardening would claim a further increase in manual effort. M. Șerban calculated that in Rumania the cultivation of one hectare of maize necessitated about three times as much labour as one hectare of wheat. Likewise the interesting study of Dettweiler on manual labour in agriculture established that an ordinary mixed farm under grass and cereals needed 6·49 permanent labourers per 100 ha., but that a middle-sized farm with one-fifth of its area under beet would require 16·83 labourers for the same acreage. From a different angle the same point has been proved by certain German regional calculations; they found that the number of permanent agricultural labourers per 100 ha. varied between 12·54 in Eastern Prussia, 20·59 in Bavaria, and 32·38 in the Rhineland—three regions which display a typical transition from large farms to small holdings, and from corn-growing to dairy farming and market gardening. Dr. Felix Bornemann having inquired into the detailed organization of twenty-two German peasant farms, together covering an area of 140 ha., found that ninety-three persons were employed on them (members of the household and wage-earners), i.e. 66·4 workers per 100 ha. This figure, of course, was exceptionally high and suggested a considerable wastage of labour. But the inquiries relating to large farms probably listed individuals who performed solely agricultural labours, whereas some of the persons included in the above figure no doubt gave at least part of their time to domestic work. One should perhaps make some allowance, too, for the time and energy which the peasants themselves spend in repairing buildings, fences, carts, and most other things about house and farm, when the large farmers would have recourse to masons and other artisans.

For all these reasons smallholders everywhere display a characteristic preference for crops which demand intensive labour but yield larger returns. To some extent that was already noticeable in Rumania before the reform. The peasants' choice was then cramped by excessive demands made on their labour under the old contract system. But the following table shows that in 1911 there was a significant difference between large and small cultivators in the division of the areas under various crops:

Crops	Farms above 100 hectares	Farms below 100 hectares
	%	%
Wheat	50·97	49·03
Barley	24·04	75·96
Oats	33·03	66·97
Maize	14·01	85·99
Cereals total	30·44	69·56
Oil and textile plants	68·80	31·20
Potatoes and pulse	44·98	55·02
Industrial and other plants	32·31	67·69
Of these, tobacco	—	100·00
Market gardening	2·72	97·28
Fodder crops and hay	31·28	68·72
Orchards, &c.	5·87	94·13 [1]

One can trace the phenomenon in the *Agricultural Statistics for 1911–15* (1918), which contain on p. iii a table showing the area covered by various crops in the periods 1906–10 and 1911–15. The comparison is useful because the line dividing the two periods roughly coincided with the moderate increase in smallholdings after the rising of 1907, the yearly averages showing the following changes:

Crops	1906–10		1911–15	
	Hectares	%	Hectares	%
Cereals	5,075,544	85·2	5,040,054	84·2
Oil-bearing and textile plants	66,707	1·1	94,491	1·6
Pulse	82,794	1·4	95,908	1·6
Industrial and other plants	19,497	0·3	25,074	0·5
Artificial grazings	116,868	2·0	175,263	2·9
Natural ,,	437,678	7·3	390,830	6·5
Vineyards	86,460	1·5	86,395	1·4
Plum orchards	70,600	1·2	75,522	1·3
Total	5,956,148	100·0	5,983,537	100·0

Besides an average increase of 27,400 ha. in the cultivated area, there was a decline in the area under cereals and natural grazings, while the more profitable crops—industrial and leguminous as

[1] M. Şerban, *op. cit.*, p. 28.

well as artificial grazings—showed a proportional increase. The following figures, referring to certain specific crops, clearly bring out the differentiation between large and small farms:

	Small	Large
1911–15		
Tobacco, mustard, chicory, &c.	10,400–13,700 ha.	200–250 ha.
1915		
Leguminous and tubers	73,731 ha.	30,926 ha.
Market gardening	24,943 „	810 „
1919		
Leguminous and tubers	34%	16%
Industrial crops	96·48%	3·52%

None of the crops requiring intensive cultivation and special care appear to have been in favour with the large farmers.

But quite apart from technical considerations, a change in the repartition of crops was needed for a progressive re-organization of Rumanian agriculture. In her case this was not merely the automatic consequence of the transfer of land from large to small owners. Nor was it claimed solely by the interests of the new peasant proprietors; nor brought about accidentally by the State's interference. Price restrictions, export duties, and other extraneous factors played their part in causing the area under wheat to be reduced, as will be shown later on when discussing the country's economic policy. But the important point is that the needs of more progressive farming urgently demanded the adoption of a system of crop-rotation. The following comparative figures for 1911 indicate how unsatisfactory had been the allocation of the cultivated area in Rumanian agriculture:

		Cereals	Pulse and vegetables	Industrial plants	Fodder crops	Fallow and grazing (on arable)
Rumania	ha.	5,182,424	111,467	118,390	171,420	620,300
	%	83·35	1·89	1·90	2·76	10·10
Germany	%	62·38	17·65	0·71	10·39	8·87 [1]

From the time when the Crimean War opened a wide road for the export of Rumanian corn, and thereby stimulated a change from a quasi-pastoral to a cereal phase, wheat-growing

[1] M. Şerban, *op. cit.*, p. 23.

spread rapidly and somewhat one-sidedly, as may be judged from the figures below:

Year	Yearly average under wheat
	Hectares
1840 about	360,000
From 1862 to 1866	697,220
„ 1867 „ 1871	877,200
„ 1872 „ 1876	1,060,340
„ 1886 „ 1890	1,509,700
„ 1891 „ 1895	1,438,200
„ 1896 „ 1900	1,590,000
„ 1909 „ 1913	1,851,000
In 1914	2,111,730 [1]

There were farms of 1,000, 2,000, 3,000 ha. on which nothing but wheat was sown; and frequently the large cultivators grew wheat on the same land during several years in succession. For these reasons most Rumanian agricultural economists consider that wheat has reached the maximum extension—about one-third of the arable area—compatible with a rational exploitation of the soil; and that henceforth any increase in the total crop must come not from a wider acreage, but from more intensive farming and stronger returns.

Quite apart from the land reform, therefore, the scientific problem of maintaining the fertility of the soil, as well as the economic requirements of the home market, necessitated a fresh transition from the cereal phase to a system of crop rotation. The land reform has merely quickened this second and belated development in the country's agricultural organization. Before noting this particular aspect of its effects, one must take into account the slight rearrangement in the disposition of the land which the addition of new provinces, with differing geographical characteristics, caused to the Old Kingdom. An indication of that change can be seen in the table on p. 290.

It will be seen that within her new frontiers Rumania had a lesser percentage of cultivated land and proportionately more forest, pastures, &c.; as is natural because of the more mountainous character of Transylvania and Bucovina. The greater percentage of fallow was probably a temporary phenomenon

[1] N. Xénopol, *La Richesse de la Roumanie.*

Area occupied by	Old Kingdom 1915		Greater Rumania 1922	
	Area in hectares	% of total area	Area in hectares	% of total area
Crops	5,549,280	40·2	10,338,289	35·1
Fallow	564,803	4·1	1,891,800	6·8
Vineyards and Plum orchards	161,410	1·2	290,022	0·9
Meadows	391,704	2·8	1,408,057	4·8
Pastures	1,075,643	7·8	2,858,500	9·7
Forests	2,497,632	18·1	7,308,600	24·6
Rivers	854,990	6·2	5,329,132 (Rivers, Buildings, etc.)	18·1
Buildings, roads, and other barren land	2,694,838	19·6		
	13,790,300	100·0	29,424,400	100·0 [1]

during the application of the land reform. The table on p. 291 shows in detail the provincial variations in the use of the land.

A general picture of the two periods of transition in Rumania's agricultural organization, to which we have referred, is presented in the following table:

Use of land	1860		1912		1923	
	Hectares	%	Hectares	%	Hectares	%
1. Cultivated area [2]	2,494,220	19·91	5,413,281	41·68	12,330,088	41·8
2. Vineyards and plum orchards	105,130	0·84	161,557	1·23	313,000	1·1
3. Grazings	1,046,610	8·35	552,289	4·20	1,597,300	5·4
4. Permanent grazings [3]	3,043,130	24·29	1,871,763	13·78	2,858,485	9·7
5. Forests	2,223,200	17·82	2,422,290	18·44	7,094,056	24·1
6. Waterways, roadways and building land	3,607,310 (6 and 7)	28·79 (6 and 7)	1,365,949	10·40	5,303,794 (6 and 7)	17·9 (6 and 7)
7. Surface unutilized			1,348,615	10·27		
	12,519,600		13,135,744		29,496,723 [4]	

Within the new frontiers arable land has lost some of its importance, at the expense of pastures and forests, and for this reason the percentage under cereal crops remains the same (the 1912 figure does not include fallow, while that for 1923 does). But the effect of the peasants' advent is clearly discernible in the

[1] G. Ionescu-Siseşti, *Structure Agraire*, 1922, p. 7.

[2] 1860 and 1912 without fallow.

[3] 1860 and 1912 include fallow.

[4] Figures for 1860 and 1912 after M. Şerban, p. 57. Those for 1923 from *Buletinul Statistic*, 1924, No. 2.

DISTRIBUTION OF THE LAND IN 1927

Classification	Old Kingdom		Bessarabia		Bucovina		Transylvania		Rumania	
	Area	% of arable land	Area	% of arable land	Area	% of arable land	Area	% of arable land	Area	% of arable land
	Hectares		Hectares		Hectares		Hectares		Hectares	
I. Arable land . .	6,496,948	100·00	2,677,761	100·00	293,814	100·00	2,979,749	100·00	12,448,272	100·00
1. Cereal . . .	*5,726,003*	*88·14*	*2,330,162*	*87·02*	*178,665*	*60·81*	*2,305,526*	*77·37*	*10,540,356*	*84·67*
2. Artificial grazings and fodder crops .	*332,357*	*5·11*	*72,283*	*2·69*	*38,834*	*13·22*	*213,084*	*7·16*	*656,558*	*5·28*
3. Other crops and fallow . . .	*438,588*	*6·75*	*275,316*	*10·29*	*76,315*	*25·97*	*461,139*	*15·47*	*1,251,358*	*10·05*
(a) Market gardening	*174,682*	*2·68*	*82,460*	*3·08*	*59,384*	*20·21*	*118,678*	*3·98*	*435,204*	*3·49*
(b) Industrial plants .	*111,751*	*1·72*	*182,206*	*6·82*	*4,927*	*5·08*	*60,577*	*2·03*	*569,461*	*2·97*
(c) Fallow . .	*152,155*	*2·35*	*10,650*	*0·39*	*12,004*	*0·68*	*281,884*	*9·46*	*446,693*	*3·59*
II. Natural grazings and pastures . . .	1,409,958	—	448,309	—	171,723	—	2,034,685	—	4,064,675	—
III. Plantations and orchards	273,094	—	149,140	—	7,400	—	193,691	—	623,325	—
IV. Forests . . .	2,927,459	—	234,204	—	460,009	—	3,627,315	—	7,248,987	—
V. Buildings, roads, waterways, quarries, mines, barren land .	2,777,751	—	669,836	—	173,284	—	1,483,070	—	5,103,941	—
Total area . .	13,885,210	—	4,179,250	—	1,106,230	—	10,318,510	—	29,489,200	—

relatively rapid expansion of artificial grazings, notwithstanding a simultaneous and considerable addition to natural grazings and pastures (the figure for 1912, it should be noted, including fallow). Artificial grazings and fodder crops are steadily increasing, as shown by the figures below:

	Hectares
1923	469,211
1924	511,837
1925	571,461
1926	605,503
1927	656,558

In 1927 they covered 5·27 per cent. of all the arable land.

The same effect, in the second place, is visible in the redistribution of crops. The time which has elapsed from the beginning of the reform is too short to have allowed any fundamental change to have taken place in the nature of the crops; the peasants' lack of knowledge and training forbade any sudden transformation. But its beginnings appear sufficiently defined in the opposite table. [1]

For the time being, cereal crops would seem to retain as large a share of the arable land as they occupied before, but not without a constant proportional diminution:

Year	Total arable land	Area under cereal crops	Percentage of arable land under cereal crops
	Hectares	Hectares	
1906–10 (average)	5,956,149	5,075,544	85·21
1911–15 "	5,983,537	5,040,054	84·20
1921	10,042,898	9,146,577	91·07
1922	10,338,289	9,444,357	91·36
1923	11,062,073	9,657,168	87·29
1924	11,731,420	10,210,561	87·04
1925	12,269,362	10,471,630	85·27
1926	12,276,807	10,405,047	84·75
1927	12,448,272	10,540,356	84·67

[1] From the *Statistica Agricolă a României pe Anii 1911–15*, 1918, and the *Statistica Agricolă pe Anul 1926*, 1927. The table has been worked out as an illustration of the new trend in the utilization of the soil, but it can lay no claim to complete accuracy. Besides the deficiencies to which attention has been drawn, Rumanian agricultural statistics suffer from a too frequent regrouping of the various items. Headings adopted for one period no longer mean quite the same in the next, and comparisons are thereby rendered difficult and tentative. In the above table the figure which refers to the total area cultivated in 1911–1915 includes natural grazings, on arable land, whereas the corresponding figure for 1926 does not include natural grazings but includes 412,632 ha. fallow.

Year	1 Total area of country	2 Total cultivated area	Cereal crops	Pulse and vegetables	Industrial and other plants	Fodder crops	Natural grazings
A. Yearly average 1911–15 Ha.	13,017,700	5,983,537	5,040,054	95,908	119,565	175,263	390,830
Per cent. of (1)	—	46·0	38·7	0·7	0·9	1·3	3·0
Per cent. of (2)	—	84·2	—	1·6	2·1	2·9	6·5
B. 1926 . . . Ha.	29,489,200	12,276,807	10,405,047	418,439	435,186	605,503	1,330,934
Per cent. of (1)	—	41·6	35·3	1·4	1·5	2·1	4·5
Per cent. of (2)	—	—	84·7	3·4	3·5	4·9	10·8

The decline of the area under cereal crops is too regular not to suggest a settled tendency. It tells, however, only part of the story, and a more significant change is revealed by the displacement in the ranks of the principal crops, as shown by the table below:

Year	Total area under cereal crops	Wheat	Maize
	Hectares		
Yearly average 1910–4	5,040,054	1,936,527 38·4%	2,072,653 41·1%
Yearly average 1923–7	10,256,952	3,118,394 30·4%	3,847,116 37·5%
1927	10,540,356	3,101,153 29·4%	4,219,423 40·3%

While the area under wheat has been subject to marked oscillations, and after a temporary recovery is again in decline, the area under maize has expanded steadily from year to year:

	Hectares
1923	3,404,492
1924	3,621,751
1925	3,930,780
1926	4,059,432
1927	4,219,423

One can discern in these several tables unmistakable signs of that incipient reorganization of agriculture which, as we have suggested, was to be expected from an increase in peasant farming. Industrial crops, market gardening, &c., are relatively gaining ground; the progress in cattle-breeding and dairy-farming is indicated by the expansion in grazings and fodder crops, and not less by the change in the distribution of the principal cereals. Maize has been for a long time the staple food of the peasant population, but only in the Old Kingdom, and its maintaining now its importance could not be explained on that score alone, just when the consumption of wheaten bread is rapidly spreading to the villages. But maize is equally important as a fodder crop, and the fact that its cultivation continues so extensively is further evidence for the likely development of cattle-breeding. The argument is justified by the serious fall in wheat-growing; so that the relative constancy in the position of maize, in comparison with the decline of wheat, must in a certain measure be counted as another step away from the cereal phase in which Rumania's agriculture has lingered hitherto.

In the light of this incipient transformation, one is tempted to reflect on the claims M. Garoflid put forward on behalf of the large farmers, that at least they deserve credit for having forced the peasants to cease being shepherds and become cultivators. That perhaps they did, yet in a period when European corn-growing was already losing to overseas competition—a competition which in the end is forcing the European farmers to become shepherds again. In Rumania that intermediate stage extolled by M. Garoflid ruined a whole class, exhausted the soil, and destroyed a flourishing branch of farming, which under the wing of the new reform the peasants are now struggling to revive. Wheat-growing is not likely to prove more popular with them than with the peasants of western Europe.

Section 2

The Present Position and the Future of Wheat and Maize

(*a*) *Wheat.* From the general problem of Rumanian agricultural production the decline of wheat-growing has been singled out for special attention, no doubt because wheat formed the backbone of Rumania's export and was therefore the element by which Rumania's agricultural prosperity was commonly measured abroad. Hence the disappearance of Rumanian wheat from the western markets has caused much apprehension concerning the effects of the reform on Rumanian agriculture as a whole. Wheat-growing and the export of wheat will probably not play the same predominant role in Rumanian economic life again, and it is therefore of general interest briefly to survey the circumstances on which that assumption is based.

Wheat has been subject to all the conditions which have caused an all-round decline in Rumania's harvests. In addition, however, wheat-growing has been affected by a number of specific circumstances, first of all by a considerable reduction in the area under this crop. In the Old Kingdom, before the War, wheat had spread itself out over nearly two-fifths of the cultivated area, but in Greater Rumania, at present, its cultivation covers less than 30 per cent. of the area brought yearly under the plough. We have pointed out in the preceding pages that in general the peasants give preference to crops which demand greater labour on a lesser area; maize, &c., is relatively gaining ground at the expense of wheat. Wheat, moreover, is difficult to grow on the exiguous strips which compose most of the peasant holdings. To the decline both in wheat-sowings and harvests the exhaustion of the wheat-lands by the wanton exploitation of the former large cultivators has further contributed. Certain regional variations in wheat-growing are explained by local circumstances. In Moldavia, where the winter lasts longer, spring wheat must be sown during a short period towards the end of March or the beginning of April, which involves greater risks than in other regions and with other crops. Wheat used to be grown there mainly by large farmers, on their own account, and the fall in its cultivation is considerable. On the other hand, in Muntenia and Oltenia, wheat can be sown towards the end of February: in these regions wheat was cultivated largely in *métayage*, and, figures calculated and privately communicated by M. Emil Marian show that the area under wheat has remained practically unchanged in the various counties since 1876. There have been variations from year to year, but not over longer periods.

An additional set of facts which affected the growing of wheat sprang from the peasants' strained circumstances after the War. Out of their poverty they had then to rebuild their economy, and also to pay for the land they had received at the resettlement. Now seed for a hectare of wheat costs about 1,600 lei, which mounts up with cleaning to 2,000 lei, and with interests of 30–40 per cent. on this money to 2,600–2,800 lei per ha.; a sum which is frequently beyond the means of the peasants or in any case a sum which they are disinclined to lay out. They prefer to sow maize, which gives them food for themselves and for their animals, and the seed for which only costs about 150–160 lei per ha. At the same time the peasants have been tempted by the disorganization of railway transport to employ their teams for carting the harvest, &c., instead of sowing autumn-wheat, which forms the main wheat-crop—2,839,636 ha. autumn-wheat as against 261,517 spring-wheat in 1927, the latter mainly in Bessarabia; they could obtain from carting high profits and ready cash enabling them to meet the higher cost of keeping animals. The Rumanian railway system was designed to link up the towns and not, in the first place, to tap the richest agricultural regions, and the effect has been to tempt the peasants in emergencies with opportunities for safe and easy gain, which disturbs their farming, and in the same measure the temper of the economists concerned with the furtherance of agriculture.

These conditions are probably of a lasting nature. They make it unlikely that the area under wheat will ever reach its former proportion, quite apart from the influence of more transient factors, to which we shall presently refer, connected with the state of the market and with official policy. Of the twelve and a half million ha. arable land the peasants now hold some eleven million ha. But among the peasants only those can grow wheat who own more than 3·5 ha. In the present organization of their household and farming the peasants require 18 hectolitres maize for their own food, 8–10 hectolitres for birds and farm-yard animals, and 700 chile for 2 oxen—a total quantity for which the 3·5 ha. just suffice. If one divides the expropriated area by the number of peasants who have received land, the average holding distributed works out at about 3·5 ha. each, and if to that area one adds the land distributed in former re-settlements, the conclusion is that an area of roughly 9,500,000 ha. is divided into very small holdings on which it is not practicable to grow wheat. The remaining 3,000,000 ha. arable land about equal the area which has in fact been put under wheat.

Hence the area which can be devoted to wheat-growing is likely to remain stationary, while internal requirements are greatly increased. In some of the new provinces, especially in Transylvania and Bucovina, the population eat bread and potatoes, but maize not at all. Together with the growth of urban centres, that circumstance demands a higher contribution in wheat from the Old Kingdom, as Transylvania and Bucovina could not satisfy their own needs even before the War and had to draw supplies from other parts of the Austro-Hungarian monarchy. More important still is the effect of changes which are taking place in the habits of peasants in the Old Kingdom. The long period spent in the army has accustomed many of them to eat bread, and being on the whole better off after the reform, their improved standard of living finds expression in a change from a maize to a wheat diet. The figures below purport to indicate the internal consumption of the principal cereals, but they are approximate to a degree, being obtained simply by the deduction of exports from the total production (inclusive imports):

	1910–15		1920–4		1925		1926	
Cereals	Total in quintals	Per head	Total in quintals	Per head	Total in quintals	Per head	Total in quintals	Per head
Wheat	12,098,897	1·63	21,151,656	1·27	28,677,613	1·71	27,459,903	1·63
Rye .	324,103	0·05	1,902,834	0·11	2,027,811	0·12	2,588,422	0·16
Maize	18,164,942	2·46	30,578,446	1·82	35,801,546	2·14	53,936,484	3·23
Barley	2,675,657	0·36	8,003,050	0·48	8,366,492	0·50	11,039,439	0·65
Oats .	2,975,684	0·41	8,242,181	0·49	7,195,035	0·43	10,979,617	0·65
	36,239,283	4·91	69,878,167	4·17	82,068,497	4·90	106,003,865	6·32

In the following table an attempt has been made to work out more accurately the figures for wheat, by deducting from the total internal consumption the quantity estimated to have been used as seed (at 180 kilogram per ha.), in quintals:

Year	Internal consumption	Used for seed	Used for food	Per head
1910–15	12,098,897	3,476,064	8,622,833	1·16
1920–4	21,151,656	4,688,325	16,463,331	0·98
1925	28,677,613	5,941,596	22,736,017	1·35
1926	27,459,903	5,989,476	21,470,427	1·28

The figures have only relative value, for in years of rich harvests the unexported surplus appears fictitiously as part of the figure representing the internal consumption. This explains the high figures for 1926. On the other hand, 1924 gave a very poor wheat crop; wheat had to be imported probably in the following spring, and that import went to swell the already large total for 1925. Wheat imports were as follows:

	Quintals
1920–4	4,383
1925	482,438
1926	0

The consumption of wheat is small in comparison with that of other countries, as the peasants in the Old Kingdom mainly eat maize. In the case of maize it is impossible to separate the quantity used as food from that used as fodder. The general consumption of cereals is considerable, being paralleled only by that of Hungary; therefore the new habit of eating bread is likely to absorb any future increase in the wheat-harvests.[1]

That phenomenon is being experienced in other countries as well, even in the Far East, in China and especially in Japan. Wheat exports to the Far East have increased on an average tenfold since the beginning of the century. During the eleven years before the War India exported an average of 45 million bushels yearly, but after the War it fell to 13 millions, as a result of greater home demands; and consumption, it is considered, is likely to keep pace with the rise in production.[2]

The change in Rumania is illustrated more tellingly even than in figures by the appearance in almost every village of at least one baker, whereas formerly bread could only be got from town. 'A loaf of bread' was generally the treat which the peasant brought his family when he went to market. The population of the new provinces, on the other hand, will not take to maize,

[1] Average consumption per head of cereals, including quantities used for seed and as fodder, 1920–9, in quintals.

	Hungary	Yugo-slavia	Bulgaria	Poland	Czecho-slovakia	Rumania
Wheat	1·54	1·03	1·61	0·35	0·65	1·27
Rye	0·68	0·11	0·30	1·43	0·87	0·11
Maize	1·55	1·90	0·86	0·02	0·19	1·82
Barley	0·58	0·22	0·39	0·42	0·68	0·48
Oats	0·36	0·22	0·22	0·86	0·86	0·49
	4·71	3·48	3·38	3·08	3·25	4·17

(After Ionescu-Sisești; from data in the Year-book of the International Institute of Agriculture.)

The table discloses certain national preferences: Rumania and Jugoslavia consume a good deal of wheat and maize; Bulgaria less maize and more wheat; in Poland maize is of no importance, rye coming first and after it wheat; in Czechoslovakia wheat and rye are consumed about equally. Czechoslovakia and Poland are great consumers of potatoes, as food and as fodder, hence they consume less cereals.

[2] *India as a Producer and Exporter of Wheat,* Wheat Studies of the Food Research Institute, Stamford University (California), July 1927.

even in times of stress. After the disastrous failure of the 1928 harvest in many parts of Bessarabia, the Rumanian Red Cross set up canteens in the distressed area and distributed to the villagers free of charge the traditional *mămăligă* (*polenta*), but later bread had to be substituted, although the cost was twice as high, because the population, though famished, could not be induced to eat maize. In general, the internal consumption of wheat before the War was estimated at 50,000 to 60,000 wagons, 80 per cent. of which were consumed in the towns. Now, the internal requirements have been estimated in a Memorandum of the *Cercul de Studii Economice* at 150,000 wagons, i. e. a proportion of 1 : 3, whereas the population has increased merely as 1 : 2. If the population of the Old Kingdom should come to consume wheat at the same rate as the inhabitants of the new provinces, internal requirements would rise to 200,000 wagons yearly, plus 50,000–60,000 wagons for seed; which means that the pre-war export could not be equalled until the harvest reached 400,000 wagons yearly. The problem of internal supplies is aggravated by the fastidiousness of the townspeople who insist on the whitest of wheaten bread. Even immediately after the War, when wheat and flour had to be imported on credit, and when France, Germany, and other countries restricted the consumption of white flour, restrictions were imposed in Rumania only upon the producers, but not at all upon the consumers.

To meet the growing internal demand in the face of a tendency to diminish supplies, the Government resorted from the beginning to measures calculated to bring them popularity by keeping the price of wheat low. Just when Rumania needed to increase her exports for purposes of reconstruction, and when Rumanian wheat could have found a ready outlet in the depleted markets of Central Europe, the Rumanian Governments prohibited the export of wheat and of wheat-flour. The neighbouring countries —Yugoslavia, Hungary, Bulgaria—had a lesser production, yet none of them resorted to similar prohibitions. In the second place, the Governments fixed maximum prices for wheat, and as this had the effect of deterring the farmers from sowing wheat, the authorities finally resorted to a system of requisitions. Any article liable to be requisitioned tends to disappear from the market, and as all these restrictive measures were applied to wheat alone, they created a psychological preference for other crops which did not expose the farmers to the constant interference of the authorities. Afterwards, when the export of wheat was left free, the Government endeavoured to ensure a plentiful internal supply by imposing high export taxes. This effectively ruined all the chances of the wheat export, the more so as prices were falling rapidly from year to year, and Rumania's restrictive measures coincided with the special efforts made by France, Germany, Italy and other countries towards the raising of their own wheat production. Nor had the Rumanian measures any other effect beyond diminishing the supplies which reached the market. As a consequence, internal wheat prices rose ever higher, even if the rise was usually one year late, keeping step in each case with the change in official restrictions. As might have been expected, restrictions on wheat caused its price to remain absolutely or relatively below that of other cereals; in 1916 wheat prices were 33·35 per cent. higher than maize, but in 1924 only 17·35 per cent. Oats and barley fetched during the first five years after the War higher prices than wheat. This led to a considerable wastage of the more valuable corn; large and small cultivators preferred to use wheat on their own farms for the feeding of animals. In the towns, too, bread being cheaper than maize or other foodstuffs, chickens, and even pigs and horses, were fed on bread. During the past few years, on account of low prices and indifferent quality, whole cargoes of wheat have gone to Denmark as fodder.

These disturbing circumstances were aggravated by the severe money crisis and the consequent high rates of interest. Wheat-growers were unable to obtain from the official prices even the equivalent of the money invested in seed and labour. In an interview published in the Bucarest *Plutus*, on November 11, 1923, the director of a big Transylvanian flour mill admitted that things had come to such a pass that they were able to buy wheat at almost any price, if they paid ready cash. Small traders and speculators used to assemble at country stations and buy wheat from the peasants by the sack, at prices which gave them no more than 50,000–60,000 lei per wagon. When its restrictive measures had completely unbalanced the cultivation of wheat, the Government tried to counteract them by other official decrees. In 1921, it was decided that wheat with a weight of more than 76 kilograms per hectolitre and only 1 per cent. impurities should be exempted from requisition and from export restrictions. But such a quality was extremely rare and above the limit fixed before the War as entitling to exemption from requisition. Some of the large cultivators may possibly have reached it, but it was most improbable that the peasants, who could not afford selected seed, would come anywhere near it. The decree was in danger of being interpreted by them as one more favour for the 'boiars', and of making them still more reluctant to grow wheat.[1]

Later on the Government offered a premium of 200 lei for each hectare sown with wheat, but that represented merely a small part of the loss which wheat-growers were suffering through price restrictions and export taxes, and the measure was therefore fated to remain ineffective. The *Argus* calculated that, with a normal harvest, wheat in 1924 gave a return of about 2,000 lei per ha., which was the same as other crops; whereas in the absence of restrictions wheat could give a return of 4,000 lei to the 3,000 lei of other cereals.

It will be seen that the decline in wheat-growing cannot be altogether attributed to the transfer of land from large to small cultivators; a fact which a later chapter, dealing with the agrarian policy of the State, will bring out more clearly. There is much justification for the point of view of the Rumanian writer who accused the State of having treated wheat 'as a philanthropic crop'. The upshot of all these circumstances is that an indifferent harvest can hardly satisfy the country's own needs. This was the case in 1928, when, immediately after the harvest, the flour-mills of Transylvania bought 2,000 wagons of wheat from Hungary.

The obstacles the State placed in the way of wheat-growing

[1] According to the *Frankfurter Zeitung* for May 9, 1925, the average weight per hectolitre of wheat was 74·6 kg. in the Old Kingdom, 73·7 in Bessarabia, 76·2 in Transylvania and 75·2 in Bucovina, giving an average of 74·9 for the whole country.

were the more unfortunate as wheat could maintain itself in Rumania only by means of more intensive methods and returns. M. Harold Wright mentioned in his book on Population that during the first year of the War the area under wheat increased by about 19,000,000 acres. Wheat-growing has made such rapid strides in America and Canada that some of the inferior qualities are now used by the Canadian farmers for the fattening of cattle, in preference to the more exportable cereals like oats and barley. M. Ionescu-Siseşti has calculated that with the present cost of production and average returns, the Rumanian farmers suffer a loss whenever the production remains below 1,200 kg. per ha. That, however, means a continuous loss, the average normal production being at present below 1,000 kg. per ha., and about 900 kg. per ha. in the case of peasant cultivators. The remedy, in his view, was to turn from wheat-growing to the production of high quality animal, dairy, and garden produce. M. Garoflid, too, warned the country some years ago that they must familiarize themselves with the idea that soon Rumania may no longer count among wheat-exporting countries. And if American wheat supplies continued to increase, the replacement of wheat by maize was, in his opinion, not necessarily a great loss. The best use to be made of the excess of fodder produced by the peasant holdings was to devote it to the breeding and fattening of cattle. Though representing the large farmers, M. Garoflid admitted that perhaps the only means of raising Rumanian agriculture might be 'to replace the export of wheat and oats and barley by the export of meat and fats'.[1]

(*b*) *Maize*. Both from the point of view of extension and of production, maize-growing remains the chief crop of Rumanian agriculture. Its prominent position is not merely accidental, but the result of a soil and climate favourable to its growth. Maize requires a rich soil, and a warm climate from sowing to harvesting, with rain during the period of vegetation and a dry autumn to facilitate ripening and harvesting. Such conditions are seldom encountered elsewhere in central and western Europe, and maize is hardly to be found except in the Neckar valley, among vines. Maize for fodder, which is harvested green, is

[1] *Argus*, October 12, 1923.

extensively grown in Germany, of course, much of the seed for it being imported from the Banat.

Maize was introduced in Rumania in the eighteenth century, through Italian and Turkish traders. It soon became a favourite with the peasants, displacing almost completely, within a short space of time, the growing of millet, which had from olden days been the main ingredient in the peasants' nourishment. Since its introduction, maize flour has formed the bulk of the peasants' food.

The advantages which the peasants find in maize are many and varied. It is well adapted to Rumania's soil and climatic conditions. In normal years, and when carefully grown, maize gives a richer harvest than any other cereal. It is a much safer crop than wheat as it requires only about half the quantity of water needed by wheat, so that it resists drought better and longer than other cereals; only once during the past fifty years did the maize crop fail altogether and maize have to be imported from America. The labour it requires, especially sowing, does not interfere as regards time with other agricultural labours. Maize needs only a small quantity of seed in proportion to the acreage, the cost being about 150–160 lei per ha.; and there is a further saving in that it is sown six months after wheat.

No other plant produces within such a short period of vegetation so much carbohydrates, sugar and fats, in as readily digestible a form. The maize-porridge and the maize-cake supply the peasant with a complete food, comprising starch, sugar, albuminoids, and fats. Its composition in this respect is more satisfactory than that of wheat, which contains only 1·9 per cent. fats while maize contains 4·4 per cent. The preparation of the *mămăligă* is a trifling labour compared with the baking of bread; it can be cooked easily and freshly over any kind of fire, which is of great importance to the peasants, who often have to work far from home, living with wives and children for days and sometimes weeks in and about their carts.

Maize stalks and leaves supply a much better fodder than straw; it serves to feed all the animals, from oxen to chickens; it fattens pigs better than anything else. And what remains over after foddering the animals with the stalks makes a fuel which is especially valuable in the timberless Rumanian plain.

Finally, maize-growing has a useful place in crop-rotation. Though a cereal, it is treated like a root, and the hoeing it requires prepares the ground for wheat and other cereal crops. The rotation maize-wheat was the first step in the improvement of Rumanian agriculture.

The main disadvantage of maize-growing is that it demands a great deal of labour in hoeing and ridging. That is why the large farmers seldom grew it, except in *métayage*; and in eastern Galicia, e.g., before the War, maize was the only crop cultivated in *métayage* (*naspuken*). This drawback does not, however, deter the peasants, as so far they do not count their own labour among the factors which determine how far a crop is profitable or not. Another disadvantage, of a social nature, has been the endemic prevalence of pellagra among the Rumanian peasantry, attributed to the consumption of maize. In so far as that was true it was no doubt due to the fact that the quality of the maize had been allowed to degenerate. Besides this the peasants sold the best and consumed the worst of what they reaped. As long as *mămăligă*, often of poor quality, was almost the only food of a population on a low level of physical fitness, it was likely enough to produce bad consequences. To-day the peasants still live mainly on *mămăligă*, but from all accounts pellagra has practically disappeared, owing to a general improvement in their standard of living.

Because maize plays only a negligible part in the food of the urban population, the Governments did not inconvenience its cultivation and export with all the heavy restrictions which they imposed upon wheat. This helped to maintain the popularity of maize with the peasants, as proved by the figures for the first few years after the War:

	Area in hectares
1920	3,295,418
1921	3,443,990
1922	3,403,854
1923	3,398,059

During the worst period of reconstruction, therefore, the area under maize remained almost stable, whereas the area under other cereals was first restricted and then gradually extended from year to year. The peasants, that is, first directed their efforts to the revival of maize-growing, and only afterwards took steps to increase the other crops. From the figures given in the preceding section one may safely deduce the fact that maize is, if anything, gaining ground, and that it is likely to remain by far the most important crop in Rumanian agriculture. As a food it may gradually be replaced by wheat, in some measure, but that will probably be more than compensated by a growing demand for it as fodder, through the development of cattle-breeding.

Section 3

The Effect of the Reform on Production

Rumania's agricultural production registered a serious decline after the War, from every point of view. The cultivated area diminished; total production, as well as the yield per ha., fell; and the quality of the produce was also worse. As that decline occurred during the period which saw the application of the land reform, it was simple enough to assume that the first was the result of the second; especially as most critics took it for granted that the change had meant a transfer of land from well-equipped large farmers to a mass of unprepared peasant cultivators. How little foundation there was for such an assumption has been shown in the preceding chapter. It was seen that the large farmers had depended almost completely on the teams and implements and labour of the peasants; and that was bound to be still more so at the end of the War.

Certain writers, therefore, hold that far from depressing production, the reform has actually saved it. 'The agrarian reform has saved our agricultural production', says M. Ionescu-Sisești, 'and has protected us against the irremediable collapse of our national economy. . . . Only by offering the peasant the land for which he was craving has he been induced to make an effort beyond man's normal powers.' The writer finds support for his view in the significant fact that 'the area which has remained untilled from expropriated land is much smaller than that from land not liable to expropriation'.[1] Taking all the arable land, the part uncultivated amounted to 24·21 per cent., whereas from the expropriated land 16·93 per cent. remained untilled in 1919 and 12·86 per cent. in 1920. The writer appears to overlook the obvious retort that it was just the expropriation which made it difficult for the large cultivators to secure the necessary labour for their farms, but this merely qualifies without invalidating the truth of the observation. It was not an isolated experience. 'The Polish landowners, ruined by the devastation of the War, are most of them unable to farm their estates. Apart from measures of land reform, therefore, it was found necessary to pass special legislation for placing that surplus area in the form of compulsory leases at the disposal of the peasants.' [2]

When trying to discover the real reasons for the fall in production, one is first of all faced, as was pointed out before, with the insoluble difficulty of how to disentangle the effects of the reform from the effects of the War. The latter made themselves felt in most of the civilized countries, says the report of an inquiry made by the International Institute of Agriculture. 'Essentially the present crisis is a marketing crisis. In many countries the prices of agricultural products do not cover the cost of production, nor compensate for the considerable increase in the prices of manufactured goods.' Certain general propositions may safely be laid down, however. Seeing that the agricultural deficiency became mitigated as the War years receded, it is reasonable to assume that, to some degree, it represented a consequence of the War. And as that improvement became accentuated as the application of the reform advanced, it is evident that not all the trouble could have been caused by the reform itself. Nevertheless, there is no doubt that during the first years production was disturbed by the application of the reform. But here, again, we are faced with a complicating circumstance which makes it impossible to reach any nice conclusion. During the period of transition agriculture was abnormally harrassed by the economic policy of the Govern-

[1] *Reforma Agrară şi Producţiunea*, 1925, p. 23.

[2] G. Daszinska-Golinska, *La Réforme Agraire en Pologne*, Warsaw, 1921, p. 87.

ments who ruled Rumania during the ten years after the War. That policy, to be detailed later, must for ever leave open the question as to whether that agricultural decline would have come to pass at all, or with such virulence, had the State not shown itself conspicuously indifferent to the recovery of the country's paramount industry.

Some of the most severe effects of the War have been mentioned in discussing the change in the technical means of production. Two-thirds of the more fertile land of Old Rumania remained for two years under the occupation of an enemy desperately in need of food-stuffs. From December 1, 1916, till April 4, 1918, the Central Powers took away 1,840,352 tons of cereals, fodder, and oil-bearing seeds; including in that total 1,223,340 tons of wheat and 227,522 tons of maize. They were also able to carry away almost the whole of the 1918 harvest. Muntenia was completely denuded of supplies at the end of the War; oats and barley had to be brought in from Moldavia, and wheat, for consumption and seed, from abroad.[1] The loss in men and animals, in machines and implements would have been sufficient severely to disturb agricultural production, especially as economic conditions at the end of the War made the replacement of the destroyed materials difficult and often impossible. Everywhere one can still see old agricultural implements, which had been long discarded, summarily mended and brought into use again. Haphazard cultivation during the War, even more careless than before it, has left behind rich crops of weeds. At the end of the War, moreover, there was a serious shortage of seed, as the needs of the country and the enemy requisitions had used up all the supplies of corn; wheat and flour had to be imported in 1919 even for internal consumption. To all these circumstances one must add the collapse of the transport system; roads and railways are not even yet in a fit state to cope with a normal harvest.

(*a*) *The Cultivated Area.* The manner in which the reform was applied made the agricultural confusion still worse. The Govern-

[1] For a description of the enemy occupation see G. Antipa, *L'Occupation Militaire de la Roumanie*; and on the state of agriculture during the War, G. Ionescu-Siseşti, *L'Agriculture Roumaine pendant la Guerre*—both in this series.

ment decreed the immediate expropriation of large property, and the whole operation was carried out during the winter 1918–19. This hasty application of the first part of the reform made it impossible to transfer the land to the peasants with the same speed. Of the 2,200,000 ha. expropriated during the first year over 600,000 ha. remained in the hands of the State; 250,000 ha. were rented by the former owners, while 350,000 ha. remained uncultivated, the State having to pay for them a yearly rental of 20,000,000 lei. The landowners were not unwilling, as they got a safe income without any trouble and effort; in Dobrogea, e.g., they received the full regional rent though the land remained untilled, whereas the unexpropriated land could not be let even at half the regional rent. In Constanța county, 65,000 ha. were taken over by the State, but only 12,000 ha. could be distributed to the peasants, and that with great difficulty. In his Memorandum to the King, M. Garoflid estimated at six to seven hundred million lei the loss resulting from that unutilized area. In general, the mechanic basis of the expropriation hampered production in the first two years, i.e. just when the former owners had to reorganize their exploitations, because no one knew precisely how much land would be left him; and this was made worse by the continuous talk of a second expropriation which, as we know, was in fact decreed. The joint effect of the War and of the application of the reform is seen in the following figures relating to the area under wheat:

	Hectares
1914	2,111,730
1918	1,616,042
1919	696,680

or according to the kind of farms:

	1914	1919–20	
	Hectares	Hectares	
Large farms	927,529	131,300	
Village associations	—	270,449	559,282
Small farmers	1,184,138	288,833	

Compared with the average of the preceding five years, the area cultivated in 1919 dropped by about 1,400,000 ha. in the Old Kingdom. Hence, the loss could have been due only in a minor part to the reform, as merely about one-fourth of it concerned the expropriated land:

	Hectares	%
Cultivated area before the War	5,941,264	
Area left fallow during 1919	1,438,902	24·21
Area expropriated in 1919	2,244,741	
Part of it which remained untilled in 1919 . .	374,982	16·93
Part which remained untilled in 1920 . . .	284,428	12·86

The decline in the cultivated area was in any case short-lived. The table below suggests that the farmers themselves did all that depended on them towards the recovery of agriculture:

Year	Old Kingdom	Bessarabia	Bucovina	Transylvania	Rumania
1920	—	—	—	—	8,658,480
1921	—	—	—	—	10,042,792
1922	—	—	—	—	10,338,289
1923	—	—	—	—	10,734,420
1924	—	—	—	—	11,388,144
1925	6,137,309	2,863,875	287,875	2,589,105	11,878,164
1926	6,455,100	2,611,400	274,539	2,935,768	12,276,807
1927	6,496,948	2,677,761	293,814	2,979,749	12,448,272

It is interesting to note how the cultivated area was distributed between large and small property, in the several provinces:

Bessarabia

Year	Large Property	Small Property	Peasant Co-operatives	Total
1925	223,968	2,422,673	217,204	2,863,845
1926	174,963	2,343,444	92,993	2,611,400
1927	176,269	2,418,762	82,730	2,677,761

Bucovina

Year	Large Property	Small Property	Peasant Co-operatives	Total
1925	75·362	212,513	—	287,875
1926	74·838	199,701	—	274,539
1927	75,464	218,350	—	293,814

Transylvania

Year	Large Property	Small Property	Peasant Co-operatives	Total
1925	260,570	2,303,962	24,573	2,589,105
1926	261,251	2,662,712	11,805	2,935,768
1927	261,609	2,712,467	5,673	2,979,749

Old Kingdom

Year	Large Property	Small Property	Peasant Co-operatives	Total
1925	917,240	5,141,499	78,570	6,137,309
1926	1,121,397	5,273,319	60,384	6,455,100
1927	991,016	5,454,716	51,216	6,496,948

Rumania

Year	Large Property	Small Property	Peasant Co-operatives	Total	Fallow
1924	1,414,228	9,491,798	482,118	11,388,144	343,276
1925	1,477,140	10,080,647	320,347	11,878,134	391,228
1926	1,632,449	10,479,176	165,182	12,276,807 }	include
1927	1,504,358	10,804,295	139,619	12,448,272 }	fallows [1]

One should perhaps explain that most of the land listed as being exploited by peasant co-operatives was really in the hands of the temporary village associations, formed to take up expropriated land until its final measurement into individual lots; therefore the constant fall of the figures in that column indicates the progress that was being made with the measurement of the holdings, and not a decline of agricultural co-operation.

In 1922 the cultivated area still remained about 10 per cent. below the pre-war average. By 1925 the pre-war area was equalled, and the expansion slowed down, but has not come to an end; so that in extent, at any rate, the reform would seem to be furthering the exploitation of the soil. The demand is still considerable among the peasants, but there is relatively little land left that might be brought under the plough.

The chief available land reserve consists of some 400,000 ha.

[1] The tables are compiled from the yearly statistics of the Ministry of Agriculture. The figures for 1924 and 1925 do not include fallow; afterwards fallow land was included in the general total.

alluvial soil in the low-lying regions along the Danube, Dnyestr and Pruth. This land is liable to flooding, and in 1912 a law provided for the reclaiming of the land by means of a system of dams; work begun on the strength of those preliminary studies was interrupted by the War. The agrarian laws gave the owners a respite of ten years during which they might carry out the work on their own account; otherwise the land is to be taken over by the State. A commission of experts appointed by M. Mihalache when he took over the Ministry of Agriculture, in 1928, reported that only a little had been done so far, and that the work lacked a basis sufficiently scientific to ensure its reliability and permanence. New plans are being drawn up for the early reclaiming of what is bound to prove an extremely fertile addition to the country's arable land.

(*b*) *Production.* The decline of production has, unfortunately, proved more persistent. The cultivated area now exceeds the pre-war extent, but as the yield per hectare is generally less, the total harvest remains deficient in that proportion.

Certain peculiar circumstances have contributed to that unfortunate result. In certain parts of the country, to begin with, the rainfall would seem to have changed in the post-war years. The average rainfall for the hundred years before the War had reached 605 millimetres. During the first five years after the War, the general average rainfall amounted to only 565 millimetres. In 1924, the average was only 540 millimetres. In that year the Black Sea basin recorded a rainfall of merely 388 millimetres, the basin of the Dnyestr 378 millimetres, and the basin of the Pruth 341 millimetres. These quantities hardly sufficed for a mediocre harvest. Autumn has become a dry season, rain being recorded mainly at the end of spring and in summer. In 1928 there was hardly any rain from the end of May till the snow season. The usual direct steamer traffic from Vienna to Giurgiu could be maintained only by repeated transfers into boats with a lesser draught, and for a time it was feared that the traffic would have to be stopped altogether. One reason for the change is probably the merciless cutting down of forests, especially during the War. There is a great difference in the rainfall received by the various regions, which ranges from 1,200 millimetres in the highlands to 800 millimetres in the hills and less than 400 millimetres in the lowlands, but so far no system of irrigation has been devised to draw advantage from these variations. Moreover, while the quantity of rainfall might be sufficient, it generally is concentrated on a small number of days, 65–8 in the year. This relative dryness is enhanced by the action of the strong north-easterly winds, which sweep across the flat corridor between the Dnyestr and the Carpathians. Lately the winters have been marked by severe cold setting in before the fall of the snow. As a consequence, about one million hectares had, e.g., to be ploughed and sown afresh in the spring of 1929.[1] These climatic conditions explain some of the apparently careless habits of the peasants. It is an old tradition among them to leave in the autumn the fields insufficiently cleaned and harrowed, because in the absence of plantations there is thus a better chance of holding the snow on their fields. It often happens that wheat sown between maize, and kept back by the maize harvest, is in the spring more forward than wheat sown in time and properly cared for.

Another circumstance, of a general character, which has affected agricultural production after the War, is the shortage of labour on the land. Of the resettled peasants many are reluctant to engage as labourers, provided that they can get sufficient sustenance out of their own holdings, with the help of carting, &c. The younger men, on the other hand, who have remained without land, migrate to the towns to seek domestic or industrial work. The census attempted two years ago had to be scrapped as its results were obviously false, but even without a measurement in figures one can establish everywhere a rapid growth in the urban population; and the shortage of labour during the period of agricultural activities leaves no doubt that there is a slight movement of population away from the land. To some extent this may be induced by the change in the succession laws, which prohibit the splitting-up of peasant holdings below a minimum of 2 ha.; younger sons have therefore a lesser chance of inheriting any land at all.

[1] For a discussion of the relation between climate and vegetation in Rumania see the paper read by M. Jean Camarașescu before the International Agricultural Congress held at Bucarest in June 1929.

Some of the circumstances which have influenced production adversely may be attributed directly to the reform. Some million hectares having passed into the hands of small owners, it might have been expected, in the Old Kingdom, at least, that they would strain their powers in order to get more out of that area than had its former large owners. These hopes have not been, and could not be, realized so far. For if extensive cultivation gave such poor results when applied on a large scale, its achievements were bound to be still frailer when the system was applied over a mass of scattered fields. Of course, such results as used to be obtained by tenants and bailiffs, through putting the screw on the peasants, will henceforward not be obtained. But in the Old Kingdom, at any rate, one could hardly speak of a decline in knowledge brought into the service of agriculture; though the lack of a certain guidance which used to be exercised by the large cultivators no doubt makes itself felt. All these, however, are temporary effects, due not so much to any deficiency in the new methods as to the survival of the old ones.

A less transitory drawback to the reform springs from the distribution of land, not to those best able to cultivate it, but to the peasants who most suffered in the War. The principles of the reform, and probably still more the abuses committed during their application, have, moreover, placed fairly considerable extents of land in the hands of artisans, gypsies, petty officials, &c., thereby aggravating that defect. More serious still, the reform has inevitably increased the chaotic scattering of the peasant fields; in the absence of all measure of consolidation, additional fields were given where land was available and not where the recipient already, perhaps, had some land of his own. That means not merely a wastage of time and an uneconomic repetition of the various labours, but also a considerable wastage of land. The lines of demarcation between these innumerable small fields are alleged in certain places to occupy as much as 30 per cent. of their total extent; and they are also hotbeds of weeds, as the people concerned are not anxious to remove the only things which make the boundaries conspicuous.

Whatever the part, great or small, played by each of these factors—most of which are transitory—production still lags behind the pre-war level. In comparing the two periods one must be careful to keep in mind the change which has taken place in the distribution of the various crops; for, as the table[1] below shows, the extent of each of the five principal cereal crops is at present vastly different, though the area they jointly cover is similar to the corresponding pre-war area:

Average area 1909–13, in 1,000 hectares

Crop	Old Kingdom[2]	Bessarabia	Transylvania	Bucovina	Total
Wheat .	1,852	728	930	21	3,531
Rye .	128	202	125	29	484
Barley .	534	672	136	32	1,374
Oats .	447	71	308	47	873
Maize .	2,084	765	974	67	3,890
	5,045	2,438	2,473	196	10,152

Average area 1923–27, in 1,000 hectares

Crop	Old Kingdom	Bessarabia	Transylvania	Bucovina	Total
Wheat .	1,491	732	869	26	3,118
Rye .	69	100	86	23	278
Barley .	871	698	153	28	1,750
Oats .	692	173	276	45	1,186
Maize .	2,161	737	789	60	3,747
	5,284	2,440	2,173	182	10,079

The post-war averages show a decrease of over 400,000 ha. in the area under wheat, 200,000 ha. for rye, about 150,000 ha. for maize; and an increase of over 300,000 ha. for oats and of nearly 400,000 ha. for barley. The increase of the total area under the five crops in the Old Kingdom almost exactly equals the area they occupied in the two counties of southern Dobrogea, Durostor, and Caliacra, acquired in 1913. It is more difficult to explain the considerable drop in Transylvania, unless the figures of M. Jasny refer to a somewhat larger territory than that occupied by the present Rumanian province, perhaps because part of the Banat went to Jugoslavia.

With the aid of the same sources one may attempt a general

[1] The first part of the table after the article of N. Jasny in *Wirtschaftsdienst*, Kiel. November 30, 1928; the second part from the statistics of the Ministry of Agriculture.

[2] Without the Quadrilateral.

comparison of the total production of these five crops, in the same periods:

Average production 1909–13. In 1,000 tons.

Crops	Old Kingdom[1]	Bessarabia	Transylvania	Bucovina	Total
Wheat .	2,389	585	980	28	3,982
Rye .	119	194	130	24	467
Barley .	544	623	166	30	1,363
Oats .	422	81	328	67	898
Maize .	2,730	860	1,351	88	5,029
Total .	6,204	2,343	2,955	237	11,739

Average production 1923–7. In 1,000 tons.

Crops	Old Kingdom	Bessarabia	Transylvania	Bucovina	Total
Wheat .	1,300	476	825	29	2,630
Rye .	51	71	79	23	224
Barley .	622	399	147	25	1,193
Oats .	503	87	221	47	858
Maize .	2,472	795	961	86	4,314
Total .	4,948	1,828	2,233	210	9,219

For purposes of general comparison a third table is given below, showing the average yield per ha. of the principal crops during the five years before Rumania entered the War and during the five more or less normal agricultural years after the War. In the absence of figures for the pre-war yield in the new provinces, the comparison is limited to the Old Kingdom:

Year	Wheat	Rye	Barley	Oats	Maize
1911–15 . .	11·5	8·7	10·4	9·5	13·2
1923–27 . .	8·7	7·4	7·1	7·2	10·9

If these figures could be taken to represent a normal and stable relation, the fall in the average yield would justify some concern as to the effect of the reform on agricultural production. But production figures are in our case not reliable enough, unless calculated over very long periods. The crops were, and are, left altogether at the mercy of an erratic climate, with the result that their yield oscillates violently from year to year, and has

[1] Without the Quadrilateral.

never yet risen above a very low level. Among the following European countries, Rumania took the lowest place before the War, when her agriculture was thought to be flourishing:

Average yield in quintals. 1911–14

	Wheat	Barley	Oats
Belgium	25·3	27·5	23·8
Holland	25·2	26·2	20·6
Denmark	31·6	22·5	18·7
Germany	21·7	20·9	19·9
England	21·3	18·7	18·0
Sweden	21·0	16·9	14·9
Austria	13·9	15·6	13·1
France	13·5	14·1	13·1
Hungary	13·4	14·3	11·5
Rumania	11·5	10·4	9·5

The Rumanian figures refer to the period 1911–15. Had we taken the figures for 1911–14, as in the case of the other countries, the Rumanian averages would have dropped to 11·3, 10·1 and 9·3; while if we had considered the period 1910–13, they would have risen to 13·2, 10·7 and 9·8. The year 1910, namely, gave a bumper crop, with a wheat average of 15·5 quintals per ha., while the 1914 yield was merely 6·0 quintals per ha.—a level to which production has not descended even in the worst year after the War.

As a further illustration of how difficult it is to draw general conclusions from short-period averages, we give below certain averages for the periods 1920–7 and 1923–7:

Average yield per ha. in quintals

Region		Wheat	Rye	Barley	Oats	Maize
1. Old Kingdom	1920–7	9·0	7·5	8·3	8·3	10·8
”	1923–7	8·7	7·4	7·1	7·2	10·9
2. Rumania	1920–7	8·5	7·9	7·7	7·9	10·9
”	1923–7	8·4	8·0	6·8	7·2	11·2

The averages for the whole period are, in general, higher than the averages for the more normal period 1923–7. The conclusion would seem to be that production is getting worse in the measure in which War damages are made good and the disturbance caused by the application of the reform becomes attenuated—which obviously is an impossible paradox. The correct explanation would seem to be rather that in the second set of averages the year 1924, in which the harvest failed badly, plays relatively a larger part; wheat averages fell to 6·8 and 6·1, rye to 4·1 and 5·6, barley to 3·6 and 3·6, oats to 5·4 and 4·9, maize alone keeping a normal level with 10·9 and 10·9 quintals per hectare.

Maize, which happens to be the specifically peasant crop, is the only crop in the above table which maintained its average after the War; which appears to invalidate the idea that peasant cultivation is responsible for the generally lower yields. There is, however, one other possible interpretation of the fact that the averages for 1920–7, including the worst period of reconstruction, are better than the averages for 1923–7. During this stretch of time the total cultivated area has steadily increased, and the increase was limited exclusively to peasant farming; hence it might be argued that average yields fell lower in the measure in which peasant farming expanded. As the argument is as likely as not to be raised it is as well to refer to it here; but we doubt there is any substance in it. For if it were correct, then we should have to find lower average yields recorded just for those crops the cultivation of which is extending, whereas in reality the very opposite happens to be true. In the case of none of the other cereal crops has the cultivated area increased as constantly and considerably as in the case of maize; yet maize, we have seen, is precisely the crop which at the same time has best maintained its yield.

M. Crum Mihăilescu, of the Ministry of Agriculture, has worked out a table which shows that in the two years, 1926 and 1929, which gave good harvests after the War, the peculiar

Crop	Compared with pre-war average (=100)		Compared with average of 1923–7 (=100)	
	1926	1929	1926	1929
	%	%	%	%
Wheat	75	60·3	103	100
Maize	100	93	130	119
Rye	93·5	104	120	133
Barley	100	118	158	184
Oats	148	153·5	148	154

peasant corn crops have equalled or even surpassed the pre-war yield; and in a general way the table proves that the lower yield is not a settled phenomenon. The table compares the yield of the 1926 and 1929 harvests, first, with the pre-war average and, secondly, with the average of the period 1923–7.

It will hardly be possible to appraise the peasants' achievements or failures under the new system until they have had a chance to acquire equipment, and to dispose of their produce, without impediments from the State, but rather with such assistance in the way of credits, transport facilities, training and research, &c., as their role in the country's economic life entitles them to receive. For a preliminary estimate one must await the collection of figures showing the contribution of the various categories of farmers to agricultural production. An attempt to differentiate the production of large and small farmers was made by the Ministry of Agriculture in 1928, but the results turned out to be plainly erroneous. The various figures given here may, of course, be taken to be generally applicable to peasant farming, which now represents about 90 per cent. of the total cultivation.

Detailed tables giving the extension of the various crops, their total and average production, their distribution among the several provinces, &c., will be found on pp. 332–7. They are not such as to need further elucidation, and we will not burden the text by repeating them here. Their most characteristic feature is the steady increase of the area under artificial grasses and other fodder crops:

Year	Cultivated area (hectares)
1923	469,211
1924	511,837
1925	571,461
1926	605,503
1927	656,558

There is a similar expansion among semi-fodder crops—oats, barley, &c., and, above all, maize. Likewise, an increase in commercial crops, market-gardening, &c. This greater variety, besides the technical advantages which have already been discussed, has the merit of reducing the risks connected with the former one-sided growing of cereals, and it should thereby help to maintain more steady general returns.

As regards individual crops, wheat-growing was affected by the change to small cultivation, by the fiscal and customs policy of the State, as well as by unpropitious weather; the three autumns of 1921–3 were exceptionally dry, which interfered with the sowing of autumn-wheat, as the peasants show a bias against sowing 'in the dry' between 'St. Mary's days'.

Rye was already losing ground before the War. In the Old Kingdom it covered 200,000 ha. in 1896–1900, merely 128,000 ha. during the period 1909–13, and fell to 75,613 ha. by 1915. It shows no sign of improving. Barley lost its favourable position in the market between 1924–6, but recovered later; compared with an average area of less than 550,000 ha. in the years before the War, its cultivation reached an average of 870,882 ha. in the Old Kingdom in the period 1923–7. Oats expanded rapidly after the reform, till 1924, then began to decline, with a slight recovery in 1928. Maize has captured the interest of the peasants, and at present occupies one-half as much land again as wheat.

Maize is also the crop which has best maintained its yield, especially if one takes into account the peasant custom of sowing beans between maize, a custom exercised in 1927 on about one-fifth of the total area under maize; pumpkins and melons are likewise extensively grown as subsidiary crops with maize. This does not mean that the yield is anything like satisfactory yet. Rumania is now the largest maize-grower in Europe, but the methods of cultivation are still those in use a hundred years ago. The maize varieties have been allowed to degenerate, and while technically selection is easy, it is difficult in practice without proper guidance, as maize being a self-fertilizer the selection must be carried out by all the farmers of a region. Instead of finding two cobs on each stem, as is normal elsewhere, the Rumanian crop gives an average of one cob to each two stems. The average yield moves at present round 1,000 kg. per ha., whereas the American maize-growers obtain from 3,000 kg. upwards.

Before the War, Rumania found it so easy to dispose of crops produced with the sweated labour of the peasants that her only concern was how to increase her harvests. World conditions have greatly changed since then. Markets are overflowing, and it is only now that those in whose hands lies the guidance of Rumanian agriculture begin to realize that quality must come first and quantity where it may. Everything remains to be done in that direction. Wheat still contained in 1927 impurities averaging 8·27 per cent. for the whole country, though it is true that grain from Dobrogea with 34·10 per cent. impurities was largely responsible for that bad average. The best corn came from Transylvania, with 2·44 per cent. impurities; in the Old Kingdom the best results were obtained in the lowlands of the Baragan and Bugeac, with only 2·62 per cent. impurities, and the worst in the Moldavian-Bessarabian plain, with 6·10 per cent. Rye provided the largest proportion of foreign matter with 3·9 per cent.; actual impurities, which really cause the quality of flour to deteriorate, only amounted to about 1·1 per cent.

However a certain improvement is taking place. The average weight of wheat in 1927 was 77·3 kg. per hectolitre. From 295 samples obtained by the Central Agronomical Station the wheat crop was classified according to quality as follows:

Excellent corn weighing over	80 kg. per hl.		12·0%
Very good corn „ „	78–80	„ „	35·6 „
Good corn „ „	76–78	„ „	33·4 „
Medium quality „ „	74–76	„ „	9·7 „
Poor quality „ below	74	„ „	9·3 „ [1]

In other words, about one-half of the harvest was of very good quality; and about one-third of good quality; the percentage of poor quality corn was similar to the percentage of corn of very good quality in the preceding years. The provenance of the corn, according to quality, was as follows:

Region	Excellent	Very Good	Good	Mediocre	Poor
	%	%	%	%	%
Transylvanian plain	3·7	7·0	2·9	0·4	0·8
Transylvanian highlands	2·0	7·4	4·9	0·4	1·2
Danube lowlands	3·3	4·7	6·7	0·4	0·4
Muntenian highlands	0·4	2·0	2·9	0·4	—
Moldavian highlands	0·4	1·6	1·6	1·2	—
Moldo-Bessarabian plain	—	7·0	6·6	3·3	2·4
Baragan and Bugeac	2·0	4·7	4·9	1·6	2·3
Dobrogea	0·4	1·2	2·9	2·0	2·0

[1] This and the following tables from a study on 'The Quality of Corn in the Agricultural Year 1926–7', by the Director and Staff of the Central Agronomic Station, *Buletinul Agriculturii,* Sept.–Oct. 1928.

The best quality came from the main corn-growing districts; the poorest from the Moldo-Bessarabian lowlands, which, in fact, suffered from the worst drought. The following two tables may be of interest, as showing, first, the variations in weight of the 1927 samples from the several districts, and, second, a comparison of the average weight per hectolitre during the years 1922–7:

Weight per hl. in kg. in 1927

District	Maximum	Minimum	Average
Transylvanian plain	80·9	75·4	78·3
Transylvanian highlands	80·1	76·1	78·1
Danube lowlands	82·5	76·8	78·9
Muntenian highlands	79·8	74·1	77·6
Moldavian highlands	78·6	75·3	78·7
Moldo-Bessarabian plain	78·5	75·1	77·0
Baragan and Bugeac plain	78·9	74·0	76·7
Dobrogea	76·5	73·8	75·1

Average weight in kg. per hl. during 1922–7

District	1927	1926	1925	1924	1923	1922
Transylvanian plain	78·3	74·2	76·3	75·2	79·2	77·3
Transylvanian highlands	78·1	75·8	77·3	77·1	78·3	77·7
Danube lowlands	78·9	72·4	76·0	75·1	77·5	77·3
Muntenian highlands	77·6	70·6	74·6	73·5	76·0	76·5
Moldavian highlands	78·7	73·3	76·8	74·4	77·2	77·7
Moldo-Bessarabian plain	77·0	73·8	73·7	74·8	77·9	75·0
Baragan and Bugeac plain	76·7	70·0	76·2	72·4	78·7	76·7
Dobrogea	75·1	71·3	75·1	69·6	76·9	75·1

The 1927 harvest showed an all-round improvement in the quality of the principal cereal crops, which in some measure made good the deficiency in quantity:

Weight per hl. in kg.

	Wheat		Rye		Barley		Oats	
	1927	1926	1927	1926	1927	1926	1927	1926
Old Kingdom	77	72·3	69	67·3	62	59·9	45	42·1
Bessarabia	74	69·9	69	67·8	63	59·9	49	47·5
Bucovina	77	77·7	73	72·9	68	62·5	48	47·5
Transylvania	76	75·0	72	70·9	64	63·8	43	45·4
Rumania	76	73·7	70	69·7	64	61·5	46	45·7 [1]

[1] *Statistica Agricolă pe Anul 1927*, Part I, 1928.

There is every reason to hope that, with better opportunities, the peasants will henceforward perform the various agricultural labours more carefully and at the proper time. But this would not suffice to stop the downward trend of the average yield. There is as great a need for the use of selected seed, with a choice of the varieties best suited to the Rumanian soil and climate; and a still greater need for a more merciful exploitation of the soil. At present, some twelve tons of selected wheat seed are put on the market yearly by private farms, and some 550 tons of wheat and 600 tons of other cereals from the State's model farms. The seed-selecting stations of the National Agricultural Society and of the higher Agricultural Schools are endeavouring to produce new varieties, best fitted for the country's conditions. Excellent work is being done by the 'Sămânța', a private company formed after the War, whose seed-selecting stations are bringing increasing supplies each year into the market as well as producing new varieties; e.g. as the new maize variety 'Regele Ferdinand', which would seem to give a much richer yield in dry years than the American and other varieties. So far, of course, the supply of selected seed is still trifling, and the seed used by the peasants generally consists of a bewildering mixture, as one may see in looking over a wheat-field before it is cut.

The use of manure is hardly known, extraordinary as this statement may sound. The large farmers never kept enough animals before the War to produce farm-yard manure in usable quantities; and the inquiry of M. Șerban, cited in an earlier chapter, established in fact that none of the estates which he investigated, and which were of the best, made any use of farm-yard manure. The peasants, though having it, have never learnt its use; their animals are in fact not kept under shelter, unless in very bad weather, and such farm-yard muck as then gathers is generally burnt or left to rot at the edge of the village, or at best used after sun and rain have worn out its goodness—so indifferent has extensive cultivation made every one to the needs of the soil.

Equally rare is the use of artificial manures. The monograph prepared by the International Institute of Agriculture on the use of artificial manures gave the yearly consumption before the War as 1,471 tons of phosphates and 177 tons of potash. The book gave no figures referring to the post-war years. Internal production and consumption from the 'Mărăşeşti' factory was as follows:

	Superphosphates	
	Internal output	Total consumption
	tons	tons
1921	1,900	470
1922	1,900	1,340
1923	1,450	1,370
1924	1,450	1,800
1925	4,000	3,750
1926	5,300	4,000

The same company imported 120 tons of potash salts in 1926. Customs returns indicate that 615 tons of phosphates were imported in 1925, which should be added to internal consumption; while in the same year 9,926 tons of phosphates produced in Rumania were exported. The import of Chili saltpetre amounted

In 1921 to 58 tons
„ 1922 „ 109 „
„ 1923 „ 161 „
„ 1924 „ 745 „
„ 1925 „ 353 „

Ammonium sulphates are used on a relatively larger scale, as well as cyanamide of calcium. The latter is produced in a Transylvanian factory with a capacity of 30,000 tons yearly; the present output is about 12,000 tons, most of which is exported. The output of ammonium sulphates could reach 1,860 tons, but only 905 tons were produced in 1926. Hence, even this incipient production is still much higher than the demand. One might mention that in the opinion of M. Ionescu-Siseşti, expressed before the Agricultural Congress, Rome, 1927, the soil of the steppe region is still so rich that its fertility would not be raised by treatment with chemical manures. This would refer to about one-third of the arable area; the chief problem of that region is water supply. The remaining two-thirds of the arable area are in need of manuring.

In the light of all these circumstances one may sum up the reasons for the decline of production as being, first, damage and disorganization caused by the War, which have not been made good during the first decade of peace; in a measure that disorganization has been rendered more acute by the application of the reform; and in a much greater measure the work of reconstruction has been prevented by the State's policy. Hence, the drop in total and average production was in no way an unnatural phenomenon. The drawbacks which in this respect might be attributed directly to the reform are a less economic cultivation of the cereal crops, through the splitting-up of the land, and, from the same cause, greater wastage in harvesting the crops. This deficiency will have to be made good by better organization; but it fades into significance compared with the general evils which remain to be remedied as springing from antiquated methods and from an utter lack of care for the soil's fertility. What the soil can give is shown by the performance of Varjas, a commune in Banat, where by experimenting with the most suitable varieties, the best distance for planting the seed, &c., the local peasant farmers have caused maize to yield up to 5,000 kg. per hectare. But that is a remote ideal. The reform of land tenure has been carried to a very far point, but the reform of agriculture is not even begun.

(*c*) *The New Provinces.* In considering the effect of the reform on production, a somewhat sharp line must be drawn between the Old Kingdom and the new Provinces, especially Bucovina and Transylvania. In the Old Kingdom the reform could not have a deep influence on agrarian economy, as the dispossessed large owners and cultivators were doing nothing to improve the soil and its output, or the live and dead stock. The situation was different in Bucovina and Transylvania, and in the latter province, more especially in Banat. M. Garoflid, as Minister of of Agriculture, readily admitted this. After speaking, in his *exposé de motifs* to the Transylvanian law, on the norms of expropriation, he added:

> 'We shall make certain exceptions in the application of those expropriation measures. There are in Transylvania intensive agricultural undertakings—true agricultural factories. It would be a great loss to our national organization if we should destroy them; and the strict application of the norms mentioned above would destroy them. With a view, therefore, to protecting them, I have provided that those landowners may be allowed to retain up to 500 jugars.'

The owners or tenants of these large farms usually lived on the spot. Most of them possessed adequate agricultural knowledge and capital. The farms were properly equipped with buildings and machines; the live stock was numerous and of good quality, and cultivation was based on a sound variation and rotation of crops.

In practice the application of the reform went farther than M. Garoflid first intended, and there is no doubt that many of those large farms have been thereby irreparably damaged. Valuable material has been destroyed altogether. Farm buildings are falling to pieces, where they have not been pulled down by their owners and the material sold. Machines, tractors, &c., especially the heavy ploughs, are useless for the small holdings, and many of them were simply abandoned, a total loss. A company was formed after the reform for the export of unusable large machines. Great damage has also been done to cattle-breeding which, in Transylvania, had been furthered in a systematic manner. The province was divided into breeding districts, each of them specializing in a breed best adapted to local climatic and other conditions; the co-operatives or individuals who kept bulls had to conform to the arrangement, no other breed being allowed in the district. A considerable part of that breeding material was supplied by the large owners. Now that whole organization has simply vanished. Fine animals had to be sold to the butcher. The production of meat and dairy products has likewise depreciated, in quality or in quantity, with the disappearance of these large farms; some of them, like the estate of Count Cskonics, at Jimbolia, were organized to supply a whole chain of towns with pasteurized and bottled milk. Even now the only decent supply of milk, butter, &c., to reach Bucarest comes from Sibiu. The Transylvanian critics admit, therefore, that in the Old Kingdom the reform may actually have had a good influence on the large farmers; it has induced them to farm better, for fear of a new expropriation, and it has also enabled them to do so, by leaving them often 500 ha. arable land. But they assert that in Transylvania the maximum left has seldom exceeded 125 ha., which means that the destruction has been greater just where the economic organization was better. They consider that a different system should at least have been applied round the main towns, so as to ensure a plentiful supply of good and cheap food-stuffs for the urban population.

The authors and supporters of the reform do not deny either that well-organized large farms had existed in Transylvania or that the activity of these farms has been crippled by the reform. Yet they have a two-fold reply to make to the criticisms summarized above. In the first place they point out that once land was granted to the peasants in the Kingdom and in Bessarabia, it was impossible to treat those of Transylvania less well. The argument is, of course, unanswerable. There is no doubt that any attempt to maintain the large Transylvanian estates in a province eaten up with nationalist animosities, would have meant the risk of a still more destructive expropriation by popular violence. There is one reserve, however, to be made on this point. If the expropriation of the large estates was unavoidable, the expropriation of smallholders is inexcusable. Herr Fritz Connert, in his article, gives examples of Transylvanian peasants who have been deprived of some of their land, though their small or medium holdings were below the 100 ha. generally fixed as the utmost limit in every part of the country. And these were intelligent, prosperous farmers, who could serve as model to the new holders; themselves and their holdings representing just that type of rural middle-class, the absence of which is deplored by all agricultural authorities in the Old Kingdom. The second argument in reply to the Transylvanian criticism questions the ability of the large estates to keep going in the changed economic and social conditions after the War. Formerly the landlords controlled the political machine; all legislation, as well as its application, was subservient to their interests. An article in the *Argus* (November 18, 1922) contended that land-tax was paid in Hungary on the basis of a survey made a hundred years earlier. The consequence was that 'a hen laid enough eggs to pay the taxes of a big estate'. Count Besan owned 2,615 jugars and paid about four shillings; Baron Koranyi, a former Minister of Finance, paid something incalculable, but much less than about a farthing per jugar, and so on. Again, the law on agricultural labour was applied by the *pretor* (the sub-prefect), who depended altogether on the goodwill of the local landlord; a labourer stood no chance of getting other work in the neighbourhood if he left against the wish of his employer. At present, the whole legal and social situation is altered; many labourers, moreover, have got some land of their own. Even a less radical reform would have made the former labour relations impossible, and it is unlikely that the large estates would have continued to pay with a higher wage bill and a diminished hold on the labour supply.

The argument merely describes a labour evolution which is being experienced throughout agricultural Europe. If it may fairly be applied to Transylvania, it must be infinitely more fitting for Bessarabia. The Bessarabian landowners had nothing like the equipment of the Transylvanian, and the Bessarabian peasants were much less tractable after the War than the others. To state, therefore, as does M. P. V. Synadino in his article, that the average cultivated area has fallen by 4·4 per cent.—from 2,304,902 ha. in 1902–11 to 2,203,366 ha. in 1920–4—and average production (wheat, rye, barley, oats, and maize) by 16·5 per cent. —from 1,968,660 tons in 1902–11 to 1,645,000 tons in 1920–4— is to show that the post-war output is not as good as the pre-war output; but it by no means proves that things would have been better if the land had been left in the hands of the large owners. Quite a number of circumstances which have nothing to do with the scale and technique of farming have contributed to that decline; and it is worth while noting that Bessarabia, which beyond doubt has suffered most from the faults of omission and commission in the State's policy, is also the province in which total and average production has most fallen off from its pre-war standing.

The tables of averages reveal indeed the interesting fact that in almost every case the averages relating to the Old Kingdom are higher than the averages for the whole country, which means that the addition of the new provinces has influenced the average yield unfavourably. Bessarabia, especially, remains in every case behind the national average. Transylvania has in most cases higher averages than the whole of the country, while Bucovina has the highest of all, often exceeding the national averages by fully one-third. The average production per ha. of the five principal cereal crops during the period 1920–7 was as follows, in quintals:

Old Kingdom	8·8
Bessarabia	7·6
Bucovina	10·7
Transylvania	9·1
Rumania	8·6

Bessarabia has had bad luck, both with the weather and with the administration. It was the province whose agriculture most needed care during the period of reconstruction, as farming had been disorganized worse than elsewhere by the forcible taking over of the land, and all its transport and trade links cut off with the severance from the Russian Empire. Yet from the time of its union with Rumania up till the end of 1928 the province had been left to its own devices. The failure of the 1928 crop produced famine conditions in large areas of the province.

But it is more difficult to explain the great difference in the returns from Transylvania and Bucovina. The two provinces have similar geographical features, and they have both been under Austro-Hungarian administration. But the agrarian situation in the provinces differed in two respects; in Bucovina landownership was nicely graduated, with a good proportion of medium and small property, and the reform had only to be applied on a limited scale; Transylvania, however, had a high percentage of large property, and in consequence the reform was applied on a more thoroughgoing scale. Are the better returns in Bucovina due to a healthier distribution of property of old standing, or are the inferior returns of Transylvania due to the greater encroachment of the reform? The answer would be instructive from more than one point of view, if it could be given reliably, but for this we should need more detailed material than we possess. Nevertheless, the post-war difference is perhaps sufficiently explained by the pre-war averages. According to the table used in his article by M. Jasny, Transylvania produced before the War 2,955,660 tons of wheat, rye, barley, oats, and maize, on 2,473,000 ha., which gave an average of 11·9 quintals per ha.; Bucovina produced 237,000 tons on 196,000 ha., which meant an average of 12·1 quintals per ha. Again, in the table given on an earlier page and showing the average yield of wheat, barley, and oats in ten European countries during the period 1911–14, Austria (which included Bucovina) had for the three cereals together an average of 14·2 quintals per ha., and Hungary (which included Transylvania) of 13·1 quintals; the figures for the period 1906–10 being 13·8 and 11·5 quintals per ha.; or if we take the two periods together, the Austrian average was 14·0 and the Hungarian 12·3 quintals. The difference between the averages of the two countries, in the period 1906–14 was, therefore, 1·7 quintals; the difference between the averages of Transylvania and Bucovina in the period 1920–7 has been 1·6 quintals. The first figures refer to whole countries and to three cereal crops, the second figures merely to isolated provinces and to five crops, but the approximation is too near not to have some meaning in it. In whatever way the reform may have affected the agriculture of the two provinces, it has not altered the relation of their average production; and as the reform went a great deal farther in one province than in the other, its effects are not greatly elucidated by a difference in average yield which remains much as it was before. M. Jasny's table, it is true, would give merely a negligible difference of 0·2 quintals per ha. in favour of Bucovina, apparently for 1909–13, but, then, it would allow an average of 12·3 for the Old Kingdom—i. e. 0·2 more than Bucovina and 0·4 more than Transylvania—which contradicts the more specific figures of the international table, as well as general experience. This simply proves that almost any conclusion might find support in figures relating merely to a few crops and a few years. At any rate, the effect of the reform on Transylvanian production could not have been so disastrous as the destruction of a few model farms may suggest, as the average Hungarian yield was little above the Rumanian before the reform, and as it, too, has declined after the war.

Production in the new provinces has certainly gone down, but how great a share circumstances unconnected with the reform may have had in bringing this about will appear in discussing official policy. The slow execution of the reform would certainly seem to be responsible for a transitory decline of the cultivated area in Transylvania and of its output, as nearly one million ha. expropriated land were still let out to improvised tenants in 1924. Other damaging effects have been of a local and peculiar nature. A Transylvanian deputy complained in the Chamber (April 12, 1927) that the inhabitants of certain communes in the Satu-Mare district were unable to raise any crops, because of the increase in wild animals. The communes in question are surrounded by forests, which used to belong to the Karolyi family; they were then hunted regularly, and the landlord was obliged to indemnify the peasants for any depredation of their crops. But the expropriation has transferred the forests to the State; wild life has increased, and the peasants are receiving no compensation at all for the damage caused to their crops. Still more peculiar was the case brought to the notice of the Chamber by Dr. N. Lupu (May 24, 1927). He asserted that the peasants in the Maramureș county were in desperate straits, and had constantly to appeal for gifts of food. The region had always been poor, but the present trouble was due to the peace settlement. 'When the new frontiers were drawn, a hundred mountain slopes with pastures used and owned by the neighbouring communes were left to Czechoslovakia. Now these people have no means of making a living; the number of animals has fallen to one-fourth, as the villagers sold them to buy maize.'

The only conclusion emerging with any clarity from all these facts and arguments is the unhelpful one that the problem of the reform's effect on the agriculture of the new provinces, as of the Old Kingdom, bristles with complexities. The industry is passing through a period of transition, in which the action of the reform happens to be intermingled with the activity of other forces. Only in the long run, and other things being equal—as the set economic phrase goes—would it be possible neatly to disentangle the first from the remainder.

CHIEF CEREAL CROPS IN THE YEARS 1920–7

Year	Old Kingdom			Bessarabia			Bucovina			Transylvania			Rumania		
	Area	Production		Area	Production		Area	Production		Area	Production		Area	Production	
		Average per hectare	Total		Average per hectare	Total		Average per hectare	Total		Average per hectare	Total		Average per hectare	Total
	Hectares	Quintals	Quintals	Hectares	Quintals	Quintals	Hectares	Quintals	Quintals	Hectares	Quintals	Quintals	Hectares	Quintals	Quintals
Wheat															
1920	781,680	8·6	6,746,933	571,038	9·1	5,207,600	12,095	9·5	114,974	657,897	7·0	4,616,367	2,022,710	3·3	16,685,874
1921	1,161,186	10·1	11,761,287	536,027	4·4	2,376,059	15,207	10·8	164,963	775,915	9·1	7,079,175	2,488,335	8·6	21,381,484
1922	1,208,813	10·2	12,246,356	574,634	10·2	5,915,283	22,415	11·3	253,594	843,778	7·8	6,625,233	2,649,670	9·5	25,040,465
1923	1,225,027	10·7	13,138,439	769,682	11·5	8,834,117	19,663	12·4	243,983	675,969	8·2	5,576,201	2,260,341	10·3	27,792,730
1924	1,392,047	6·8	9,436,097	854,415	4·1	3,650,882	29,847	7·7	231,026	895,793	6·5	5,847,439	3,172,102	6·1	19,165,444
1925	1,588,393	9·5	15,164,403	781,084	2·8	2,257,721	24,652	11·7	288,746	906,778	11·9	10,795,177	3,300,887	8·6	28,506,047
1926	1,717,296	8·5	14,722,483	648,276	7·9	5,158,998	26,528	13·2	350,664	935,387	10·6	9,945,468	3,327,487	9·1	30,177,613
1927	1,532,632	8·4	13,011,202	607,154	6·4	3,911,664	29,033	10·9	318,048	932,334	9·7	9,086,158	3,101,153	8·5	26,327,072
Average 1920–7	*1,325,884*	*9·0*	*12,028,400*	*667,786*	*6·9*	*4,664,040*	*22,430*	*10·9*	*245,749*	*827,981*	*8·9*	*7,446,402*	*2,844,081*	*8·5*	*24,384,591*
Average 1923–7	*1,491,079*	*8·7*	*13,004,524*	*732,118*	*6·5*	*4,762,676*	*25,944*	*11·0*	*286,493*	*869,252*	*9·4*	*8,250,088*	*3,118,394*	*8·4*	*26,303,781*
Rye															
1920	65,327	7·9	520,372	149,202	8·0	1,197,316	18,490	7·1	131,825	82,584	6·6	549,707	315,603	7·6	2,399,220
1921	72,259	7·0	510,143	99,326	5·0	495,100	16,659	8·9	143,024	138,332	8·4	1,158,475	326,576	7·1	2,306,742
1922	59,316	7·9	469,871	90,691	10·5	955,088	25,031	9·5	237,608	91,485	7·4	675,881	266,523	8·7	2,358,447
1923	66,057	9·0	595,270	118,177	9·2	1,087,075	20,271	10·1	204,858	65,951	8·4	552,981	270,456	9·1	2,440,189
1924	59·895	4·1	248,178	100,139	4·9	499,470	24,609	6·5	159,228	86,811	7·0	607,796	271,454	5·6	1,512,674
1925	64,471	7·6	495,456	95,260	3·5	334,503	20,347	11·7	238,327	90,403	10·6	963,191	270,481	7·5	2,031,477
1926	83,532	8·1	676,925	93,056	9·4	877,903	23,331	10·8	252,206	95,704	10·9	1,048,652	295,623	9·7	2,855,736
1927	69,271	8·0	549,909	94,654	8·1	768,126	27,116	10·8	293,019	90,214	8·4	757,082	281,255	8·4	2,368,136
Average 1920–7	*67,516*	*7·5*	*508,266*	*105,063*	*7·3*	*776,822*	*21,981*	*9·4*	*207,518*	*92,685*	*8·5*	*789,220*	*287,246*	*7·9*	*2,281,577*
Average 1923–7	*68,645*	*7·9*	*513,148*	*100,257*	*7·1*	*713,415*	*23,134*	*9·9*	*229,537*	*85,816*	*9·1*	*785,940*	*277,853*	*8·0*	*2,241,642*

Barley															
1920 . . .	495,344	11·6	5,757,126	751,546	10·3	7,741,325	29,508	8·6	253,931	123,775	7·8	967,120	1,400,173	10·5	14,719,502
1921 . . .	709,576	8·1	5,767,872	658,901	3·9	2,563,586	32,681	8·8	289,567	168,215	7·3	1,231,898	1,569,373	6·3	9,852,923
1922 . . .	873,039	12·5	10,877,806	681,963	11·7	7,997,827	32,195	10·2	326,975	140,257	8·6	1,215,456	1,727,454	11·8	20,418,064
1923 . . .	926,900	6·7	6,297,234	762,789	7·3	5,533,790	38,143	9·7	370,381	150,559	7·6	1,141,507	1,878,391	7·1	13,252,912
1924 . . .	888,713	3·6	3,228,665	778,921	2·9	2,289,868	30,895	4·7	147,240	152,202	6·8	1,031,244	1,850,731	3·6	6,697,017
1925 . . .	847,728	7·9	6,743,040	680,653	2·2	1,516,965	28,101	8·1	228,409	147,583	11·7	1,704,864	1,704,061	5·9	10,193,278
1926 . . .	825,138	10·3	8,506,629	554,379	11·4	6,339,174	19,646	13·4	263,487	152,404	11·4	1,740,179	1,551,567	10·9	16,849,469
1927 . . .	865,938	7·3	6,392,475	715,641	5·9	4,280,587	21,297	10·1	216,014	161,384	10·1	1,718,126	1,764,260	7·1	12,617,202
Average 1920–7 .	*804,046*	*8·3*	*6,685,105*	*698,099*	*6·8*	*4,782,890*	*29,858*	*9·0*	*262,000*	*149,547*	*8·9*	*1,343,799*	*1,680,751*	*7·7*	*13·075,045*
Average 1923–7 .	*870,882*	*7·1*	*6,215,608*	*698,476*	*5·7*	*3,992,076*	*27,616*	*8·8*	*245,106*	*152,826*	*9·6*	*1,467,184*	*1,749,802*	*6·8*	*11,921,975*
Oats															
1920 . . .	516,820	11·5	5,954,364	197,386	9·4	1,870,845	27,741	9·0	252,250	224,445	8·2	1,843,438	966,393	10·3	9,920,897
1921 . . .	661,055	9·0	5,981,669	249,257	4·5	1,124,120	35,542	9·1	323,921	293,151	7·5	2,201,974	1,339,006	7·8	9,631,684
1922 . . .	752,351	10·3	7,760,405	312,677	11·7	3,676,471	35,456	10·3	365,708	233,038	6·7	1,561,918	1,333,522	10·1	13,364,502
1923 . . .	787,988	6·7	5,319,661	247,079	6·9	1,698,901	47,982	8·9	468,872	262,353	6·1	1,608,526	1,345,402	6·7	9,095,960
1924 . . .	697,981	5·4	5,779,507	218,351	2·0	438,697	45,925	9·4	432,207	274,325	5·2	1,447,804	1,236,580	4·9	6,098,215
1925 . . .	691,450	6·6	4,599,250	168,988	1·4	237,893	49,298	10·6	526,324	275,111	7·4	2,037,158	1.184,847	6·2	7,400,625
1926 . . .	658,692	10·2	6,746,554	98,662	11·7	1,161,444	40·339	12·9	521,044	280,766	11·2	3,161,245	1,078,419	10·7	11,590,287
1927 . . .	625,954	7·3	4,682,288	129,204	6·2	804,007	40,037	10·6	424,394	289,213	9·6	2,770,745	1,084,408	8·0	8,681,434
Average 1920–7 .	*674,036*	*8·3*	*5,602,962*	*202,700*	*6·7*	*1,376,547*	*40,290*	*10·2*	*414,340*	*266,550*	*7·8*	*2,079,101*	*1,196,072*	*7·9*	*9,472,950*
Average 1923–7 .	*692,413*	*7·2*	*5,025,452*	*172,456*	*6·0*	*868,188*	*44,716*	*10·6*	*474,568*	*276,353*	*7·9*	*2,205,081*	*1,185,931*	*7·2*	*8,573,304*
Maize															
1920 . . .	1,888,223	14·8	27,904,087	713,572	14·1	10,085,388	62,190	15·9	990,102	631,433	12·9	8,154,522	3,295,418	14·1	46,238,468
1921 . . .	1,861,367	8·5	15,915,910	748,403	7·1	5,324,031	60,653	10·1	612,603	773,567	8·1	6,251,164	3,443,990	8·1	28,103,708
1922 . . .	1,976,717	7·0	17,445,153	652,853	8·8	5,942,346	52,617	7·5	654,713	721,667	9·0	6,379,737	3,403,854	8·9	30,421,949
1923 . . .	1,965,803	12·3	24,306,596	678,924	10·8	7,096,408	57,967	14·8	858,710	731,798	8·4	6,196,779	3,404,492	11·3	38,458,493
1924 . . .	2,131,189	10·9	23,432,110	692,776	8·3	5,809,968	58,234	13·8	805,744	739,255	12·7	9,440,869	3,621,454	10·9	39,488,691
1925 . . .	2,308,543	11·1	25,783,004	763,473	5·5	4,211,910	64,359	16·9	1,087,680	794,471	13·2	10,508,873	3,930,780	10·5	41,591,467
1926 . . .	2,345,657	13·3	31,144,283	805,156	19·5	15,763,154	58,787	14·4	845,721	849,832	19·6	13,081,206	4,059,432	14·9	60,834,364
1927 . . .	2,554,607	7·4	18,923,688	776,699	8·8	6,880,953	58,861	12·3	722,928	829,256	10·6	8,803,893	4,219,423	8·4	35,331,462
Average 1920–2 .	*2,129,013*	*10·8*	*23,106,853*	*725,232*	*10·5*	*7,639,269*	*59,200*	*13·8*	*822,275*	*758,909*	*11·3*	*8,602,130*	*3,672,355*	*10·9*	*40,058,575*
Average 1923–7 .	*2,161,159*	*10·9*	*24,717,936*	*737,405*	*10·7*	*7,952,478*	*59,628*	*14·4*	*864,156*	*788,962*	*12·1*	*9,606,324*	*3,747,116*	*11·2*	*43,140,895*

CEREAL CROPS IN 1910–15

and their Distribution between large and small farms.

Crops	Area in hectares						Average production per hectare in quintals.						Total production in quintals.					
	1910	1911	1912	1913	1914	1915	1910	1911	1912	1913	1914	1915	1910	1911	1912	1913	1914	1915
Large property (above 100 hectares)																		
Wheat	1,044,101	983,827	1,043,451	758,582	927,592	852,022	16·8	14·5	12·6	15·2	6·7	14·4	17,562,295	13,699,326	13,129,368	11,512,899	6,262,447	12,271,120
Rye	29,879	12,004	9,017	7·737	6,325	7,023	14·0	11·4	10·6	12·7	6·6	11·9	421,137	137,964	95,810	98,055	41,520	83,306
Barley	140,717	121,928	115,158	146,276	123,120	122,121	13·2	12·2	10·8	12·3	10·4	12·3	1,863,795	1,490,972	1,243,366	1,807,011	1,283,674	1,506,328
Oats	150,254	132,583	116,483	192,577	131,063	146,905	10·9	7·9	10·0	12·0	8·9	11·6	1,648,452	1,058,109	1,162,813	2,304,047	1,161,410	1,698,533
Maize	266,845	292,135	266,716	264,341	235,732	222,188	14·3	17·0	16·2	16·5	15·4	12·8	3,821,140	4,987,408	4,310,242	4,355,170	3,627,361	2,836, 155
Millet	4,582	8,572	12,120	18,284	9,944	12,029	7·7	10·7	8·2	8·0	8·4	7·9	35,585	92,428	105,885	147,008	83,693	95,651
Small property (below 100 hectares)																		
Wheat	904,116	946,337	1,025,969	864,523	1,184,138	1,052,227	13·9	13·0	10·9	13·2	5·3	11·5	12,600,104	12,334,235	11,204,963	11,400,441	6,337,298	12,164,910
Rye	143,982	119,792	98,227	82,846	77,748	68,590	10·7	9·2	8·2	10·0	5·7	9·4	1,551,659	1,110,384	800,696	830,356	448,521	645,133
Barley	408,674	385,273	384,727	416,263	445,302	432,779	11·2	10·8	8·6	10·1	9·1	11·1	4,577,309	4,195,550	3,314,417	4,215,706	4,082,606	4,814,469
Oats	296,506	268,832	265,302	329,572	296,443	284,058	9·4	11·0	7·1	9·7	8·4	9·3	2,815,217	2,958,375	1,877,827	3,210,289	2,512,574	2,646,166
Maize	1,719,414	1,793,116	1,812,504	1,882,630	1,829,834	1,885,101	13·9	13·7	12·9	13·9	13·0	10·7	23,942,744	24,663,851	23,522,011	26,354,050	23,838,314	20,307,000
Millet	22,639	30,837	32,155	36,642	28,087	38,698	8·6	10·0	8·2	7·4	8·4	8·0	196,274	308,730	264,620	273,342	234,917	312,362

THE AREA UNDER CEREAL CROPS 1920–7
(in million hectares)

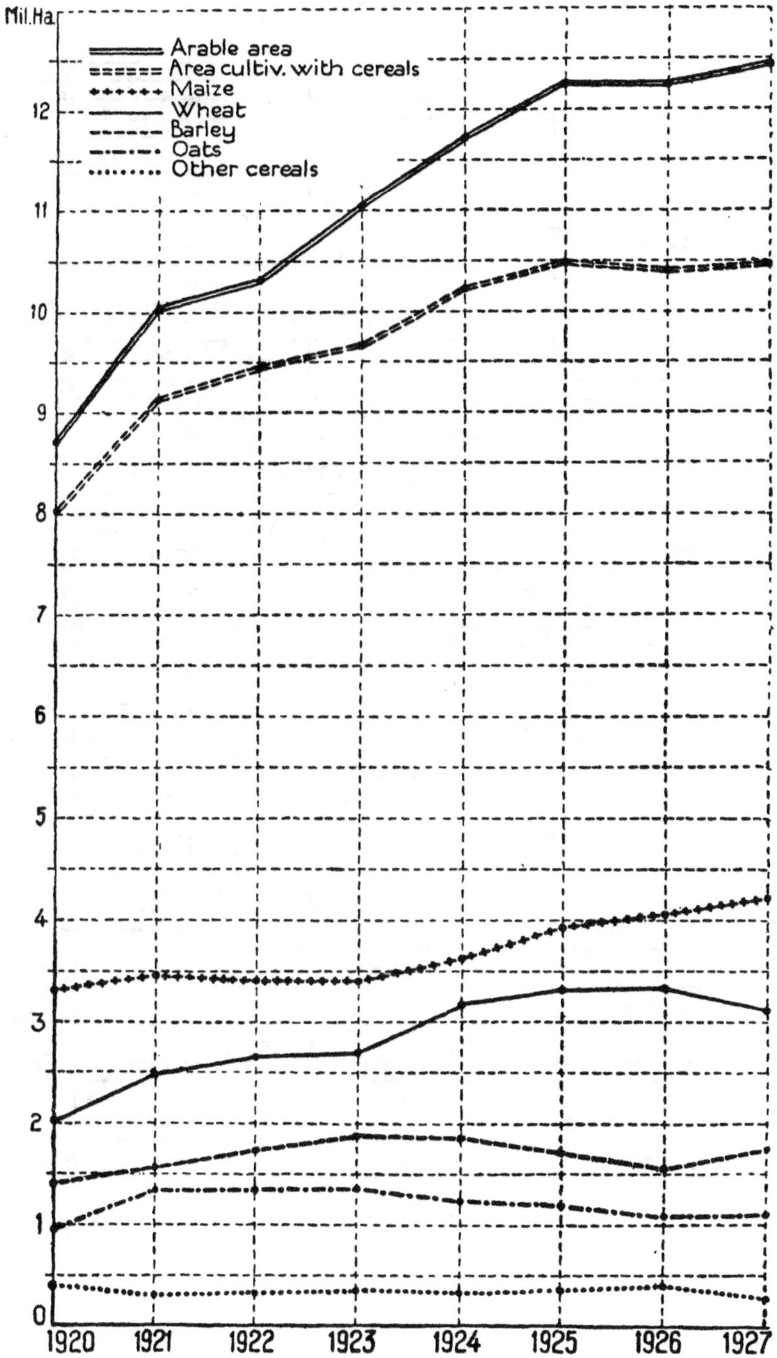

TOTAL PRODUCTION OF THE CHIEF CEREAL CROPS IN 1920–7
(in million quintals)

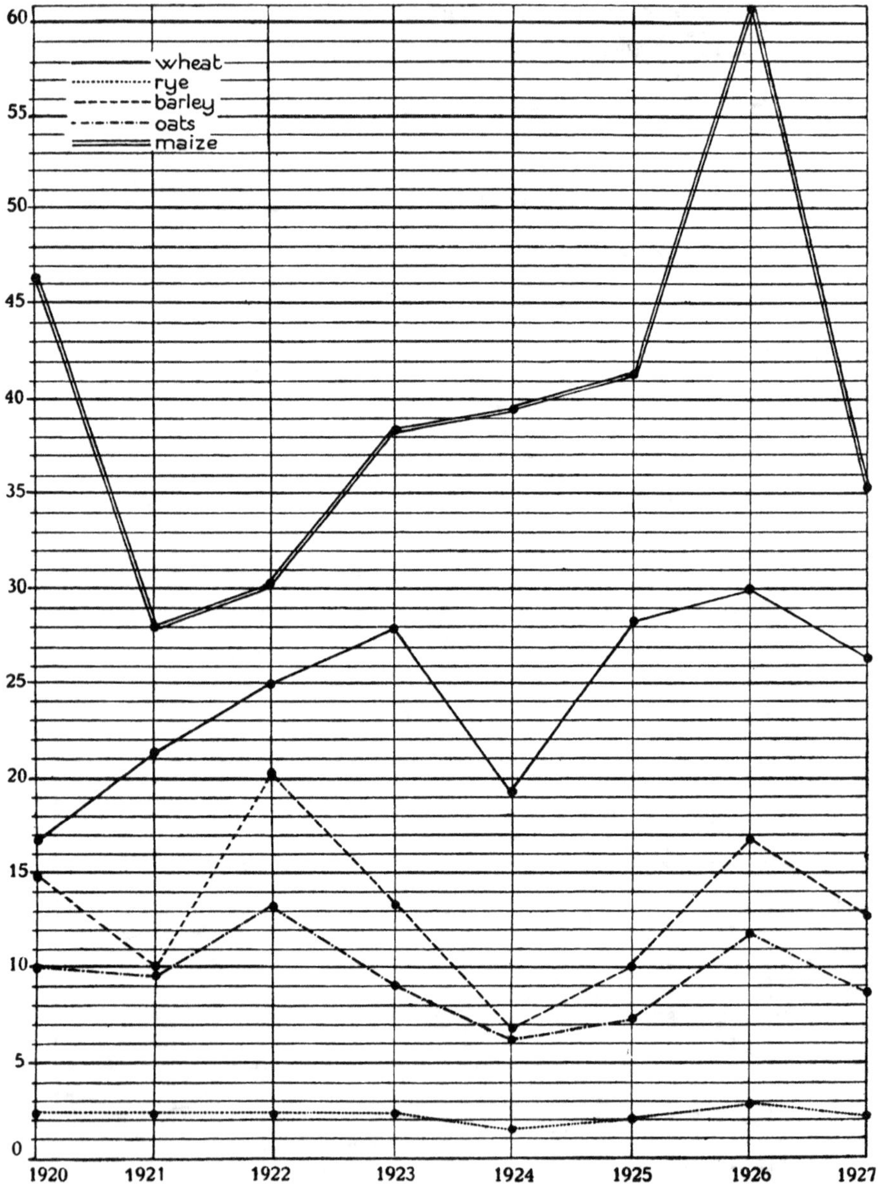

AVERAGE PRODUCTION OF THE CHIEF CEREAL CROPS IN 1920–7

(in million quintals)

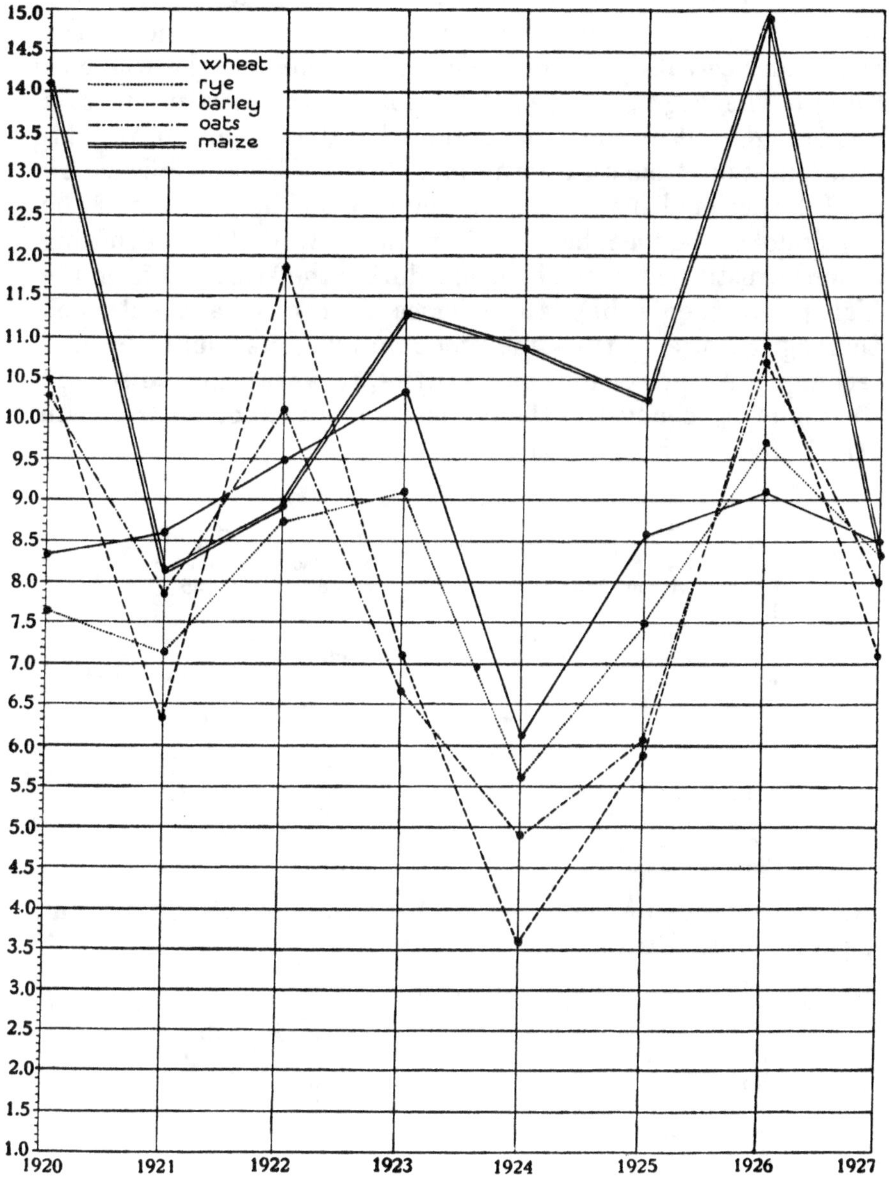

SECTION 4

THE EFFECT OF THE REFORM ON PRICES AND EXPORT

(*a*) *Export.* The lower yields in Rumanian agriculture may possibly be attributed to the land reform, but the reduction of the area under cereals is part of a tendency common to the whole of Europe. The geographical redistribution of agricultural production was brought out by the data contained in the volume on Agricultural Questions from an International Point of View, published by the International Institute of Agriculture on the occasion of the economic conference held at Geneva in May 1927. Profiting from the disturbed state of Europe during the War, and from the high prices then ruling, the oversea countries have extended their agriculture, at the same time improving its equipment and technique. An index-figure based on the pre-war averages disclosed the following changes in the production and export of the main cereal crops during the period 1921–5:

WHEAT

	Production	Export (incl. flour)
Europe (without Russia)	88	35
Russia	51	4
North America	132	241
South America	137	—
Australia	137	183

RYE

	Production	Export
Europe (without Russia)	79	31
Russia	88	69
North America	233	6,377 [1]
North America	254	536

All the American export went to Europe, and 485,100 quintals even to Soviet Russia.

BARLEY

	Production	Export
Europe (without Russia)	88	91
Russia	46	5
North America	117	319
South America	248	347
Australia	161	—

[1] Before the War, 164,000 quintals; in 1921–5, 10,459,000 quintals.

Oats

	Production	Export
Europe (without Russia)	85	108
Russia	56	3
North America	121	271
South America	112	75
Far East	227	—

Australian production showed a falling off, but that of the Far Eastern countries increased.

Maize

	Production	Export
Europe (without Russia)	84	85
Russia	234	9
North America	105	179
Central America	107	5,200
South America	—	109
Far East	108	—
Africa	129	436

In Australia maize-growing is insignificant.

Rumania's contribution weighs but lightly in the scale of world production. Yet she generally has an important exportable surplus, because of the small density and frugal habits of her population. Her place in the list of exporting countries is therefore higher than in that of producing countries. Corn-growing for export did not assume any importance till the middle of the nineteenth century. Until the beginning of that century the population was sparse and foreign trade was a monopoly of the Porte. In that phase Rumanian agriculture had a predominantly pastoral character. Freedom of trade was obtained through the Treaty of Adrianople (1829), and the Crimean War brought Rumanian agriculture into prominence. From that time onwards her corn exports never ceased to grow until the War. Enemy requisitions and war damage placed the country in the position of having to import corn and flour in 1919. In that year agricultural export was nil. It resumed its course in 1920, but with considerable differences in quantity and in kind from the pre-war export.

The export of corn has gradually been rising, but with strong variations from year to year, in keeping with the variations in production, a bad harvest reacting unfavourably on the export of the following year. Price conditions and, especially, the fiscal policy of the Rumanian governments have, however, had a strong influence on the export of agricultural products after the War; and the change in the distribution of crops has been responsible for parallel changes in the nature of exports. Rumania's corn export amounted in the twenty-five years before the War to 54 per cent. of the total harvest. During the period 1920–4 it reached a yearly average of 135,179 wagons (wheat and flour, barley, oats, and maize), i.e. about 16·5 per cent. of the production of these four cereals. The bad harvest of 1924 brought the export down to 81,563 wagons in 1925; it improved in 1926 to 187,284 wagons (including 1,395 wagons millet and 8,808 wagons beans); and the bumper maize crop in 1926 enabled the 1927 export to reach the exceptional figure of 305,658 wagons—the highest level since the War and equal to Rumania's pre-war export.

The decline of the Rumanian corn export is due, therefore, in the first place, to a fall in her own production, and, in the second place, to a redistribution of the world's production and trade. In point of production, Rumania retains a prominent position with regard to maize, being third in the list after the United States and the Argentine; she takes fifth place among the producers of barley and oats; but only seventh place among the producers of wheat, if we consider merely the countries which have an exportable surplus, and falls to the tenth place if we include Spain and Italy, who consume the whole of their own production. Generally speaking, therefore, one might say that Rumania has maintained her position among producing countries in respect of maize, oats, and barley, but has lost it in the case of wheat.

Her exportable corn surplus is greatly reduced in comparison with that before the War. That is due partly to the addition of the new provinces. Before the War, apparently, Transylvania[1] had a small surplus of wheat (65,000 tons) and oats (35,000 tons), but was short of barley (40,000 tons) and maize (32,000 tons), which means that she just about covered her needs in corn; and

[1] L. Michael, *Agricultural Survey of Europe* (The Danubian Basin), Part I, Washington, 1924.

this was also the case with Bucovina. As the production of both provinces has fallen off after the War, the deficiency has had to be made good from the surplus of the Old Kingdom. During the period 1902–11 the average yearly corn production of Bessarabia was 1,968,663 tons, from which, according to M. Synadino, 999,360 tons were available for export. During 1920–4 the average production amounted to 1,675,000 tons. Putting the internal needs of the province at 1,261,390 tons—in view of the increased population since 1902–11, and of the larger number of animals the new peasant holders kept—M. Synadino concluded that, during the period 1920–4, Bessarabia offered for export only 38,000 wagons of corn as compared with 99,000 in the period 1902–11. During the period 1911–15 the Old Kingdom produced an average of 5,989,791 tons (wheat, oats, barley, and maize) yearly; and, during 1923–7, an average of 4,896,352 tons, i.e. 1,093,439 tons less yearly. The average yearly export of the four cereals during the period 1911–15 reached 2,444,914 tons, this means that the surplus in the second period was of no more than about 1,350,000 tons, or rather one million tons, making allowance for the greater internal demand. During this period, therefore, the Old Kingdom and Bessarabia—the two provinces which have a corn surplus—could not spare more than 130,000 wagons of corn, in round figures, for export; and this roughly coincides with the actual amounts that were sent out of the country, as may be seen from the table below:

In quintals

Crop	1911–15	1920–4	1925	1926
Wheat and flour	10,002,330	865,928	310,872	2,717,710
Rye and flour	—	297,336	10,419	267,390
Maize	1,102,935	5,989,118	5,800,872	6,897,880
Barley	10,434,207	4,985,242	1,826,927	5,810,030
Oats	2,909,670	1,380,286	207,274	610,670
Total	24,449,142	13,517,910	8,156,364	16,303,680

The violent oscillation of the corn export is illustrated in the diagram on p. 342.

Among European countries exporting corn before the War Rumania took second place, after Russia. Through the virtual

THE EXPORT OF CORN

(in million tons)

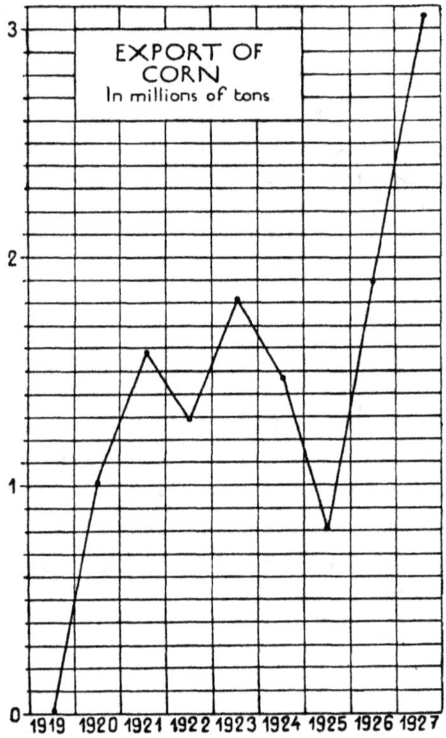

(*After G. Ionescu-Sisești*)

disappearance of Russia, for the time being, from the market, Rumania has advanced to the first place in Europe, without, however, being able to keep proportionately in step with the advance in the export of the oversea countries. The following tables show the interesting changes of position which have taken place since the War in the export of the main cereal crops:

1. *Wheat*

Period 1909–13	quintals	Period 1921–5	quintals
1. Russia	41,174,000	1. Canada	56,334,000
2. Argentine	24,249,000	2. United States	38,232,000
3. Canada	20,152,000	3. Argentine	33,179,000
4. United States	14,271,000	4. Australia	20,160,000
5. Rumania	13,314,000	5. India	4,381,000
6. India	13,176,000	6. Hungary	4,580,000
7. Australia	11,430,000	7. Russia	1,787,000
8. Bulgaria	2,130,000	8. Yugoslavia	949,000
9. Serbia	1,010,000	9. Bulgaria	505,000
		10. Rumania	502,000

2. *Oats*

Period 1909–13	quintals	Period 1921–5	quintals
1. Russia	10,683,000	1. Canada	4,894,000
2. Argentine	6,170,000	2. Argentine	4,621,000
3. Canada	1,803,000	3. United States	1,772,000
4. Rumania	1,559,000	4. Rumania	1,374,000
5. Algeria	568,000	5. Algeria	540,000
6. United States	567,000	6. Russia	249,000

3. *Rye*

Period 1909–13	quintals	Period 1921–5	quintals
1. Russia	5,341,000	1. United States	8,751,000
2. Rumania	894,000	2. Russia	3,706,000
3. United States	145,000	3. Canada	1,708,000
4. Argentine	69,000	4. Hungary	575,000
5. Canada	19,000	5. Poland	485,000
		6. Argentine	405,000
		7. Rumania	250,000

4. *Barley*

Period 1909–13	quintals	Period 1921–5	quintals
1. Russia	36,998,000	1. United States	4,633,006
2. Rumania	3,568,000	2. Rumania	4,476,000
3. India	2,295,000	3. Canada	4,235,000
4. United States	1,628,000	4. Russia	1,855,000
5. Algeria	1,067,000	5. India	1,396,000
6. Canada	1,029,000	6. French Morocco	1,201,000
7. French Morocco	652,000	7. Czechoslovakia	837,000
8. Tunisia	618,000	8. Argentine	771,000
9. Argentine	166,000	9. Poland	618,000
		10. Algeria	598,000
		11. Hungary	113,000

5. *Maize*

Period 1909–13	quintals	Period 1921–5	quintals
1. Argentine	29,401,000	1. Argentine	31,929,000
2. United States	10,264,000	2. United States	18,313,000
3. Rumania	9,837,000	3. Rumania	6,149,000
4. Russia	7,544,000	4. South Africa	4,040,000
5. Bulgaria	2,067,000	5. Yugoslavia	3,352,000
6. Serbia	1,075,000	6. Bulgaria	1,092,000
7. South Africa	952,000	7. Russia	706,000

These tables show that Rumania has maintained second place in the export of barley, with an increased total; she has kept the fourth place in the export of oats, but with a reduced quantity; she has retained the third place in maize export, with a reduced quantity; but she has dropped severely in the wheat table from fifth to tenth place, and from 133,000 to 5,000 wagons yearly.

More significant still, she has been outdistanced in wheat export by her neighbours—Hungary, Yugoslavia, and Bulgaria. Bulgaria and Yugoslavia suffered as much as Rumania, or more, through the War; and Yugoslavia, too, carried out an extensive land reform, greatly misapplied; their being able to better Rumania's wheat export could, therefore, only be explained through the restrictive policy adopted by the Rumanian Governments for the purpose of accumulating cheap internal supplies.

> 'The exported quantities', wrote M. Ionescu-Siseşti, 'could have been larger, in spite of the present lowered production, if the outflow of the available surplus towards export were stimulated by means of appropriate measures. Proof that we are not doing this may be found in the fact that our internal consumption, reckoned per head, appears to be high, although our population is frugal and the seed ration given to our animals is very reduced in relation to the manner in which animals are fed elsewhere. We do not consume, we waste.' [1]

The yearly consumption of cereals per head was 417 kg. during the period 1920–4, which includes food for men and animals as well as seed; in the same years it was 308 kg. in Poland, 319 in Czechoslovakia, 339 in Bulgaria, 348 in Yugoslavia and 471 in Hungary. Hungary was the only country to exceed the Rumanian average, but she exports a larger number of fattened animals. Internal consumption has been quicker in

[1] *Participarea României la Producţiunea şi Comerţul Mondial de Cereale,* 1928, p. 28.

reaching the pre-war level than either production or export. The average consumption during 1910–15 was 490 kg. per inhabitant; this was equalled by the figure for 1925.

Rumania's corn export has fallen, therefore, from a yearly average of 40·8 per cent. of the total harvest (wheat, barley, oats, and maize) in 1911–15 to 18·7 per cent. in 1923–7; and the proportion of wheat in those figures from 16·7 per cent. to 2·2 per cent. The general fall in corn export is accompanied by a change in the kind of cereals exported. This is seen in the following table, referring to the export of the four main cereal crops:

Crop	Percentage of their production		Percentage of total export	
	1911–15	1923–7	1911–15	1923–7
Wheat	45·2	7·4	41·0	11·6
Maize	37·0	20·6	42·6	52·9
Barley	52·0	42·9	11·9	30·3
Oats	26·7	10·3	4·5	5·2

Wheat has been replaced by spring cereals—barley, maize, oats and even millet—which were allowed to be exported and brought higher returns with less effort and expense. The maize export represented 9 per cent. of the world trade in 1925, 8·6 per cent. in 1926 and 15·4 per cent. in 1927, when Rumania came next after the Argentine.

These changes in the nature of the agricultural export are concomitant with the variation in the distribution of crops described earlier in this chapter. Indeed, one can trace the effect of the transition to small-scale farming more clearly in the change of exports than in the change of crops. The two tables on pp. 355–6, which give the quantity and the value of all Rumanian exports during the years 1919 to 1926, supply all the details to illustrate that transformation. Here we will merely give a table showing the proportion of the various products in the value of exports during 1921 to 1926, in comparison with the corresponding figures for 1913 (see p. 346).

The export of live stock was valued at $579,000 in 1913; it reached $6·5 millions in 1923, over 14 millions in 1924 and nearly 20 millions in 1925. Compared with a percentage of 3·2 of the total value of exports which animal products of all kinds represented in 1913, this category rose during 1921–6 to an average value of 14·31 per cent.—a significant increase. The percentage oscillated between 8·7 in 1923 and 20·84 in 1925; and it is worth

Products	1913	1921	1922	1923	1924	1925	1926
Live stock . . .	0·4	9·06	13·33	7·0	10·19	14·50	8·10
Animal food-stuffs .	1·7	2·93	0·63	0·9	2·62	4·52	4·20
Hides and skins . .	0·5	0·27	0·23	—	0·32	0·58	0·42
Wool	0·4	1·10	0·16	0·2	0·26	0·55	0·39
Animal waste . .	0·2	0·30	0·49	0·6	0·82	0·69	0·41
Cereals . . .	67·0	50·48	38·31	49·6	43·90	25·22	36·95
Vegetables, seeds, and fodder . . .	5·1	4·64	4·87	7·4	4·87	4·83	4·10
Vegetable oils . .	—	0·06	0·01	—	0·53	0·51	0·40
Timber and derivative products . . .	3·5	5·73	18·52	16·5	19·25	21·50	15·03
Oil	19·6	22·41	18·53	13·1	12·13	19·85	25·00
Other products . .	1·6	3·02	4·92	4·7	5·11	7·25	5·00
	100·00	100·00	100·00	100·00	100·00	100·00	100·00

noting that it falls and rises with some regularity inversely to the rise and fall of the percentage for cereals. The export of animal products, that is, now plays a compensatory part to the export of corn, which illustrates in a concrete manner the suggestion we made before, that the risks of agriculture would be lessened by the greater variation in the nature of farming.

The lesser decrease in the value of agricultural exports, as seen in the tables on pp. 355–6, compared with the sharper drop in total quantities, suggests, moreover, that the change in the varieties exported has not been unprofitable in itself. But, in general, there is no doubt that Rumania has lost the place which she occupied in international trade before the War. The first volume of the *Memorandum on the Balance of External Trade*, 1913–25, published by the League of Nations, states that the total value of international trade has increased during that period by 5 per cent. At the same time Rumania's share in it has decreased by one-half, although the country's area and population have doubled. In 1913 exports amounted to $130 millions, equal to 0·71 per cent. of the world trade; in 1925 this share had fallen to 0·47 per cent. Imports represented

0·58 per cent. in 1913 and 0·46 per cent. in 1925 of the world's trade. The same conclusion is drawn from the figures indicating the value of Rumania's export per head of population. From $16·98 per head in 1913 it has fallen to $8·23 in 1925, Rumania passing from the eleventh to the twenty-first place among European exporting countries; and if for a precise comparison with 1913 the figure $8·23 is divided by the index 1·575, representing the depreciation in the purchasing power of the dollar, the value of exports would be merely $5·20 per head of population. World trade, in short, has increased by 5 per cent., whereas Rumania's share of it is 3·5 times smaller. And that is not a general phenomenon among the countries of south-eastern Europe. Hungary exported $18 per head in 1913 and $17 in 1925; Bulgaria $3·7 and $8, and Yugoslavia—though exceptionally hard hit by the War—has increased its exports from $3·5 in 1913 to $12 per head in 1925. Once again one must draw attention to the fact that Yugoslavia has carried out a land-reform as extensive as Rumania's, which has not prevented an increase in her exports. One might also note the obviously greater powers of recovery of a country of peasant holders, such as Bulgaria, in comparison with, for instance, Hungary, a country in which large estates predominate.

Up to a point, the fall in Rumania's corn export may be laid directly at the door of the land reform. The corn trade, especially for export, was bound to suffer more than production from the break-up of the large estates, with their centralized administration, with established connexions with traders, banks, and shipping firms. The division of the harvest among a large number of small farmers has made its collection, storing, and transport more difficult and more costly than it was before; likewise more difficult to estimate the available quantities quickly, and in consequence the surplus that might be offered for export. The same circumstance has reduced the uniformity of the product, the peasants using the more expensive selected seed even less than the large farmers. The large cultivators, too, usually supply a mixture of varieties, and primitive means of harvesting and handling are causing a considerable percentage of impurities to appear in the peasant corn. Though Rumania's soil and climate can produce the best wheat in Europe, with which alone the Russian could compete, the denomination 'of Danubian origin' has acquired a derogatory meaning in the international corn trade.

(*b*) *The Value and Prices of Agricultural Products.* The drop in the value of Rumanian exports is in part due to the fact that the price of agricultural products has increased less than that of industrial goods. The general nature of Rumania's exports and imports has remained the same as before the War, but notwithstanding—or, rather, because of—the drop in the value of exports, more has proportionately to be sent abroad than before. Whereas 3·7 tons of exports corresponded to one ton of imports before the War, after it the relation changed as 5:1, i. e. five tons were exported for each ton imported. During the pre-war years the average value of exports was estimated at $28 per ton,[1] and of imports at $105, which characterizes the difference between the nature of the goods bought and those sold. That relation of values was upset after the War, to Rumania's disadvantage. During the period 1919–23 the average value of exports remained unimproved at $28 per ton, but the value of imports has more than doubled, rising to $250 per ton. During the following years, 1924–6, the average value of exports remained about the same at $29 per ton, while the moderation in the price of industrial goods brought the value of imports down to $160 per ton.

The gross value of the agricultural production[2] has been

[1] The average export value of corn was estimated at $28 per ton in 1914; it rose to $95 in 1915, an increase which has brought the general average for the pre-war years to $32. See L. Colescu, *Comerţul Exterior al României Inainte şi după Războiul Mondial*, 1928, p. 564.

[2] An estimate of the total value of Rumanian production in 1923 has been given by M. I. I. Georgianu in the *Analele Statistice şi Economice*, vol. viii, Nos. 1–2, 1925.

	(in million lei)
1. Agricultural production	75,000
2. Timber products	5,000
3. Live stock and animal products	50,000
4. Mineral products	6,000
5. Industrial production lei 29,000, of which the actual finishing contribution made by industry was	14,000
Lei	150,000

estimated on the basis of internal wholesale prices as follows (in million lei):

Crop	1923	1927
Cereals	53,539	47,207
Alimentary crops	6,635	6,781
Industrial crops	3,484	5,958
Natural and artificial grasses	5,238	4,816
Orchards and vineyards	6,633	7,102
Total	75,529	71,864

No official figures are available to show the yearly increase in the value of live stock, &c. Exports amounted to 7,662 millions (without animal products) in 1923, i.e. about 10 per cent. of the gross value of the harvest; the improvement in 1927 was exceptional, due to the large maize crop of 1926. In the years after the War, the market value of agricultural products, though higher than before the War, has not risen in the same degree as the value of industrial and manufactured goods. Agricultural industry has thereby everywhere been placed at a disadvantage which is one of the main causes of the severe agricultural crisis now experienced by most European and oversea countries. How much worse, therefore, must the situation be for the Rumanian farmers, who have been prevented from obtaining even the relatively low prices ruling in the world markets? They have not had, like farmers elsewhere, an opportunity of laying aside handsome profits during the War years; and price and export restrictions, export taxes and other government measures spoilt their chances of making the most of the boom which followed immediately upon the War. Hence they had no reserve of profits to enable them to hold out in the bad years that were to come. The ton-value of agricultural exports, we have seen, remained the same as before the War, because the difference was absorbed by export taxes; and as the pre-war values were naturally those ruling in the world markets, it has meant that post-war prices had to be lower than world prices by at least the amount of the export tax, if Rumanian corn, &c., was to reach those markets at all. The accompanying diagrams show indeed that the internal prices of wheat and maize, as in fact of all other

WHEAT PRICES

The price of wheat per quintal, in gold francs, in the principal markets

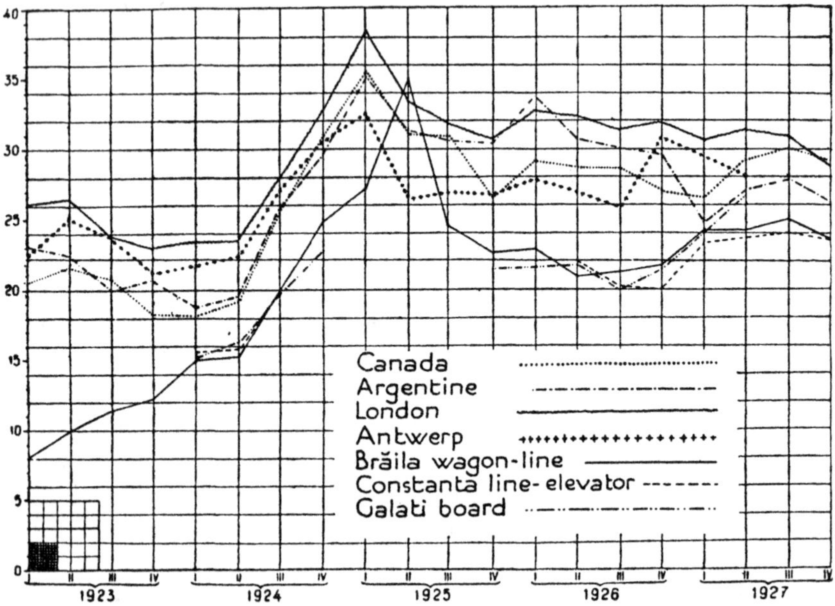

MAIZE PRICES

per quintal, in gold francs, in the principal markets

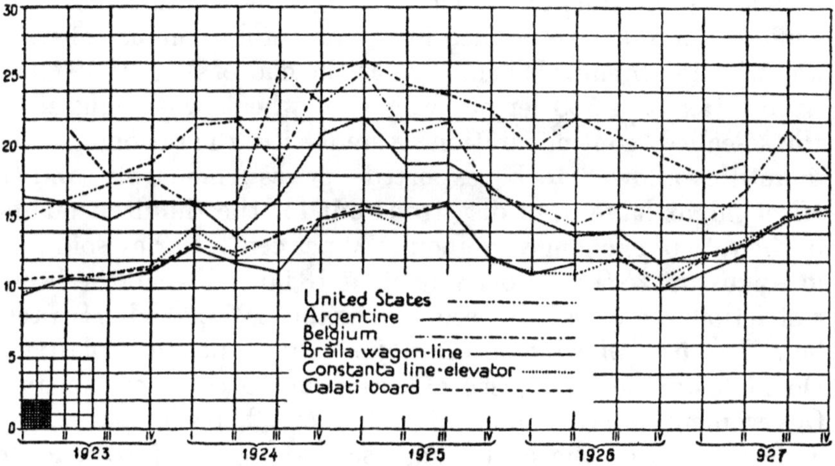

(*After G. Ionescu-Sisești*

crops, have been lower than in any other of the world's markets; and not merely in the importing European markets, which would have been natural, but also in the overseas exporting markets.

The price-fixing and taxation policy of the government was bound to discourage production. Taken in conjunction with the violent fluctuations in the exchange, it was even more likely to confuse and hamper exports. Frequently, indeed, exports moved inversely to production, which only a general confusion of prices could explain, as exemplified in the wasteful home use of wheat and wheaten bread. When, e.g., at the end of August 1921, the export tax of 1,500 lei per wagon of wheat was replaced by the so-called 'contingentation' system—i.e. the rationing of the quantities allowed to be exported—prices fell sharply, and the effect might have been disastrous but for the simultaneous fall in the rate of exchange by about 100 per cent. Barley sold early in August at 25,000 lei per wagon, at 18,000 lei when the export restrictions were first imposed, and again at 25,000 lei in September. In the following autumn just the opposite took place. Deluded by the high price of corn, due to the low exchange, Government imposed an export tax of 20,000 lei per wagon. Soon afterwards the exchange rose and prices of corn fell to such a level that towards the middle of August 1922 the export tax absorbed about half the value of the corn in the producer's hands. Had the exchange continued to rise, it might soon have absorbed the whole value of the corn. The incident disclosed what fantastic effects could result from a high fixed tax, when a tax sliding with the value of corn might have achieved some stability. The opposite table, giving the prices of the four main crops in the principal Rumanian ports, shows how strongly they oscillated from quarter to quarter.

Before the War, Rumanian wheat naturally sold at world prices. In 1924 the wagon of wheat at Craiova, one of the chief collecting centres, was 75–80,000 lei, while in Paris it was worth 115,000 lei. Altogether the dearness of agricultural products in Rumania was merely apparent, because of the continuous fall in the exchange. Whereas before the War the wagon of wheat had a value of 2,000 Swiss francs, M. Garoflid calculated that towards the middle of 1922 it valued only 700 Swiss francs;

while the cost of production had remained as high as before in gold values. Another consequence of that official interference with the market was that during the first years of the exceptional régime wheat prices were almost level with those of maize and barley. By some writers this was fallaciously attributed to excess

(In gold francs)

Year		Wheat	Maize	Barley	Oats
1923	I .	8·10	9·67	9·71	10·33
	II .	11·27	10·53	10·57	11·22
	III .	11·19	10·73	9·50	9·87
	IV .	12·05	11·23	10·43	9·06
1924	I .	15·26	13·19	12·82	10·41
	II .	15·91	12·03	11·73	9·68
	III .	19·68	13·84	16·72	11·97
	IV .	24·11	14·61	17·98	14·69
1925	I .	27·25	15·58	17·95	17·80
	II .	33·00	14·90	15·35	19·10
	III .	23·05	15·62	14·56	15·37
	IV .	21·21	11·85	12·37	12·14
1926	I .	21·96	11·34	12·04	11·85
	II .	21·16	11·50	13·30	12·24
	III .	20·91	11·99	12·55	11·35
	IV .	20·88	10·02	13·34	10·73
1927	I .	23·64	11·80	16·95	13·73
	II .	23·88	13·00	19·14	16·42
	III .	24·04	15·09	19·03	16·14
	IV .	23·58	15·22	19·48	17·57

of world-production; but Rumania was a closed market, yet internal wheat prices were lower by half than those ruling in the world markets. In October 1923, indeed, Rumanian wheat sold at 45,000 lei, while American wheat offered at about 90,000 lei at Constanța. Nor did these circumstances stabilize the internal market. Because of the shortage of money and of export prohibitions for white flour, Rumanian mills only purchased what they strictly needed at the time. The same reasons forced cultivators to offer as much as they possibly could; with the result that at a certain moment, in 1923, wheat actually fell by about 15,000 lei below the inadequate maximum prices fixed by the Government.

There is no saying, of course, whether the economic and fiscal policy of Rumania's governments would have been different without the reform. But, faced with such a policy, the large cultivators could hardly have survived it, unless permitted to go to the uttermost limit in squeezing the soil and sweating the peasants. All improvement in agricultural methods and equipment would have been out of the question. In fact, those better equipped farms in the Kingdom and in Transylvania, which spent relatively more on labour and on the upkeep of live and dead stock, would have been the first to succumb. For a time corn production may have kept up the flourishing appearances of the past. Large owners and tenants could not have shown their resentment in the form of passive resistance, which the wellnigh self-sufficing peasants can afford to practice. Through their efforts the export of corn might have risen to better figures, without necessarily implying a better yield. Exports from bad cultivation and in spite of bad policy could only have been ravaging in the extreme. In the state in which the War left the country, such exports would have been sucked out of the very marrow of the soil and of those who tilled it, to the lasting impoverishment of both. There is, at least, hope in the new agrarian organization. Though it has not been prolific in the adverse conditions which have surrounded its infancy, it may bear the seed from which a real agriculture will spring. In any case, it cannot fail to deal more kindly with the soil and its labourers.

EXPORTS. 1919–26

VALUES

Nature of Exports	1919		1920		1921		1922		1923		1924		1925		1926	
	Thousand lei	% of total	Thousand lei	% of total	Thousand lei	% of total	Thousand lei	% of total	Thousand lei	% of total	Thousand lei	% of total	Thousand lei	% of total	Thousand lei	% of total
Cereals and cereal products	2,320		2,316,725		4,171,662		5,378,485		12,187,077		12,214,415		7,319,002		14,090,215	
Vegetables, flowers and plants	15,459		227,436		382,256		682,971		1,860,225		1,355,383		1,401,561		1,492,951	
Fruit	1,168		19,795		98,597		148,369		227,090		401,156		594,635		485,342	
Wines	1,949		8,079		2,954		32,045		4,144		7,296		17,884		11,930	
Total	*20,896*	*20·01*	*2,572,035*	*74·60*	*4,655,469*	*56·34*	*6,241,870*	*44·46*	*14,278,536*	*58·02*	*13,978,250*	*50·24*	*9,333,082*	*32·16*	*16,080,438*	*42·07*
Live animals	2		803		744,202		1,870,895		1,717,155		2,835,594		4,207,463		3,070,247	
Animal products	1,992		26,382		242,888		101,507		265,315		730,015		1,387,242		1,942,991	
Total	*1,994*	*1·91*	*27,185*	*0·79*	*987,090*	*11·95*	*1,972,402*	*14·05*	*1,982,470*	*8·06*	*3,565,609*	*12·81*	*5,594.705*	*19·27*	*5,013,238*	*13·12*
Timber	*5,162*	*4·95*	*124,552*	*3·61*	*473,001*	*5·73*	*2,599,972*	*18·52*	*4,091,405*	*16·62*	*5,356,504*	*19·26*	*6,239,080*	*21·50*	*5,899,369*	*15·43*
Oil	*63,173*	*60·52*	*657,767*	*19·08*	*1,858,671*	*22·49*	*2,601,579*	*18·53*	*3,222,426*	*13·09*	*3,375,370*	*12·13*	*5,759,729*	*19·84*	*9,480,922*	*24·80*
Various	*13,160*	*12·61*	*66,308*	*1·92*	*288,778*	*3·49*	*623,473*	*4·44*	*1,035,631*	*4·21*	*1,547,868*	*5·56*	*2,098,360*	*7·23*	*1,749,553*	*4·58*
General total of exports	*104,385*	*100·00*	*3,447,847*	*100·00*	*8,263,009*	*100·00*	*14,039,296*	*100·00*	*24,610,468*	*100·00*	*27,823,601*	*100·00*	*29,024,956*	*100·00*	*38,223,520*	*100·00*

EXPORTS. 1919–26

QUANTITIES

Nature of Exports	1919		1920		1921		1922		1923		1924		1925		1926	
	Tons	% of total	Tons	% of total	Tons	% of total	Tons	% of total	Tons	% of total	Tons	% of total	Tons	% of total	Tons	% of total
Cereals and cereal products	2,744		983,042		1,538,865		1,212,372		1,777,617		1,414,330		884,527		1,791,707	
Vegetables, flowers and plants	10,694		84,000		94,472		99,907		144,033		77,235		112,344		139,230	
Fruit	622		5,630		18,257		16,963		13,233		31,132		43,902		26,522	
Wines	556		1,538		534		5,665		932		706		1,547		737	
Total	*14,616*	*13·39*	*1,074,210*	*73·22*	*1,652,219*	*60·90*	*1,334,907*	*32·80*	*1,935,815*	*39·49*	*1,523,403*	*32·19*	*1,002,320*	*21·72*	*1,958,196*	*32·08*
Live animals	1		573		51,638		107,254		58,423		69,796		78,909		65,796	
Animal products	242		2,860		7,267		2,919		7,935		13,461		24,590		34,692	
Total	*243*	*0·22*	*3,433*	*0·23*	*58,905*	*2·17*	*110,173*	*2·71*	*66,358*	*1·35*	*83,257*	*1·76*	*103,499*	*2·24*	*100,488*	*1·65*
Timber	*22,562*	*20·67*	*92,234*	*6·28*	*570,818*	*21·03*	*1,968,575*	*48·37*	*2,224,935*	*45·39*	*2,483,377*	*52·49*	*2,467,197*	*53·47*	*2,220,870*	*36·39*
Oil	*44,014*	*40·33*	*249,097*	*16·98*	*377,328*	*13·91*	*435,526*	*10·70*	*416,025*	*8·49*	*437,914*	*9·26*	*787,617*	*17·07*	*1,501,911*	*24·61*
Various	*27,705*	*25·39*	*48,144*	*3·29*	*53,868*	*1·99*	*220,782*	*5·42*	*258,516*	*5·28*	*203,533*	*4·30*	*254,033*	*5·50*	*322,210*	*5·27*
General total of exports	*109,140*	*100·00*	*1,467,118*	*100·00*	*2,713,138*	*100·00*	*4,069,963*	*100·00*	*4,901,649*	*100·00*	*4,731,484*	*100·00*	*4,614,666*	*100·00*	*6,103,675*	*100·00*

CHAPTER X
THE EFFECTS OF THE REFORM ON RURAL ECONOMY

Section 1

The Effects of the Reform on Live Stock

The rearing and breeding of animals acquired a much greater importance in Rumanian agriculture after the reform than it ever had before. It is the branch of rural economy which more than any other favours smallholders; and to them animals are valuable both as a help in cultivation, and as objects in the production and sale of which the peasants can easily hold their own against competition from large producers. The fodder crops grown by smallholders in connexion with the live-stock industry introduce a greater variety in farming, and the nature of their cultivation helps to clear the soil of weeds and to prepare it for cereal crops. Rumania's agricultural production, moreover, can no longer be maintained on a competitive basis without the regular use of manures; which gives additional value to numerous live stock.

Before the dispossession of the peasants the rearing and breeding of animals formed the main source of wealth for the Rumanian Principalities. Their animals were famous for number and fine quality. Among the Greeks of the Eastern Empire 'Valachian' was synonymous with 'owner of rich herds'. Rumanian horses were sung in the 'Niebelungen Saga' (Part XXII, verse 4), and Germany for a long time bought them as remounts for her calvary; hence, the old name of 'Wallach', given to geldings, first appearing towards the middle of the sixteenth century. Moldavian mares were introduced into Denmark for the improvement of the native breed. The Turks greatly appreciated Rumanian horses and Moldavia's yearly tribute included 'forty good Moldavian horses'. They even had a proverb that 'there is nothing to beat a Persian peasant and a Moldavian horse'.

Almost as renowned were Moldavian horned cattle, which used to be exported to all the neighbouring countries. But with the growing encroachment of the landlords upon peasant lands, the area under grass constantly lessened; while the economic importance acquired by corn-growing, and the political influence acquired by corn-growers, caused the neglect of that branch of rural economy which had made the prosperity of the peasants. For the peasants it became increasingly difficult to keep good animals, and the new extensive farmers had no interest in keeping them. The State took so little interest in the maintenance of the former wealth in live stock, that advantages were secured for the export of corn in the trade convention concluded with Austro-Hungary, while the export of animals, formerly so extensive and profitable, was at the same time restricted.

The consequent stagnation, or rather relative decline, in the quantity of live stock, is made evident by the table below, though it cannot also show the much greater relative decline in the quality of the stock:

Year	Oxen and Buffaloes			Pigs		Horses	
	Total number	Per 100 inhab.	Per 100 hectares arable	Total number	Per 100 inhab.	Total number	Per 100 inhab.
1860 .	2,751,168	70·2	41·06	1,088,737	27·8	512,839	13·1
1873 .	1,886,990	43·3	—	836,944	19·2	433,593	9·9
1911 .	2,666,945	37·7	33·34	1,021,465	12·8	828,962	10·4

(After M. Şerban, *op. cit.*, p. 120)

Notwithstanding this decline in cattle rearing, Rumania still took a high place among European countries with regard to the number of animals. But the quality of her live stock was poor, as may be inferred from the following table, in which the proportion of animals is given side by side with the proportion of grass and pasture:

Country	Year	For 1,000 inhabitants			Year	% Arable	% Grass-Pasture
		Horses	Oxen	Sheep			
Germany . .	1900	74	336	173	1900	50·4	16·9
Austria-Hungary	1910	85	346	264	A 1911	37·6	25·4
					H 1910	45·6	22·9
Belgium . .	1910	43	253	25	1910	52·3	20·0
Bulgaria . .	1913	110	370	1,994	1910	45·2	16·6
France . .	1908	81	330	367	1910	47·8	20·1
Italy . . .	1914	28	184	332	1911	51·9	21·2
Great Britain .	1900	31	169	578	1911	30·6	63·5
Rumania . .	1910	146	434	949	1905	60·2	15·1

(After Dr. N. D. Cornăţeanu).

The country, therefore, had a large quantity of animals, but devoted a small area to their feeding; the village commons were of poor quality, and most of the year the animals had to subsist on straw and maize stalks. Moreover, most of the large animals were draught animals, an excessive number, in fact, because the prevailing customs forced the peasants to supply working teams and because the circumstances in which the work was performed, and the poor quality of the stock, made it necessary to employ a proportionately large number of animals.

The distribution of live stock before the reform was remarkable for the enormous disproportion in the number of animals owned by each of the two categories of cultivators. Immediately before the War, the peasants owned 85 per cent. of the total number of horses, 87 per cent. of all the horned animals, 76 per cent. of the sheep and 83 per cent. of all the pigs. In Muntenia the number of traction animals the peasants owned reached 92 per cent. of the total. The total number of owners was 1,238,677 in 1916, which meant that 69 per cent. of all the households owned animals as compared with 67 per cent. in 1900; on the land this percentage, of course, was higher, about 83 per cent. of all the households, the number of rural owners being 1,163,458 in 1916. The other side of the picture was represented by the considerable extent of grazing and pasture in the hands of the large owners: according to the 1911 statistics they held 174,805 ha., while 385,396 ha. were owned by the peasants. A great deal of these peasant grazings were of the poorest quality. The reform having given the peasants land for cultivation, it also increased the number and size of communal grazings, thereby emancipating the peasants from their dependence on the large owners and tenants. But in this case also the reform has not moved beyond the first step of granting land. As the peasants' knowledge and experience is greatest in connexion with the keeping of animals, the best way to intensify production would have been to encourage and give scope to that special ability. The peasants as cultivators prefer maize and barley to wheat, and from that standpoint, too, elementary economic considerations should have suggested that it was preferable not to let those raw materials be exported as such, but rather transformed into meat and fats. Official policy, however, has remained as indifferent as before to the fate of this potential source of wealth.

Nothing has been done so far for the growing of fodder crops on the land set aside for the keeping of cattle. Most of these fields are buried in weeds. There is hardly any clover to be seen, and beet and turnips are equally rare. Almost nothing is known about the preparation of fodder, about the importance of a regular feeding-time, or about the gradation of food according to the animals. Grass and pasture remain the backbone of cattle-rearing in Rumania, and the land reform has rightly provided considerable land for this purpose. But very little of it has been sown with grass, and most of these fields which had been formerly under maize have not been levelled yet. No arrangements are in force to regulate the seasonal use of such grazings.

That indifference is proving especially costly to Transylvania and Bucovina, both of them well adapted and accustomed to cattle-breeding. In some parts of these provinces, cultivation is in fact subordinate; between them they have grass and pasture covering about 2·5 million ha. They still had a useful amount of breeding material at the end of the War, but the destructive price and tariff policy the Rumanian governments adopted has threatened ruin to the live-stock industry of the new provinces, as that of the Old Kingdom was ruined before.

If, nevertheless, the country has been able to make good within a few years the decrease in live stock caused by losses and over-consumption during the War, that is altogether due to the efforts of the peasants. This is admitted by every one. The large owners and tenants had neither the experience nor the means, in capital, buildings, grazing, &c., of rebuilding the country's depleted stock. The Old Kingdom alone is supposed to have lost over six hundred thousand horses and more than a million horned animals during the War. The following table indicates these losses, as well as the rapidity with which they were replaced in the Old Kingdom:

	1900	1916	1919	1921	1927
Horses	864,324	1,218,563	603,075	793,108	967,706
Horned cattle	2,588,266	2,937,877	1,862,744	2,579,941	2,680,265
Sheep	5,656,444	7,750,809	3,306,327	5,413,850	7,052,738
Pigs	1,709,205	1,402,187	822,453	1,309,408	1,509,347

The yearly increase in the two periods 1900–16 and 1921–4 is shown below:

	Yearly increase 1900–16	1921–24
Horses	28,390	58,532
Horned animals	21,834	546,272
Pigs	(decrease)	66,646 [1]

It should be mentioned that the post-war figures represent only estimates—and those for the first peace year probably a rough estimate—as a census of animals has not been taken since the War. Nor has their distribution among the various categories of landowners been taken account of, though it is clear that the proportion owned by the peasants must be even greater than before.

The total number of animals, according to the official Statistics of Domestic Animals, has changed after the reform as follows:

	Horses	Oxen	Sheep	Pigs
1921	1,686,728	5,520,914	11,194,047	3,159,591
1922	1,802,051	5,745,534	12,320,569	3,146,806
1923	1,828,129	5,553,871	12,480,967	2,924,603
1924	1,843,208	5,398,704	13,611,902	3,133,144
1925	1,814,804	5,049,078	12,950,212	3,087,869
1926	1,877,285	4,798,384	13,581,869	3,167,722
1927	1,939,438	4,552,166	12,941,051	3,075,782

According to figures published in the *Buletinul Agriculturei* (No. 2, 1925), the total number of domestic animals in the provinces composing Greater Rumania changed as follows:

Before the War	28,836,000
After the War	23,364,000
Minus difference	5,472,000

All European countries have seen their live stock reduced through the War. For purposes of comparison we give below the figures for the same periods relating to three of Rumania's neighbours:

	Before the War	After the War	Minus difference
Hungary	8,773,000	6,370,000	2,403,000
Yugoslavia	29,005,000	16,892,000	12,113,000
Poland	22,116,000	18,444,000	3,672,000

[1] R. C. Stere, *Buletinul Agriculturei*, April–June, 1927.

It would seem that Rumania stands alone among the countries which have been involved in the War in having been able to increase her live stock, between 1919 and 1925, by 42 per cent.; thereby coming nearest to making good the War damage. Among sixty-nine countries compiling animal statistics, Rumania took twelfth place in 1925 with 79·2 animals to the square kilometre, Denmark being first with 148, Uruguay second with 143, and England third with 139.

The number of domestic animals in proportion to the size of the country and of its population has undergone the following changes:

	Horses		Oxen		Sheep		Pigs		Animal units [1]	
Year	Per 1,000 inhab.	Per sq. km.	Per 1,000 inhab.	Per sq. km.	Per 1,000 inhab.	Per sq. km.	Per 1,000 inhab.	Per sq. km.	Per 1,000 inhab.	Per sq. km.
1900	144	7·0	431	20·0	942	43·0	285	13·0	703	32
1916	154	9·0	371	21·0	989 [2]	57·0	175	10·0	648	37
1927	111	6·6	261	15·4	741	43·9	176	10·4	478	28

All species of animals, except horses, were continually decreasing in number before the War. In the first peace years the peasants made a strong effort to make good the War losses, with considerable success, but the movement lost impetus after a few years, and a fresh decline set in instead. From 1926 to 1927 horses only increased in number with 62,153 or 3·31 per cent. The number of oxen has never ceased to diminish since 1922; from 1926 to 1927 it was reduced by 246,218, or 5·4 per cent. Sheep decreased by 640,818 or 4·71 per cent. Pigs only lost 91,940 or 2·9 per cent. The number of animal units fell from 29 to 28 per square kilometre, and from 501 to 478 per thousand inhabitants.

The stagnation in the keeping and breeding of live stock is

[1] Animal units are calculated by taking each large animal as=1, and each small animal as=1/10.

(The figures for 1900 and 1916 from the 1916 census of domestic animals; those for 1923 from the official statistics for the year.)

[2] The figure for 1900 was compiled in November, before the lambing season—that for 1916, compiled after lambing, included 2,346,381 lambs; as the absolute increase was merely 1,013,437, there was in reality a relative decrease.

undoubtedly due, above all, to the price and tariff policy pursued by the State since the War. While the Ministry of Agriculture was striving to improve the stock, and obtained a credit of 100,000,000 lei for buying bulls from Switzerland, other departments adopted a policy which rendered the breeding and fattening of cattle altogether unprofitable. Their line of policy was to keep the cost of living low; to that end the export of animals was allowed only with special permits, which merely led to an abusive traffic in such permits and to excessive slaughtering of young stock. From June 1923, the export of live stock was left free, but a tax of 10,000 lei was imposed on each large horned animal, irrespective of its weight, which tax was later reduced to 8,000 lei. Under the system of permits the export tax had amounted to about 2,000 lei per animal. The export of pigs and fowl was still prohibited. Animals destined for export had to pay 50 per cent. above the ordinary freight on the railways. The first effect of these measures was to check the export of live stock just during the years when the price of meat was much higher in all the neighbouring countries. The Transylvanian breeders were especially hard hit; before the War they used to send about 10 per cent. of their stock yearly to Vienna and Prague. The position in the Old Kingdom was almost piquant. Rumania had never ceased to complain against the tariff barrier with which Austria-Hungary shut off the import of animals and meat; yet as soon as she was freed of that obstruction, Rumania proceeded to set up a similar barrier herself against her own live-stock industry. Still more strange was the policy which permitted the export of bran and oil-cakes in return for a tax of no more than 4,000 lei per wagon; animals, that is, were forcibly kept in the country, but their food was allowed to go abroad. A Transylvanian breeder asserted in an interview with the *Adevèrul* (September 25, 1924), that foreign cattlemen were buying animals in Yugoslavia, where the duties were ten times lower, and then fattening them on Rumanian fodder.

The effects of these restrictions were reflected in the export figures shown in the table on the next page. 44,363 head of large cattle were exported during the first quarter of 1923, when export permits were required; during the first quarter of 1924, when export was 'free', subject to the tax of 10,000 lei per animal, the number of horned animals exported fell to 31,367. In 1911 Hungary exported 400,000 oxen and 500,000 pigs, valued at 16 milliard lei. In 1923 Greater Rumania

	1922	1923	1925
Horned cattle	151,000	102,000	86,000
Pigs	176,320	90	162,050
Sheep	25,281	none	48,463

exported animals to the value of 5–6 milliards and in 1924 of only 1 milliard.[1]

If one reckons that with 5 kilogram corn one can produce 1 kilogram meat whose export value is double, it is easy to estimate the potential income which was sacrificed by that restrictive policy. Its second effect was inevitably to depress internal prices to such an extent as to make the fattening of cattle unprofitable. In fact, fattening was abandoned by most growers, as is proved by the extravagant slaughtering of calves, lambs, and sucking-pigs; during the winter 1927–8 I purchased sucking-pigs in Rumania and Yugoslavia at half-a-crown apiece. This fact and its serious consequences were emphasized at the meeting of the Agricultural Syndicates, in June 1925, when it was pointed out that the export tax amounted to 25 lei per kg. of beef and 31 lei per kg. of pork, which was just about the value of the meat itself. Internal consumption not unnaturally increased under such conditions; it rose from 11 per cent. to about 18 per cent. of the stock yearly. In 1923, one million large horned animals were killed out of a total stock of 6,000,000. The total number of animals killed in public slaughter-houses during 1925 was 1,029,139 oxen or 19·34 per cent. of the stock; 1,883,474 sheep, or 14·13 per cent.; and 481,315 pigs, or 15·5 per cent. of the stock. These figures do not include animals killed by peasants for their own consumption. The meat consumption was estimated in 1925 at 51·35 kg. per urban inhabitant and 3·50 kg. per rural inhabitant. In itself that increase in home consumption is desirable, both for the advantage of the industry and for the better feeding of the rural population. But

[1] *Argus*, November 19, 1924.

at present it represents a considerable wastage, as a large number of animals are killed very young, which is bound to compromise future chances of export, under a wiser policy.

The possibility of exporting animals and meat, especially to the Balkan countries, is still considerable. Even Poland is exporting cattle to Greece, transporting them across Rumania, where they are shipped at Constanţa. Through Transylvania and Bucovina, again, Rumania is favourably linked with the markets of Central Europe, which once preferred to be supplied from those two provinces. 'If we were to put agriculture on the only road which could at present lead to its intensification—that of breeding and fattening cattle—we could very soon satisfy the requirements in animal products of all the Central European countries.'[1] Rumania enjoys a privileged position as long as Russia cannot supply these countries with fodder as she did before the War. The land reform, and the consequent reduction in corn-growing, has helped Rumania to avoid the crisis through which the corn-exporting countries are generally passing. The larger losses in population and the reduction in purchasing power which the belligerent countries have suffered, coupled with the extension of the area under cereals in the oversea countries, have resulted in the overproduction of corn, which is responsible for the present crisis. The number of animals, on the other hand, has materially diminished everywhere, while the consumption of meat is increasing. A reasonable economic policy—which would have taken account of the general decline in live stock and of the power of absorption of the neighbouring markets—might have turned to great advantage this placing of Rumania's agriculture upon the shoulders of the peasants.

Section 2

The Effect of the Reform on Rural Industries

(*a*) *Domestic Industries*. The Rumanian State, from the year of its establishment, in 1866, up till the time of the land reform, had shown itself indifferent, if not actually hostile, to the fate of the industries the peasants practised in their homes. The chief

[1] Article by C. Garoflid, *Argus*, May 12, 1927.

ambition of the national leaders having been to create a national factory industry, they showered privileges upon all those who made attempts in that direction, at the expense of the older but simpler activities. Domestic industries were excluded from the benefits of the law for the encouragement of national industry, which granted many transport, customs, and other facilities. Raw materials destined for large-scale industry were exempted from customs duties, but those imported for the use of domestic industries did not enjoy the same favour. Similarly, the commercial treaties concluded by Rumania only protected the large-scale industries. 'Why not grant domestic industry the same protection?' protested Dr. Antipa, 'if it finds it possible to maintain itself when placed on an equal footing with large-scale industry? . . . The small industries have a place of their own to fill in our social and economic life.'[1] As late as 1921, when a congress of large-scale manufacturers prepared a project of expropriation for the benefit of industry, an article had to be added, apparently in response to official wishes, which specifically excluded domestic industry from all share in the proposed arrangement. Not till July 1923 was the first act of grace shown to domestic industry, when Articles 34 and 35 of the fiscal law passed in that year allowed certain reductions and exemptions from taxation.

That disregard for domestic industry is the more difficult to understand as in Rumania such activities were needed on the land for more than one reason. A peasant holding 5 ha. spends about sixty days in farming them, with the prevailing system of cultivation, as Dr. Cornăţeanu has calculated in detail. M. Alimăneşteanu has allowed 120 days for all the work which a peasant has to do, so that even taking into account the unconscionable number of religious and other holidays, the peasant has a surplus of 159 working days for which he must find an occupation. There is room for much improvement in the methods of farming, and every step forward will take up more of the peasant's time. But there are on an average four months in the year when snow and cold stop all work in the field, and which the peasant and his family must spend in demoralizing idleness

[1] Dr. Gr. Antipa, *Problemele evoluţiei poporului Român*, p. 329.

unless they ply some handicraft. Further, a large number of peasants own merely 2–3 ha., which are not sufficient for their upkeep. They depend, therefore, on some additional income, which they could only get from domestic industry if it happens that there is no factory, mine or quarry in the neighbourhood. It lies within the power of domestic industry, therefore, to solve the problem of existence for a number of peasants, and to give to most of them the means for a better standard of living. National economy, too, would stand to benefit enormously from any development which during the long winter months could harness the peasants' labour power to some productive work. The social and moral aspects of the problem are not less important for the nation's progress.

The association of farming with some handicraft, one alternating with the other, is to be found in many parts of Europe—in Switzerland, Belgium, Saxony, &c. In the Rumanian regions it is of old standing, especially in the highlands, where sometimes whole villages are in winter and during spells of bad weather engaged in the manufacture of all kinds of wooden and earthen articles, which they sell to traders or take into the neighbouring fairs. In general, village industries have developed in Rumania out of home industries, and have only rarely been transplanted from the towns. The manufactured objects vary from mere spokes to elaborate musical instruments. The Transylvanian village industries, especially those plied by men, are more varied, but those of the Old Kingdom play a more important part in national production. Almost 75 per cent., e. g., of the butts which the makers of wine and spirits and even the town merchants require, are supplied in the Old Kingdom by the village cask-makers of Muntenia and southern Moldavia. The northern Moldavian districts put on the market large quantities of the long coarse-spun overcoats (*şuman*) worn by the peasants, both the cloth and the tailoring being done in the villages.

Information concerning Transylvanian village trades was contained in the statistics on house and village industries (*Haus- und Volksindustrie*) published by the Hungarian authorities a few months before the outbreak of the War. These statistics distinguished between trades plied professionally, i. e. solely with the intent of selling the articles, and the more peculiarly domestic industries, when the articles were manufactured for household use and were only occasionally sold. The following figures, for 1910, refer to the districts now included within Transylvania:

Number of persons engaged in agriculture and gardening	Number of persons engaged in domestic industries			
	As principal occupation		As accessory occupation	
	Men	Women	Men	Women
1,797,259	15,757	6,192	18,341	21,179 [1]

The large number of women recorded as engaged in some accessory occupation was due to the wide margin allowed by these statistics; M. Anastasiu considers that many of the occupations noted in them did not fill even thirty days of the year's work, and that in consequence they could hardly be counted among regular accessory occupations. The table, it should be noted, did not include tinkers, window-menders and other itinerant traders, who were counted separately.

The largest proportion of village traders was in the county of Turda-Arieş, where out of a total working population of 54,357, 1,167 men were engaged in some trade as a main occupation and 1,597 as an accessory occupation—the total of 2,764 representing 7·1 per cent. of the working male population. The county includes the famous *moţi*, workers of wooden objects, with whom agriculture is the accessory occupation, as they cannot make a living out of the poor soil of the high district in which they live. After working a certain quantity of objects they load them into carts or on the backs of horses, and visit fairs and villages until they have disposed of their goods. Generally they are several weeks away from home, returning with a load of wheat, maize or rye. Another interesting group of itinerant traders are the Rumanian tinkers from the county of Solnoc-Dobâca, whom one meets on every road up and down the country. These men manufacture only a small part of their goods at home, and the bulk on the way, according to local demand. The land is worked by the members of the family who

[1] After O. A. Anastasiu, pp. 56–7.

remain at home, with the help of those trading, when one of the more important field labours is to be done. One might mention also the women engaged in weaving and embroidering in the county of Sibiu, which has the largest number of them; some of them used to take their goods as far as Bohemia, or actually settled there for a while, renting a room and a loom.

Bucovina has the usual run of village trades. The making of sheepskin jerkins (*cojoc*) is widely practised. The Bessarabian women spend most of their winter-time in weaving carpets. Men's coats now come mostly from factories. There are in Bessarabia a number of village presses for extracting oil from flax and sunflower seeds.

In the statistics for the Old Kingdom accessory rural occupations belonging to industry, commerce, and transport are grouped together. According to M. Anastasiu, the relation between industrial and commercial occupations was 5:1, i.e. the persons engaged in rural industries were five times as numerous as those engaged in commerce. The total number was in 1913 as follows:

Total number of working persons in the villages		Number of villagers engaged in some accessory occupation	
Men	Women	Men	Women
1,736,202	1,579,315	80,580	5,735[2]

These figures do not include the villagers who were engaged in mining as an accessory occupation, nor those with whom agriculture was the accessory occupation.

Village industries are more developed in some of the counties, especially in Muscel, Prahova, and Bacău, all of them belonging to the mountainous and hill regions. Their character is sometimes determined by the nature of local productions, cask-making being naturally in demand in the vine-growing districts; in other cases the industry is traditional, as the highly finished manufacture of embroidered cloths and garments in Muscel. The use of 'national' costumes has been brought into fashion again, and

[1] These are the sums of the detailed figures given by M. Anastasiu on p. 58. He then gives on p. 66 the totals 3,091,129 for the working population, and 75,294 men and 6,375 women as engaged in accessory occupations.

there are several societies now acting as distributors for this domestic industry. An attempt is also being made to develop the manufacture of 'national' modern furniture and of pottery, but many of the objects thus put on the market are rough abominations which have nothing of the usual peasant taste and workmanship about them. Among the more curious industries one may mention the manufacture at Slăvuța (Dolj county) of the tall and narrow rectangular wattle-cages in which the peasants store their maize to dry; they are made in winter and taken in carts about the neighbouring districts. Thanks to that industry the village is one of the wealthiest, and boasts three co-operative banks. The county of Neamț has a large village industry for the manufacture of peasant overcoats, which successfully compete with factory products. At Cleja (Bacău county) some of the villagers manufacture and sell straw-hats. In the same county, the villagers of Nadiș have specialized in the manufacture of the *cobză*, a musical instrument similar to the Russian *balalaika*. Some 2,000 are made each year; before the War they used to be sent as far as Russia. The result is that the village has 'well-built houses, all of them nicely fenced in'.[1]

The highland villages are all of them older than those in the lowlands, and this explains why village industries have a more traditional character in the highlands. The lowland villages, besides being of more recent origin, often had to be abandoned or moved to other parts, during the stormy periods of Rumania's history. Moreover, during the many wars and military demonstrations which took place on Rumanian soil, until the last quarter of the nineteenth century, the lowland villagers were frequently obliged to provide transport for the various armies; carting has remained the main accessory occupation of the lowland peasants in winter time and has caused them to neglect the more stable village industries.

The existence of numerous village industries has enabled a denser population to live in the highlands than could have existed on the produce of the soil. This has its importance for corn-growing in the plain, as peasants from the hills and from the mountainous districts, after finishing their own limited

[1] O. A. Anastasiu, p. 69.

agricultural labours, come down in the summer in groups to help as labourers in haymaking and harvesting.

No information is available from which one might estimate the influence of the land reform on domestic industry as a whole. As the peasants, taken altogether, have more land than they had before, and as they are developing more intensive crops, they must give proportionally more time to agricultural work and have less time to spare for accessory occupations; but their time is yet far from being fully claimed by farming. The effect upon individual domestic industries varies from place to place. In some cases the reform has facilitated the supply of raw materials, such as timber and hides, and the village industries using them have profited by it. In other cases, especially in the case of materials coming from abroad, the supply has become more difficult owing to the fall in the exchange, high customs duties, and changed relations between the value of agricultural and industrial products. The high price of cotton, e.g. has caused home-weaving to be neglected in many parts. It is only now that authorities and economic experts are beginning to discuss a constructive policy for the encouragement and development of domestic industries.

(*b*) *Agricultural Industries.* The dominant role agriculture plays in Rumania's economic life is only shown the more clearly by a consideration of the country's industry. Before the War, of course, industry was altogether insignificant; the total mechanical power utilized industrially amounted to merely 1·37 h.p. per square kilometre. The addition of Transylvania has raised that average to 1·61 h.p. (it is merely 0·27 h.p. in Bessarabia) by 1923. But the distribution of power among them shows that most of the industries depend on agriculture for their raw

	h.p.
Engineering	55,587
Timber and paper	66,581
Chemicals	56,526
Food-stuffs	98,584
Textile	13,530
Tanning	7,750
Pottery	39,779
Flour Mills	117,102
Printing &c.	2,112
	457,551

materials. According to a table published in the *Argus* on October 21, 1923, the large-scale industries used at that time 457,551 h.p., distributed as in the foregoing table.

It would be interesting to work out in detail the extent to which Rumanian industry depends on agriculture, but the material for such a study is not yet available. The trials through which Rumanian industry has passed since the War suggest that the undertakings which are using agricultural raw materials have the best chance of surviving and prospering. The two tables below give some indication of the development, since the reform, of the factories which are engaged in the transformation of agricultural raw materials.

I. The Food Articles Industry in 1915

Nature of Industry	Number of factories	H.P.	Capital in land, buildings, and installations	Value of Production	Personnel (administration and workers)
			Lei	Lei	
1. Mineral waters	6	231	2,580,177	106,927	270
2. Breweries	5	5,810	8,388,514	10,031,320	688
3. Spirits and champagne	7	155	498,664	1,937,384	187
4. Distilleries (alcohol)	15	1,271	4,067,125	10,943,717	805
5. Chocolate and sweets	20	353	1,149,151	8,058,181	959
6. Coffee (chicory)	1	22	376,034	840,600	69
7. Meat and vegetable preserves	15	394	2,917,115	5,475,992	1,690
8. Glucose	2	185	2,370,311	794,457	138
9. Vinegar	8	339	1,308,816	1,061,419	169
10. Steam bakeries	13	755	2,937,990	6,401,695	472
11. Flour products	6	177	914,594	1,153,083	198
12. Corn cleaning	5	1,405	1,698,127	6,224,261	80
13. Flour milling	96	22,120	41,000,000	115,476,500	2,913
14. Dairies	4	417	444,137	230,905	23
15. Vegetable oils	19	977	4,987,813	9,403,448	424
Total	222	34,611	75,638,568	178,139,889	9,085
Industries not coming under the law of 12/2/12					
16. Sugar factories	4	7,500	32,000,000	37,000,000	2,066
Total[1]	226	42,611	107,638,568	215,139,889	11,151

[1] The figures for 1915 refer only to factories enjoying the advantages of the law of February 1912 for the encouragement of national industry.

II. The Industry in 1926

Nature of Industry	Number of factories	H.P.	Capital in land, buildings, and installations	Value of Production	Personnel (administration and workers)
1. Sugar factories .	12	25,023	71,304,123	3,748,870,000	13,622
2. Flour mills . .	231	48,819	72,943,102	6,098,844,385	5,531
3. Alcohol factories .	178	7,074	33,065,512	1,065,518,394	3,573
4. Breweries . .	53	6,672	33,152,736	937,457,077	2,912
5. Chocolate and sweets	64	2,821	8,960,289	717,139,479	2,531
6. Distilleries . .	49	157	3,719,068	295,178,529	644
7. Meat preserves .	51	842	2,707,353	258,353,901	656
8. Fruit preserves .	22	715	3,419,288	149,645,587	1,341
9. Vegetable oils .	67	4,587	6,825,619	919,482,577	1,279
10. Flour products .	19	476	1,357,006	61,203,772	429
11. Dairies . .	10	229	812,355	97,616,463	303
12. Biscuit factories .	12	104	489,716	54,903,177	226
13. Vinegar . .	26	62	1,052,842	44,438,173	218
14. Coffee substitutes .	5	148	590,100	67,789,050	200
15. Corn cleaning .	22	1,648	2,983,550	85,394,600	205
16. Steam bakeries .	19	445	4,075,463	357,535,597	1,002
Total 1926 .	840	99,822	247,458,122	14,959,370,761	34,672

The progress made by these industries is satisfactory, but it is overshadowed by the growth of co-operative peasant undertakings.

Section 3

The Effect of the Land Reform on the Co-operative Movement

(*a*) *History of the Movement.* Through the transformation of land tenure in Rumania into a system of smallholdings the whole meaning of the problem of co-operation has changed. It has now to be regarded not merely as a means of giving aid to individual farmers, but as an essential complement to the reform. Experience has proved that everywhere the peasant farmer is able to hold his own in the process of production. It is only when he emerges from it and enters the market that he finds himself at a disadvantage, in competition with the large producers, and at the mercy of traders and other intermediaries. The initiative of the American and Canadian farmers has shown, however, that co-operation can free the small farmer from that subjection to the market. In the more progressive peasant countries co-operation has brought within reach of the smallholders those advantages which larger farmers may have had in matters of purchase and sale, as well as in matters of production. Those who fought for the Rumanian land reform had clearly in their minds the problem which they would have to face after its achievement. 'The parcellation of the arable land', said M. Mihalache in introducing his Reform Bill in 1920, 'does not do away with the idea of joint cultivation when the technique of farming demands it. But it begins by creating those individual rights which are the cement of all lasting association.'

If in Rumania rural co-operation lacked vitality hitherto, that no doubt was due in the main to the oppressive economic and social conditions under which the peasants lived. For the past shows that among the Rumanian people the spirit of co-operation has at all times been strong and widespread. Even now ancient forms of economic association are still to be found among the rural inhabitants of the Rumanian lands. In various parts of the Carpathians the peasants who engage in the rearing of sheep still do so largely on a joint basis. They keep their animals together, engage shepherds for all of them jointly, rent jointly pastures in the mountains and in the Danubian lowlands, whither they send the sheep for the winter. All expenses are borne in common, and the produce is sold as a whole, profits being shared between them according to each partner's contribution in animals or in special services. Any traveller in the Carpathians must have come across one or more of the *târle*, the log huts in which the shepherds spend the whole summer, grazing the sheep and making cheese. Likewise one finds numerous and old 'companies'—that is, associations of fishermen—in the regions of the Danube and of the Black Sea, where the catching of large fish necessitates the working of a number of experienced and well-equipped men together. Tasks and profits are divided among them according to each man's experience and ability. It is interesting to note that almost always these associations do not rest on a written agreement, but simply on mutual trust and 'the sacredness of the given word'.

From the description of the old village organization, in the first part of this study, one could see how much of it rested on joint ownership and joint use. Tradition, as much as economic advantage, plays a great part in the constant demands of the peasants for village grazings rather than for individual grass fields. Almost everywhere the social duty of helping one's neighbour still finds expression in the *clacă*, when the villagers work together for each other in turn; and in the *şezători*, held of a winter night, when under the stimulus of the singing and joking of the young men, the women and girls of the village diligently carry out together some piece of handiwork.

The evolution of rural life after the emancipation of the serfs did not offer a congenial soil in which those old traditions could strengthen and develop. Economically the peasants were continuously depressed, and their social role in the country's life was altogether nullified. The lack of education meant a lack of all contact with the doings of the West; and the towns, which elsewhere had formed the channel through which the idea of co-operation reached the country-side, were not ripe for such a function in Rumania. The excessive cheapness of living right up to the War left no inducement for the establishment of co-operatives of consumption; and in the absence of large-scale industry, the artisans were never driven by competition into associating with each other. That, no doubt, is the reason why the propaganda begun by M. P. S. Aurelian about 1870, under the stimulus of western experiments, for the establishment of co-operatives in the towns, fell on deaf ears. A few co-operatives of artisans were set up in some of the towns, between 1882 and 1892, but none of them could take root. In Bucarest and elsewhere it was for some years a fashion for certain luxurious grocery stores to parade the label of co-operatives, though they had nothing of the co-operative system in their organization and working. Quite different, of course, was the position in the villages. Owing to the increasing misery of the peasants the co-operative idea seemed to offer a way of saving them from utter ruin—an opportunity which, unfortunately, the uneducated and impoverished peasants were hardly capable of using unaided.

The co-operative idea penetrated into the Rumanian

provinces by way of the Saxon colonists in Transylvania. Being in close cultural contact with Germany, they were from the beginning informed of the co-operative attempts connected with the names of Schulze and of Raiffeisen, and followed them up with similar experiments in Transylvania, from 1852 onwards. It was the example of one of these Saxon co-operatives of credit which induced a distinguished Transylvanian leader, Visarion Roman, to create at Răşinari, in 1868, the first 'society for deposits and loans'. Without being itself a strictly co-operative undertaking, this was the precursor of the many so-called popular banks which later spread over Transylvania and Rumania. In 1872 the Răşinari society and other similar institutions were attached to the bank 'Albina', the first Rumanian bank to be founded in Transylvania. The bank, under the directorship of Roman, acted at first as a guide and supporter of the small credit societies, so that nineteen of them had come into being by the end of 1872. The bank, however, being a commercial undertaking, disliked the tendency among the credit co-operatives to emancipate themselves; it, therefore, withdrew its support, thereby causing the collapse of the subordinate institutions. The last of them, the Society of Sibiu, closed down in March 1875. Another such popular bank, however, the 'Aurora' of Nasăud, established in 1873 independently of the central bank, was able to survive and to prosper; on the eve of the War it had 1,700 members, about £7,000 paid up capital, and as much again reserves. The 'Aurora' possesses historic importance for the Rumanian co-operative movement, because it served as a model for the establishment of the rural co-operative banks in the Old Kingdom from which the whole co-operative system, as it now exists, has sprung.

The despair to which their ever-growing misery was driving the peasants vented itself in the risings of 1888 and 1891. In their need the peasants had sold themselves hand and foot to the large owners and tenants, pledging their labour for years in advance. Such money as they still could obtain in times of stress they had to get from publicans and usurers, at rates of interest which not infrequently rose to 500 per cent. The first to realize the need of doing something to improve the peasants' material situation were the village teachers. Education was like good seed scattered on barren rocks as long as the peasants lived in such utter misery. After the rising of 1888, the rural teachers began in their professional gatherings openly to speak of the need of setting going a movement for self-help amongst the peasants. They were as good as their word, and some of them soon laid the foundation of the co-operative movement, not without considerable risk and danger to themselves from the mistrustful authorities. Some of the pioneers had to pay for their initiative with persecution and imprisonment.

With information obtained from Transylvania, and guided by the statutes of the 'Aurora,' the first popular bank in Rumania, the 'Dumitra' was founded in the village of Dara in October 1891. It began with a membership of thirty-four and a capital of 127 lei (£5 sterling). A second popular bank was formed in March 1892, with thirty-nine members and a capital of 1,328 lei (£53 sterling); it was followed in the same year by five others. There are a few points of interest to be noted in connexion with these first co-operative attempts. The first organizations were those designed to supply credit, in the form of popular banks. That was the form which had best succeeded in Transylvania, and its was most in keeping with the needs of the Rumanian peasants at the time. Truly co-operative action was hardly meant; it was merely a measure of self-defence against usury. That was indeed about the only kind of action which the peasants and their friends, the village teachers, could then undertake with their own means. The second point of interest is that all of the early popular banks were established in mountain districts; that is, where the peasants were relatively better off and enjoyed a less dependent social standing than their fellows lower down in the cornlands. Thirdly, one may note that the first co-operatives began with an extremely limited membership and capital; this indicates the difficulties with which they had to contend, due either to the poverty or to the distrust of the participants. Yet this point may serve as an example of the decisive part which self-help plays in co-operation. It is characteristic that of those first seven co-operatives only one was unable to survive—and that was the one which began with the largest membership and the largest capital. The others have persisted to the present day and are among the most prosperous popular banks.

After a short period of stagnation the movement recovered its impetus, and from 1896 developed apace. Its growth was especially rapid between 1900 and 1902, thanks to the enlightened encouragement of two Ministers of Education, Dr. C. Istrati and M. Spiru Haret. Realizing the economic and social blessings which the movement might bring to the peasants, Istrati and Haret used their authority for protecting the village teachers against persecution on account of their co-operative activities. They took steps in fact to assist them therein. They entrusted one of the teachers with making the principles and methods of co-operation known among village teachers, and several young teachers were sent to study the co-operative movement in various foreign countries. That first period saw the establishment of 711 popular banks inside ten years, on the initiative and with the means of the villagers. During this period were laid the foundations of the co-operative essays which were promising to transform village life even in the uncongenial conditions which prevailed before the reform.

(*b*) *The Legal Status of Co-operative Societies.* The first legal provisions referring to co-operative societies were contained in the Commercial Code of 1887. They were neither sufficiently definite nor sufficiently simple, so that many of the early co-operative credit associations kept to the form of simple mutual associations, and did not register as co-operative companies in the sense of the Commercial Code. The rapid growth in the number and membership of the popular banks made it necessary to give them the possibility of organizing themselves legally, on a basis which should be systematic and at the same time simple enough to be adaptable to the circumstances of varying institutions. That legal basis was supplied by the Law of March 28, 1903, 'Concerning Rural Popular Banks and their Central Office,' which was the work of M. C. Stere. The main purpose of the law was to co-ordinate the movement and link up the popular banks among themselves. In reality its main effect was to place the whole movement under State control. The newly established Central Office was not so much a joint institution of the popular banks as a State organization destined to control, to guide, and to provide with funds the popular banks in the villages. For this purpose the State place 20,000,000 lei at the disposal of the Central Office.

The new law allowed three types of popular banks:

1. A type of independent banks (Article 7), which were not bound to assume any of the established co-operative principles.

2. A type of approved banks (Article 31), which were allowed to work with the Central Office on fulfilling certain conditions, such as the obligation under which its members were to reside in the commune in which the bank was established or at most in neighbouring communes.

3. A type of banks whose capital did not consist of shares, the members being jointly and fully responsible for all the activities of the bank (Article 34). The Central Office was entitled to advance to this type of bank the necessary working capital, after satisfying itself that the members possessed sufficient means to cover any eventual risks. In their case the Central Office could impose certain norms for the administration of the bank.

The law, it may seem, wanted to favour the last type of banks, as Article 9 provided that they should be exempted from the payment of the business tax provided they fulfilled the following conditions:—(*a*) the members should be jointly and fully responsible; (*b*) at least 50 per cent. of the net annual profits should go to create a reserve fund, and that fund should not, in case of liquidation, be distributed among the members, but used for purposes of communal utility; (*c*) the administration of the bank, with the exception of book-keeping, should be voluntary. As the law, however, exempted from the business tax all banks whose capital was less than 20,000 lei (those with a larger capital paying only half the business tax), and as all banks were exempted from stamp duty for all transactions not exceeding 300 lei, it does not appear that the law was specially bent on encouraging the creation of banks of the Raiffeisen or of a similar type. The only advantage reserved for popular banks of that type was a grant of 50 per cent. from the profits of the

Central Office, in proportion to the transactions which had taken place between it and the particular bank; but that was an indifferent concession in comparison with the other advantages in which all the banks shared. In fact, the amendment introduced in 1908 equalized the situation of the various types of banks; it provided that all popular banks of whatever kind which did not charge a higher rate of interest than 10 per cent. for members and 13 per cent. for non-members, were to be exempted from the payment of the business tax. The author of the law admitted in introducing his Bill that 'these provisions are perhaps not such as would encourage the creation of nothing but mutual societies, in the strict sense of the word'.

The law concerning popular banks did not lay down any norms regarding the composition of the business capital, except in the case of those banks whose capital was not formed of shares. For the rest, the law left in force the provisions of the Commercial Code (Articles 225 and 226), which laid down that no member of a co-operative society may participate with more than 5,000 lei, or hold shares of a nominal value above that sum; they also decided that the nominal value of each share could not be greater than 100 lei or smaller than 25 lei. The absence of specific directions had this result: that, in most cases, the principle of joint and unlimited solidarity of the members was not applied. It seems that in consequence the popular banks did not endeavour to satisfy, as fully and as cheaply as possible, their members' need for credit, but rather pursued the accumulation of profits.

Subsequent modifications and amplifications of the law extended its provisions to co-operatives of production and of consumption, as well as to peasant co-operatives for the holding and purchase of land. Until the end of 1918, the whole movement was under the guidance of the central organization known as the 'Central Office of the Popular Banks and Village Co-operatives'. The decree-law of January 3, 1919 enlarged and transformed this institution into 'The Central Office of Peasant Co-operation and Resettlement'. The new institution consisted of five sections. Three of them were to guide and control the co-operative movement: the Central Office of Popular Banks; the Central Office of Village Co-operatives of Production and Consumption;

and the Central Office of Village Associations and of Agricultural Exploitations. The other two sections were charged with the execution of the agrarian reform: the Directorate for Land Questions and Mortgage Credit, and the Directorate of the Survey. Each of the first three sections was autonomous, with its own capital and an administrative council, half of whose members were appointed by the State and the other half elected for three years by a congress of the co-operative societies. In their turn, each of the three section councils appointed two representatives to a general council which co-ordinated the work of the whole institution.

The Co-operative Code of July 12, 1928, abolished the Central office of Village Associations and Agricultural Exploitations, leaving only two Centrals: the Central of the Popular Banks and the Central of the Co-operatives, the latter including the functions of the abolished section.

The joining of the co-operative movement and of the technical and financial execution of the land reform into one State organization has been regarded as a fresh proof of the State's intention of keeping the whole co-operative movement under its control. The capital of the several central offices is of mixed origin, yet these offices lack real financial and administrative autonomy, as they are dependent on the support of the National Bank, and because the appointment of their personnel is in the hands of the State. These circumstances perhaps explain why a genuine co-operative movement has not yet developed in the Old Kingdom; numerically the growth of co-operative societies has been rapid enough. The protection of the State has not encouraged that spirit of enterprise and initiative which would aspire of its own strength to the solving of those problems which are facing the movement.

(*c*) *The Credit Co-operatives.* The object of the law of 1903 was to encourage the foundation and development of the banks by two methods: (i) by measures exempting them from the ordinary legal obligations; (ii) by the setting up of a credit organization which was to be at the same time a supervisory body.

Among the first there should be mentioned:

1. Simplification of the formalities for the establishment of

such banks: the founders merely had to deposit the terms of constitution with the justice of the peace, without payment of any tax or initial charges (Arts. 4 and 5).

2. Recognition of each popular bank as a trading company with limited or unlimited liability, joint and several or otherwise, on a share basis, &c. (Art. 2).

3. Recognition of the banks as corporate bodies (Art. 3); consequent powers of purchase of real and personal property.

4. Exemption from stamp duties (Art. 9), reduction of the cost of legal proceedings (Art. 13), certain facilities with respect to the carrying on of credit operations (Art. 36).

5. Exemption from the taking of special licences.

The supervisory institution contemplated by the law took the name of Central Office of the Popular Banks. It was a State institution intended to assist the popular banks and to act as a check upon them. The Office was placed under the management of public officials, but an administrative council was shortly afterwards set up in addition, which, by the fact of remaining in office for seven years, offered a guarantee of continuity and was secured in some measure against political influence, although most members of the Council were nominees of the Government. The Minister of Finance had a right of veto. The Central Office, as a credit institution, had at its disposal a fund of 20,000,000 lei, provided by the State, and a current account with the National Bank. These resources could be used to assist village banks, but not indiscriminately. The law itself made no distinctions concerning the grounds of exemption from the ordinary legal obligations referred to above. But it was intended that only those banks should enjoy the credit of the Central Office which offered exceptional guarantees or conformed more closely to the co-operative ideal. The law enumerated all the conditions which a popular bank must satisfy before relations with the Central Office can be established. For example, membership and the granting of loans must be restricted to persons residing in the same commune (or, with the authorization of the Central Office, in a neighbouring commune), and persons who were already members of another bank could not be admitted to membership; the interest charged was not to exceed a certain rate fixed by the

Central Office; the members of the administrative board were to be jointly and severally liable; no modification of the rules could be made without the sanction of the Central Office, &c. The existence of an initial capital was not made an essential condition. The Central Bank had power to advance capital, provided that there were among the members of the bank twenty farmers residing in one and the same commune and having assumed full joint and several liability. The object of this provision was to encourage banks which approximated to the Raiffeisen type. Such banks had only been established in villages in which there were a number of peasant proprietors, and in 1918 there were only forty-six of them.

In its supervisory capacity the Central Office was called upon to exercise a continuous control over all the popular banks, without making any distinction between those with which it already had business relations and others. The banks were expected to communicate their statutes and their balance sheets. These powers of inspection were very wide, the Central Office being even entitled to change the managing board.

Years	Number of banks	Membership	Paid up capital	Reserves	Balance-sheet total
			Lei	Lei	
1904	1,625	121,786	6,850,976	—	10,168,811
1905	1,849	198,411	13,665,824	—	16,703,135
1906	2,021	240,253	18,509,519	792,614	27,275,474
1907	2,223	295,325	27,746,241	1,262,418	41,153,303
1908	2,410	246,707	37,851,898	2,857,612	58,670,708
1909	2,543	402,938	49,034,211	3,659,160	75,708,924
1910	2,656	454,187	61,016,395	4,350,172	93,567,883
1911	2,755	510,118	79,592,265	5,409,042	121,477,347
1912	2,862	563,270	99,067,743	6,249,380	157,135,008
1913	2,901	583,632	107,142,203	7,925,815	170,790,003
1918	2,965	641,359	186,438,528	16,387,093	325,265,138
1919	3,114	678,061	243,863,256	19,228,872	482,217,716
1920	3,194	702,864	301,850,404	22,709,226	646,304,101
1921	3,211	705,150	348,068,894	26,861,301	813,459,882
1922	3,213	717,507	398,974,921	—	1,064,029,394
1923	3,747	875,879	478,915,265	—	1,507,952,912

Co-operative agricultural credit was thus organized by law in two grades: the popular banks and the Central Office. Two years later an amending measure of March 15, 1905, authorized the popular banks of a particular district to combine into federations or unions. In this way three grades of agricultural credit were created: popular banks, federations, and the Central Office. The Central Office was to deal exclusively with the unions or federations of popular banks. Generally speaking, a grouping by districts was followed.

In virtue of the decree-law of January 3, 1919 the Central Office became the first section of the new Central Office of Peasant Co-operation and Resettlement.

The development of the popular banks since the passing of the Law of 1903 is shown in the table on p. 383.

At the end of 1927 4,773 popular banks were working with the Central Office. The total membership of 952,997 included:

	1913		1927	
		%		%
Farmers	530,460	90·9	818,091	85·8
Artisans	14,956	2·5	49,051	5·2
Officials	14,196	2·5	32,850	3·4
Business men	10,382	1·8	21,722	2·2
Landowners	4,724	0·8	15,832	1·7
Schoolmasters and priests	8,914	1·5	15,451	1·7

Classified according to the number of their members, the 4,593 banks which had sent in their balance-sheet for 1927 were divided as follows:

987 with less than 100 members.
1,781 „ „ „ 100–200 members.
988 „ „ „ 200–300 „
644 „ „ „ 300–500 „
172 „ „ „ 500–1,000 „
21 with more than 1,000 members.

The banks with a limited membership predominated, and that was true also with regard to their paid-up capital:

Banks	Paid-up capital
	Lei
2,843	Up to 200,000
1,184	From 200,000– 500,000
369	„ 500,000–1,000,000
138	„ 1,000,000–2,000,000
57	„ 2,000,000–5,000,000
2	Above 5,000,000

From the point of view of their individual shares the

members of the popular banks were distributed as follows at the end of 1923:

Members		Capital Shares	Total Capital	
	Per cent.	Lei	Lei	Per cent.
245,431	28·0	2 to 50	8,987,435	1·87
169,314	19·3	51 „ 100	13,431,475	2·8
120,790	13·8	101 „ 200	21,430,945	4·47
105,970	12·1	201 „ 500	43,149,930	9·1
96,398	11·0	501 „ 1,000	72,745,980	15·16
68,423	7·8	1,001 „ 2,000	100,431,935	20.93
69,553	8·0	2,001 „ 5,000	218,737,565	45·67
875,879	100·0		478,915,265	100·00

It will be seen that the members who contributed up to 500 lei formed 73·2 per cent. of the total membership, but held merely 18·18 per cent. of the total capital, whereas 26·8 per cent. of the members, with shares above 500 lei, held 81·82 per cent. of the total capital. That unequal distribution was interpreted by Peasantist critics of the Rumanian co-operative movement as showing that 'although nominally one-half of the rural population belongs to the popular banks, the immense majority of the members only participates with infinitesimal shares—with the first payments they make on becoming members, for the purpose of obtaining a loan. Hence one cannot say seriously that co-operation has taken root among our rural population, as one might be led to do from a superficial glance at statistical data.'[1] In reply, it has been pointed out that the unequal distribution of capital shares was a natural consequence of the poverty in which the bulk of the peasantry lived hitherto. Further, it was considered natural that in the backward state of rural life confidence in banks should grow but slowly, and that most of the members should therefore contribute little, even if they could have contributed more. Nevertheless, it is admitted as striking that for an equal amount of capital contributed by the two extreme categories, there should be one hundred members in the lower to one member in the higher category.

The co-operative movement would seem to have increased in favour since the War, whether due to the greater confidence of the peasants in it or to their greater affluence after the agrarian re-

[1] Madgearu and Mladenatz, *Reforma Cooperației*, p. 9.

form. The number of members holding less than 500 lei capital decreased while that of members holding more increased; the tendency, therefore, being for a majority of the members to control a majority of the capital. The following table shows that change to have begun before the War and to have become accentuated after it:

	1909	1910	1911	1913	1919	1920	1923
Percentage of members with less than 500 lei capital	94·9	94·6	93·4	91·9	83·0	81·1	73·2
Percentage of capital held by them	48·04	46·44	43·54	40·9	26·16	22·24	18·8
Percentage of members with 501–2,000 lei capital	5·1	5·4	6·6	8·1	17·0	18·9	26·8
Percentage of capital held by them	51·96	53·56	56·46	59·1	73·84	77·76	81·82

The percentage of members contributing between 500 and 2,000 lei increased 5·25 times within a period of fourteen years; during the same period the part of the total capital held by them only increased 1·5 times, which means 3·5 times less than the increase in their number. The tendency, therefore, is clearly towards a concentration of membership in a middle category, as there was a simultaneous decrease in the percentage of members contributing between 2,000 and 5,000 lei. Deposits showed similar differences in the size of the contributions, and a similar improvement since the War. They amounted to 430,476,732 lei on 31st December 1923—when the Central Office published the last report based on the norms in use since 1904—a sum equal to the total capital of the popular banks, although interest on deposits was merely 8 per cent. when the interest which could be obtained for private loans was anything from 20 to 30 per cent. Deposits were distributed according to individual amounts as follows:

Depositors	Size of deposit	Total amount of deposits	
	Lei	Lei	%
95,875	1 to 50	7,935,948	1·84
45,190	51 ,, 100	4,145,189	0·96
69,435	101 ,, 500	30,543,839	7·10
31,820	501 ,, 1,000	29,431,506	6·83
29,830	1,001 ,, 2,000	49,849,321	11·58
91,492	2,000 and upwards	308,570,929	71·69
363,642		430,476,732	100·00

These figures indicate a perceptible decrease in the percentage of small depositors and a considerable increase in the bigger deposits. On December 31, 1920, the number of those who deposited more than 2,000 lei did not reach even 9 per cent. of the total number of depositors, the sums deposited by them being 56·13 per cent. of the total; at the end of 1923 the number of such depositors was over 22 per cent. and the sum of their deposits 71·69 per cent. of the total.

From a total of 1,184,065,557 lei outstanding loans on December 31, 1923, 24,906,051 lei had been lent to various co-operatives and 1,159,159,506 lei to individual peasants. The sum of 25,000,000 lei lent to the peasant co-operatives is relatively modest, but it is equal to the State's contribution to the Central of the popular banks. On the other hand the sum of 1,159,159,506 lei lent to the peasants is considerable, especially as it represents really a balance, the total sum of the transactions carried out by the banks during 1923 amounting to 4,739,035,675 lei. Loans were classified as below, according to the security on which they were granted:

	Lei	Per cent.
Loans on personal security . . .	209,712,877	18·09
Loans on bills of exchange . . .	472,943,635	40·80
Loans on pledges	385,142,153	33·23
Loans on mortgages	91,360,840	7·88

Or, according to the persons benefiting:

	Lei	Per cent.
Loans to members	798,455,366	68·88
Loans to non-members	360,704,139	31·12

The majority of the loans were granted against bills of exchange, and would seem, therefore, to have been taken up by cultivators lacking working capital. That view is confirmed by the use to which the money was put:

	Lei	Per cent.
For food and forage	96,450,320	13·38
For purchase of live stock and implements	310,460,324	43·07
For rent	116,450,415	16·15
For land purchase	145,320,310	20·17
For other purposes	52,079,224	7·23
	720,760,593	100·00

The remainder of 438,398,912 lei, similarly distributed, had been granted in previous years. The bulk of the loans therefore were taken up for productive purposes, and only relatively a small amount for consumption.

The total number of borrowers was 735,452 in 1923. They were classified according to the size of their loans as follows:

Borrowers		Size of loan	Total amount of loan	
	%	Lei	Lei	%
185,450	25·2	Up to 100	14,564,320	1·26
150,320	20·5	101 „ 500	69,435,260	9·99
184,260	25·0	501 „ 1,000	185,630,420	12·02
97,630	13·3	1,001 „ 3,000	204,650,340	17·65
74,328	10·1	3,001 „ 6,000	333,443,530	28·76
43,464	5·9	6,001 and upwards	351,435,630	30·32
735,452	100·0		1,159,159,500	100·00

Most of the borrowers, 70·7 per cent., took out loans of less than 1,000 lei and only 5·9 per cent. larger loans of 6,000 lei and more. This suggests that the small peasant undertakings are run with a minimum of capital; but it is probable that the popular banks were unable to satisfy all the demands of their peasant customers. Loans are not easily granted, and the banks almost always give less than the borrowers demand, because of insufficient capital. Subsequent years have shown an improvement in that respect. The total amount lent reached 2,173,898,442 lei on December 31, 1925, 29,404,479 lei being lent to co-operatives and 2,144,493,963 lei to individual peasants. On the same date the transactions of the banks reached a total of 6,926,608,503 lei.

The total capital of the banks increased materially from 478,915,265 lei at the end of December 1923 to 745,455,714 lei on December 31, 1925. Likewise, deposits rose from 430,470,732 lei to 591,280,976 lei during that period. The fact that capital and deposits for fructification together represent 63 per cent. of all liabilities proves that the popular banks were existing on their own resources.

Loans raised by the popular banks from various institutions were distributed as follows on December 31, 1923:

	Lei
From the Central	32,946,009
From the Federals	281,374,345
From various banks and institutions . .	16,492,840
	330,813,194

Loans from the federals of popular banks represented 85 per cent. of the total, which shows that the connexion between the banks and their federals is becoming closer; this must tend to increase the autonomy of the movement. The development is encouraged by the Central; it grants loans direct to the popular banks in exceptional cases only, its usual policy being to place at the disposal of the federals the funds which they may need.

In 1923 the profits of the popular banks amounted to 54,840,333 lei; the average dividend paid by them was 6 per cent., which was more than modest considering the state of the money market at the time.

The Federals of the Popular Banks. Most of the popular banks of a district, and some of the other co-operatives, are grouped in federals which act as a link between the co-operative societies and the Centrals. The federal banks control and guide the individual societies, supply them with credits, and look after co-operative propaganda. They are administered by a board elected at the general meeting of the affiliated societies. The development of the federals is shown in the following table:

Year	Number of the federals	Number of the federated banks	Share capital	Reserves	Deposits in the banks	Balance-sheet total
			Lei	Lei	Lei	Lei
1908	2	20	37,060	768	1,500	328,866
1909	3	30	96,000	1,530	25,034	530,327
1910	4	65	175,152	3,056	42,221	620,215
1911	6	119	174,128	4,702	421,875	1,979,760
1912	8	211	586,279	10,858	335,996	4,758,615
1913	14	330	963,458	21,011	479,151	6,801,044
1914	17	431	1,093,385	35,384	432,510	7,861,917
1915	22	630	1,333,799	54,752	787,992	9,899,817
1916	32	1,351	2,705,275	239,962	3,978,513	34,891,979
1917	35	1,492	3,187,845	353,553	11,962,721	37,392,633
1918	45	1,984	7,878,549	732,230	57,908,787	113,389,475
1919	45	2,640	14,418,739	1,541,165	59,916,578	197,771,834
1920	46	2,864	15,061,480	1,681,719	53,034,919	272,841,881
1921	45	2,919	18,922,018	2,451,060	35,521,102	396,044,836
1922	42	2,972	21,189,892	3,014,110	32,211,587	392,536,204
1923	53	3,334	27,716,045	6,289,734	29,209,846	558,527,452

At the end of 1923 there were in the whole country 44 county and 9 regional federals. To the federals were affiliated 3,334 popular banks—that is almost 90 per cent. of the banks working with the Central as well as 58 co-operatives for the holding or purchase of land, and 726 various co-operatives. On December 31, 1925, there were 50 county and 9 regional federals, with 3,370 popular banks, 49 land co-operatives and 653 various co-operatives affiliated to them. While, therefore, the number of affiliated co-operatives has decreased, the number of affiliated popular banks has increased with more than 500.

The transactions of these federals amounted in 1925 to 3,400,000,000 lei; their assets and liabilities at the end of 1925 to 1,107,657,890 lei—a considerable increase from the 558,527,452 lei at which they stood at the end of 1923.

As the bulk of the assets consisted of loans granted to co-operatives and of advances against crops, the federals would seem to have been fulfilling the function for which they were created. On the other hand, by far the largest item on the liability side represented loans obtained from the Central, which implies that the federals were not yet self-supporting, but served rather as outposts of the Central to facilitate its financial relations with the large number of popular banks. This view is supported by the inadequate capital of the federals, as well as by the small size of the deposits they received. The two items amounted at the end of 1923 to 56,925,892 lei, which was merely about one-seventh of the loans contracted by the federals from the Central. The activities of the federals improved so much that the sum of the loans they granted reached 904,181,213 lei at the end of 1925. That improvement was due above all to the increase in loans contracted from the Central, which reached 844,674,913 lei; there was a slight increase in the paid-up capital to 48,549,346 lei, but a decrease in deposits, which fell to 25,612,910 lei.

Profits at the end of 1925 were, however, 2·5 greater as compared with 1923, amounting to 10,277,990 lei.

The Central of the Popular Banks. The law of 1903 created a Central Office of the Popular Banks and Village Co-operatives as a parent body to the movement. In 1919 the office was merged into a new and broader institution called the Central Office for Co-operation and Resettlement, which consisted of five sections. The first of them is the Central of the Popular Banks, which guides, controls, and finances the credit co-operatives; almost all the rural and urban popular banks and their federals being connected with it. The Central is an autonomous, incorporated institution. It was originally attached to the Ministry of Agriculture, but with the creation, in 1923, of a 'Ministry of Labour, Social Insurance and Co-operation', the Central of the Popular Banks as well as the Central of the Co-operatives of Production and Consumption became separate sections in the new department. Urban and rural co-operation were linked up at the same time.

The Central is the organ through which the State participates in the co-operative credit movement. It is administered by a board of fourteen members, seven of whom are elected for three years by the congress of popular banks and their federals, the others being nominated by the National Bank and various Ministries. Properly speaking, the Central is a Central Credit Bank for the popular banks, as well as for the other groups of co-operatives which it finances through the intermediary of their own Centrals. It works through the district or county federals, and its activities are strictly limited to credit operations. Only in the case of co-operatives specially established for the purchase of land does the Central supply credit direct, in the form of mortgages.

The balance-sheet of the Central closed on December 31, 1923, with a total of 589,700,773 lei and two years later with a total of 1,233,088,314 lei. Its capital consisted of 25,000,000 lei contributed by the State, of contributions from the popular banks and their federals amounting to 6,593,440 lei in 1923 and 9,717,090 lei in 1925; reserves amounted to 35,491,115 lei in 1925, so that the total capital of the Central was 70,208,205 lei at the end of 1925. A large part of it was immobilized—as of the assets 35,598,232 lei were in public stock, 3,477,878 lei in buildings, and 268,502 lei in furniture—altogether 39,344,613 lei.

The Central carried out, in the main, current account transactions with the federals. These amounted to 990,000,000 lei in 1925, based probably on the credit granted to the Central by the National Bank. The Central was in practice little more than an intermediary between the National Bank and the co-operatives. One effect of that excess of bureaucratic links was that loans granted by the National Bank at 3–4 per cent. reached the popular banks at 12–13 per cent. Effective loans granted to popular banks only reached 127,000,000 lei in 1925, almost all of them against mortgages. That proves that the Central suffers from a shortage of capital and cannot adequately feed the co-operative movement.

The above figures show that the co-operative credit movement is growing apace, though the figures relating to capital and to balance-sheets must be corrected with the index representing the depreciation of the Rumanian exchange. While the currency lost 97·5 per cent. of its value from 1918 to 1926, i. e. was reduced to a value forty times smaller, the capital of the popular banks only doubled during that period, which means that it increased by 100 per cent. whereas it should have increased by 4,000 per cent. The increase in capital followed a normal ascent. The number of banks having risen by 30 per cent. and the membership by 50 per cent. it was but natural that their capital should increase by 100 per cent. That was in general the relation in which the three factors progressed since 1904. The considerable increase in the total of the balance-sheets is explained, in the first place, by the growing need for productive credit among the new peasant proprietors and, in the second place, by the fall in the exchange, which influenced the size of the individual loans.

The weakness of the credit co-operatives is clearly disclosed by a comparison of their working capital in 1913 and 1927:

	(In thousand lei) 1913	1927
Paid-up capital	107,142	1,265,465
Reserves	9,033	111,563
Surplus	—	170,095
Total capital	116,175	1,547,123
Deposits	18,390	964,849
Other liabilities	11,124	1,159,597
Total borrowed funds	29,514	2,124,446

The relation between the banks' capital and borrowed funds was, therefore, 3·9 : 1 in 1913 and 1 : 1·4 in 1927. Or take the following figures:

	1913	1927	
	Lei	Lei	Gold lei
Average value of individual share	183	1,327	33·20
Capital per member	198	1,623	40·60
Borrowed funds per member	51	2,229	55·70
Total working capital per member	249	3,852	96·30

This comparison proves that financially the credit co-operatives are much weaker than before the War, especially if one takes into account not only the diminished purchasing power of the lei, but the wider functions which the co-operatives have to fulfil after the land reform.

Another feature which discloses the weakness of the co-operative movement is the inadequate connexion between the popular banks and the co-operatives of production and consumption. The 25,000,000 lei the popular banks placed at the disposal of the various co-operatives shows the latter to be still in an embryonic stage.

(*d*) *Consumers' Co-operative Societies and Co-operative Societies for Production.* A great variety of co-operative societies has come into existence and multiplied in Rumanian country districts. These include consumers' societies, societies for joint sales or purchases, for working of forests, mines or quarries, vine-growing societies, co-operative bakeries, dairies and societies for fishing or market gardening, &c. Legally they have the same standing as the popular banks, as the provisions of the Law of 1903 were extended to them in 1905. As regards direction, control and credit, they depended until 1919 on the Central Union of the Popular Banks. In that year a Central of the Co-operative Societies for Production and Distribution was set up as an autonomous section of the Central Office for Co-operation and Resettlement. Its organization resembles that of the other Centrals. From that time these co-operatives multiplied rapidly.

These groups of co-operatives have had a chequered career, as some of the societies were started without sufficient exploration of the field of activity into which they ventured and of the economic problems they would have to face. As a result quite a number of them were unable to keep going and closed down, not only losing capital thereby, but at the same time shaking the confidence of the peasants in the movement as a whole.

Nominally, there were 755 such co-operatives in existence at the beginning of 1919, but two-thirds of them were in liquidation or weakened to such an extent by the War that they soon had to close down. With the end of the War, however, and with the beginning of the reform, the movement entered upon a period of rapid development. In 1919 there were founded 230 new consumers' societies with 13,500 members and 1·6 million lei paid-up capital; 400 more started life in 1920, with 33,000 members and 4·5 million lei paid-up capital. The table below gives the number and distribution of these co-operatives at the end of 1921:

Kinds of society	Number of societies	Members	Paid up capital	Goods sold
			Lei	Lei
Distribution and joint sales	1,945	133,083	32,834,403	238,802,413
Forest working	730	34,999	17,007,237	18,679,675
Dairies	28	721	67,555	991,755
Mills	31	1,519	369,601	56,743
Bakeries	17	994	175,779	1,702,900
Fishing	25	1,268	640,818	423,748
Mining and quarrying	11	326	646,644	158,757
Miscellaneous	96	4,457	1,156,126	3,381,146
Totals	2,883	177,367	52,898,163	264,197,137

These figures refer to the whole country. In the Old Kingdom, there were 1,500 societies at the end of 1921, with 145,000 members, and 2,593 at the end of 1924 with 236,713 members.

The main characteristic of the post-war evolution would seem to be the preference which most villages show for a mixed type of co-operative society, belonging in principle to the co-operatives of production, but engaging also in common purchases, because of local needs, smallness of available means, and difficulty in finding sufficient leaders. Common purchases include most of the implements and materials required for production—manure, seeds, machines and implements, fodder, &c.—which the large farmers obtain through their agricultural syndicates, but also, frequently, the kind of household articles which generally fall within the scope of consumers' co-operatives.

Similarly, joint sales are often combined with the partial preparation of the produce for the market, as is the case with fruit-drying, &c. These co-operatives, therefore, engaged in satisfying practically any and every need of the villagers, and this is no doubt the reason why their type is so popular with the peasantry. The Central of the co-operatives of production and consumption has drawn up a form of Statutes, adopted by all these co-operatives, which makes it obligatory for them to distribute a bonus to the purchasing members; it imposes the creation of reserve funds for cultural and social purposes, as well as the establishment of an indivisible and untransferable fund which shall ensure the continuity and autonomy of the society.

The Central is also acting as a wholesale distributing society, which may have contributed to the success of these mixed co-operatives, as it would seem that most of the co-operatives were forced to close down before the War just because of the absence of co-operative distributing centres. To facilitate distribution, the Central usually deposits considerable supplies with the more important co-operative societies. Finally, the Central has acted as an intermediary for the sale and, especially, for the export of the produce which the affiliated village co-operatives are getting together.

The development of this mixed type of co-operative society has the advantage of making unnecessary the establishment of a whole number of small co-operatives in the same place. Their popularity is proved by the fact that there were 2,593 of them in existence in 1924, besides 200 in Bessarabia and 600 in Transylvania. From the balance-sheets which 1,890 of these co-operatives supplied to their Central it appears that their membership rose from 133,883 in 1921 to 236,713 in 1924. Their capital increased during the same period from 43,622,433 lei nominal and 32,834,403 lei paid-up, to 109,147,866 lei and 89,747,033 lei. These sums indicate a considerable increase in the subscribed capital, and at the same time in the proportion of paid-up capital, from 74 per cent. in 1921 to 82 per cent. in 1924. At the same time by December 31, 1924, these co-operatives set aside reserves amounting to 20,637,873 lei and a fund for cultural and social purposes of 5,739,210 lei. Their transactions increased even more rapidly. Between 1921 and 1924 the value of goods purchased rose from 236,895,250 lei to 886,978,545 lei, and the value of goods sold from 238,802,413 lei to 817,104,704 lei. Even assuming that the fall of the exchange has doubled the price of goods, it still appears that the total business transacted in 1924 was twice as large as that of 1921.

The balance-sheet of these co-operatives closed on December 31, 1924, with a total of 476,818,161 lei. The main assets were goods purchased on commission, representing a value of 219,593,043 lei; the chief liabilities were debts amounting to 230,000,000 lei, a sum which suggests that the co-operatives still suffered from insufficient means of their own.

(*e*) *The Co-operatives of Production.* The mixed co-operatives of purchase and sale are associations of producers, each of the members having an agricultural exploitation of his own. Those branches of activity which require a larger capital, a greater division of labour, and so on, have led to the establishment of real co-operatives of production. Some of these societies, especially those for the exploitation of forests, have progressed remarkably well, while others have had a less satisfactory history.

The development of the co-operatives for the exploitation of forests after the War bids fair to equal the success of the Land Holding Societies before the War. In the view of certain Rumanian writers, their importance reaches beyond the economic field.

> 'They have solved in our highland districts two problems: one social and the other national. The peasants used to be robbed of their possessions, and their labour exploited, by forestry companies working all along the Carpathian Mountains; to-day, thanks to the forestry co-operatives, the peasants are regaining their ancient rights of ownership and they find it possible to emerge from serfdom and to become masters of their property and labour. . . . Moreover, those rapacious timber companies were largely in foreign hands. Through the forestry co-operatives, therefore, the peasants are finding the means of restoring to the country a national possession, and at the same time of giving that important branch of our national economy its Rumanian character again.' [1]

Unfortunately, some of these co-operatives are not affiliated with a Central, so that the figures which follow do not include

[1] N. Ghiulea, *Asociaţiile Ţărăneşti*, pp. 185–6.

all of them. Of the 730 forestry co-operatives in existence in 1921 only 347, with a membership of 34,999, communicated their balance-sheets to the Central; in 1924 the total number was 843, of which 476, with a membership of 57,935, supplied balance-sheets. These figures show an increase in the number of co-operatives, a proportional increase of those affiliated to the Central, and a relative increase in membership from 100 for each society in 1921 to 122 in 1924.

The total balance-sheet of the 476 forestry co-operatives amounted to 568,783,451 lei on December 31, 1924. The fully paid-up capital on that date was 51,580,387 lei and reserves 46,409,320 lei. The work of the co-operatives depended a great deal, however, on financial support from the Central and other credit institutions, to which the co-operatives owed 234,669,296 lei; of the latter sum 51,143,982 lei was due to popular banks and their federals. The assets of these co-operatives consisted of installations, machines, tools, &c., valued at 72,297,905 lei, of real estate valued at 36,729,284 lei, of current work valued at 30,008,078 lei, and, especially, of timber—standing, at the mills or in warehouses—valued at 137,159,528 lei. Assets further included 97,758,518 lei advanced by the co-operatives as deposits.

Other varieties of co-operatives of production included at the end of 1924 the following:

25 fishing co-operatives.
26 mining co-operatives.
21 co-operative bakeries.
26 co-operative flour mills.
22 co-operative dairies.
24 agricultural co-operatives.
9 co-operatives for the exploitation of mineral waters.
1 co-operative pottery, and
88 various co-operatives.

Of these 242 co-operatives, 132 with a membership of 12,148 communicated balance-sheets which amounted at the end of 1924 to 97,930,796 lei. Their capital was 40,302,012 lei and reserves 4,907,119 lei. They had debts amounting to 15,845,671 lei at the Central and 21,797,333 lei outstanding with merchants and banks. Assets consisted of machines, tools, installations, &c. valued at 12,345,789 lei, real estate 10,288,565 lei, manufactured goods 3,640,592 lei, other goods 9,034,012 lei, and raw materials 2,832,418 lei.

The Central of the village co-operatives of consumption and production has varied and extensive activities. It acts as the banker as well as the wholesale purchaser and salesman of the affiliated societies; it procures supplies in emergencies for local authorities and for the army; it builds houses for the colonists who are being settled in the frontier regions; it exports corn, &c., &c. The Central plays thereby an important part in the country's economic life. Its balance-sheet reached the formidable sum of 1,722,607,199 lei at the end of 1924, though capital and reserves merely amounted to 175,000,000 lei. The Central, therefore, still dominates the whole activity of the co-operatives of production and consumption.

(*f*) *Agricultural Co-operation.* Agricultural co-operation as such has been influenced by the agrarian reform more than any other branch of the co-operative movement. The main groups of co-operatives falling within this section were the landholding and land-purchasing societies. They expressed the peasants' great need of land, which was characteristic of the pre-reform conditions, and they flourished especially after the rising of 1907, when the new legislation imposed upon State and corporations the duty of letting their estates to peasant co-operatives only. It was in the nature of things that there should be much less scope for this type of co-operation after the reform. Large property is reduced to a shadow, and is, so to speak, on its trial; few landowners therefore let their estates, and the land which comes into the market for sale is insignificant in extent.

The reform is, on the other hand, responsible for a new type of agricultural co-operatives. It was to be hoped that having secured land and economic autonomy, the peasants would join together for the purpose of mutual help in production and sale. It so happens that the impetus for this development has been given by the reform itself, though not directly or deliberately. It will be remembered how the land expropriated in virtue of the decree-law of December 1918 was handed over not to individual peasants, but to the so-called 'associations of re-settlement'. They were village associations modelled on the landholding co-operatives, and were meant to carry on for a few years only, until the individual holdings could be measured and distributed. For political reasons these associations were already disbanded in 1920, and the land divided among the peasants. But during their short existence these associations had planned, and in part realized, various arrangements for the furthering of their interests. Some of them had purchased machines or had contracted for the supply of implements and seeds, others had made arrangements for co-operative selling, or had established breeding stations, and so on. When these associations were broken up, their members found themselves in a quandary. It was not always easy to divide the materials which had been acquired jointly and for joint use; and in some cases the former associates realized that it would not be to their advantage to interrupt the activities they had started in common. Many requests for advice reached the Central, with the result that wherever possible the break-up of an association of resettlement was made the occasion for the establishment of an association with a co-operative character, free of all connexion with the land reform. This type of society received the name of 'farming co-operatives'.

Co-operative Land-holding Societies. The first co-operative landholding societies were formed on the basis of the ordinary law; certain small alterations introduced in 1904 into the Code of Procedure and into the law of documentary evidence, facilitated the formation and the working of these societies. In March 1908, a clause was added to the Act of 1903 on popular banks, by which all the privileges granted to popular banks were also secured to these landholding societies. An important step was taken in 1908, when a new act provided that land belonging to the State or to corporate bodies could only be leased to co-operative landholding societies, unless it had been shown on inquiry that it was impossible to form such a society. Since the State owned a large number of estates and the property held in mortmain was also very considerable, circumstances favoured the development of these societies. Finally, the new reform laws (except that for Bessarabia) lay down the rule that the remaining large estates may not be let out for more than seven years, and that under equal conditions preference must be given to peasant co-operatives or to agronoms. But though the law facilitates the formation of these societies it does not leave them to work without control, as the risks of inexperience and abuse are too great. Every co-operative landholding society, therefore, must have its rules approved by the Central Bank and is for the whole period of its existence liable to inspection by that body; while the provisions in respect to societies to which the Central Bank grants loans are even more stringent. The Central has the right to appoint an agricultural expert as administrator of the undertaking, and to introduce into the rules of the societies any modifications that may be thought necessary.

While the land is held jointly, cultivation is always individual; excepting occasional reserves for the growing of fodder or seed. The management of the farming is entrusted to an agricultural expert as administrator; the rotation of crops is fixed by these experts on the most economic lines; the assignment of portions of land is decided by lot, the area of each being in relation to the working capacity and the number of persons in the family of the member. The maximum accorded to one member has been 10 ha. The joint purchase of all farm requisites, such as seed, live stock, machines is also usual. These co-operatives, which combine individual initiative of the members with expert control, have proved not only the means of bringing about an improvement in the economic position of the small cultivators, but also an excellent instrument for training them professionally.

The considerable expansion of co-operative landholding activities before the War is shown in the table on p. 401.

The rent paid by the members is fixed annually and is calculated to cover all expenses and to leave a surplus as reserve. After the land reform the activity of these societies was necessarily curtailed. In accordance with the law of July 1928 the management and supervision of the co-operative landholding societies have been removed from the Central of Popular Banks and given to a newly-formed Central of Co-operatives.

Societies for Co-operative Land-purchase. Land purchasing co-operatives were first made the subject of legislative provisions in March 1908. Their constitution and working are governed by the same rules as those for landholding societies. Their purpose

Years	Co-operative landholding societies	Members	Paid-up capital	Area	Annual rents
			Lei	Hectares	Lei
1903	8	—	—	4,940	94,785
1904	168	—	—	10,557	256,025
1905	37	—	—	30,358	626,144
1906	18	—	—	54,681	1,611,428
1907	103	11,118	409,258	37,344	2,183,822
1908	172	23,236	852,162	133,227	3,628,063
1909	273	36,371	1,286,524	190,521	5,574,531
1910	347	45,583	1,954,118	248,340	7,762,871
1911	378	62,009	2,386,433	283,381	9,220,806
1912	587	65,170	3,944,068	369,922	12,404,085
1913	495	76,678	4,289,981	374,891	12,497,081
1918	496	82,293	—	406,664	17,235,115
1923	102	16,472	—	51,554	7,071,693
1924	125	18,498	—	47,810	13,761,162 [1]

is to purchase estates or part of estates and then divide the land among the members, according to each one's ability to buy. A recent measure provides that no division of the land shall take place except in accordance with a scheme approved by the Central. These co-operatives have necessarily a transitory character, as they cease to exist when the price has been fully paid and the land divided up among the members. The members pay for their shares in annuities calculated to extinguish the purchase price within a given number of years. Until the liquidation of the society the estate purchased by it is considered as undivided and can be used to guarantee loans and other obligations.

There is no means of judging how much land had passed into the hands of peasants through the instrument of such societies. Their activity was hampered by lack of credit; therefore it was usual for the peasants first to rent an estate and by accumulating profits to try later to buy it. Up to the end of 1924 the number

[1] In his interesting article published in No. 2 of *Agrarna Probleme*, M. Timov sees in the rapid expansion of these societies a 'proof of the continuous development of the process of differentiation [among the peasants]; this is confirmed by the reduction of the average per head in the rented area. . . .' Do not the facts support rather the contrary inference? The reduction of the individual average suggests the growing intrusion of the village proletariat among the mass of farming peasants, i. e. a process of levelling and not of differentiation. This effect, moreover, has a permanent character. For the more land is absorbed by village proletarians, through the means of landholding and land purchasing co-operatives, the less land remains to be snapped up by the wealthier peasants—those *kulaks* so unpleasant to the eyes of Communist theoreticians.

of land purchasing societies liquidated was fifty-three, with 5,079 members, and they had purchased 15,682 ha. at a price of 20,779,351 lei.

The reform laws gave the State a right of pre-emption on all land sales involving more than 50 ha. in the Old Kingdom and 25 ha. in Transylvania. The intention apparently was that land obtained in that way should then be placed at the disposal of land purchasing co-operatives, but the financial crisis has prevented the State from applying that policy on a large scale. Nevertheless, the movement is remarkably active, in view of the small number and size of the estates which are put up for sale:

Year	Number of co-operatives	Area purchased	Purchase price
		Hectares	Lei
1923	104	46,401	134,324,954
1924	177	61,109	370,484,980
1927	337	81,134	—

The Agricultural Co-operatives. This type of society, inaugurated after the agrarian reform, is meant to do for the peasants what the agricultural syndicates are doing for the large farmers in Rumania and elsewhere. Their activity is varied, developing in one direction more than in another according to circumstances. In fact, these co-operatives are supposed to satisfy every need of the small cultivator, in the way of buying and selling. But in addition they are concerned with the improvement of the land and of the methods of farming, with insurance and bookkeeping, and with 'any and every operation destined to contribute to the advancement of agriculture'. Hence these co-operatives have at the same time professional, commercial and cultural ends in view. There was nothing like them before the reform, except where landholding co-operatives fulfilled some of these functions. In the state of economic dependence in which they then lived and laboured the peasants were not able to organize themselves. Now that 90 per cent. of the land is in the hands of the peasants, these agricultural societies may become the chief factor in the solving of the technical and economic problems raised by such a vast extension of small-scale agriculture.

Though the character and activity of this type of society comes very near to that of the landholding co-operatives, the law provides altogether special rules for its government. The first agricultural co-operative was founded early in 1921, and was soon followed by others, their number increasing as shown below:

	Societies
1921	71
1922	165
1923	222
1924	659
1927	825

With 52,009 members, the total paid-up capital on December 31, 1924 was 15,617,563 lei. The balance-sheet total was 92,542,634 lei. The activity of these societies is on a moderate scale so far. That is not surprising, seeing that their Central has no capital of its own, the finances it requires being obtained from the Central of the Popular Banks. The budget of this Central is entirely contributed by the State. Like the popular banks, these societies are grouped in federals; at the end of 1924 there were nine of them, with 453 affiliated societies and a paid-up capital of 3,345,738 lei.

(*g*) *Co-operation in the New Provinces. Bessarabia.* At the time of Bessarabia's union with Rumania the co-operative societies of that province were governed by the Russian Co-operation Code of March 20, 1917 which had removed the restrictions and the State control imposed by the laws of 1871, 1895 and 1904. After the union these societies continued to work independently of the movement in the Old Kingdom. Some new co-operatives were, however, established on the basis of the provisions ruling in the Kingdom, as a result of propaganda carried on by the central offices of Bucarest, which set up regional branch offices at Chișinău.

According to the study of MM. Madgearu and Mladenatz. the number of credit co-operatives on January 1, 1920 was 446. Data could be gathered only from 339 of them, with 228,781 members, giving an average of 674·8 per society. It would seem that 78 per cent. of the rural population took part in the co-operative movement. The volume *Cooperația în România Intregită* published by the Ministry of Agriculture in 1920 gave the following estimate of the number of Bessarabian societies: 28 in 1904; 253 (245 credit) in 1910; 357 (349 credit) in 1914; and 1,056 in 1919, 429 of which were credit societies, 9 credit unions, 603 consumers' co-operatives and consumers' unions. During and after the War, therefore, trading operations exceeded credit operations; the co-operatives played an important part during the War in supplying the army, especially with bread. The total membership was given as about 600,000. The total of the balance-sheets was: credit societies, 52,600,000 lei; credit unions, 38,500,000 lei; consumers' societies, 7,900,000 lei; consumers' unions, 3,850,000 lei.

There are in Bessarabia 25 German co-operatives of consumption with 4,160 members, gathered into one union; and a union of Jewish co-operatives, including 40 credit societies with 27,170 members, 70·3 per cent. of whom are merchants and artisans.

Transylvania. Transylvanian co-operatives functioned on the basis of the Hungarian law XXIII of 1898. All the societies could affiliate to the Central Credit Co-operative Society of Budapest, obtaining in return certain exemptions from taxation and rates. The Budapest Central supplied credit and controlled the activities of the affiliated societies. Public authorities had a right of control over them, which explains why the Rumanian inhabitants preferred the limited company form of association.

On the strength of that law 675 societies, affiliated to the Central, were founded in Transylvania, with a membership of about 200,000. In 1918 their capital was 17 million Hungarian crowns; reserves, 55 millions; deposits, 88 millions; and credits from the Budapest Central, 23 millions. Rumanian statistics gave the number of societies on January 1, 1921 as 1,677, namely, 459 credit societies, 196 co-operatives of production, 377 consumers' societies, 25 societies for joint sales, and 620 credit and saving societies. After the union with Rumania, a number of new popular banks were founded, grouped round five centrals. In 1920 the legislation valid in the Old Kingdom was extended to Transylvania. In that year the consumers' societies, connected with the Central 'Hangya' of Budapest, founded a Central of their own at Aiud, under the name of 'Central of the Hangya

Co-operative Societies'. The progress of the affiliated societies is shown in the following table:

	1920	1921	1922	1923
Number of societies	536	535	537	514
Membership	145,733	148,412	149,965	131,522
Capital	5·4 mill.	5·8 mill.	6·2 mill.	10·2 mill.
Reserves	2·9 ,,	3·85 ,,	4·2 ,,	4·6 ,,
Turnover	90·7 ,,	101·00 ,,	151·1 ,,	204·4 ,,

In addition to the popular banks affiliated to the Bucarest Central, there were at the end of 1923 two other groups of credit co-operatives in Transylvania. A group of 330 popular banks with 98,218 members, affiliated to the Co-operative Alliance of Cluj; at the end of 1923 they had a paid-up capital of 5,728,084 lei and deposits 38,349,568 lei; loans had been granted to the amount of 32,710,344 lei. The second group consisted of 182 Saxon Raiffeisen societies, with 18,201 members, affiliated to the Saxon Union at Sibiu. The total capital was only 223,015 lei, but reserves amounted to 1,805,454 lei and deposits to 46,047,061 lei. Up to the end of 1923 they had granted loans amounting to 28,775,521 lei. The Suabian union at Timișoara included thirteen societies with 2,000 members.

Bucovina. Co-operation was governed by the law of 1873, which allowed the movement complete freedom of action. In Bucovina there were four co-operative groups. The Rumanian group consisted of 156 co-operatives affiliated to the Central Office at Bucarest. They had 23,216 members, a capital of 543,798 lei and deposits 10,680,248 lei. The total amount of the loans granted by them reached 10,680,248 lei. The German group had 67 co-operatives affiliated to a Central of the German Credit Societies at Cernăuți. Their membership at the end of 1923 was 11,011, their capital 152,970 lei, deposits 15,156,811 lei, and loans granted reached 14,460,551 lei. The Ruthenian Group consisted of 41 credit co-operatives affiliated to a Central of the Ruthenian Societies. (In 1928 the Ruthenian union was in liquidation.) They had 6,414 members, a total capital of 64,363 lei and deposits 1,250,976 lei. Loans granted in 1923 amounted to 1,354,850 lei. The Polish Group consisted of 12 Raiffeissen Societies and one Schulze-Delitzsch, affiliated to a Central of

Polish Societies with 2,576 members altogether. Their total capital was 89,443 lei, deposits 1,352,335 lei and loans were granted for 1,221,669 lei.

These various societies, which did not work with the Central Office of Bucarest, raised the total number of credit co-operatives by 789, membership by 159,636, balance-sheets total for 1923 by 153,250,152 lei, and the sum of the loans granted to individual peasants and their associations by 89,259,634 lei.

(Comparative tables for the several branches of the movement are given on pp. 412–13.)

(*h*) *State and Co-operation.* To sum up, the land reform has given immense scope to co-operation in Rumanian agriculture, and the opportunity is being seized eagerly enough by the peasants and by the leaders of the movement. In contrast, however, to this real emancipation of the peasants, through the abolition of landlordship, a more elaborate State control has been imposed upon their associations than they ever suffered before. No other question is so arduously debated within the movement as that of its relations with the State. It is admitted all round that co-operation would not have reached its present extent had it not enjoyed the fostering care of the State to a degree unparalleled in any other country of the world. It is clear, indeed, that at present the movement could not dispense with the help of the State without risking the disappearance of some of its members. Yet this prospect does not dismay the leaders. There must be something seriously wrong with the system when even one of its chief executive officials—M. T. Măndru, director of the Central of Agricultural Co-operatives—openly declares that progress depends on one essential condition: 'The State shall no longer use artificial means for keeping alive societies without vitality or initiative, for they compromise the whole movement.' [1]

Such a statement gives substance to the widespread conviction that official tutelage has acted as a disguise for political interference. The first essays of co-operation were viewed with suspicion and hostility by Liberals and Conservatives alike.

[1] T. Măndru, *Cooperativele Agricole*, 1925, p. 12; a pamphlet provided with the significant motto: 'True co-operation does not beg, it conquers.'

Afterwards some of the Liberal leaders became its warm supporters—Dr. C. Istrati, M. Spiru Haret, and, later, M. I. Duca. But the nominated personnel of the various boards is itself evidence that political bias has not been foreign to their appointment; and with a partisan administration, it is not difficult to believe what is widely asserted, that political allegiance frequently determined the success of an application for a loan or for such other favours as the co-operatives could dispense. The critics of the present system further point out how altogether out of proportion has been the relation between the control the State has claimed and the assistance it has granted. Every activity of the co-operative societies is subject to close official control and is dependent on the financial support of the Centrals, which are themselves tied to the National Bank. Yet in 1923–4 the total credits allowed by the Bank to the co-operative movement as a whole amounted merely to 600 million lei, whereas private banks received some 7,500 million—though the co-operative movement is supposed to satisfy the needs of a class of producers who have in their hands 90 per cent. of the land and of the means of production.

These criticisms, therefore, rest as much on grounds of principle as on grounds of circumstance. The intervention of the State may gradually lose its political bias, in the measure in which the great public institutions cease to be the monopoly of one political party. But it is clear that the leaders of the co-operative movement object to State tutelage on principle, and not merely because of the political colour it wears at the moment. The 1921 Report of the Union of Raiffeisen Societies of Sibiu denounced the 'co-operative laws of the Old Kingdom as most reactionary', and the same organization, which stood outside political squabbles, lodged a complaint against those laws with the International Co-operative Alliance. In a guarded opinion on the whole position Mr. Diarmid Coffey, at that time Librarian of the Dublin Co-operative Library, considered it 'likely that the movement in Rumania has not the spontaneity which should characterize true co-operation, and that it must lean heavily on State institutions.'[1] The standpoint of the co-operative leaders was formulated before the Chamber by one of the pioneers of

[1] Diarmid Coffey, *The Co-operative Movement in Jugoslavia, Rumania . . .*, p. 69.

the movement, M. Stan Morărescu. He pointed out that originally the law concerning popular banks had left untouched the principle of the movement's autonomy. To that phase of the legal régime corresponded the most flourishing period which Rumanian co-operation has known from its foundation to the present day.

> 'Official tutelage may have had its usefulness, but to-day it has become an obstacle in the way of the normal development of co-operation, and its abolishment is an indispensable condition for progress. The Co-operative movement demands to be given its freedom and to be relieved of executive control, remaining like the companies with limited liability, only under the control of the judiciary. . . . Nothing to-day could justify the maintenance of the co-operative movement in the position of a minor placed under a guardian, especially after the union with the new provinces, when side by side with the controlled co-operation in the Old Kingdom we have the autonomous co-operation of Bessarabia, Transylvania, and Bucovina. . . .' [1]

Rightly or wrongly, the leaders of the movement seem to fear that official interference, which hitherto has been used for political ends, might in future be made an instrument of financial exploitation. That suspicion finds some support in the financial policy which has ruled in Rumania since the War, and which has been justified, exactly in its relation to co-operation, by a writer who has endeavoured to endow the practice of the rulers with a theoretical apology. Rumania, in his view, is still in the phase of struggle against 'destructive capital'—i.e. foreign capital—and the battle can only be won by the creation of a national capital. Could this be achieved direct through co-operation, and not through capitalist banks? The writer, accepting the Marxian dialectics, resolutely answers 'No'. The soil is therefore not yet ripe for real co-operation. It is a delusion to think that one could organize co-operation 'with peasants who have barely emerged from feudal servitude, and who naturally lack those spiritual qualities out of which co-operation blossoms'. These qualities are produced only by a prolonged and racking trial at the hands of capitalism—only in the school of capitalist production. Capitalism must come first, co-operation afterwards. The one will bring the other. [2]

[1] *Aurora*, Bucarest, November 25, 1925.

[2] St. Zeletin, *Cooperaţie Română?*, 1925, pp. 13–14.

This is not far from the conviction of the opposite camp that the one intends to batten on the other. Hence they cannot believe that a political machine which has been openly under the influence of banking finance during the last decade could be a trustworthy foster-mother for the co-operative ideal. There are some who believe, indeed, that the fear of their being successful played its part in the hasty abolition of the village associations in 1920.[1]

'There is no doubt', exclaimed the Peasant leader, M. Mihalache, 'that co-operation as we understand it will have to wage a heavy battle with the forces of capitalism'. . . . 'For it is the purpose of such co-operation completely to emancipate the small producer from the domination of capital, and so to secure for him the whole profits of his labour'. . . .[2]

The land reform, therefore, besides having set new problems for co-operation to solve, is also changing its whole background. Those who speak for the new peasant proprietors seem intent upon transforming into a movement what hitherto has been largely an organization.[3]

[1] See article in the able review *Societatea de Mǎine*, Cluj, January 6, 1924.

[2] *Aurora*, December 5, 1923.

[3] In November 1928 the National-Peasant Party for the first time came into power, and at once set to work to adapt the country's economic laws and policy to their views. A new Co-operative Code, prepared under the supervision of the Minister of Labour and Co-operation, M. Ion Rǎducanu—himself an old leader of the movement—was passed in the summer of 1929. M. Rǎducanu justified the early reform of the Code introduced by the Liberal Government in July 1928 on the ground that it was imbued as much as ever with the idea of State control, which it extended to the autonomous co-operative groups in the new provinces. The strenuous opposition of the Raiffeisen societies of Transylvania only gained them a respite of ten years within which they had to adapt themselves to the legal system enacted for the Old Kingdom. M. Rǎducanu also contended that the apparently fine figures relating to the co-operative movement were in fact hiding a triple crisis—moral, technical and material. Partly as a result of the War, and partly of the economic conditions prevailing after it, the movement had attracted many people who cared only for the gains they could make through it; many societies were inefficiently run; and inflation had shaken the economic strength of the movement just in the period when it should have played an important part in the progress of the new peasant farmers. To that weakness of the co-operatives the State replied not with more generous aid, but with sterner control. The new Code was devised to create a legal frame within which the movement might evolve freely.

That legal frame, said the *exposé de motifs* to the new Code (from which this summary is made), 'should be sufficiently wide to offer to the co-operatives the freedom of movement which every economic enterprise needs if it is to develop in the present economic world. But, at the same time, the legal provisions must be sufficiently definite to make sure that a society calling itself "co-operative" fully conforms to the co-operative idea, and is clearly differentiated from other kinds of companies.' The law's first two chapters therefore indicate that within its meaning 'co-operatives are societies formed by an unlimited number of people for the realization of common ends, by means of a common economic enterprise'.

The Co-operatives. The new Code leaves it to every co-operative to fix through its Statutes the geographical and material limits of its activities. In the belief that the Rumanian movement suffers from an excessive division of forces, the Code facilitates the fusion of two or more of the existing societies.

The Code maintains the provision according to which capital shares must be equal; but it no longer lays down a maximum, merely providing that the rules of each society shall prescribe how many parts a member may take up, fifty being the highest limit. The societies may be based on the limited or unlimited responsibility of their members. The Code allows the establishment of co-operatives with a share capital, the shares of course named, as a form of association likely to be more suitable for certain co-operatives of production, with an industrial character, needing a more stable capital. To prevent a sudden decrease in capital the Code allows societies to lay down a term, not exceeding ten years, during which no member may withdraw.

In order not to hamper the work of the societies the Code allows them to state in their Statutes that they may also work with non-members. But the law prevents this from being made a source of additional profit for the members. Of the gross profits, at least 10 per cent. must go to reserves; at least 5 per cent. to a fund for social activities; at the most 10 per cent. for the administrative board, if the Statutes provide for their remuneration; part of it as a dividend for the paid-up capital, not exceeding the maximum fixed for each kind of co-operative by the National Office; while the remainder is to be distributed among the members according to the part which each has played in the realization of the surplus. If non-members also played a part in this, their proportional share from the net profits must in no case be distributed to the members, but must be added fully to reserves. Hitherto, according to M. Răducanu, contributions in labour, consumption, &c., were almost never remunerated, all the profits being distributed as dividends to capital.

The co-operatives are ultimately under the control of the general meetings, in which the members have equal votes, and which elect from among themselves the managing and controlling personnel.

The Federals. Like every other economic movement, co-operation needs to concentrate its forces. This end is served by the Federals. 'In our conception, the Federal is purely and simply a co-operative whose members are the co-operative societies. Hence, though we attach overwhelming importance to the federal organization of co-operation, our project only contains very limited provisions for the constitution, organization and activity of the federals.' They come within the general provisions of the law.

The activity of the federals must be concerned solely with satisfying the needs of the associated co-operatives.

The Code leaves the federals altogether free to decide through their Statutes the geographical extent of their operations, which may cover the whole country, as well as the kind of co-operatives which they will accept as members.

'In our view, the future lies with the regional federal banks, grouping together all the various co-operatives; with a national wholesale society for supplying the co-operatives, especially by way of imports, with all the goods for domestic use; with national federals for the disposal of agricultural produce, fruit, &c., collected through local or regional co-operatives. But, we repeat, the law must do no more than provide the legal frame necessary for the functioning of these organizations, and not to impose detailed schemes which would merely cramp and choke the life of the federals.'

The Unions. The framers of the new Code believe that federal organizations must make a clear cut between economic and social activities. Hence, side by side with the federals, which are to be purely economic societies, the Code contemplates the setting up of Co-operative Unions.

The Unions are, in the first place, to exercise as a compulsory function legal control over the affiliated societies.

In addition, their Statutes may contemplate such other activities as the protection of the movement's interests, technical and legal aid, provision for co-operative propaganda and teaching, and any other activity concerning the cultural side of co-operation in the region in question. 'We see in the organization of such unions, as the organs of self-government of the movement, the means of creating a real co-operative life.'

The law sets no geographical limit to the constitution of the Unions. Nor does it make the organization into Unions compulsory, as not all the co-operatives may be able at present to face the outlay involved. But if a majority of co-operatives in a certain region demand it, the National Office may make affiliation compulsory for all co-operatives in that region.

The Central Co-operative Bank. The old Centrals, financed and controlled by the State, were in their turn exercising the double function of financing and controlling the co-operatives. Their activities, centralized and bureaucratic, were somewhat mixed, and discontent with their conduct was widespread. The new Code attempts to segregate the functions of the central bodies.

In place of the Central of the Popular Banks the new Code sets up a Central Co-operative Bank. It is to enjoy wide autonomy, but as it cannot for the time being dispense with State support, its Statutes require the approval of the Government.

The Central Bank's capital is to consist of 500 million lei contributed by the State, and of social parts of 10,000 lei each, every affiliated co-operative having to take up at least one social part. The Bank is to grant credits in whatever form to the affiliated societies and to carry out on their behalf any and every banking operation.

The board of management will consist of three delegates of the State, one of the National Bank and five delegates elected by the general meeting of the affiliated societies. At the general meeting the voting power of the State is limited to one-third of all the votes cast at the meeting. Disputes between the State and the Central Bank are to be settled by an arbitral commission presided over by a member of the supreme Court.

The Directing Co-operative Bodies. The task of guiding and co-ordinating the activities of the various branches of the movement is entrusted to a General Council of the Co-operative Movement. It consists of fifteen members, of whom six are to be elected by the general congress of the co-operative societies, five are appointed by the government, one is delegated by the supreme Court, while three are to be co-opted from among recognized experts on co-operative questions.

The functions of the General Council are carried out through the National Office of the Co-operative Movement. The Office will guide and control the activity of the Unions, laying down the rules for the control of co-operative societies. At the same time, the Office serves as a court of appeal from decisions taken by the Unions. Finally, the Office will represent the interests of the movement in its relations with the political power. 'We regard the National Office as a central of the Co-operative Unions. When we have Co-operative Unions in the whole country, the National Office will disappear, to be replaced by a Central of the Unions.'

The new Code has been accepted by the co-operative organizations of the national minorities in the new provinces, which had hitherto preferred to continue working on the strength of the pre-war laws. Now the Hungarian and German co-operatives are represented on the General Council, and thus the process of unifying the co-operative system is actively under way.

Meanwhile, a Co-operative Central for Import and Export has been founded, with financial aid from the Central Bank. It is a limited company, its members consisting of those co-operative societies which desire to carry out joint purchases and to arrange for the joint sale of their produce. The Central undertakes any commercial operation falling within the needs of its members. It has begun to work at once, for the sale of this year's abundant harvest, and for supplying to the peasants agricultural machines and implements, as well as selected seed, &c.

THE EVOLUTION OF RUMANIAN RURAL CO-OPERATIVE SOCIETIES BETWEEN 1921 AND 1928

(From figures supplied by the Directorate of Agricultural Statistics, Ministry of Agriculture)

Societies	1921	1922	1923	1924	1925	1926	1927	1928
I. Credit Co-operatives.								
A. *Credit Co-operatives (Popular Banks).*								
Number of Societies	3,211	3,213	3,747	3,956	4,207	4,413	4,766	4,810
Number of Members	705,150	717,507	825,879	847,217	886,844	915,388	962,515	973,641
Capital	348,062,894	398,947,921	478,915,265	584,034,833	745,356,714	971,744,812	1,265,465,317	1,510,790,577
Reserves	26,861,301	31,346,607	41,712,706	51,293,643	65,925,616	85,633,423	111,563,247	136,884,662
Loans Granted	572,847,255	817,975,765	1,184,065,557	1,535,557,167	2,173,885,908	2,834,820,027	3,452,234,914	4,142,686,897
Net Profits	29,521,375	41,271,884	54,840,333	69,186,269	93,562,625	126,014,157	170,145,521	201,655,324
Total balance-sheet	813,459,882	1,064,029,394	1,507,952,912	1,894,469,321	2,709,844,978	3,608,670,423	4,414,494,398	4,902,269,266
B. *Federals of Popular Banks*								
Number of Federals	45	42	53	57	59	59	56	56[1]
Number of Affiliated Societies	3,395	3,513	4,118	4,401	4,572	4,853	4,822	4,773
Capital	18,922,018	21,189,892	27,716,045	35,648,159	48,549,346	58,577,686	75,177,071	89,909,245
Reserves	2,451,060	3,014,110	6,289,734	5,389,813	8,303,487	10,983,789	14,303,199	16,808,223
Loans Granted	118,737,966	193,316,228	349,956,443	484,578,720	904,181,213	1,152,769,882	1,223,812,521	1,492,749,859
Total balance-sheet	396,044,836	392,536,204	558,527,472	667,368,406	1,117,657,890	1,349,175,053	1,404,148,682	1,797,096,880
C. *Central of Popular Banks*								
Affiliated Societies	—	—	—	—	—	—	—	2,610[1]
Capital: State Contribution	12,000,000	12,000,000	25,000,000	25,000,000	25,000,000	25,000,000	25,000,000	25,000,000
Capital: Contribution of Popular Banks and Federals	6,549,459	6,593,190	6,593,440	6,593,440	9,717,090	9,639,190	9,638,990	9,889,490
Reserves	10,166,223	10,939,593	15,104,115	26,069,921	35,491,115	53,046,000	77,244,000	97,725,495
Loans Granted	142,850,901	246,880,420	450,666,025	647,662,007	1,104,726,558	1,326,340,087	1,510,629,001	1,925,717,137
Total balance-sheet	378,254,310	473,565,366	589,700,773	788,210,589	1,233,088,314	1,458,568,381	1,682,994,963	2,046,504,412
II. Co-operatives of Supply and Sale								
A. *Societies of Supply and Sale*[2]								
Number of Societies	1,950	2,248	2,912	3,252	3,339	3,339	2,886	2,623[1]
Number of Societies which sent in their balance-sheets	1,287	1,413	2,062	2,584	2,520	2,443	2,250	2,394
Number of Members	133,883	147,854	245,718	293,802	280,666	284,972	257,592	240,881
Capital	43,622,434	54,069,565	101,899,865	125,507,421	147,366,144	164,957,358	187,315,006	198,771,372
Reserves	3,595,323	3,474,631	13,639,767	22,312,846	28,672,885	38,255,019	43,816,758	45,357,586
Turnover	238,802,413	260,190,294	815,535,891	817,104,704	1,095,529,374	1,146,222,107	1,174,002,774	763,101,803
Profits	—	14,410,974	19,529,098	19,277,539	17,088,206	13,944,224	3,412,065[4]	19,831,436[4]
Total balance-sheet	173,769,956	218,792,237	391,759,143	562,571,151	853,317,732	967,970,214	1,034,296,464	1,171,818,025
B. *Federals of Supply Societies*								
Number of Federals	—	23	20	19	12	13	13	6[3]
Number of Affiliated Societies	—	822	758	736	513	468	319	225
Capital	—	6,185,660	5,788,081	4,251,178	3,892,837	3,813,162	4,207,392	1,552,825
Reserves	—	770,499	4,571,221	5,170,020	7,226,836	1,542,996	2,596,205	1,472,534
Turnover	—	112,332,020	93,823,973	45,046,049	88,753,494	69,178,729	61,949,201	39,475,329
Profits	—	2,606,162	58,662	5,176,418[4]	673,737	11,749,275[4]	20,660,034[4]	1,980,435[4]
Total balance-sheet	—	207,312,501	211,282,700	211,890,943	151,331,567	192,590,492	170,884,755	54,369,819

C. *Societies of Production*								
Number of Societies	—	—	—	242	257	321	405	344[1]
Societies which sent their balance-sheets	107	103	113	131	141	151	139	156
Number of Members	9,313	11,131	9,176	12,148	15,464	16,139	16,189	14,570
Capital	2,503,093	6,339,490	7,711,614	14,779,982	29,449,419	36,470,070	40,969,478	38,921,004
Reserves	388,343	395,749	1,006,764	1,395,201	2,779,096	5,612,020	5,671,410	6,238,551
Turnover	6,715,084	11,385,278	28,714,083	71,185,595	115,074,494	142,935,212	158,937,848	106,488,358
Profits	—	1,172,593	1,279,560	3,255,555	423,864	6,356,519[4]	6,872,690[4]	8,008,755[4]
Total balance-sheet	10,121,651	28,509,725	36,089,253	97,930,796	148,593,966	211,048,857	249,180,180	169,442,522
D. *Central of Societies of Production and Distribution*								
Number of Societies Affiliated	367	448	751	837	981	1,063	1,168	1,214
Capital	26,041,906	27,194,088	32,660,878	36,998,814	46,494,380	51,940,976	58,311,469	87,245,880
Reserves	22,246,465	55,126,096	75,613,279	89,262,497	130,066,644	139,169,443	171,617,383	176,554,143
Turnover	226,596,541	90,495,689	205,987,960	143,686,269	182,478,826	161,465,825	64,827,059	42,735,250
Profits	12,497,373	19,891,511	25,398,000	7,633,392	5,587,487	4,954,404	3,879,290	9,245,412
Total balance-sheet	584,889,109	637,254,651	790,977,385	730,241,115	1,002,305,731	997,560,415	1,069,711,594	1,224,592,234
III. LANDHOLDING AND FORESTRY SOCIETIES								
A. *Landholding Societies*								
Number of Societies	—	—	102	125	123	119	100	104[1]
Number of Members	—	—	16,472	18,498	16,607	14,769	12,079	12,569
Rented Area (in hectares)	—	—	51,554	47,810	53,936	42,451	37,181	37,220
Yearly Rent (in lei)	—	—	7,071,693	13,761,192	21,851,307	21,431,015	17,246,298	14,088,548
Capital	—	—	2,294,038	3,250,500	3,943,609	4,587,995	3,338,976	3,963,485
Reserves	—	—	880,243	824,095	1,502,125	2,066,313	2,423,900	2,245,107
Total balance-sheet	—	—	13,080,420	17,568,279	22,853,964	22,904,880	22,814,980	22,859,930
B. *Land Purchasing Societies*								
Number of Societies	—	—	104	177	220	297	343	402[1]
Number of Members	—	—	15,346	22,069	25,778	32,149	35,559	39,485
Purchased Area (in hectares)	—	—	46,401	61,109	69,084	78,637	82,876	90,898
Purchase Price (in lei)	—	—	134,324,954	370,484,980	576,163,593	772,663,964	985,005,144	1,228,565,563
Capital	—	—	65,796,148	27,765,256	77,138,212	76,307,886	91,057,062	98,431,650
Reserves	—	—	1,122,994	6,853,089	5,030,238	6,324,361	7,064,964	7,435,266
Total balance-sheet	—	—	152,789,344	243,672,307	414,048,241	567,029,949	625,796,205	668,974,623
C. *Societies for the Exploitation of Forests*								
Number of Societies	730	823	830	843	889	918	862	819[1]
Number which sent their balance-sheets	347	403	484	476	473	480	452	451
Number of Members	34,999	42,922	56,054	57,935	57,048	63,103	61,754	56,196
Capital	24,136,822	30,782,929	53,418,118	69,543,860	93,136,364	108,474,938	125,614,287	123,102,001
Reserves	1,553,954	1,893,460	7,114,611	13,639,630	18,915,859	24,599,772	28,019,050	25,217,145
Turnover	18,679,675	2,258,205	198,434,091	235,492,832	299,875,758	382,326,045	333,561,322	400,273,586
Profits	—	11,762,679	27,081,844	12,891[4]	36,277,363[4]	25,860,880[4]	37,076,902[4]	48,107,263[4]
Total balance-sheet	200,876,034	342,892,361	377,714,896	568,783,451	91,317,450	855,267,569	888,056,861	905,817,998

[1] Provisional figures.
[2] Including Agricultural Societies.
[3] Of the 14 federals in existence in 1928 only 6 sent in their balance-sheets; 5 were being wound up; 3 did not send in any statement of accounts.
[4] Loss.

CHAPTER XI

THE EFFECTS OF THE STATE'S ECONOMIC AND FINANCIAL POLICY UPON THE WORKING OF THE REFORM

In the preliminary remarks to Chapter IX it was pointed out how difficult it was to estimate the effects of the reform on production, because of the variety of other agents which, side by side with the reform, acted and reacted upon the country's rural economy.

There were in the first place the consequences of the War. They were not peculiar to Rumania, but a universal phenomenon, which saddled every country with the arduous problem of reconstruction. Reconstruction meant in the main the renewal of the factors of production—human and mechanical—destroyed or damaged in the service of the War; as well as a partial readaptation of the economic machine to altered conditions of supply and demand. In Rumania the land reform changed the whole structure and direction of rural life, and readaptation became as large a part of the problem of reconstruction as renewal. The whole task, therefore, was heavier and more complex than elsewhere, comparable rather to the problem of reconstruction in Russia than to that in western Europe.

Such as it was, the recovery of agriculture was the pivot on which the whole problem turned; because agriculture had been and was bound to remain the country's chief field of production, and because, if one excepts a minor contribution from the oil industry, agriculture alone was capable of giving a surplus wherewith to pay for the considerable imports urgently required to renew the means of production. In the proportion in which that task rested upon a particular branch of industry, that branch patently deserved assistance and encouragement from those who directed the country's policy. In this particular case such goodwill was also calculated to have inestimable psychological value as a stimulus to the millions of new peasant owners, to set to work with a will and get the utmost out of their new holdings.

To fulfil that function, the land reform would have had to be complemented by a helpful agrarian policy. If agriculture was to carry the State on its shoulders, it would first have to be helped itself on to its feet again. Rumania's internal resources were exceedingly low at the end of the War, and nothing in the way of bountiful favours could have been expected by any class of producers. But as the aid which the State could give the farmers was scanty, all the more cause was there to remove out of their way anything which might hamper their own effort to make good. Then, only, would the land reform have had a clear chance to prove what forces, good or evil, it was bringing in its train. That test will never be applicable now to the first decade of the new agrarian régime. Instead of pursuing these simple ends, which in this case were truly obvious, official policy harnessed itself to the ambitious ideal of achieving national self-sufficiency. Not only did it strive to conjure up a national industry—which, it was clear, could not have satisfied the country's immediate needs, let alone give a surplus with which to purchase indispensable goods from abroad—but it tried to reach that difficult goal without any foreign help. Taking 'through ourselves' as its motto, it attempted to create an artificial industry with such means as could be squeezed out of a worn-out country, and, moreover, at the same time to revalorize the exchange. The inevitable result was a harrowing stringency of money. Being the only field of production which could spare any means at all for the carrying out of these plans, agriculture was made the Cinderella of the industrial and financial fields of activity, in which an extreme economic nationalism wanted to raise its flag. And as, in its new state, agriculture rested more than ever on the shoulders of the peasants, it was the peasants who after having been subjected to the landlords were presently subjected to industry and finance.

In such conditions the farming class could not rise to the great occasion the land reform offered it. In fact, it found it hard even to reach its former precarious state and repair the damages which the War had done to its material equipment. The paradoxical policy which produced that situation must therefore be described, however briefly, if the birth and infancy of

Rumania's new agrarian organization is to be presented in its true light.

SECTION 1

THE RESOURCES OF AGRICULTURE

(*a*) *The Working Capital of Agriculture, as affected by the Reform.* After the War, and with the beginning of the land reform, agriculture found itself faced with exceptional circumstances. Its task was formidable indeed. Having always been undercapitalized it was bound to feel the effects of the War even more than it did in other countries. But, in addition, Rumanian agriculture was now primarily in the hands of several million smallholders, who had to find at one and the same time means wherewith to organize their farms as well as ready money for heavier taxes, and, especially, for paying for the land they had received. The large owners, in their turn, if they wanted to stay on the land, had to meet the new situation by reorganizing their farms for intensive cultivation, with equipment of their own. This at a time when they had suffered through the reform a considerable loss in revenue, receiving in return bonds on a very low scale of compensation and which at once lost half of their nominal value when offered on the Stock Exchange.

The compensation allowed by the decree of December 1918 was below the actual value of the expropriated land, but was still substantial enough. After the second expropriation it became merely nominal. The index had indeed been raised from twenty to forty times the rent fixed in 1916, i.e. the amount of the compensation was doubled, but in the meantime the currency had depreciated to 1/30–1/40 of its nominal value. From data collected from tribunals in the Old Kingdom, it appears that during the period 1911–16 whole estates were sold at an average of 988 lei per ha. in the lowlands and 684 lei per ha. in the hills—which was equal, roughly, to £39 10*s*. 0*d*. and £27 10*s*. 0*d*. at par.[1] The compensation granted to the expropriated landowners in the Old Kingdom varied from 1,200 to 3,000 lei, the average being about 2,000 lei per hectare; and

[1] V. O. Popovici-Lupa, *Viaţa Agricolă*, March, 1921.

only half of this at the first expropriation. This was worth about £20 towards the middle of 1919, not more than about £6 10*s*. 0*d*. at the time of the second expropriation, in the summer of 1921, and soon afterwards, when the bonds were handed over, less than £2; the actual value being about £2 10*s*. 0*d*., at the rate at which the Rumanian exchange has now been stabilized, or, roughly, one pound per acre. That, however, was the nominal compensation; the actual amount the landowners received was still less. For the bonds in which the State paid them never rose above 61 on the Stock Exchange, and fell to an average of 50. As the price has been calculated at forty times the rent fixed in 1916 and currency has depreciated to 1/32 of its gold value, it means that the nominal value of the bonds represent 5/4 and their actual market value merely 5/8 of the yearly rent for one hectare fixed officially in 1916. And, further, the yearly rent representing normally 1/20 of the object's value, it follows that the compensation given to the landowners was equal to $5/8 \times 1/20 = 1/32$ or about 3·2 per cent. of the pre-war value of their land.[1] The State was admittedly short of resources, and it is no concern of ours to inquire whether it could have taken a greater burden upon itself. What we are trying to clarify in this chapter is the State's attitude towards the agricultural classes, and the point is, therefore, not whether the State could have offered a fairer price to the landowners, but whether it treated other classes and groups who had some claim upon it with equal stringency. The only other measure comparable in character and scale to the land reform was the expropriation, so to speak, of the owners of Russian rubles and of Austro-Hungarian crowns; and they, one must note, were given twice the market value of the object they had to surrender. The transaction cost the State nearly 7·5 milliard lei,[2] the nominal

[1] M. Synadino, on p. 9 of the article quoted before, states that in Bessarabia the compensation was fixed at 750 lei per ha. Before the War one ha. was worth 1,000 Swiss francs and gave a net revenue of 42–5 francs yearly; the compensation given by the State represents about 7 Swiss francs, which will give a yearly income of 7 centimes.

[2] The stamping and withdrawal of Austro-Hungarian notes, which according to the Treaty of Saint-Germain was to be done at once, was not begun till late in the summer of 1919 and not carried out till August 1920. The 'primitive technique of the stamping process, and the reprehensible delaying of the measure of unification for

value of the currency being less than 14 milliard lei. According to estimates, dating back to 1924, cited by M. Cioriceanu, the total price of the expropriated arable land was slightly above 15 milliard lei, and together with forests expropriated later in Transylvania, about 17 milliard lei. The Central Resettlement Office puts the cost of the arable land at only 12 milliard lei (6 million ha. at an average of 2,000 lei); the actual value being about 145 milliard lei (estimated at the moderate average of 800 gold lei per ha.). The State, in other words, spent 7·5 milliards in buying back 14 milliards' worth of notes, and it offered 12 milliards for land worth 145 milliard lei. In the first case, moreover, it actually spent 7·5 milliards, without any further profit to itself; in the second case it has incurred merely a long-term debt of 6 milliards—one-half of the price being paid by the peasants—and it has got in return all the mineral wealth of the subsoil, as well as 175,000 hectares land, reserved for public purposes, at the expropriation price of one pound an acre.

Agriculture was urgently in need of capital in the first years of peace, having to make good the damage caused by the War and to equip itself for the more intensive phase in which it was entering. Most of the large owners, as we have seen, had to

nearly two years' (Dr. Netta, p. 464) gave rise to an extensive contraband in these depreciated notes. About 4 milliards Austro-Hungarian crowns were presented for stamping in August 1919, but about 8·5 milliards for exchange in August 1920. Moreover, the rate of exchange was 'unreasonably and unjustifiably' (Dr. Cioriceanu) fixed at 40–100 per cent above the market value: Austro-Hungarian crowns=0·50 lei, Romanoff rubles=1 leu, Lwoff rubles=0·30 lei, while notes issued by the Central Powers during their occupation through the Banca Generală were valued at par:

	Withdrawn	Valued at Lei
A–H crowns . .	8,580,089,979	4,290,044,988
Rubles . . .	1,289,039,590	1,001,842,785
Banca Gen. notes .	2,170,000,000	2,170,000,000
Total		7,461,887,773

The explanation commonly offered for this excess of generosity is that the State wished to deal liberally with the population of the new provinces. Yet no such liberality has been evinced in the treatment of owners of land in the new provinces, as that vouchsafed to the owners of bank-notes. It is probable, in fact, that at the time the rate of exchange was fixed, the bulk of the cancelled notes were in the possession not of individuals but of banks, and many more in the possession of banks in the Old Kingdom than in the new provinces. (See Dr. Xenofon Netta 'Politica Monetară a României', article in *Buletinul Institutului Economic Românesc*, Sept.–Oct. 1928, pp. 463–6; and Dr. I Cioriceanu, *La Dette Publique de la Roumanie*, Paris, 1927, pp. 69–72.)

purchase anew everything in the way of live and dead stock. But the compensation they were to receive from the State for the expropriated land was made over to them very slowly. The formalities for securing the bonds were complicated; two years after the first expropriation the owners had not received even a payment on account. The first bonds were not issued till November 1922, though bearing a coupon for May 1922.[1] The bonds are payable within fifty years. No sinking fund has been provided so far, and no amortization-draw had taken place until the end of 1928. According to the Director of the Central Resettlement Office, Professor A. Nasta, the total cost of the expropriated area would be approximately as follows:[2]

Old Kingdom . .	2,726,346 ha.	at 2,215 lei	=	6,038,856,390 lei
Bessarabia . .	1,491,916 „	„ 782 „	=	1,166,678,312 „
Bucovina and Transylvania	1,671,447 „	„ 2,180 „	=	3,643,754,460 „
				10,849,289,162 lei

An estimate communicated by the Central Resettlement Office puts the total cost at 12,016,194,000 lei (6,008,097 ha. at 2,000 lei). Until the end of June 1929, bonds have been issued for 6,176,710,200 lei. In addition, bonds to the value of about 600 million lei have been issued to cover mortgages of the Rural Credit Institute of Bucarest and the Urban Credit Institute of Jassy.[3]

[1] Payment for the first expropriation in the *Old Kingdom* was authorized by the Minute No. 625 of the Cabinet Council, published in the *Monitorul Oficial* of April 6, 1922; payment for the second expropriation by the Minute No. 517 published on March 27, 1927. The same Minute authorized payment for the land expropriated in *Bucovina*. For *Bessarabia* payment was authorized by the Cabinet Minute, No. 1459, published on July 3, 1923; for *Transylvania* by the Minute No. 92, published on January 31, 1924. Payment for the forests expropriated in Transylvania and Bucovina was authorized by the Cabinet Minute No. 3322, published on November 7, 1926. By arrangement with the Ministry of Finance, the Central Resettlement Office was entrusted with the transference of the bonds to the expropriated owners, as well as with the collection, through the usual fiscal channels, of the payments which the peasants had to make. The owners received 80 per cent. of the sum as soon as the price was settled, and the remainder after the final measurement of the expropriated land; interest at 5 per cent. was calculated, however, from the day when the land was taken over. Half of the capital was to be paid by the State, the other half by the peasants; interest was altogether a charge upon the State.

[2] *Reforma Agrară*, 1926, p. 7.

[3] M. Cioriceanu puts the value of bonds issued up to the end of 1926 at 11 milliard lei. These did not include payments for the Bessarabian estates expropriated from French and British citizens, the value of which was fixed at about one milliard lei, payment being made in 4 per cent. consolidated bonds of a total value of £1,103,000 (*op. cit.*, p. 72).

The total value of bonds issued up till the end of June 1929 was, therefore, less than the money paid in cash to the holders of foreign notes in 1920. Expropriation bonds were neither exempted from taxation nor qualified to be accepted as security by the National Bank, like other State papers; a disability which has affected their market value.

The newly settled peasants were to pay one-half of the price of the land, plus surveying expenses, about 200 lei per ha. (The State had advanced about 800 million lei for purposes of surveying.) These sums could nominally be claimed at once, but provided the new holders paid one-fifth of the sum on receiving the land, they could pay the other four-fifths within twenty years. According to the estimates of the Central Resettlement Office quoted above, the total amount to be paid by the peasants would be one-half of the price of the arable land, i. e. about 6 milliard lei, plus about 1·2 milliard lei surveying costs. Until the end of 1928, they had been debited with 4,094,105,074 lei.

The governments having adopted a strongly deflationist policy, they were anxious to bring back into circulation the notes which the peasants were supposed to be hoarding; and, like all Chancellors of the Exchequer, the Ministers of Finance were bent upon increasing as much as possible the immediate revenue without troubling about the effect of such a course upon national economy and upon subsequent budgets. Hence the subordinate local authorities were instructed to make propaganda among the peasants for the immediate payment of the full price of the land they had received. Whether the peasants were really hoarding money is an open question, but official propaganda certainly met with considerable success. It was, of course, to the advantage of the peasants to pay their debt while the currency was so depreciated, but psychological motives probably influenced them more than financial calculations. Past experience has made the peasants extremely reluctant to incur debts, and they do not feel secure until the proper 'papers', i. e. the title-deeds, are in their hands. Hence, a large number of them made the effort to pay the price in full, as may be seen from the table at the end of the volume. Up to the end of 1925 the peasants had paid nearly one-and-a-half milliard lei. It seems probable that payments kept pace with the debiting of the peasants, at least, until 1925; the position of the farmers was then getting steadily worse.

The law obliged the governments to devote all moneys thus collected from the peasants to the amortization of expropriation bonds. In fact no payment of this kind had been made up to the middle of 1929. The position, therefore, of the agricultural industry with regard to available capital was briefly this: both large and small farmers were in need of all possible resources, as many of them were about to start farming on their own account, and all of them had some shortage of stock and equipment to make good. The former landowners, however, were receiving the sums due to them from the State but tardily and in depreciated bonds; while the peasants were pressed to part with their money just when they most needed it. Nor was this money returned to the industry, as the legislator had intended that it should be. Hence, during the first ten years the State has actually turned the reform into a source of revenue; it has handed out long-term bonds for what it had to pay, while it has cashed ready money for what it had to receive—the operation having the effect of a forcible loan limited to the agricultural industry. The income which the State derives from the subsoil has compensated in part the smaller sum which the State has to defray by way of interest on the bonds, and the first might have come near to balancing the second if the State's possessions had been properly administered. At any rate, even if the expropriated owners had sold all their bonds—which they could not have done without severe loss—it is evident that the liquid capital they might have collected would hardly have exceeded the sums which the peasants handed over to the State. At best, that is, one section of the farming community might have balanced, at a great sacrifice, the capital outlay of the other section; in reality, the working out of the process of compensation has probably resulted in a diminution of such liquid capital as the agricultural industry possessed. (For detailed figures, see Appendix III.)

(*b*) *The Supply of Agricultural Credits.* The depreciation of the currencies has enabled owners of real estate to rid themselves easily of mortgages and other debts. Rumanian landowners have benefited, as have landoners elsewwhere, from that condition,[1] yet much of their advantage was lost through untoward events. They did not, to begin with, have the opportunity, enjoyed by the farmers of other countries, of making money during the War. During the first two years of the War, the frontiers were closed, and during the other two years two-thirds of the land was in enemy occupation, while in the other third production and marketing were stringently controlled by the Government. In both parts of the country, live and dead stock deteriorated. The indebtedness of the large owners was wiped out after the War, but so was their revenue. Few of them possessed any capital values beyond land, and this they lost against a nominal compensation. For the large farmers it was even more difficult to complete their stock than it was for the peasants. The peasants were not greatly in debt, but neither had they any capital. Their possessions may have represented a considerable sum in the present depreciated currency, but they have not a corresponding revenue and still less reserves in cash. After the War many of them had to get land, most of them had to get some stock; and if they got the one cheaply, they had to pay heavily for the other, so that altogether they had a difficult task before them, requiring fair facilities in regard to credit and general conditions.

If the pecuniary needs of small and large farmers were great, the possibility of satisfying them was inversely limited. The shortage of money was general, and the guarantees which farmers could offer were inadequate. The large owners had relatively little land left which might constitute a basis for credit, and in

[1] According to S. Nenițescu (cited by Antim in *Chestia Socială*, p. 183), land property was burdened before the War with the following mortgage debts:

Private mortgages	133,948,621 lei
Credit institutions	254,227,136 „
Mortgages without interest	43,746,021 „
Total	431,921,778 „

The yearly interest amounted to 26 million lei. To that were added taxes, and interest on current debts, so that altogether agriculture had to pay yearly in taxes and interest 50,951,260 lei out of a total income of 211,930,346 lei, i.e. 26·04 per cent. In the view of certain Rumanian economists, the landowners were rapidly approaching the point where they would have had to sell out; and that had some part in the ease with which they accepted the reform, in 1917.

Total deductions from land tax amounted to 138,586,750 lei yearly during 1923–7; allowances for mortgage debts were merely 1,285,275 lei, the total interest payable being 6,555,855 lei. (Dr. Creangă, *Veniturile și Averea României Mari*, p. 14.)

the first years after the War the probability of a second expropriation weakened that basis still more. Smallholders were in a still worse position, as the law forbade them to sell or mortgage the lots received through the land reform. The whole enormous area transferred to the peasants was thus removed from use as a basis for credit—and this in a country in which land played a more important role as a credit factor than trade and industry taken together, and at a time when agriculture was quite peculiarly in need of credit. Moreover, the reform had rendered equally difficult the obtaining of credits on produce. As the peasants' corn is not standardized it cannot be stored in elevators; each lot is stored separately, in the flimsy buildings or open yards of country stations, and credit operations are, of course, not possible on the strength of such individual and uncertain warehousings.

The main difficulty arose, however, not from the special circumstances of agriculture, but rather from the financial and economic policy of the State. Even before the War the nation's capital was barely sufficient to finance agriculture and the elementary industry then in existence. Only a small portion of the public loans could be covered at home. Notwithstanding the crisis from which all the branches of production have suffered, economic activity, compared to the size of the population, is greater in new Rumania than it was in the Old Kingdom. New industries and trades have sprung up, the exchange of goods between the several provinces is more varied, and in consequence the general requirements in money and credit are proportionately higher than they were in 1914. Yet the actual supply, both national and foreign, has been much smaller, because until the end of 1927 the governments pursued a policy of revalorization. In 1914 the Old Kingdom had a paper circulation of 500 million gold lei for a population of eight millions, which allowed 63 gold lei per inhabitant. At the end of 1928 the total paper circulation was 21 milliard lei, worth in round figures 660 million gold lei. The number of inhabitants being now eighteen millions, it follows that the circulation amounted to 36·5 gold lei per inhabitant. The parallel is not complete without taking into account the loss in the purchasing power of the *leu*, estimated at about 30 per cent. To bring the paper circulation not only up to its earlier numerical equator, but to its pre-war purchasing power, the country would have needed 82 gold lei per inhabitant, and for the whole population 1,476 million gold lei or 48 milliard paper lei.

In the second place, the State pressed these diminished resources into the service of a policy aiming at the rapid development of new industries and at the nationalization—in a political sense—of existing industrial undertakings. The banks were forced to use their resources to that end. The capital invested in large-scale industry varied as follows:

	In gold lei
1901	314,646,903[1]
1915	361,226,733[1]
1926	852,105,482

All the big banks, and the public exchequer, plunged knee-deep into that industrialist current. At one moment Messrs. Marmorosch, Blank & Co. alone controlled two-fifths of the country's industry. But none of the banks took a direct interest in farming enterprises. Nor were individual investors tempted to do so. The insecurity of land property and the State control of agricultural marketing, on the one hand, coupled with the favours granted to banking, industry, and trade, caused in the supply of capital a regular 'flight from the land'. Almost all the bills discounted by the National Bank were industrial and commercial, while agriculture was starved of credit. A former Governor of the Bank, M. Oromolu, admitted this himself, in 1925. In his report to the Board of Governors he remarked that Rumania had a National Bank to help trade and a Society for Industrial Credit to help industry, but no organized agricultural credit on a similar scale, though the bulk of the producers were farmers.

The credit at the disposal of farmers nowhere came near what they needed for reconstruction and for the improvement of production. Rumania possessed only one institution of agricultural credit for large-owners, the First Rural Credit Society of Bucarest. Until 1924 its transactions were based on mortgages; in that year a banking section was attached to it. Mortgage loans

[1] The first two figures refer only to the industries in the Old Kingdom enjoying the benefits of the Law for the Encouragement of National Industry.

were given by means of bonds on rural property; in 1925 the total issue amounted to 133,600,000 lei. Loans granted by that Institution and its branches in 1926 amounted to 484,000,000 lei, the rate of interest being 12 per cent. In addition, in 1925 the State created with the aid of the National Bank, the Lending Banks Against Guarantees, which never disposed of important resources, the loans outstanding in 1926 amounting to 246,861,250 lei. In 1923 the State had created, likewise with the support of the National Bank, a Society for Industrial Credit. In 1924 its transactions reached one milliard lei, those of the Lending Banks only about 95,000,000 lei. For the 4,171 farms above 100 ha. which existed before the War 'a sum of 600 million gold lei was provided by mortgage loans alone up to 1913, while to-day . . . the whole of the agricultural industry is able to obtain from mortgage banks, co-operatives, and other agricultural institutes no more than 4·25 milliard paper lei, that is, 100–30 million gold lei; of this 3 milliards is obtained through rediscounts from the National Bank. Mortgage bonds have up to now been impossible to place on the market. . . .'[1]

The small cultivators depended on the co-operative Popular Banks. These were supplied with funds from a central institution established with State assistance, the latter being in its turn assisted by the National Bank. The credits which the National Bank allowed to the whole co-operative movement did not exceed 600 million lei in 1923–4. According to the *Buletinul Agriculturei* for April–June 1927 (p. 126), the credits granted by Popular Banks to the peasant farmers in 1925 amounted to a little over 2 gold lei per ha., which included investment credits as well as working credits. Altogether, the credit offered to agriculture by the various public and semi-public institutions has not exceeded 3 per cent. of the pre-war total.[2] This assertion would certainly appear to hold good for Bessarabia. Investments, loans, &c., of all the Bessarabian banks had amounted to 503 million rubles in 1913, equal to 45 milliard paper lei; at the end of 1927 they were merely 1–1·2 milliard lei; the peasants, who now hold nearly all the land, are supplied by the Popular Banks.

[1] C. Stoicescu, in the *Manchester Guardian's Rumanian Supplement*, May 1927.
[2] S. Timov, in *Na Agrarnom Fronte*, No. 9, 1925.

In 1913 there were 337 banks with assets of 29 million gold lei, making an average of 962 lei per head of inhabitant. At present the number of banks is 557 with assets of 632 million lei, or 211 lei per inhabitant; this sum includes 356 million lei worth of corn distributed in 1925–6 for food and seed, because of the failure of the harvest, which was a measure of temporary assistance, so that the assets actually available for banking transactions were only 87 lei per head of inhabitant.[1]

The situation was hardly better in other parts of the country. The director of a big bank in the once prosperous Banat declared to a correspondent of the Bucarest *Plutus* (November 6, 1923), that 'Rumanian peasants—some of them worth millions—are daily coming to me to demand credits of 20–30,000 lei, for somewhat longer terms, and we cannot do it'. Because of that neglect of agricultural credit large farmers were forced to borrow from private banks, at 25–30 per cent. interest, while the peasants were abandoned to the usurers, to whom they had to pay anything up to 100 per cent. The predicament in which farmers, large and small, found themselves was disclosed in an interview with M. I. Prohaska, the director of the greatest Banat flour-mill, published in the *Plutus* of November 11, 1923: 'To-day', he said, 'we can buy wheat at any price—I am giving away a professional secret—provided we pay in cash, so great is the shortage of money among farmers.'

A law for the creation of an Agricultural Credit Institution was passed by the National Peasant Government in the summer of 1929. The initial capital is to be of 500 million lei, one half of which is to be subscribed by the State, with permission to increase the capital gradually to five milliard lei, by issuing bonds. Loans are to be granted on mortgage for shorter or longer terms; the law allowing the mortgaging also of land distributed under the agrarian reform. It was hoped to start the new institution in the autumn of 1929.

[1] Teofil Ioncu, 'Creditul în Basarabia', article in *Dreptatea*, Bucarest, January 13, 1928.

Section 2

The Burdens of Agriculture

(*a*) *Taxation.* Direct taxes on rural property were made uniform for the whole country by the law of 1923, which also established an income tax for the first time. The assessments made on rural property on that occasion were to remain valid during the quinquennial period 1923–7. The basis of assessment was to be either the rental value per hectare, or the taxable income of the property. Because of the conditions then prevailing the Ministry of Finance decided that the assessments should not exceed certain moderate limits. In fact the assessment of that year fixed the taxable revenue of the 13,685,921 ha. arable land at 3,038,768,028 lei, that is 220 lei per ha., or 4·40 in gold lei as against the pre-war estimate of 30 lei per ha.

In virtue of the fiscal law passed in December 1926, the basis of assessment was raised, as from 1927, to three times the taxable income fixed in 1923.

The taxable income of the various categories of land was estimated as follows:

	Taxable revenue	Extent	Taxable revenue per hectare
	Lei	Ha.	
Grass lands	738,860,717	5,226,731	141·36
Market gardens	76,354,104	178,680	427·32
Orchards	62,258,422	234,524	265·46
Vineyards	114,417,141	236,910	482·95
Forests in exploitation	360,134,514	2,387,898	150·81 [1]
Other taxable properties	29,100,826	121,821	238·88
Arable land	3,038,768,028	13,685,921	220·00
Total	4,419,893,752	22,072,485	200·00

The small property gives the highest taxable revenue per ha. of arable land, with an average of 224 lei for the whole country:

Properties up to 5 ha. (1,437,917,614 : 6,400,109)=224,67 lei
„ above 5 „ (1,600,850,423 : 7,255,549)=219,72 „

A more detailed segregation into categories is not possible for

[1] The low taxable income of forests in course of exploitation is due to the assessment at 75 lei in Transylvania and 145 lei in Bucovina—a rate which is altogether inexplicable when compared with the 1,078 lei fixed in the Old Kingdom and the 1,195 fixed in Bessarabia.

the whole country because the tables for Transylvania were set up in jugars and those for the other provinces in hectares, so that the categories do not coincide. For this reason the figures for Transylvania have been left out of the table below, which gives the taxable income per ha. of the various categories of property in the Old Kingdom, Bessarabia, and Bucovina:

Properties up to 5 ha.	(1,082,403,315 : 4,723,816) =	229·16 lei
,, of 5 to 10 ,,	(428,990,419 : 1,964,501) =	218·37 ,,
,, of 10 ,, 50 ,,	(294,238,865 : 1,473,920) =	199·63 ,,
,, of 50 ,, 250 ,,	(239,474,049 : 1,059,566) =	226·01 ,,
,, over 250 ha.	(259,349,285 : 1,167,784) =	222·08 ,,

Medium-sized property of 10–50 ha. gave the lowest taxable income, which is explained by its being situated in the less fertile regions of the hills.

These low assessments caused the revenue derived from the land tax to remain on a very moderate scale. Land farmed by its owner was taxed at 12 per cent., land farmed by tenants at 14 per cent., and land whose owner lived abroad at 24 per cent. The gross yield of the tax was as follows:

	Lei	Per cent.
Old Kingdom	420,041,473	47
Bessarabia	166,093,389	19
Bucovina	29,789,124	4
Transylvania	284,064,524	30
Total	899,988,511	100

The net yield was still lower, as certain allowances were made for mortgages, for large families, &c. The allowance for mortgages was as below:

	Sum of interest on mortgages	Tax deductions
	Lei	Lei
Old Kingdom	4,340,138	900,837
Bessarabia	20,615	4,123
Bucovina	477,880	85,959
Transylvania	1,717,220	294,354
Total	6,555,853	1,285,273

The table indicates the limited size of the mortgages resting on land property, as mortgages on the expropriated area were paid off in expropriation bonds. Very striking is the case of Bessarabia, where mortgages were practically wiped off, no doubt because the lending institutions resided in the former Russian Empire.

A second category of allowances benefited the landowners who were not liable to income tax, having an annual income of less than 10,000 lei; they were entitled to deductions ranging from 10 to 20 per cent., according to the number of their children. Finally, landowners whose property did not bring in more than 2,000 lei yearly income were entitled to an allowance of 25 per cent. The average income having been assessed at 220 lei per ha., it followed that most of the peasants with holdings up to 10 ha. could claim that allowance—in fact 2,944,074 out of 3,612,745 or 81 per cent. The amounts deducted on these two grounds were as follows:

	For large families	For incomes below 2,000 lei
	Lei	Lei
Old Kingdom	22,389,910	42,493,399
Bessarabia	13,457,022	20,204,368
Bucovina	1,806,083	2,888,981
Transylvania	13,669,843	20,386,568
Total	51,322,858	85,973,316

The total amount of these deductions was 138,586,450 lei, leaving a net yield of the tax of 761,402,060 lei. This sum was apportioned between various public authorities as follows:

	Lei
State Treasury	459,419,550
Counties	187,440,210
Communes	114,542,300
Total	761,402,060

The yield of the tax represented 17 per cent. of the taxable value, a rather high proportion, but the burden was not severe, as the taxable value had been assessed with great indulgence. M. Ionescu-Siseşti considered that during 1923–7 the State claimed only one-fifth of what the land tax might have yielded. Revenue from land tax was estimated at 1,300 million lei in the 1927 budget and at 1,200 million lei in the 1928 budget.

It would have been useful for the purpose of this study to

compare the incidence of taxation as between agriculture and industry. Without such a comparison it is not possible to judge whether the smallness of the land tax was truly a favour shown to the farmers or merely part of a generally over-lenient fiscal policy. But such a comparison is notoriously hard to make, and in Rumania's case it cannot even be attempted, because only a small portion of the public revenue has been derived from direct taxation. Moreover, in a country with a lax administration the collection of taxes may affect their incidence as much as or more than the assessment of those taxes. One can, therefore, merely note two facts mentioned by M. Anastasiu in his article on the direct taxation of limited companies.[1] He states that the direct taxation levied upon the co-operatives of production in the Old Kingdom represented 5·6 per cent. of their capital, and that levied upon large industrial undertakings about 2·3 per cent. (4 per cent. in Transylvania); though the writer is careful to point out that the validity of the comparison depends on too many elements to be accepted at its face value (pp. 605–6). By putting side by side various figures he established, however, another significant fact. During the four years 1923–6 the currency had lost about 38 per cent. of its internal purchasing power; during the same period the capital of limited companies in the county of Ilfov (which includes Bucarest, the seat of most Rumanian companies) had almost doubled. Yet the amount of income tax those companies paid decreased during that interval in absolute and, especially, in relative values. One would assume, wrote M. Anastasiu, that the collecting authorities would try their hardest to check an abuse tolerated by the assessment commissions. 'The truth is just the opposite. The fiscal authorities of Bucarest have indeed instituted numerous taxation appeals during 1927 and 1928, but only against small individual firms or minor limited companies; they did not lodge a single taxation appeal against any limited company with more than 100 million lei capital' (p. 595).

M. Ionescu-Sisești gives figures showing that the yield of the land tax accorded with its assessment.[2]

[1] In *Buletinul Institutului Economic Românesc*, Nov.–Dec. 1928.

[2] *Repartiția Propietăților Agricole* . . ., p. 185.

Before discussing import and export duties, which have constituted the main tribute levied on agriculture during the past few years, brief reference might be made to another peculiar tax which is imposed solely upon the rural population, and, in fact, only upon the peasants. Though servitudes were abolished long ago, one of them has remained in being to the present day—namely, the obligation of rural inhabitants to give five days' service for the upkeep of public roads. If a peasant has no animals he works with his hands, otherwise he has to give five days' work with his cart and oxen. Landowners residing in the towns pay 30 lei road tax there and nothing else in the country, no matter how many animals they may keep on their estates. It may happen that some peasants cannot or will not perform that servitude; in order to constrain them to it a departmental decision of the authorities concerned fixed the equivalent of five days' manual labour after the War at 150 lei, and of five days' labour with the ox-cart at 500 lei, plus a fine of 50 lei.[1] During subsequent years the money value of the servitude rose to 660 lei for five days' manual labour, 1,100 lei with one ox, 2,200 lei with two oxen, 2,990 lei with four, and 4,500 lei with six oxen. The average peasant holder, owning 5 ha. land and two oxen, may have had to pay, therefore, merely 100 lei in land tax during the period 1923–7, but unless he performed the work he would have had to pay 2,200 lei as road tax, which amounted to a tax of 440 lei per ha., levied only upon his class. Some illuminating facts concerning this peasant servitude came to light during the sitting of the Ilfov County Council, on the 17th of February 1929.[2] The Prefect of the county expressed the wish on behalf of the new Peasant Government that part of the debt the peasants had incurred on account of the road servitude should be remitted. He based his proposal on two reasons. He first remarked that the county's finances were flourishing and that in consequence they need not be guided by the attitude of other counties, which implied that the Government's generous suggestion was not extended to the whole country. And, secondly, he pointed out that often the obligation had remained unperformed through no fault of the

[1] Statement of M. Mihalache before the Chamber, June 23, 1921.

[2] See *Dreptatea*, Bucarest, February 19, 1929.

peasants, but through lack of work, it being difficult to obtain supplies of gravel. Hence it would seem to be the rule to charge the peasants with the money tax even when they are willing to perform the work, but there is no need for it. In other words, an obligation to perform certain public work, when needed, of course, appears to have been transformed simply through a departmental decision into a permanent tax with a class incidence. The discussion further disclosed how heavy was the burden which that obligation laid upon the peasants. The Prefect's proposal, namely, was that arrears dating from the period 1908–20, amounting to 2,975,478 lei, should be cancelled altogether; while of the arrears from the years 1921–7, amounting to 72,320,085 lei, one half should be remitted. The figures are revealing, indeed. For they show that some peasants—quite a number, in fact, as the first sum must have included a proportion of gold lei—have been unable to pay this imposition throughout a period of twenty years; and that in one county alone, and that one of the richest, arrears from this obligation grew within seven years to the formidable sum of seventy millions, notwithstanding the land reform. The Ilfov County Council adopted the Prefect's suggestion, and it also slightly reduced the money value of the road servitude for the current year. But seeing that a Peasant Government had come into power in the autumn of 1928, the surprising thing is that neither the Prefect nor any of the elected Councillors thought of suggesting that this medieval servitude should be abolished altogether. Though during 1929 a peasant in the county of Ilfov, owning 5 ha. land and two oxen, would have to pay only 297 lei in land tax,[1] he had to pay in addition 1,750 lei as the equivalent of the road servitude. Between themselves these two taxes—and there are a good many others—would levy over 400 lei per ha.

(*b*) *Tariffs.* 'The real fiscal contribution which agriculture has made towards the needs of the State has been not the tax on agricultural land, but the export tax,'[2] says M. Ionescu-Sisești.

[1] The threefold taxable value of 1923, i. e. 660 lei per ha., multiplied by 5 = 3,300 imposed at 12 per cent. = 396, minus the allowance of 25 per cent = *297 lei*.

[2] *Repartiția Proprietăților Agricole*, p. 185.

To which one must only add that import duties played as effective, if less spectacular, a part in hampering the progress of farming. The stringency of capital and credit described above left the farmers to their own devices in the phase of reconstruction. Thanks especially to the efforts of the peasants, the live stock was replenished rapidly enough. It was obviously more difficult for the farmers to equip themselves with machines and implements adequate in quantity and quality for intensive farming.

Import Duties. Even before the War the equipment of agriculture had been of the poorest. The creation of many new smallholdings meant that much more had to be done than merely to make good the War damage; that is, if large and small farmers were to have the means for increasing production. The governments, however, were bent upon protecting national industry—in this case really consisting of one factory only, the 'Reşiţa', which had been nationalized, in a political sense, under Liberal auspices. Home production could not in any case satisfy all the urgent needs of agriculture. Nevertheless, high import duties were placed on agricultural machines and implements. Ploughs had to pay on an average 20–30 per cent. of their value, and, in general, agricultural machines paid an average of 1·20 lei per kg., while light motor-cars and commercial vehicles, which are not manufactured in the country at all, were only imposed with about 0·60 lei per kg. The table on p. 434 compares the import duties imposed in 1916 and 1927 on agricultural machines with those levied on motor-cars.

One should perhaps point out that if the comparison is to be real, one must take into account commercial vehicles, rather than luxury cars for private use. One will note, further, the enormous increase in the duties on agricultural machines—fifteenfold in the case of ploughs—while the duties on commercial vehicles and the lighter private cars have been reduced by one-half to one-third. The result was that during 1925–6, when Rumanian imports reached the highest figure recorded up to that date, the import of motor-cars (10 million dollars) far exceeded the import of agricultural machines (less than 3 million dollars in 1925 and less than 6 millions in 1926).

As the prices of agricultural machines and implements, as of all industrial products, stood at a very high level at the end of the

	Tax per 100 kg.	
	1916	1927 [1]
	Gold lei	Gold lei
AGRICULTURAL IMPLEMENTS:		
Pickaxes	1·00	20·00
Hoes, scythes	5·00	10·00
Spades, grubbing-axes, pikes	5·00	17·50
Forks	5·00	12·50
Ploughs, harrows, rollers, ridge-ploughs, extirpators, grubbing-axes, completely fitted up, and their accessories	1·00	15·00
Planting, binding, threshing- and harrowing-machines; grape-crushing-machines and fodder presses worked by hand	1·00	20·00
Steam ploughs or ploughs moved by gas or electricity, fertilizer-scattering-machines; harvesting- and mowing-machines; sorters; sowing-machines, fodder-presses moved by cattle or engine; dairy machinery weighing more than 25 kg.; fodder-binding-machines with iron frames; potato and beetroot-harvesting-machines; maize-threshing-machines with engine; cereal-harrows with engine; vineyard-watering-machines	2·00	7·50
Threshing-machines worked by hand or cattle	2·00	20·00
Threshing-machines worked by engine:		
(*a*) the drum 800 mm. long	2·00	32·00
(*b*) the drum more than 800 mm. long	2·00	7·50
MOTOR-CARS:		
Passenger-cars with open coach-work		
(*a*) under 1,000 kg.	45·00	16·00
(*b*) 1,000–1,200 ,,	30·00	24·00
(*c*) 1,200–1,500 ,,	30·00	74·00
(*d*) above 1,500 ,,	30·00	124·00
Passenger-cars with closed coach-work (coupé, sedan without partition)		
(*a*) under 1,000 kg.	45·00	50·00
(*b*) 1,000–1,200 ,,	30·00	74·00
(*c*) 1,200–1,500 ,,	30·00	100·00
(*d*) more than 1,500 kg.	30·00	150·00
Motor lorries, trucks, tanks, fire-engines, &c.		
(*a*) 3,000 kg.	30·00	15·00
(*b*) 1,500–3,000 kg.	30·00	12·50
(*c*) less than 1,500 kg.	30·00	10·00

War, the addition of such heavy duties made their purchase well nigh impossible, especially for the millions of small peasant

[1] Minimum and Maximum tariff. There is a 'maximal' tariff, which is not applied, however, as Rumania has tariff conventions with all the countries which come into question.

farmers who had in their hands 90 per cent. of Rumania's arable land. As recently as the end of 1928, when the prices of many industrial products had reverted to the pre-war level, or had even fallen below it, retail prices for agricultural implements and machines were still exceptionally high in Rumania, as may be seen from the following table: [1]

Agricultural implements (Average types)	Pre-war prices	Post-war prices	
	Lei	Lei	(gold)
Thrashers	12,000	400,000	(12,500)
Grading- and sorting-machines	150–300	15,000–20,000	(465–625)
Binders	800	50,000	(1,560)
Sowers	600	30,000	(935)
Mowers	250	16,000	(500)
Rakes	200	10,000	(310)
Reapers	450	18,000	(560))
Ploughs	50–70	2,200–4,500	(70–140)
Harrows	40–60	1,200–1,700	(38–53)

These prices refer in the main to the products of American industry, which predominate in the Rumanian market and which, relatively, have more rapidly approached to pre-war prices than corresponding English and German products.

One might mention also the duty of 28 lei which farmers have to pay for each imported sack, which means about 50 per cent. of its value, as the price of the object at the frontier is 55–70 lei. As the annual requirements amount to some 2,000,000 sacks this duty adds some 50 million lei yearly to the impositions which agriculture has to support.

Export Duties. Most of the belligerent countries adopted measures after the War tending to moderate internal consumption while intensifying exports. The governments concerned were naturally anxious to nurse their depleted supplies, and at the same time to recover at the earliest possible moment a favourable balance of trade. The Rumanian Government followed a contrary line of action. Its chief ambition was to ensure a liberal and cheap supply of food for the urban population.

In pursuance of that policy the State had recourse to a variety of measures—control of retail prices, restriction or prohibition

[1] By courtesy of 'Plugul' Cy., Ltd., Bucarest.

of exports, imposition of heavy export duties—against which the Central Union of Agricultural Syndicates protested in a Memorandum, in 1921. It warned the Government against the effect of such a policy on production, and urged instead the following measures: complete freedom for internal corn trade; export should be free, on payment of a tax; a bonus for wheat growing as long as wheat export was taxed, the necessary sums to be obtained from the proceeds of the export tax. The whole was to form a transitional program which after a year was to give way to a régime freed of all restrictions and control.

The views of the Agricultural Syndicates failed to influence the State's policy. After a short period, during which the export of agricultural products was altogether prohibited, the State introduced in 1920 the so-called system of 'contingentation', i.e. a rationing system which permitted producers and traders to export certain quantities on payment of a low tax, if they supplied at the same time a determined quantiy for internal consumption. That system merely led to an abusive traffic in export permits, which was demoralizing trade and administration. Even the banks, in a Memorandum presented to the Government in 1921, pleaded for its replacement by a system of export taxes, however much they disliked all interference with trade. Maize remained subject to the 'contingentation' system till 1922 and wheat till 1924. The free export of oil-bearing grains was not allowed till 1925. But beginning with 1922–3 the 'contingentation' system was abandoned in principle in favour of high export duties. By this means the State hoped to prevent an excessive export of food supplies, and consequently a rise in the cost of living, and to secure simultaneously fresh revenue for the Treasury. The table on pp. 437-8 gives the full list of export taxes payable for the principal cereals.

As the duties were assessed on quantity they equalled on occasion through the play of the exchanges 50 per cent. of the value of the produce. To obviate possible losses to the Treasury from the continuous fall in the currency, these taxes were made payable from the middle of 1922 in stabilized foreign exchanges, on the basis of £, as noted in the last column of the table. We have already mentioned the export tax of 10,000 lei, after-

Product	Unit of taxation	Date of decree	Tax in lei	Tax in £	s.	d.
	kg.		Lei	£.	s.	d.
Wheat	100	1 Aug. *1915*	6	—		
	,,	7 June 1919	20% ad valorem	—		
	10,000	14 July 1920	12,000	—		
	,,	9 Dec. 1923	25,000	—		
	,,	8 Aug. 1924	45,000	44	0	0
	,,	17 July 1925	30,000	29	5	0
	,,	7 Mar. 1926	18,000	18	0	0
	,,	1 Aug. 1926	13,000	13	0	0
	,,	23 Jan. 1927	5,000	5	0	0
	,,	19 Oct. 1927	3,500	3	10	0
	,,	24 Jan. 1928	2,800	3	10	0
Maize	100	1 Aug. *1915*	4	—		
	,,	7 June 1919	20% ad valorem	—		
	10,000	18 May 1920	13,500	—		
	,,	2 July 1920	7,800	—		
	,,	27 July 1920	2,200	—		
	,,	24 June 1921	1,500	—		
	,,	6 Sept. 1922	30,000	—		
	,,	13 Nov. 1923	20,000	—		
	,,	14 Oct. 1925	12,000	12	0	0
	,,	21 Mar. 1926	10,000	10	0	0
	,,	23 Jan. 1927	5.000	5	0	0
	,,	19 Oct. 1927	3,500	3	10	0
	,,	24 Jan. 1928	2,800	3	10	0
Rye	100	1 Aug. *1915*	6	—		
	,,	7 June 1919	20% ad valorem	—		
	10,000	14 July 1920	10,000	—		
	,,	21 Oct. 1920	10,000	—		
	,,	9 Dec. 1923	20,000	—		
	,,	7 Mar. 1926	10,000	10	0	0
	,,	17 Jan. 1927	5,000	5	0	0
	,,	19 Oct. 1927	3,500	3	10	0
	,,	24 Jan. 1928	2,800	3	10	0
Barley	100	1 Aug. *1915*	5	—		
	,,	7 June 1919	20% ad valorem	—		
	10,000	26 June 1920	10,800	—		
	,,	2 July 1920	6,000	—		
	,,	27 July 1920	2,200	—		
	,,	5 Aug. 1920	4,200	—		
	,,	27 Sept. 1920	3,500	—		
	,,	21 Oct. 1920	3,500	—		
	,,	24 June 1921	2,000	—		
	,,	9 July 1922	20,000	—		
	,,	14 Oct. 1925	12,000	—		
	,,	21 Mar. 1926	10,000	10	0	0
	,,	23 Jan. 1927	5,000	5	0	0
	,,	19 Oct. 1927	3,500	3	10	0
	,,	24 Jan. 1928	2,800	3	10	0

Product	Unit of taxation	Date of decree	Tax in lei	Tax in £ s. d.
Oats	100	1 Aug. *1915*	5	—
	,,	7 June 1919	20% ad valorem	—
	10,000	4 June 1920	11,500	—
	,,	2 July 1920	7,200	—
	,,	27 July 1920	2,200	—
	,,	5 Aug. 1920	3,000	
	,,	22 Sept. 1920	3,500	—
	,,	21 Oct. 1920	3,500	—
	,,	30 Aug. 1921	2,000	—
	,,	9 July 1922	20,000	—
	,,	7 Mar. 1927	10,000	10 0 0
	,,	23 Jan. 1927	5,000	5 0 0
	,,	19 Oct. 1927	3,500	3 10 0
	,,	24 Jan. 1928	2,800	3 10 0

wards reduced to 8,000 lei per head of large horned animals. Sheep paid 400 lei per head. Quality cheese paid 45 lei per kg. and dried fruit 20 per cent. ad valorem. It is suggestive that wheat paid 45,000 lei per wagon, but white flour, which had a much greater value, only 25,000 lei, because steam-mills were considered as part of the national industry. Petrol of whatever strength only paid 8,000 lei per wagon. The receipts from customs duties are shown in the table below:

	Import duties	Export duties	Various	Total
1922 April–Sept.	599,354,037	383,786,257	141,971,889	1,125,112,183
1923	1,531,245,485	4,979,445,369	468,917,459	6,979,608,313
1924	1,988,435,220	5,112,053,981	547,377,881	7,647,867,082
1925	3,003,468,689	3,906,883,124	616,956,167	7,527,307,980
1926 (first seven months	1,540,341,739	2,201,460,105	330,175,554	4,071,977,398

Export taxes brought therefore considerably higher revenues than import taxes from the time when the 'contingentation' system was abandoned. In 1926 customs duties produced 9 milliard lei—i.e. more than one-third of the State's total revenue—and about one-half was the yield from export taxes. As the bulk of the exports consisted of corn and agricultural products, the bulk of the export taxes was inevitably paid by agriculture, which is confirmed by the customs returns:

Year	Export taxes on agricultural produce	Percentage of taxes from value of agricultural export
	Lei	Per cent.
1919	3,597,630	15·71
1920	387,498,424	14·90
1921	469,147,382	8·31
1922	1,436,564,947	17·48
1923	4,364,679,096	26·87
1924	4,375,732,991	24·01

For 1925 and 1926 figures are available only for some of the principal agricultural exports:

1925	2,486,038,000	20·83
1926	2,488,020,000	14·26

These taxes represented a very high percentage of the total value of the agricultural export.[1] During 1923 and 1924, for which the figures are final, the revenue which the State derived from these taxes was about six times higher than the total amount of the land tax.

Two characteristics of the above table of export taxes should be specially noted. The changes in the various rates, it will be seen, were frequent; and, moreover, they were not decreed simultaneously for all kinds of grain. Taken together with the frequent and considerable variations in exchange, these circumstances made all transactions for future delivery a sheer gamble and altogether rendered the corn trade chaotic. The trade, in fact, was almost wiped out; there were many failures, among old-established firms. None of the flour-mills worked full time; those put up for sale could find no purchasers.

The effect was bound to be mercilessly expressed in the figures relating to agricultural exports. During the half-year which followed the imposition of the 45,000 lei tax no wheat was exported at all. Considering that in 1925, when the country had plentiful stocks, exports suffered a further serious decline, it seems clear that the phenomenon was caused by abnormal conditions in the corn

[1] Export duties on manufactured products (many of them the produce of agricultural industries) represented during the same period a much lower percentage of the value of those exports:

1922 1·28 per cent. 1923 7·39 per cent.
1924 7·46 per cent. 1925 6·29 per cent.
1926 3·85 per cent.

trade rather than in corn-growing. The imposition of export duties reacted more directly on farmers producing for the market than on those who, like most of the peasants, produced primarily for their own consumption; until 1927, therefore, the peasant farmers would seem to have been less severely hit by the export tariff. This assumption, made by several Rumanian writers, may have merely the appearance of truth. It is true that corn exports came mainly from the large farms, but internal requirements in corn, and especially in animal and dairy products, were satisfied mainly by the peasants; and the tariff policy was of course designed chiefly for the purpose of depressing the internal price of agricultural produce. In Transylvania 65 per cent. of the horned animals were milch-cows, giving milk valued at 10 million lei. Transylvania exported in 1921 27,000 kg. butter and 400,000 kg. cheese; in 1924 the export was 70 kg. butter and 150 kg. cheese. Yet the import of foreign cheese did not diminish, the duty of 24 lei per kg. being more than offset by the burdens which Rumanian farmers had to bear. Condensed milk for export was taxed with 40 lei per kg., while Dutch condensed milk sold at 39 lei per kg. in Hamburg.[1]

At any rate, the State's fiscal policy was reversed in 1927. The land tax assessments, as we have mentioned, were increased threefold, while export duties were reduced as below (decree of January 2, 1927):

Cereals, leguminous and oil-bearing seeds	5,000 lei per wagon
Wheat flour	4,000 ,, ,, ,,
Large horned animals	2,000 ,, per head
Pigs	300 ,, ,, ,,
Sheep	100 ,, ,, ,,

In view of its higher value, wheat paid relatively, under the new tariff, a lower tax than other cereals. Customs receipts assumed from that date a more normal aspect, in the distribution of import and export duties:

	(in milliard lei)	
	1927	1928 (six months)
Import duties	3·14	3·24
Export duties	1·34	0·36
Various	0·26	0·05

[1] From an article in the *Argus*, June 24, 1925.

But agriculture was cheated of the benefits which the change was supposed to give it, because in the meantime the exchange had risen, while the price of agricultural products in foreign markets had fallen.

In relation to the total extent of the arable area, the burden the agricultural industry had to bear in export duties during 1924–5 was in effect equal to a tax of 325 lei per ha. An article from the pen of a prominent agricultural expert published in the *Argus* of October 17, 1923, maintained that farmers had to disburse over 40 per cent. of the total value of their production in direct and indirect taxes, while having themselves to bear all the costs and risks of that production. Manufacturers, traders, bankers, &c., paid taxes on their net income, but the farmers, complained the writer, had to give the State almost half the value of their gross revenue, without regard to the harvest's return per ha. and to cost production. In fact, as another writer pointed out, the agricultural producers were doubly hit by the State's tariff policy: while it prevented them from selling to the best advantage what they produced, it forced them to use the products of an excessively protected national industry. The State derived therefore the further advantage that, by depressing the cost of living, it could continue to underpay its officials, as well as to buy cheaply supplies for the army.

(*c*) *Transport.* Rumania's transport system has never been adequate for the needs of her agriculture. Roads and railways are much below the requirements of an industry whose trading chances depend on the possibility of transporting great bulk safely and cheaply within a short space of time.

Most of the country roads are mere tracks, built and summarily maintained by compulsory peasant labour. In bad weather they quickly become unusable, except for slow and heavy ox-carts. Figures obtained from the Ministry of Public Works at the end of 1928 put the total length of classified roads at 87,500 kilometres. Of these, 16,500 kilometres were in relatively good condition; 28,400 kilometres were fit to be used only in dry weather; and 42,600 kilometres were natural roads, without any hardened surface at all. Only 10,886 kilometres in the first category had a proper macadamized surface. None of these roads are fit for the present-day fast and heavy mechanical traffic. Since the War, almost nothing has been done to improve the quantity and quality of the roads; in fact, even the few principal arterial roads, like that which runs along the highly industrialized Prahova valley, have been allowed to fall into disrepair. The new provinces have received even less attention. Only now, ten years after the War, has the construction of a road been undertaken to link up Jassy, the Moldavian capital, with Bessarabia. In the latter province the condition of the roads altogether prevents any regular traffic as soon as the bad weather sets in. When a new Government came into power in the autumn of 1928, the Minister for Bessarabia went on a tour of inspection, but after only a few rainy days in an exceptionally dry year he was unable to pursue his trip and had to turn back. As most of the 15,200 bridges on these roads are built of wood they are now in a state of decay; on many of them only half the width is used for traffic. Some of the bigger bridges destroyed during the War are not yet rebuilt. The Department estimated that 600 million lei were needed yearly to keep the existing roads in tolerable repair, and 120 millions for the bridges. The actual budget allowance is 60–70 millions for the roads and 20–30 millions for the bridges.

The railway system was planned to link together the principal towns and to serve certain strategic ends, but not the needs of the country's main agricultural regions. The lack of double tracks and of sidings, besides the inadequacy of the rolling stock, provided even before the War the spectacle of mounds of sacks filled with corn stacked under the open sky in railway yards. The railways suffered severely in the War and the country's subsequent economic policy delayed their recovery. At the same time, the demand made upon them has grown with the addition of the new provinces and the expansion of the urban centres. An inquiry which the *Argus* conducted in the spring of 1924 into the retail prices of agricultural produce, disclosed differences of 40 to 200 per cent. between towns which sometimes were merely 30–40 kilometres distant from each other.[1]

If agriculture were to depend for the transport of its production solely on the railways, it is doubtful if even the internal

[1] *Argus*, May 28 and June 13, 1924.

markets could be properly served. River transport on the Danube, and to some extent on the Pruth, somewhat relieves the situation, which is only saved, however, by the peasant carts. In the neighbourhood of markets and of the ports from which grain is shipped, agricultural and other products are carried almost exclusively in ox-carts. Dr. Zahareanu calculated that in 1923 the railways could place not more than 125,000 wagons at the disposal of agriculture, which was altogether insufficient. A normal harvest of 14 million tons would leave a surplus of about 7 million tons for export, but the railways could not possibly cope with such a quantity. Nor would there be adequate accommodation for storing it. In 1926, when production and export improved, the railways proved unable to transport the whole of the harvest surplus from the interior to the ports during the autumn. When the export season was almost over the Rumanian State Railways were able to rent from the German, Polish, and Czechoslovak railway authorities 6,000 freight wagons whose delivery began early in December and which were to be used solely for the export trade. That belated arrangement was in any case but a partial solution, as railway engines were equally lacking.[1] One must keep in mind that for purposes of export rapidity of transport plays an important part in determining the prices which the agricultural produce of a country secures.

Besides suffering from that deficient railway service, agriculture was burdened with an unfavourable freight tariff. In fact, agriculture was about the only industry which paid the full tariff on the Rumanian railways. According to the Law for the Encouragement of National Industry, all factories and works to which the benefits of that law had been accorded—and this meant almost all companies employing more than fifty workers—paid only half the ordinary freight tariff both for their own goods and for machines, &c., which they imported. During the brief two weeks' spell of M. Barbu Stirbey's Government, in the summer of 1927, the oil companies secured the same privilege for the transport of their products. The timber industry already enjoyed such favoured treatment. Agricultural products alone paid the full tariff. Agriculture was represented by one member only on

[1] *Manchester Guardian Commercial*, December 23, 1926.

the governing board of the Rumanian State railways, while industry and finance had ten. That was not all. Everywhere else the railways, especially when owned by the State, concede lower freights for goods going abroad; but in Rumania agricultural produce destined for export was actually charged double the ordinary tariff after the War. One of the incidental effects of that strange freight policy was to place at a disadvantage the Transylvanian farmers, as their produce had a relatively long land journey to make before reaching the ports.

Section 3

Mercantilism or Agrarianism?

This brief outline of agriculture's resources and burdens naturally invites the question: was the State's unhelpful conduct the consequence of an error of judgement or the operation of a deliberate policy? One wants an answer to this question if merely in order to supply a conclusion to the historical sketch of the agrarian problem contained in the first part of the study; and what that answer must be can hardly be doubted, when the events of the last decade are viewed against the background of the pre-reform period. On the one side was agriculture: even with its actual primitive equipment and methods it maintained the bulk of the population, and supplied two-thirds of the exports and two-thirds of the public revenue; yet, during the trying period of reconstruction, agriculture was hampered on all sides by financial and administrative obstacles. On the other side was industry: hardly significant so far—if one excepts mining and rural industries—and probably bringing no profit to national economy; yet every other activity and comfort of the nation was subordinated to the fostering of its precarious life. All the exceptional laws passed during and after the War have been openly designed for the protection of banking, industry, and trade.

The impetuous wooing of industry after the War represented a phase widely different from the former encouragement of industry. Earlier aspirations aimed at nothing more than the establishment of an industry capable of transforming the country's raw materials. This desire was never altogether free from an undercurrent of fear, lest industrial development should hamper agricultural production and raise the cost of living. Legislation was devoted primarily to the interests of corn-growing, and the first commercial treaties to the interests of corn export. The first law for the encouragement of national industry was adopted in 1887; but in 1891 a less protectionist tariff replaced that of 1886; that of 1906 reverting to stronger protectionist measures. In 1914 there were about 850 establishments enjoying the benefits of the law for the encouragement of national industry. Their production hardly appeared in the export tables. At home, after thirty years of privileges, they were still incapable of holding their own against foreign competition without the support of high protective tariffs. The value of the raw materials used in industry in 1915 was

Native materials . .	275,702,618 lei—80 per cent.
Foreign materials . .	66,586,699 lei—20 per cent.

Of the capital invested in these State-encouraged undertakings during 1915, 40 per cent. was absorbed by agricultural and allied industries. Hence, at the end of the first long period of encouragement, the activities of Rumanian industry remained closely dependent on the development of agricultural production. Nor has that state of things essentially altered since the War. The figures relating to the use of mechanical power, cited in the previous chapter, showed that the industries relying on agricultural raw materials still predominated. Yet the character of Rumanian industry—which would be a point of interest in a discussion of its viability—is not what chiefly concerns us here. Our argument is concerned in a general way with the determined effort made to widen the sphere of industrial activity after the War, and with the fact that this effort sprang from social rather than from economic causes—a circumstance which helps to explain much of the recklessness displayed in that attempt. Through the land reform, says a Rumanian writer, 'the hitherto wealthy class, who had been dependent upon agriculture, was diverted for a livelihood to other pursuits. This class, politically predominant, was left with no other alternative than to turn its attention to the development of industry. What is more natural, therefore, than that it should begin to favour industry, even at the expense of agriculture, through every means at its disposal?'[1] This observation applies not only to the expropriated landowners but equally to the enterprising class of tenant farmers, who possessed money and great driving power. It so happened, moreover, that on being dismissed from agriculture these capitalists and entrepreneurs had a path towards industrial and banking pursuits cut ready for them through the action of political events. The virtual expropriation, under the guise of 'nationalization', of foreign-owned undertakings in the new territories, gave them the opportunity of compensating themselves in the industrial field for the influence they had lost on the land. The number and the capital of limited companies grew very rapidly:

Year	Number	Capital
		Lei
1919	929	1,982,084,376
1924	2,158	22,690,613,902
1925	2,440	27,565,510,560
1926	2,622	34,226,144,455
1927	2,817	37,480,485,955
1928	2,953	41,244,445,955

Among the benefits most of these undertakings enjoyed under the Law for the Encouragement of National Industry were the following: sale of factory sites at advantageous prices; free access to water power; exemption from customs duties for machinery, parts and accessories; various rebates on taxes; reduction in freight rates to industries importing more than 50 per cent. of their raw materials; preference in the distribution of public contracts. After the War, to these specific favours were added the prohibition of export of raw materials, and the fixing of maximum prices for fuel; not to speak of high protective duties and of various credit privileges. Yet all these attentions failed to make industry prosperous, even in the best post-war years. Data available at the Ministry of Industry and Commerce showed that

[1] Article on 'Rumanian Industry and Manufactures' in the *Manchester Guardian's Supplement*, May 1927.

a large proportion of undertakings had been unable to make ends meet:

Year	Number of undertakings	Suffered losses
1919–20 . .	929	86
1921–2 . .	1,266	176
1922–3 . .	one-fifth of the balance-sheets closed with a loss or without profits [1]	

From 1926 onwards Rumania also experienced, for the first time, the plague of unemployment, especially in the engineering industry. The remedies which a writer in the Bucarest *Argus* suggested for it are instructive as a revelation of the outlook presiding over that industrial effort. 'The Government alone', he said, 'can supply a remedy for this exceptionally grave situation. It should, to begin with, prohibit for at least six months the import of all and any metal goods which can be produced at home'; private firms should be obliged to buy their machinery from Rumanian factories, when prices are equal; and public authorities should be forbidden to import any goods which could be manufactured in the country.[2]

This determination to create a national industry called for a financial policy adapted to the purpose. Failure to introduce an effective taxation system, when it could have been done, during the War, and the financing of the War by means of paper money and Treasury bonds, had already shaken the foundations on which the country's finances rested. The 'nationalization' of economic life after the War proved to be the straw which broke the back of the *leu*. Such an enterprise could obviously not be financed with foreign money, as most public and private undertakings had been financed before the War; every internal resource was therefore mobilized to assist in carrying it out. The mirage of national self-sufficiency appears to have been so bewitching as to convince statesmen and business men that an extensive industry could be created without capital, and the currency at the same time revalorized into the bargain. Foreign holders were bought out at almost any price. The Treasury, the banks and individuals did not hesitate to assume heavy obligations towards them in strong foreign currencies. The report of the Deutsche Bank on the sale of the Steaua Română shares

[1] *Argus*, September 15, 1923.

[2] *Ibid.*, March 7, 1928.

frankly admitted that greater profits were made from that transaction than from ten years of normal banking. As the exchange continued to fall, however, these foreign obligations threatened to ruin all who were bound by them. Finally, when payment could no longer be postponed, the State intervened and concluded on their behalf long-term arrangements—in reality compulsory, if costly, moratoria—which altogether ruined the nation's money and credit.

In all these speculations agriculture had neither part nor profit; as it had nothing to do with the bitter and damaging wrangle concerning the treatment of foreign capital which ensued from that policy. Agriculture was in all these incidents a mere spectator, but a spectator who was ultimately made to pay for the broken glass. By means of a variety of fiscal and restrictive measures the farmers were prevented from raising the price of their produce in the same measure in which the currency had depreciated.

> 'That constriction of prices reduced the farmers' income in such a way as to force them to consume the whole of their profit, renouncing any improvement of stock or fresh capital investments. . . . Because of the long process of production, and of the manner in which the farmer sells his produce, the labour and worry of a whole year are staked upon the chances of one market day, and he stands to lose from exchange variations more than all other producers. . . . Agricultural production is incompatible with a fluctuating exchange.'[1]

Yet in spite of the collapse of the exchange the country's rulers did not abandon the belief that in the end they could force the foreign value of the leu to conform to its internal purchasing power; they remained, therefore, consistent in their intent to keep food plentiful and cheap, by checking the export of agricultural produce while impeding the coming in of foreign goods and money.

It was characteristic of that attitude that no restrictions whatever were imposed on the consumption of bread, as regards either quantity or quality, in the year after the War, though wheat and flour had to be imported on borrowed dollars and pounds; and it was still more characteristic that when afterwards agricultural exports were prohibited or restricted, this applied not only to wheat and meat, but also to oats and barley, to millet and—caviare. The Government's restrictive measures

[1] C. Garoflid, article in *Buletinul Institutului Economic Românesc*, February 1924.

nevertheless failed to achieve any of the results they were intended to produce. The expensive 'nationalized' industry is leading a disenchanted existence; protection from foreign competition has not presented it with a capacious internal market. The *leu* has ultimately had to be stabilized at a lower rate than either *leva* or *dinar*—though Bulgaria and Serbia had been damaged severely by the Balkan Wars and the Great War—no doubt because the two neighbouring Slav countries preferred to encourage export rather than to restrict it. Nor has the Treasury ultimately derived any benefit from that policy. For by depressing production and exports, the export duties, as well as all other taxes, though very high, have yielded little, whereas lower taxes might have produced a larger revenue in more active economic conditions. Least of all did the State's policy succeed in keeping prices low. Low prices discouraged production; this caused in certain years a shortage of agricultural supplies, in which case the State's control of prices was bound to become inoperative. All that it achieved was to reopen the doors to the speculation and profiteering which had flourished elsewhere during the War. Moreover, the cheapness of food soon became an illusory boon for the urban population, as the lack of exports and the depreciation of the currency made all imported manufactures very dear.

When these two restrictive measures—export and price control—failed to work effectively, the State did not hesitate to have recourse to more drastic means of constraint for the satisfaction of general or public needs. In the winter of 1920 the authorities requisitioned from the more successful farmers 1,000 wagons of wheat to be distributed for seed to those who had none. When in 1922 the army could not obtain wheat at the controlled price of 24,000 lei per wagon, it proceeded to requisition what corn it needed. Having learnt a lesson from the experience of the Soviets, or perhaps merely for reasons of convenience, it did not, however, requisition the corn direct from the peasants, but from millers and corn merchants. They were obliged to surrender 20–30 per cent. of the wheat they had collected—and for which they had paid 30,000 lei—at the fixed maximum price of 24,000 lei per wagon. Such requisitions had perhaps even a more disturbing effect on production and trade than the other measures, especially in a psychological sense; export taxes and controlled prices were at least known quantities, but the system of requisitions left farmers and traders at the mercy of every interference in a country ridden with unrestrained petty officials.

The cumulative action of this multiplicity of restrictive and oppressive measures inevitably had the effect of bewildering the activities of agricultural producers and of gradually depressing the whole rural industry. The passive resistance, largely unpremeditated, adopted by the peasants who refused to grow wheat, could not save agriculture from being penalized for the benefit of the other branches of economic activity. Capital drew an excessive retribution for its services in the shape of interests which doubled the original debt within three years; as did industry, in demanding for its products a much higher relative price than before the War. According to the calculations of M. D. Gheorghiu, Director of Customs, the average price per hectolitre of wheat in the period 1906–14 was 13·99 lei, equal to 18·44 lei per 100 kg. In 1927, 100 kg. of wheat were worth 850 lei, i.e. about 26·50 gold lei, an increase of about one third. We have given, when discussing the re-equipment of agriculture with machines and implements the cost of some of these goods. Unfortunately, the only available Rumanian index numbers, those collected by the *Argus*, do not extend to machines and such other goods as constitute the peasants' main purchases; apart from food, clothing was the only other general group included. Nevertheless, these figures bring out well enough the discrepancy between the prices of agricultural produce and those of manufactured goods:

	Aug. 1 1916	Dec. 20 1922	Dec. 20 1923	Dec. 20 1924	Dec. 20 1925	Dec. 20 1926	Dec. 20 1927	Dec. 20 1928	May 30 1929
Food products	100	2,207	3,355	4,370	4,990	5,276	5,268	5,842	6,177[1]
Clothing	100	3,120	4,110	5,148	5,899	6,960	7,010	7,024	7,080
Various	100	1,549	2,956	3,188	3,594	3,791	3,744	3,915	3,685
General average	100	2,292	3,474	4,235	4,827	5,342	5,340	5,593	5,647

[1] The change in the figures of the last column may be due to seasonal variations, yet it probably is not unconnected with the accession of a Peasant Government in November 1928 and the consequent reversion in the State's economic policy.

Rumania's population consists largely of frugal and almost self-sufficing peasant holders; it was only to be expected that the disproportion in value between what they produced and what they got in exchange for it would inject into the rural mass a considerable dose of indifferentism—a phenomenon which taught the Soviet Government, too, a costly lesson. Just when everything had to be rebuilt in agriculture, the peasants were reduced to poverty again; and such savings as they had were buried, as is the Rumanian peasant's habit, in sealed bottles, to save the paper money from devastation by mice. There was no inducement for them to use these savings for the purpose of increasing production, when they were forced to sell their produce at half its real value while being made to pay more than real value for such necessaries as they had to buy.

As long as the State's financial and economic policy remained so one-sidedly unpropitious, its influence was bound to frustrate the good work which the agricultural departments and their personnel were doing. They could have little chance of instilling a new spirit of enterprising optimism into the country-side as long as the peasants, rightly or wrongly, felt themselves to be as persecuted as before; and the practical work of assisting and training the farmers was necessarily cut low to fit the stinted resources which the Government placed at their disposal.

Agriculture has normally received no direct material assistance from the State, in the shape of subventions or bounties. An exception occurred when the prohibition of exports caused a serious decline in wheat-growing and the Government offered a bonus of 200 lei for each ha. under wheat. That bonus did not represent more than a fraction of the loss which the farmers were suffering through the various restrictive measures and taxes. About half a milliard lei was paid by the State in that form, without its being able thereby to check the effects of the obstacles which it was placing at the same time in the way of wheat-growing and trade. Nor has agriculture been favoured, like industry, with laws for its protection and encouragement. M. Garoflid, it is true, enacted a law in 1920 for the encouragement of mechanical cultivation; it exempted from export and price restrictions a quota of 500 kg. per ha. of wheat grown with the aid of machines, and it exempted agricultural machines from the payment of import duties. But the measure was abrogated in the following year by M. V. Brătianu on the ground that it interfered with his tariff and export policy. A number of farmers, who had already taken action on the strength of M. Garoflid's law, obtained from the Courts damages against the Government; with others the Government had to make an onerous compromise, so that the incident caused a loss to the State without any advantage to agriculture. Another measure discussed after the War contemplated the establishment of a system of elevators and standardization of corn; it would avoid the great loss resulting from the present system of piece-meal trading, and especially the useless and wasteful overlapping of transports, but so far nothing has been done to carry out the idea. The standardization of cereals was dealt with in a bill which the Liberal Government passed in the spring of 1927. The measure was denounced by the Opposition as insincere and inconsequent. When the Liberals, said M. Mihalache, at last thought of legislating for the benefit of agriculture, they began with trade and not with production. 'We are not suffering from a commercial crisis, but from a crisis of production. The problem of production must be solved first.'

The indirect aid which the State has given agriculture consists in the maintenance of schools, model farms and experimental stations. The budget of the Ministry of Agriculture and Domains amounted in 1925 to 795,000,000 lei or 2·72 per cent. of the State's total budget. In 1927 the percentage rose to 3·58 per cent., the corresponding sum being employed as follows:

	Per cent.
For agricultural teaching and research	12·78
,, model State farms	10·68
,, breeding stations	19·36
,, administration of State forests	22·99
,, application of the agrarian reform	24·26
,, state fisheries	2·80
,, other services and book-keeping	7·13
	100·00

The size of this budget hardly corresponded to the importance of the agricultural industry, or to the magnitude of the problems which the land reform had raised. Moreover, it would seem that sometimes grants formally made had failed to materialize. For instance, the Government made a profit out of the wheat which it requisitioned in 1920 at controlled prices, as the corn was paid back in kind by those who had received it and was sold by the Government at world prices; a minute of the Council of Ministers assigned the profit then made to be used for agricultural education, but nothing was actually given. Another Cabinet minute destined to the same purpose the considerable surplus which the Commission for the regulation of the corn trade had realized, but the grant was never paid. The large sums collected in export taxes have not been used in the interests of agriculture. We have already mentioned that money which the peasants paid, in advance, for the land they had received was not used, as it was meant to be, for the cancellation of expropriation bonds, but was swallowed up in the general budget of the State. Since the War, the budget of the Ministry of Agriculture has not exceeded 33,500,000 gold lei, spent largely on administrative activities; until with the advent of the National-Peasant Government the 1929 budget reached 73,500,000 gold lei, the entire surplus of 40,000,000 gold lei being spent in direct practical aids to the farmers.

It is indeed difficult to discover what steps of any importance the State has effectively taken during the last decade with a view to completing the land reform with an agrarian reform; until one comes to the founding of the Institute for Agronomical Research in 1927, the creation of M. C. Garoflid. That neglect has told upon Rumanian agriculture the more as it coincided with the strenuous efforts other countries made for the purpose of improving the equipment and output of farming. That friendly competition in the furtherance of agriculture has not been limited to the big and resourceful countries. Even in such a poor country as Bulgaria, the State found means to purchase through the Agricultural Bank machines valued at 50 million leva, which it distributed to the peasants at four-fifths of the original cost, allowing them three years in which to pay the price. In the autumn of 1927 the Czechoslovak Minister of Agriculture, M. Srdinko, expounded his Government's agricultural program before Parliament, summing up the requisite conditions for the progress of agriculture under three heads:

A higher standard of culture on the land;

The technical amelioration of the soil, and

A cheap supply of agricultural credit.

Agrarian policy in Rumania has made a beginning only with the last part of that sound summing-up; the second has not been considered so far; while the first has been widely discussed but barely attempted. Occasionally one met in villages or market towns with sporadic cultural experiments; they were due, however, not to any set national practice, but to the private initiative and zeal of individual schoolmasters or agricultural officials.

The standard of ability and of devotion to work to be found among the personnel of the agricultural services is excellent; unfortunately, it has so far served only to emphasize the shortcomings of a policy which has failed to give these virtues the scope for and the means of raising the farming community. The formal action taken to that end has been quite elaborate. A system of district committees, appointed by the Ministry of Agriculture, was set up after the War to assist the Ministry in carrying out schemes for the improvement of agriculture. These committees, of an official character, had no chance of fulfilling in their particular sphere of action an intention which was contradicted by the State's general economic policy. A fresh scheme was inaugurated in 1925, involving the establishment in each county of a Chamber of Agriculture, modelled on the Chambers of Commerce, with a mixed membership of elected and appointed members; the Chambers are entitled to elect their own members of Parliament. They were to co-operate in an advisory capacity with the Ministry of Agriculture in the promotion of better farming, being free to undertake almost any activity to that end, provided they could find the means therefore and kept within the law. Some of these Chambers have done their best to prove the useful part they could play in the solving of agricultural problems. Unfortunately, the Chambers of Agriculture were from the outset turned into party instruments, which speedily lost them their reputation and rendered them useless as critics and reformers. In November 1928, the Ministry of Agriculture instituted a new group of Committees for the improvement of agricultural production and the harmonizing of State action with private initiative. The system consists of a Central Committee under the Chairmanship of the Minister of Agriculture and including his subordinate heads of departments, representatives of other economic services and departments, of Agricultural Credit Institutions, of the Chambers of Agriculture, &c. The decisions of the Central Committee are to be adapted to local conditions by County Committees, attached to each Chamber of Agriculture; while Communal Committees of a similar mixed composition are to carry out the various plans and decisions on the spot. The success of the work, which lies all in the future, will depend on two conditions: on the ability of these Committees to safeguard their professional character, and on their functioning under the aegis of a more helpful State policy. The latter condition is being rapidly fulfilled by the new National-Peasant Government, as a result, as M. Mihalache put it, not of party bias, but of economic logic. During 1929 that Government passed a Law for Agricultural Credit, destined to facilitate the establishment of credit institutions devoted to agriculture with the help of foreign capital; a Law for Elementary Agricultural Education and for popular education, meant to educate the agricultural masses professionally; a Law for Higher Agricultural Education, meant to produce experts in the technical and social problems of agriculture, and scientists for research; the Co-operative Code, to which reference has been made in chapter X; the Law for the Improvement of the Danube regions liable to flooding, which will make possible the exploitation of immense State properties; and the Law for the Free Sale of Peasant Property, destined to help the creation of economically sound holdings. In addition the Government is now considering a bill for the creation of a proper rural survey and of ground books; a bill destined to facilitate the consolidation of peasant holdings; a bill for agricultural insurance, another dealing with irrigation, and other measures of a practical nature.

The private agrarian organizations, like the Agricultural Syndicates and their Union, and the parties representing the peasants, never had any illusions concerning the effect of the economic and financial policy described in the preceding pages. In 1920 a Congress of Agriculturists had already put forward a considered program for the guidance of agriculture into more intensive and productive channels.[1] There was in that program, of course, no suggestion of restriction or control of production, except in so far as it advocated the fixing of legal standards and obligations for the cultivation of the land. A second congress of Agriculturists convened by the Central Union of Agricultural Syndicates in November 1923 adopted resolutions insisting, above all, on the necessity of not interfering with the marketing of agricultural produce. They demanded the improvement of transport, support for export by means of trade conventions, and a healthy budget, based on the increase of direct taxation.

As regards the attitude of political parties, the Liberal Party, guided by the late Ion C. Brătianu, and afterwards by his brother, M. Vintilă Brătianu, has been in a position to dictate the economic and financial policy followed until the end of 1928. General Averescu's People's Party, at its Congress in 1925, pledged itself to distribute to the peasants who had received land, machines and implements needed for more rational farming, thus recognizing that the reform law, which it prided itself in having passed in 1921, had not solved the agrarian problem. But when the Party came into power again, in 1926, none of these good intentions were remembered. The Peasant Party's agrarian program, published in 1924, contained an elaborate plan for giving agriculture the place of honour it deserved in the country. The program recognized that this involved for the farmers duties as well as rights; it expected the State to help in creating 'an agrarian conscience', by education and special organization, but it considered that in its turn the State had a right to expect from those to whom the land had been entrusted proofs of their willingness and capacity for working it properly. The Party looked to co-operation to enlarge and improve the economic activities of the rural population. As regards industry, it admitted that it should be encouraged in so far only as it was based on the labour and raw materials which

[1] See details in *Viaţa Agricolă*, Bucarest, January 1, 1921.

the country itself could provide. That point of view was closely related to that of the National (Transylvanian) Party. In 1927 these two parties amalgamated and the new grouping, known as the National-Peasant Party, came to power in November 1928. The new Minister of Agriculture, M. I. Mihalache, promptly reasserted the Party's agrarian creed in a press interview.[1] 'Our Government', he declared, 'is not the enemy of industry. On the contrary, we believe that the interests of the two branches of national economy complement each other in the happiest possible way.' But that did not apply to industries 'which have been able to exist only out of the State's budget, and from the favours continuously granted them by governments. These are definitely parasitical industries, and they will have to disappear. The present Government has no intention of continuing the culpable support which the industries have been accustomed to receive. Its support will be directed towards agriculture. . . . We are determined to make agriculture the pivot of our whole economic life', by creating sources of agricultural credit, by professional education and by effective aid for the agricultural industry. And, in fact, the Government of M. Iuliu Maniu has at once set to work to recast the whole economic legislation and policy which have during the preceding ten years been built upon the nationalist doctrine of economic self-sufficiency.

The tendency expounded in M. Mihalache's declaration has since been reinforced by the creation of a non-party Agrarian League, in March 1929. More than twenty years ago the far-seeing statesman, Peter Carp, attempted to organize the land-owners in an Agricultural Society for the defence of agrarian interests, but the Society failed to achieve its purpose, mainly because it had left out the small farmers and because the narrow political life of the time impeded a sufficient concentration upon professional interests. The new organization appears anxious to put to profit the lessons of that earlier failure. It would seem to have sprung up almost spontaneously. In 1927 an Agrarian Cultural Association was founded in Jassy, for the purpose of co-ordinating the work of agricultural societies and institutions in the northern Rumanian provinces. In October 1928 it was

[1] *Dimineața*, December 30, 1928.

decided to extend the activity of the Society to the whole country, and the new program of the Association proclaimed, among other things, its intention of working towards the creation of an agrarian group in Parliament, which should include all representatives of agricultural interests without regard to their political colour, with a view to their permanent collaboration. To some extent, again, the impetus has come from the initiative of a group of peasants from southern Transylvania, who organized themselves into a professional society and sought the advice of leading agriculturists as to the best way of achieving their purpose. The new Agrarian League aims at bringing within its ranks all those connected with agriculture, from landowner to labourer, without regard to their political allegiance, and solely for the furthering of professional interests. The inaugural meeting of the League was held in Bucarest, on 10th March 1929. It elected M. C. Garoflid as first President of the League and passed the following resolution: 'The League is destined to encourage, to support and to realize the rights, too often overlooked, of the nation's most powerful group of producers. The Agrarian League is detached from political party interests, aiming at fulfilling the real economic interests of the country, which are ours also. . . .' The Saxon member of Parliament, Herr Fritz Connert, apparently gave the League a motto, when he said: 'Salvation will come from ourselves, let the State merely refrain from putting obstacles in our way.'

The stage thus seems set for the balancing out of mercantile and agrarian tendencies. The likelihood is that independently of the coming and going of party government, the economic experiment attempted during the last decade will not be repeated again—at any rate on such an extensive scale—partly because that experiment has failed and partly because the majority, which has suffered from it, is now roused to organize itself in self-defence. One may now expect, therefore, to see Rumanian agriculture enter definitely upon its new phase. The two main transformations which it has undergone have both been hastened in their contemporary development through the action of war. The Crimean War offered to corn-growers opportunities which induced the Rumanian farmers to abandon quasi-pastoral agriculture. The profits which then could be made from corn-growing, especially when undertaken on a large scale, tempted the politically dominant landowners to extend their possessions and to conscript the peasants' labour; and by their handling of the political machine they were able to continue making large profits even when oversea competition brought ruin upon the corn-growers of western Europe. The Great War initiated a new phase. It broke the privileges of the landed class, forcing them to hand over their land and political power to the peasants. That partition of the land created conditions which were bound at last to bring about a transition from extensive corn-growing to intensive crop-rotation. But the reform was not allowed to run its normal course. After the War, the class which had delayed the passing of extensive corn-growing, because they derived a livelihood from it, again caused a delay in the progress of agriculture by guiding all the resources of the State towards industrial activities, in which they were seeking new fortunes. The land reform, therefore, which opened the way for the transition from extensive to intensive agriculture, was made to run side by side, under unequal conditions of competition, with an effort to change the purely agrarian character of the country into that of an agrarian-industrial one. The effects of the reform were in that way interfered with and distorted by an aspiration for economic self-sufficiency, which in an agrarian country inevitably assumed a mercantile bias. The first ten years of the reform could in those circumstances hardly be accepted as offering a true reading of its ultimate development. They have to be regarded rather as an economic interlude; while the policy which has characterized them should be properly considered to have sprung, not so much from the competition of two economic imperatives, as from the struggle of two social classes for predominance—classes which, through the peculiar action of the land reform, have been ranged almost without intermixing on opposite sides of the line dividing agriculture from industry.[1]

[1] For a survey of the new agrarian current, and of the Peasant Government's views and policy, see the special supplement on Rumania published by the *Manchester Guardian* in November 1929.

CHAPTER XII

SOCIAL AND POLITICAL EFFECTS OF THE REFORM

It was characteristic of the agrarian problem in eastern Europe that it was commonly spoken of as the Peasant Question, and not, like the Western problem, as the Land Question. The major factor in the equation was the agent, not the object; and the issue hovered over the field of social policy rather than over that of economic organization. Of course, that point of view was not unanimously held. Mr. Conacher remarks with justice that both owners of corn-lands and the inhabitants of towns 'commonly conceive of agrarian land as being improperly dealt with, if disposed of in such a way as to endanger its profitable exploitation under capitalist methods for meeting a demand from elsewhere. The land-hungry peasant, on the other hand, regards land as the chief form and source of wealth, which should be equally distributed, even if such distribution were to leave each shareholder a mere subsistence. The agrarian question generally resolves itself into a conflict between these points of view.'[1] In eastern Europe, where the peasants formed the mass of the nation, agrarian measures, especially if unduly delayed, have been apt to ride roughshod over mere economic postulates, with an ease hardly to be imagined in western agriculture, and not at all in the field of capitalist industry.

Between the agrarian problems of east and west lay indeed a world of difference. There was a mere difference of degree in regard to technique, but the difference in the economic and social organization of agriculture was fundamental. A simple quantitative comparison between the productive capacity of western and eastern farming could not bring out that variation. In the view of the Russian zemstvo statisticians and agronoms, whose extensive and original labours have yielded an invaluable mass of sociological material, one must regard the two as distinct economic types—the 'capitalist' type and that of the 'wageless family economy'. Neither the criteria nor the psychology of

[1] *Agrarian Reform in Eastern Europe*, p. 9.

modern economics, with its background of 'wage labour', offer the means for gaining a true insight into the nature of the 'family *Wirtschaft*'; and modern economics cannot therefore lay down an agrarian policy which shall be universally valid. 'The usual purpose of a practical agrarian policy', writes M. Tschajanow, 'is to produce as high a rent as possible. That is the sole aim of capitalist agriculture.' But in agrarian countries, with a dense population, the practical statesman must not hesitate 'to place other ends and other criteria in the foreground, as his first duty is to secure for the bulk of the population the highest possible standard of living and the greatest possible gross income. . . . He will have inevitably to correct the economic-technical standpoint with the social, and in many respects the agrarian problem will become for him a problem of population.'[1]

If such was the correct angle from which, in eastern Europe, agrarian problems had at all times to be viewed, that angle was bound to become more acute under pressure from the popular demands which arose out of the War. Such, indeed, was the character of the ensuing land reforms that, to appraise them justly, one must reverse the order of values suggested in M. Tschajanow's phrase. The reforms left eastern statesmen with the task not of correcting the economic-technical standpoint with the social, but rather of correcting the social standpoint with the economic-technical. This task, however, was deliberately ignored, for reasons described in the preceding chapter. As a consequence nothing has materially changed on the economic-technical side of agriculture. The whole weight of the reform has been allowed to fall on the social side; and in that field one must therefore expect to find most of its modifying effects.

The one effect that stands out from all others is, of course, the virtual obliteration of the landed upper class. The wholesale expropriation of that class has been described in Chapter VII, in which it has been shown that well-nigh all the arable land, and a considerable extent of pasture and forest, have passed into the hands of the peasants. Land property, therefore, is no longer available as a source of rent or of social influence. Considering

[1] N. Tschajanow, *Die Lehre von der bäuerlichen Wirtschaft*, Berlin, 1923. The book sums up the conclusions to which the zemstvo workers had arrived before the War.

the dominant power which landowners were accustomed to wield during 'neo-serfdom', and the peasants' utter state of dependence, the social change wrought by the reform is equal to nothing less than a rural revolution. One could not describe that effect without greatly underrating its reach, merely as a change in the relative standing of the two rural classes. For, in truth, the peasants have now been left almost alone in the field, as undisputed masters of it. They have conquered the country-side decisively for their own class. Such social differentiation as may in the course of time develop among the rural population could occur only within the peasant class itself, and not above or against it. The insubstantial remnant of relatively large owners could no longer further their professional interests by oppressing or opposing the peasants, but only, on the contrary, by rallying them, by serving them, as leaders in a common cause. The new Agrarian League represents the landowners' first thought of associating the peasants with themselves. The line of social contest has shifted from the village and now runs near the boundary between land and town, between agriculture and industry. Concerning the social effect of the reform on the landed class there is therefore little to be added to what has been said so far: the work of dissolution carried to great lengths by the reform is being completed, of their own will, by these elements themselves, many of whom have sold out and embarked upon industrial and professional careers. All discussion concerning the effects of the reform on rural life and people must of necessity concentrate on the peasants, for whose benefit indeed the reform was made. As regards the life of the towns, it is altogether impossible to disentangle the repercussion of the reform from the general effect of post-war conditions; except in regard to certain peculiar philanthropic and cultural activities, which have played an important part in the welfare of the urban population, and on which the effect of the reform had been so deep and direct as to justify a brief description of it.

Section 1

Social Effects on Urban Life and Institutions

Dr. G. Caranfil has estimated[1] that a landowner whose property was worth one million lei in 1913, would have been left with only 183,852 gold lei at the beginning of 1924—taking account of expropriation, depreciation of money, and loss of purchasing power—and 151,088 gold lei in January 1928, i.e. with 15 per cent. of his original capital. One million lei invested in the best class of shares and bonds would have been worth only 53,942 gold lei in 1924 and 29,669 gold lei in 1928, i.e. 2·9 per cent. of the original capital. These figures cannot of course do more than indicate a tendency, but within these relative limits they are probably correct, and a first glance at them might suggest that the land has known how to preserve its substance at the expense of the town. But we have seen that in reality landowners and farmers were given no such opportunity; they were, on the contrary, prevented from making normal profits out of their transactions with the urban population. The official economic and financial policy left the farmers in an impecunious state; and in a country like Rumania the affluence or poverty of the farming community naturally affects the welfare of all other forms of economic activity. But if general policy, and not the land reform, was responsible for the depression of the farmers, one must with still greater reason attribute to the same cause the discomfiture of the urban traders and investors. This is proved, indeed, by Dr. Caranfil's figures. They show that the general run of the people who invested in the 'favoured' new industrial enterprises found themselves worse off at the end of the first experimental decade than did the non-favoured farmers; those, indeed, who merely failed to derive an income from such investments were the fortunate few.

One may set against this the advantages which townspeople have derived from that policy. Until two years ago, Rumania was far and away the cheapest country in Europe to live in. To this the reform may be said to have contributed, indirectly, but only in one sense. It is doubtful whether the State's policy

[1] *Argus*, March 29, 1928.

would have been the same if agriculture had remained under the influence of the powerful landed class. As it was, far from the peasants dictating their prices to the towns—as happened in many parts of central and western Europe—it was the towns which were enabled to extract a tribute from the country-side. Whether the urban population derived a lasting benefit from the State's economic policy is another matter. It could not altogether prevent a rise in the cost of agricultural products, it merely delayed it year by year; and that transitory cheapness of food-stuffs was offset by the dearness of the heavily protected manufactured goods. The truth is that, taken by themselves, retail index numbers tell nothing of the actual conditions of living until translated into real wages and salaries. This was done by M. Mihail Manoilescu.[1] In 1923 he disclosed a situation which was bad at the time, and which has never ceased to grow worse. The occupation census of 1913 showed that 1,330,132 persons, or 18·5 per cent. of the total population in the Old Kingdom were living in the towns. In Greater Rumania the larger percentage of urban inhabitants and urban occupations in Transylvania and Bucovina is offset by their lesser percentage in Bessarabia, so that in general the various proportions remain much the same. In 1913, therefore, in the above total was included:

575,223, i.e. 43 per cent., employers and their families;
332,279, i.e. 25 per cent., salaried employees and their families;
308,121, i.e. 23 per cent., workers and their families;

while the remainder, 9 per cent. were also employees, such as domestic servants, apprentices, &c. Hence 57 per cent. of the urban population were wage-earners. In almost all the European countries the real wages of these sections of the population have risen after the War. In Rumania, according to the index of the Ministry of Labour (January 1923) the wages of workers in private undertakings had risen to sixteen times the pre-war level; as the retail index of necessaries was 34·44 (June 1923), real wages remained in fact 53 per cent. lower than before the War. The situation was much worse for State employees, especially

[1] *Argus*, July 23, 1928.

for those in receipt of salaries. The memorandum submitted by the Society of Engineers in March 1923 stated that the nominal wages of railwaymen were nineteen times the pre-war level, but the salaries of railway inspectors only 3·15 higher, which meant that their real salaries were 89 per cent. lower. This latter figure was probably applicable to all civil servants. Their life throughout these years has been truly a martyrdom. Among the patients of dispensaries and sanatoria for the tuberculous State employees formed a larger percentage than that of any other occupation.

Such a pauperizing scale of remuneration has been possible because of the absence or the systematic destruction of workers' organizations; and also because of the hybrid character of much of the labour employed in industry and mining. In the Old Kingdom, a majority of the unskilled workers were peasants who took up industrial work during the periods when field work was at a standstill. Most of the skilled workers were foreigners. Rumanian workers were in a minority among industrial labour in Transylvania; the bulk of them were peasants without land, either because they came from a large family or because they had been forced to sell out; they were employed especially in mining and timber-felling. Landed peasants also were to be found in the timber industry, working in 'companies', that is in teams who contracted together for a definite piece of work and divided the profit among themselves; they usually spent about four months in industrial work. During the first years after the War numbers of villagers were attracted to the towns by the new industrial enterprises then started. During that period agriculture suffered from a shortage of labour, being unable to offer wages and general conditions as attractive as those of industry. Very soon, however, a number of factories had to reduce their activity or to close down, and the resulting unemployment checked the exodus to the towns. A proportion of workers would appear in fact to have returned to the land.[1]

[1] The inability of industry to offer a living to the workers it first attracted has increased the number of emigrants, notwithstanding the obstacles placed in their way by Rumanian authorities and by some of the oversea countries. There were few Rumanian emigrants from the Old Kingdom before the War. But Transylvania gave a considerable number, most of them peasants who first tried their luck in industry

Rapid depreciation in the purchasing power of the currency has evidently had more to do with the trials of the working and salaried classes than any excessive increase in the cost of foodstuffs. The decline of agricultural production, and consequently of exports, has certainly contributed to that depreciation. But in how far the land reform could be held responsible for that decline, and thus indirectly for the difficulty which most townspeople had of making ends meet, must be left to the reader to conjecture, from the arguments of earlier chapters. There have been a number of cases, however, in which towns or urban institutions have suffered a loss of revenue through the land reform which has affected the public service they used to render. Quite a number of the Transylvanian towns, e.g., owned stretches of arable land on their borders, and these properties have been expropriated like those of individual owners. The town of Timişoara, to give an example, owned 6,168 cadastral jugars; all of this land which lay beyond a radius of one km. from the barriers of the town, i.e. 2,587 jugars, has been expropriated, at 2,000 lei per jugar. The loss in revenue has of course to be balanced through an increase in rates and taxes.

The outstanding cases in this category of sufferers are certain prominent cultural and charitable endownments. It has been a pleasing custom with old Rumanian families to mark the favours, and also the trials, of fortune by large-hearted charity; and as all their wealth consisted of land, their charity took the form of gifts in land to churches, monasteries, hospitals, &c. The land of churches and monasteries in the Old Kingdom was already 'secularized' in 1863. The chief remaining endownments were those concerned with the maintenance of hospitals, as the Endowment of Civilian Hospitals and that of the Brăncoveanu Hospitals, of Bucarest; the Endowment of the Saint Spiridon

and then migrated to America. Rumanian emigrants numbered 7,419 in 1910. The figures for the first half-year of 1924 were, by comparison, as follows:

From Transylvania	12,223
„ Bucovina	1,782
„ Bessarabia	1,005
„ Old Kingdom	533
Total	15,543

(Dr. I. Gîrbacea, *Şomajul în Ardeal şi Emigrarea*, Bucarest, 1928.)

Hospitals and Almshouses, of Jassy, &c.; to which should be added the Rumanian Academy, whose activities were supported mainly from the revenue of land property. These institutions have been expropriated of all their arable land, receiving as compensation untransferable perpetual bonds bearing 5 per cent. interest, which means a fixed yearly income equal to about 1/16th of the revenue they derived from their arable land in 1916, without taking into account the fall in the purchasing power of the money. That consequence of the reform is the more unfortunate as hospital accommodation has always been inadequate, and as no public funds have been placed at the disposal of these endowments to enable them to close the gap in their private resources. Indeed, they have been held to the obligation of maintaining out of their funds the clinics attached to the faculties of medicine, though these were State institutions. The following extracts from the introduction to the 1927 budget of the St. Spiridon Endowment, of Jassy, give a restrained picture of the effects of expropriation on its activities.

The St. Spiridon Endowment had under its care nine hospitals, one maternity clinic, an orphanage, an asylum for old people, seven churches, and three cloisters. Its funds, collected over a period of 170 years, consisted of land properties. The revenue had been sufficient both for upkeep and improvements, because the rent from those estates had risen with every rise in the cost of living. Had the endowment remained in possession of its estates, the yearly revenue from them would have amounted in 1927 to at least 61,895,100 lei. The expropriation bonds yield a fixed revenue of 5,181,413 lei yearly. Since 1919 an effort has been made to intensify the exploitation of the Endowment's forests, but this expedient could not be pressed further without depreciating the capital itself. The 1916 budget amounted to 3,673,000 gold lei, or, roughly, to 117,536,000 paper lei; the 1927 budget was only 47,344,334 lei:

'The trustees have been forced by circumstances to introduce the system of paid hospital services, but only the surgical services are made use of by patients who can afford to pay. For the rest, we get only poor patients who have to be attended gratuitously, in accordance with the intentions of the founders. . . . Because of the insufficient revenue the

Endowment cannot pay its employees a living wage, so that it cannot claim from them devoted service. If some good employees still remain it is only because they have been many years in its service. Many of them cannot be pensioned because the Endowment has no means of paying both salaries and pensions. The heads of medical services are so badly remunerated that only one of them, who has four awards for length of service, receives in all, with various bonuses, 4,220 lei [£5 4*s.* 0*d.*] monthly; the others begin at 3,550 lei [£4 8*s.* 0*d.*].... Assistants, servants, &c., are so badly paid that we can only have the worst type of nursing personnel, and even such we cannot find. Hospital buildings have fallen into such a state of disrepair that they are a disgrace to our up-to-date State. The Endowment found itself in a most awkward position when foreign guests came to visit its largest hospital, the Central Hospital. The food served to the patients is of inferior quality. We cannot purchase even half the medicaments and surgical materials required, so that the doctors are only too justified when they complain of this state of things.... This is the situation in which the Endowment has been placed through the expropriation, besides the fact that the State has imposed upon us the obligation of maintaining the clinics attached to the faculty of medicine. As we cannot keep going with such a budget, unless the State takes upon itself to make good the loss it has caused, we shall be forced to close the clinics and some of the hospitals.'

The maintenance of the clinics cost the Endowment 8,623,494 lei in 1925, towards which the Ministry of Education contributed one million lei. In 1927 the Endowment further received a subvention of 2,000,000 lei from the Ministry of Public Health and 1,100,000 lei from the Jassy municipality.

The Bucarest Endowment of Civilian Hospitals has had to face the same difficulties. It has reduced the number of beds to 50 and then to 20 in each of its clinics; it demands from all its patients a fee of 140 lei per day, which is much above the earnings of workers or of junior clerks, and additional fees for surgical operations; its buildings are in a state of decay. The Endowment has been criticized for having reinvested its funds, with unprofitable results, in oil-shares and, in general, for inefficient administration; but be that as it may, it could only have aggravated a situation the origin of which lay in the land reform. The Peasant leaders have shown themselves unrepentant in this regard. They demanded the total expropriation of all endowments and the maintenance of hospitals, &c., out of public funds. Their one anxiety was to abolish landlordism, and in their view the Endowments had pressed the peasants as hard as any money-grabbing tenant. During the debate in the Chamber, in 1921, M. Mihalache said: 'We cannot allow a philanthropic institution first to cause disease among the people of this country by weakening them, and then to take in one here and there to nurse in its hospitals.'

The Rumanian Academy is the country's highest cultural institution, engaged in manifold and indispensable activities. The Academy, through the various sections, publishes original studies which would not find a commercial publisher; it has in its keeping Rumania's most important library; it has maintained out of its funds two agricultural schools and other educational activities. Of the 11,214 ha. land which the Academy possessed before the reform 10,569 ha. have been expropriated.[1] Most of the land, 9,447 ha., was formerly let to peasants on generous conditions; as a consequence, the Academy has received a smaller compensation than neighbouring individual owners. The expropriated area would have brought in, at current prices, a revenue of 21,138,000 lei; the fixed yearly interest on the perpetual bonds amounts to 1,343,060 lei. From the remaining land, the Academy derives an income of about 520,000 lei; the recent Elias bequest brings in 1,800,000 lei yearly; the Ferdinand Foundation contributed one million lei each year, but in 1926 this was commuted into one final gift of 5 million lei. For the maintenance of the Library the Ministry of Education contributes a subvention of 1,250,000 lei yearly. Roughly, therefore, the Academy's yearly income reaches about 5 million lei. That sum has not enabled it to carry on its former activities. It could not pay its employees salaries commensurate with the rise in the cost of living. It could not maintain its buildings in proper condition; the Academy has only recently been able to build, through the Elias bequest, a fire-proof repository for its important and irreplaceable collection of manuscripts, documents and old books. Reading accommodation is pitifully inadequate in relation to the importance of the Library and to the growing number of readers. Many useful studies submitted to the Academy are wait-

[1] These figures were kindly prepared for the purpose of this study by the Academy's administrative staff.

ing in vain to be published. The sum formerly spent on publishing was 638,954 gold lei, that of recent years only about 1,200,000 lei, or 37,500 gold lei; the proportion being still worse if one takes into account the relatively higher cost of paper and printing. One must also remember that in the meanwhile the country has doubled its size and population. The Library is suffering severely; the gap caused by the War in the supply of essential foreign publications cannot now be made good, nor can new foreign books be acquired regularly, because of the depreciated exchange. The yearly expenditure on the Library formerly amounted to 240,000 gold lei and now to 3,526,944 paper or about 110,000 gold lei. From 1928 onwards, notwithstanding the severe restriction of its activities, the Academy has had no prospect of being able to balance its budget.

Section 2

Social Effects in the Village

'If you consider the matter carefully', wrote Ion Ghica in 1872, 'you will observe that it was always only one class of our people which aroused sympathy at home and abroad—the class of those who tilled the soil; for they alone have followed the path which Providence has traced for mankind: to live by the sweat of one's brow.' Thirty-five years later, after the desperate rising of 1907, one of the big landowners, M. I. Bibicescu, recalled with much uneasiness how the peasants had toiled to protect the new State in time of stress and to maintain it in peace.

> 'Yet—I say it with the deepest sorrow—we have not shown these people much gratitude or affection. . . . The new institutions have given us rights; they could not instil into us sentiments—let alone democratic sentiments—and a sense of responsibility. So that we have been satisfied with enjoying the positions we had acquired, making use of them for the constant improvement of our comforts, but the people and their needs have been the last of our cares, if it has been one at all. . . . I look into my conscience, and acknowledge and confess myself guilty.' [1]

Just when the new State's first span of life was about to close—

[1] Preface to *In Cestiunea Agrară*.

almost a century after the restoration of autonomous government and after half a century of national independence—the deputy M. D. Patraşcanu, speaking in the Chamber at Jassy, thus assessed the gift which that political renascence had brought to the peasants:

'Consequently, of all the countries inhabited by Rumanians, it is in free Rumania that the peasant is the worst off. . . . The ultimate conclusion is still more painful: the longer one of the country's provinces has been separated from it, the more it has escaped our restrictive agrarian laws and the better is the condition of its peasants. What a terrible discovery for us!' [1]

When the peasants' land rights were for the first time restricted, by Moruzzi, in 1805, the number of big animals for which they could claim grazing was limited to 16 in Bessarabia and 12 in the Moldavian lowlands, the men with only 4 oxen forming the lowest category. By the time the peasants were emancipated in 1864 those with 4 oxen had become the leading category; in 1906, in some of the counties, only 17 per cent. of the households possessed domestic animals, while in other counties as many as 32 per cent. of the households had no large animals at all, not even a milch-cow. But while the peasant was impoverished, the country apparently increased in substance. Rumania increasingly enjoyed the reputation of being 'an eminently agrarian country', and of deriving great profit from it. Says M. Iorga:

'Our old agriculture, before the reform, only existed through an injustice, through a social evil which has now been removed. Tears, not only drops of rain, watered the furrows. In order that a certain quantity of Rumanian corn should be found, at a given time, in the world's markets, a whole population, until a short time ago, had to labour forcibly, almost under the whip; a man's own fields remaining to be cared for last and with the least advantage. And a large part of that enormous surplus which we exported came from the deprivation of the producers themselves. We were selling the bread which was denied to those who made it.' [2]

This was literally true: at the same time as the big corn export was being built up, the consumption of maize, the peasant's

[1] Speech in the debate on the bill for compulsory agricultural labours, as reported in *Lumina*, Bucarest, September 23, 1918.

[2] Article in *Plutus*, November 26, 1923.

staple food, fell from 230 kg. per head of population in 1890 to 146 kg. in 1903.

Exactly the same thing was happening in the neighbouring Russian Empire; the formidable growth of the corn export after the emancipation of the serfs, in 1861, was accompanied by a steady decrease in the average internal consumption. Such an 'eminently agrarian country' Rumania will never be again, Professor Iorga assures us: 'the free smallholders will work less for others and will eat more themselves.'

One can distinguish, in broad lines, three main periods in that downward trend in the status of the peasantry.

1. During the first the peasants enjoyed, on the whole, yeoman freedom and rights to the land until the first was formally curtailed by Mihaiu the Brave, towards the end of the sixteenth century.
2. The second period lasted till the formal emancipation of the serfs, in 1864; with a period of transition, from 1746–9 onwards, when the peasants' servile ties began to loosen and the landlords endeavoured instead to strengthen their own title to the land.
3. The third period coincided with the formal reign of constitutional liberties, when, as a consequence, the task of constraining the peasants was thrown upon the administrative machinery.[1]

During the first period, the free peasantry had the use of all the land, and if their servitudes gradually increased, it was only in so far as this was required to meet the personal needs of the landlords; until, with the creation of a standing army, the peasants were called upon to procure supplies rather than to supply soldiers. This state of things, which made the peasants servile labourers, became permanent with the loss of autonomy, when the peasants were seldom called upon to fight, but had to work all the more to satisfy demands from the ever-changing princes,

[1] M. Chebap, in his excellent though, unfortunately, unfinished monograph, divides the legal history of the peasants into two main periods: (1) from the foundation of the Principalities to 1864, and (2) from 1864 to our own days; subdividing the first into (*a*) from the foundation to 1746–9, and (*b*) from 1746–9 to our own days. The difference is not considerable, but in the light of recent research we have preferred to adopt a division in time which is formally more correct.

from the luxurious boiars, as well as, indirectly, from the suzerain Porte. When the peasants' burdens, becoming unbearable, caused them to migrate in large numbers, thereby diminishing the resources of the Treasury, Mavrocordat began the movement for the diminution of servitudes which ended with the complete emancipation of 1864. The boiars countered that tendency with successful efforts to reduce the peasants' rights to land, and to transform their own usufructuary title into one of proprietorship. The tendency, in the words of M. Chebap, was 'to enlarge the liberty of the peasants up to their full emancipation, and to restrict their rights to land up to the full emancipation of the estates of boiars and monasteries.'[1] Under the régime of the Organic Statutes the individual holdings became so small that, for the first time, the peasants found themselves under the necessity of leasing from the landlords additional land, beyond that to which they were legally entitled. In the measure, therefore, in which the landlords were losing the right to command the peasants' labour they were given means of advantageously bargaining for it. When the peasants were emancipated, in 1864, though it had to be done by a *coup d'état*, the landlords succeeded in further curtailing the land rights and in narrowing the land reserve of the peasants, a process which they afterwards nearly completed when they bought up the State domains. During the third period, moreover, when the peasants were constitutionally altogether free, the landed class, besides being owners of the land, had also acquired uncontrolled mastery over the government of the new national State. They were provided thus with a fresh instrument for securing the peasants' labour, through the use of fiscal impositions and of administrative measures; these culminated in the laws on agricultural contracts which so diluted for the peasants the wine of freedom that it was to them scarcely more tolerable than the bitter waters of serfdom.

It will be seen that throughout that evolution the one constant anxiety of the upper class had been to make sure that the peasants would work the land. If they encroached increasingly upon the peasants' land rights it was solely as a means of acquiring a hold on their labour. The landlords' action never

[1] *Regimul Legii Agrare . . .*, p. 28.

tended, like the English enclosures, to drive the peasants away, so that the landlords might extend their own farming. On the contrary, their aim was to obviate any need of doing so by tying the cultivating peasants to the village. Just as genuine feudalism was unknown in the old Rumanian provinces, so modern Rumania has never known a genuine capitalist agriculture. The landed dignitaries of the old oligarchical provinces became the landed officials and professional politicians of the new democratic State; hence the large landowners have never had the opportunity of becoming a producing class, but have remained a beneficiary class. They wanted from their land a rent, and not profits, requiring the investment of capital and a personal effort. It was not to their interest, therefore, that the peasants should become landless and perhaps abandon the village, but rather that they should be held to the tilling of the land. This led many influential landowners to put forward a curious proposal, after the rising of 1907. They realized that the existing agrarian system could not last much longer; but they did not want either to part with their land, and thus make room for independent peasant farming, or themselves to embark upon intensive capitalist farming. Instead they urged as a means of solving the agrarian problem a return to feudal agriculture: they proposed that landowners should be obliged to let the peasants farm in métayage all the area beyond 250 hectares from every estate. The produce was to be halved.[1] The proposal was characteristic of the nature of Rumanian landlordism; it would in effect have revived the servitude upon the estate, and for the peasants the obligation to pay tithe, only that now they were meant to pay one-half instead of the traditional tenth. The security and regularity of the peasant's contribution was in consequence the one purpose which influenced the policy of the upper class throughout those several periods, the difference between one period and another resulting merely from the means employed to prevent the peasants from becoming truly and fully emancipated. As long as the peasants were free of the land, their labour was tied; when their labour had to be freed, the land was tied; and when, at long last, through the latest reform, both land

[1] C. Garoflid, *Chestia Agrară*, p. 42.

and labour were finally freed, the market was tied. In the light of its own antecedents the post-war economic policy of discrimination against agriculture, now carried on almost wholly by peasants, is more easily understood; and one can understand also why the peasants, though in possession of the land, have not felt that now their existence is secure, but instead have thought it timely to organize themselves for social and political action.

The causes from which the peasants' disabilities sprang were manifold. As a husbandman he suffered from not having enough land for extensive farming, and not enough training and resources for intensive farming. At the same time he was, as a citizen, the victim of biassed legislation, and even more of maladministration of law and justice. Of these four groups of disabilities weighing upon the peasants the land reform could remove only the first, and that merely in principle, for even if it took all the land away, there was not enough of it to enable all peasant households to live from extensive farming. Correspondingly, more weight attached to the removal of the disabilities comprised in the second and third groups; yet the supply of education and working resources is at best in the embryonic stage, while discriminating legislation, though somewhat changed in aspect, remained as burdensome after the reform as it was before it. As regards administrative morals, they had little chance of improving during a time when officials could not possibly live on the salaries which the State paid them; and when every institution and section of the community—Court and Universities, army and schools—was drawn into the political contest between the old ruling groups and the new popular forces.

1. *The Economic Situation of the Peasantry.* Once more one must regret the absence of co-ordinated inquiries into the effects of the reform. Sociologically-tested material on the changes that are taking place in village life is non-existent, there being only sporadic observations, allowing glimpses into odd sections of the new habits and trends. We cannot, therefore, hope to do more, in the brief survey which follows, than point out incipient variations which appear likely to become permanent changes.

It would be natural to take it for granted that such a radical land reform would improve to a similar degree the life of the people for whose benefit it was enacted, were it not for the experience of the reform of 1864, when a great constitutional advance was nullified by means of retrogressive laws and their abusive administration. The indigence under which the peasants laboured before the reform of 1864 was transformed after it into a state of chronic poverty. From that state they could hardly fall lower, unless their holdings had been taken away from them. Yet therein lies the chief difference between the reform of 1864 and that of 1917–21: the first left the peasants with less land than they had been accustomed to till—as happened almost everywhere when the peasants were emancipated—while the second has given them possession of almost all the available arable land. The annuity they have to pay for it has been estimated by M. Garoflid to be equal to merely ten days' labour in the year. Moreover, the reduction in the size of the peasant holdings in 1864 was made the means for the conscription of peasant labour, to the detriment of their own cultivation, whereas in the present case the disappearance of large property has also done away with the means, as well as the occasion, for that exploitation of peasant labour. For the first time, therefore, since the spread of wheat-growing in Rumania, the peasants have gained that freedom of action without which an improvement in their own farming could not even begin. It is true that relatively few peasants have been endowed with economically self-contained holdings, but admitting that in consequence they may not be economically as autonomous as they could wish, it is nevertheless beyond a doubt that they are in a vastly better position than they were before. Other things being equal, a given agricultural population can obtain through peasant farming a much richer gross production than under a system of extensive large-scale farming; and if, in addition, they can in the first case retain for themselves a larger share of the produce of their labour than when the landlords had to be paid from it, it is evident that the peasants stand to benefit, under such a reform as the present, both from an absolute and from a relative increase in returns from the land.

For reasons described in the preceding chapters, the land reform has not so far been accompanied by an increase in agricultural production. The only question therefore, is whether the new agrarian régime has given the peasants a relatively larger share of the produce. The truth is, however, that large property and farming now play an altogether subordinate part; any advantage which the peasants may hope to derive must be obtained not as a class, from the other section of the rural community, but rather as a professional group, from other industries and trades. Rumanian agriculture is now practically identified with peasant farming, and what we have said concerning the fate of agriculture during the first decade of the reform described, therefore, on general lines, how the peasants fared in that period. They suffered not only through an unfriendly policy, but also through the instability and inconsequence of many measures, taken under pressure from some interested group. Too few of their grievances were redressed for them to have reached a happy economic and social position. M. Garoflid has indeed maintained that they were worse off than before. He has produced figures to show that, before the War, the income of a 'middling' peasant household, farming 5 ha., was 685 lei, and that of a 'tail-end' household, farming 3 ha., 418 lei; in 1922–3 these two categories had incomes of 15,000 and 9,480 lei, whereas to equal their pre-war income, the amounts should have been 22,600 and 13,794 lei. The incomes of these categories had thus fallen by 25 per cent. Considering that it was a fall from such mean totals as 685 and 418 gold lei, realized under the hard régime of 'neo-serfdom', what has been the use of the agrarian revolution which ruined the large owners without bringing relief to the peasants?[1] M. Garoflid based his calculation on the market prices of the principal cereals, so that probably the figures would be less unfavourable if the many other products which the peasants put on the market were included; moreover, to simplify his problem, M. Garoflid deliberately left rent out of account. But that is just the principal relief which the reform has brought to the peasants. On the basis of the customary *métayage* conditions they would have had to pay for the expropriated six million ha. arable land one-half of the produce, raised with their own seed, besides other small obligations in labour and in kind.

[1] Article in *Argus*, July 26, 1923.

The value of these subsidiary obligations was equal at least to the ten days' labour which, according to M. Garoflid, now suffice for the payment of the annuity, and to the amount of the land tax up to 1927. Therefore, that half of the produce which formerly would have been the landowners' share, roughly three million tons yearly, at the average yield of 1,000 kg. per ha., is through the resettlement a clear gain for the peasants. As the total number of peasant families is just over three millions one can estimate, in a rough and ready fashion, that through the fact of being owners and no longer tenants each peasant household has been able to add nearly 1,000 kg. corn to its yearly stock of food. During the controversy which called forth M. Garoflid's article it was pointed out by M. Manoilescu that a peasant household spent on an average not more than one-fifth of its income on manufactured goods, and that in consequence it was affected only to that limited fraction by the excessive dearness of industrial products; hence he concluded that, speaking broadly, the income of the peasants had increased 30 times but their cost of living only 13 times.[1] M. Manoilescu's point was quite justified, with two reservations. During that period many new farmsteads were created or old ones enlarged, besides which the War damage had to be made good, all this entailing a higher expenditure on industrial products than M. Manoilescu allowed for; secondly, like M. Garoflid, he had based his estimate of the peasants' income on the only available Rumanian index numbers, which relate to retail prices, and these were, of course, much higher than those which went into the peasants' pockets.

M. Garoflid also referred in the above-mentioned article to the position of the landless peasants. 'In the purely agricultural regions, with a dense population, the break-up of the large estates has restricted the demand for labour; for this reason, and because the medium-sized farms can afford to pay only reduced wages, in relation to the reduced prices obtainable for their crops, the average daily wage does not exceed 15–20 lei and food.' Had it kept level with the rise in the cost of living it should have been three times as high. M. Garoflid was writing at a time when the big farmers were complaining of a shortage of labour during the

[1] Article in *Argus*, July 16, 1923.

agricultural season, hence his figures were probably subject to considerable variations in space and in time. Four years later, in several places, we found, in fact, that the cost of labour and carting was about the same as before the reform, the lesser demand being balanced by a lesser offer on the part of the new peasant proprietors.

In the absence of systematic evidence all discussion of the peasants' material position after the reform is bound to move speculatively from one side to the other, following the writer's bias and the isolated observations on which he relies. The figures quoted in the section on co-operation, showing the percentage of members with larger shares in the Popular banks to have increased, would justify the view that the peasants are better off; while the contrary assumption could find equally strong support in the figures mentioned in the previous chapter, with reference to agricultural taxation, which showed that in the county of Ilfov many peasants had during all those years been unable to pay the road tax—an instance which was by no means isolated. Only a full statistical inquiry could establish whether these apparent contradictions were due to variations from district to district, or between one section of the peasantry and another. Also, we should then find out the reasons which have caused the peasants to get into debt again, and perhaps discover whether the worst sufferers have been the newly settled peasants, who had to organize a farmstead in adverse circumstances, or the old smallholders, who may have lost the use of some of the land they farmed before.

While the reform has made little change in the mode of farming, and in the class of farmers, it has, within the same class, transferred a great deal of land from some hands into others. To some extent there has been a change of persons, but, principally, a change in the area farmed by each household. Families with grown-up sons, for example, would hold more than before, while families without sons would hold less. Families who before may have had no means of buying land have now, perhaps, a tolerable holding of their own, while well-to-do peasants, accustomed to rent additional fields from the large owners, are probably unable to get surplus land any more. This displacement in the occupation of land has in certain villages completely upset the former state of things, and many smallholders have been left with excessive means of production, for which they are trying to find employment. This no doubt is one of the reasons why share-cultivation is still so prevalent, and, especially, why the land which came into the market after the reform has fetched such exorbitant prices. The highest prices were paid not where the land was best, but in the hill regions, where the population was most dense; and sometimes in places that were only 50 km. apart from each other the prices would vary by as much as 100 per cent. The peasant does not willingly move away from his village, and land is the only form of investment he appreciates. Being accustomed to extensive farming he values more land, and not better implements, &c.; animals, in the Old Kingdom, are kept only in such numbers as the holding can feed. Banks and other capitalist undertakings have not yet won the peasants' confidence. Hence the price of land rose continuously until two years ago; since then it has tended to decline, in some parts by as much as one half, because of the shortage of money and because of the agricultural crisis. Among the peasants who bought land several years ago, when there was literally a rush for it, many are unable to pay the debts incurred to that end; especially in the hill regions, where the land was more expensive but less productive. On the whole, the reform has had within the mass of the peasantry a levelling effect which will probably last as long as the country remains agrarian. What is left of relatively large property, if purchased or distributed, would add not even half a hectare to each of the existing peasant holdings, while land sales among peasants are an uncommon event.

If in the light of all these facts and considerations one is to venture any answer at all to the question whether the peasants are materially better off after the reform, it is perhaps safer to dissect the reply somewhat in this manner:

1. As a class, the peasants are enormously benefited, having secured almost all the arable land, at a nominal price, and in consequence a monopolistic control of agriculture.

2. As farmers, in their professional capacity, that gain is more potential than actual; as farmers the peasants are probably

not better situated than before. They have had to suffer from price and marketing restrictions; from the high cost of money, of implements, &c.; from the bad state of roads and railways; from an excess of costly formalities and administrative abuses.

3. But, for the time being, production for the market, in regard to which the peasants have been at a disadvantage, takes only second rank in their economy, production being mainly for personal consumption. As the peasants themselves produce most of the things which they consume, they are, as private householders, more comfortably provided with necessaries, being in a position to retain for their own use a larger share from the produce of their labour. It is not improbable, indeed, that if they have been unable to improve their equipment, and have found it difficult to meet taxes and other impositions, this has partly been due to their having used the new surplus in the first place for the increase of their own domestic allowance, and not for the purpose of enlarging their trading activities.[1] In brief, and speaking generally, the peasants have disposed of larger supplies than before; they have been short of money because they did not sell more; and they did not sell more because they have eaten more. Chapters IX to XI have explained why the output of agriculture did not reach the point where it might have satisfied the bodily needs of the peasants without diminishing the surplus available for the market.

2. *Material Conditions of Life.* It is the more unfortunate that one cannot establish statistically the advance the reform set going in village life, as the elements of that improvement could not, of course, be uniformly distributed either over the whole country-side or over all the aspects of the life of an individual household. There is no difficulty, therefore, in picking out isolated circumstances which will support equally well wholly contradictory views. Direct contact with the village, however, sets all doubt aside. The first blossoms of a better life are visible to every one who passes through the village street or sits in a

[1] The inquiries of Professors A. I. Chuprov and A. S. Postnikov (1897), cited by Professor Kossinsky in this series, established that in Russia the creation of peasant farms, and the increase in their yield, first of all resulted in increased consumption among the peasants, who had formerly suffered from a shortage of food. The peasants first satisfied their own needs before they turned their attention to the market.

peasant parlour, and who remembers that one must expect nothing better than poverty where blackest misery reigned only a few years before. More significant still, that change in the aspect of village life obviously reflects a change in the whole outlook of the peasant. His evident anxiety to raise his standard of living finds expression as keenly in the demand for schools as in a better provision for his bodily comforts.

(*a*) *Housing*. The reform has had a twofold effect on rural housing. It has added very considerably to the space available for peasant farmsteads, and it has initiated an improvement in the buildings found thereon.

Rumanian villages are divided into two distinct types. In the highland and hill regions, following the configuration of the land, the villages are frequently composed of scattered farmsteads; in the plain, and often even in the valleys leading to it, the village settlement is compact, stretching sometimes for one or more kilometres on both sides of the high road, side streets being found only in the larger villages. The farmsteads themselves likewise display two main types. The type found in Saxon villages, and in those influenced by them, shows a group of buildings massed round a rectangular yard, which they shut in completely and which is closed towards the street with a high and massive gate. The typical Rumanian farmstead, like the Saxon, has the house towards the street, but the other buildings are scattered round the farm-yard, and the whole is generally separated from the street merely by low wattle-fencing.[1]

The problem which the reform had to solve in this connexion differed in the various provinces. In the new provinces, especially in Transylvania, the reform found the Rumanian peasants, who formed the poorer section of the population, ill-provided with

[1] Dr. G. Banu, in his article on *Village Biology* (p. 103), makes the following comment: 'Dr. Lupaş explains this differences as follows: The Rumanian is not afraid of his surroundings. He has nothing to hide from the passer-by and builds his house with an open balcony, giving it a gay and friendly aspect, just because he feels himself at home. The other nationalities always show a retiring and unfriendly tendency, which shows that they feel themselves to be alien to this region.' One can imagine a Saxon or a Magyar writer retorting, e. g., that the Saxon and Magyar villagers had to shut themselves in for protection against the doubtful habits of the Rumanians, but that the Rumanian peasants could afford to live with open farmsteads because they had nothing to fear from their more civilized neighbours. There is no limit to the ingenuity of nationalist 'sociologists'.

farmsteads; these were either insufficient in size or badly placed, in marshy or rocky ground, on the edge of the villages. Most of the villages are mixed, but each nationality lives in a quarter of its own, so that even where better land was available the Saxon and Magyar villagers sold preferably to their own kinsmen. The reform law for Transylvania, as we have seen, permitted the expropriation of land situated within a radius of 600 metres from the edge of a village for the creation of new farmsteads. In the Old Kingdom the farmsteads suffered from overcrowding. Because of local circumstances, the peasants generally refused to sell the land they owned round the village; newly-married couples were accommodated with a house in the parental yard, with the result that in some places the yard was crowded out with the households of two or three generations. The reform has provided adequate sites at the two ends of existing villages, or sometimes it has established new settlements at some cross-roads. In addition, certain areas have been reserved for the needs of future generations. All this forms an important social aspect of the land reform.

The serious part of the rural housing problem in the Old Kingdom lay in the quality of most peasant houses. In this respect the reform could exercise a direct ameliorating influence only where new village settlements were set up, but indirectly, through the general advance in the state of the peasants, it has given a stimulus towards an all-round improvement in village building. The total number of rural buildings and households (legally married) was as follows:

Province		Buildings			Households
		Inhabited	Uninhabited	Total	
Moldavia .	(1914)	387,168	28,856	416,024	422,651
Muntenia .	,,	513,017	32,566	545,583	576,978
Oltenia . .	,,	265,160	23,002	288,162	279,448
Dobrogea .	,,	51,066	2,993	54,059	57,547
Bessarabia .	(1922)	498,191	6,271	504,462	563,910
Bucovina .	(1919)	151,325	8,164	159,489	178,157
Transylvania . . .		—	—	—	—
Total . . .		1,865,927	101,852	1,967,779	2,078,691[1]

[1] After Al. Alimănişteanu.

No peasant lived in a rented cottage. Every family had its own house which it built almost always itself with such materials as it could afford. Most houses consisted of a rough wooden frame filled in with soil and plastered with clay mixed with manure and short lengths of straw; the roof was generally covered with thatch or wooden slabs. The walls were limewashed, inside and outside, usually every spring. These houses had no foundations, the floor, like the walls, being of beaten clay. Most houses consisted of two rooms, one serving as a family bedroom and the other being reserved as a parlour, with in between a small hall in which stood the open hearth. Design and materials were extremely primitive, and, as a rule, insanitary. An inquiry undertaken by the Ministry of Agriculture, in 1906, established that the rural habitations were divided into:

3·8 per cent. underground hovels,
26·5 per cent. single-roomed cottages,
50·6 per cent. with two rooms,
14·6 per cent. with three rooms, and
4·5 per cent. houses with more rooms.

According to building materials these houses were divided as follows:

51,989 of brick or stone,
286,177 of wood, on stone foundations,
257,457, of wattle-matting covered with clay, and
242,537 were only of clay.

The last two categories were considered by Rumanian experts as unhygienic. Their prevalence would seem to be a modern phenomenon, which appeared with the loss of the peasants' timber rights; for in his monograph on Mehedinți, published in 1859, Ion Ionescu says that in the villages of that county 901 houses were built of bricks, 128,063 of wood, and only 603 of beaten clay. Most of the houses listed by the inquiry of 1906 had tiny windows, which not unfrequently were fixed and did not open at all. The census of 1912 found that the Old Kingdom still had 32,367 *bordeie*, that is, half-buried one-roomed hovels, with no windows and a low roof, of the kind in which the inhabitants of the Danubian steppe used to live during unsettled times, because the low dark roofs of these dug-outs had a better chance of escaping the eye of marauding invaders. It is characteristic that a law of 1894 prohibited the construction of such huts, and ordered the destruction of those already existing, giving their owners a respite of five years—which expired in 1899. And it is equally characteristic that these huts were most numerous in the fertile regions; the counties of Dolj and Romanați had villages in which such huts formed the majority of the habitations. The 1912 census also found 3,311 inhabited barns and stables.

If, nevertheless, these miserable habitations looked attractive to the passer-by, that was due to the peasant's innate sense of proportion, and to the fact that he only used such materials as could be found where he lived, so that the house blended harmoniously with its surroundings. The more solid and hygienic houses which are being built at present are not always an aesthetic improvement on the old, especially when they are covered with glaring tin-sheets. But the use of brick-walls is spreading, as well as of wooden floors, and windows are made on a more generous scale. It is a general trait of peasant building that every advance in well-being finds expression first in more solid and spacious farm-buildings. The peasant, that is, invests above all in the buildings with the help of which he makes his income, and is more modest with the house in which he spends it, a tendency which is more marked as the peasant ascends in the scale of wealth; so that it is easier to detect the well-being of a village from its farm-buildings rather than from its houses.

The villages of Transylvania have much better houses and buildings. In 1920 the province had 1,245,835 dwellings, with an average of five inhabitants; 220,043, or 17·5 per cent., were built of stone, 286,940 or 22·8 per cent. of brick, and 643,983 or 51·3 per cent. of timber. In the Old Kingdom the problem can be solved only gradually, and the state of things established by post-war inquiries is still extremely bad. On an average there are five persons to each room, in some villages even six, and the rooms are such that they give 4–5 cubic metres per person, instead of the 25 metres required by hygienic standards. The Central Resettlement Office is actively engaged wherever possible, with advice and help, in the form of credits, in the encouraging of higher standards in the new houses that are being built.

(*b*) *Alimentation.* During a discussion in the Rumanian Academy at the beginning of the century on the increase of pellagra, the Prime Minister, M. D. A. Sturdza, stated that from the accounts of central and county authorities it was evident that maize had to be distributed every two years to destitute peasants:

> 'In our country famine is constant, for it prevails not only when the peasant has nothing to eat, but also when he is forced to eat mouldy maize. . . . Before God there is no excuse for this: that our own kith and kin—that part of our society which works for us all and produces the country's wealth—should have to beg its daily food and to subsist mainly on a food which is poison.' [1]

The Rumanian peasant, it has been said, 'is born a vegetarian.' That was true enough, if it referred to social circumstances rather than to personal tastes. That habit was above all the result of necessity, and in part the injunction of the Church. The 'fish-days' prescribed by the Church numbered 194–200 days yearly, which most peasants scrupulously observed, without, however, being able to add more than once in a while fish, fresh or salted, to that diet. But as, from all accounts, the peasants' meatless days were usually 300 in the year, penuriousness clearly dictated them as much as religious observance. Only in regard to beef did personal taste play a part in the composition of the peasant's diet. Most peasants are animated by a feeling akin to worship for the bovine species. The ox is their companion in work and the idea of eating his flesh is repugnant to many peasants. Even in hospitals they often refuse to touch beef, and those who do it, as, indeed, most of the townspeople, always refer to it as 'cow-flesh'.

That the peasants had to subsist on an insufficient quantity of food clearly appears from the decline in the consumption of maize from 230 kg. in 1876 to 146 kg. in 1903 per head of population, without that falling off being compensated by the addition of other ingredients to the peasant's diet. Dr. A. Urbeanu affirmed, indeed, in his *Probleme Sociale*, that the quantity

[1] *Annals of the Rumanian Academy*, vol. xxiii, pp. 33–4.

of animal products consumed by the peasants had continuously and increasingly diminished since 1860. M. Vintilă Brătianu stated, in 1914, that there were villages without one single milch-cow; when milk was needed, for medical purposes, it had to be brought from the neighbouring towns. What, in such circumstances, must have been the quality of the food given by the large owners and farmers to their labourers may be inferred from the declaration of Dr. Balasian, sometime chief medical officer of the county of Ilfov. He related how he had seen peasants run away from the fields, thus losing the money due for their labour, because they could no longer stand the food. The 1907 law on agricultural contracts gave the district medical officers the right to prescribe a standard for the food which labourers received in the fields. But the law added that medical officers or village mayors could intervene for such a purpose only at the special request of the dissatisfied peasants, and this nullified the practical value of the provision, because of the degree in which the peasants were dependent on the local landlords. Moreover, the law only provided for cases in which the food was unfit for human consumption; it did not concern itself with cases when the food was merely insufficient.

During the years of the War the food of the rural population became still worse than it had been before. A large number of men were conscripted for military duties, even before Rumania entered the War, and that affected the quantity and quality of agricultural work. At the end of 1916 two-thirds of the country was occupied by the Central Powers, who during the following two years requisitioned all the available supplies, leaving the population barely enough for keeping alive. The unoccupied section of the country was crowded with the army and a mass of civilian refugees, so that the individual food allowance was scarcely better than in the occupied territory. The population continued to suffer severely until the 1919 harvest. M. Glăvan, e.g., found extreme misery in the Mehedinţi county: 'Many peasants said that on festive days, when they could not go to work and get food as part of their wages, they ate only once a day,' so that the children might have maize when the old people were at work.

The reform has exercised a more rapid and direct influence on food than on housing. It was natural that the peasant should seek to relieve his greatest want first. The two things which he did at once were: first, to replace in part his maize diet with wheaten bread, and, secondly, to stop eating maize which was unfit for food. A memorandum of the Cercul de Studii Economice[1] stated that till 1916 the home consumption of wheat amounted to 50,000–60,000 wagons yearly, 80 per cent. of which were used in the towns, leaving a yearly average of 15 kg. of wheat for each rural inhabitant. After the War, area and population of the country had doubled, but wheat consumption had trebled, and the fact that the peasants of Transylvania and Bucovina ate more bread than they produced could not altogether account for that rapid increase in wheat consumption. The change which is taking place in the habits of the peasants in the Old Kingdom is demonstrated concretely by the appearance of a baker in almost every village—an occupation which formerly was unknown outside the towns—and even of two or three, in the large villages. The extent of this change varies from place to place, but landowners of my acquaintance assert that there are villages where formerly maize alone was consumed, but in which the regular diet now includes as much bread as *mămăliga*.

The proof that the peasants no longer consume bad maize lies in the practical disappearance of pellagra. The pre-war inquiries had in general established that the peasant's diet was deficient in albuminoids, that it contained a minimum of fats and an excess of carbohydrates. During the last few years a number of inquiries made by medical students, under the direction of Professor Mezincescu, have established a marked improvement in the feeding of the rural population. The sixty inquiries relating to villages in the several geographical regions have concluded that the lowland villages enjoyed a more abundant food-supply than the villages in the hills and mountains, these latter being on nearly the same level. This observation is suggestive, because the lowland villages, while situated on the most fertile soil, had been before the reform among the poorest. The figures

[1] Reported in the *Argus*, January 20, 1923.

collected during these inquiries in a number of villages for each of the regions gave the following daily individual averages:

Regions	Albuminoids	Fats	Hydrocarbons	Total calories
	gr.	gr.	gr.	
Lowland	104	179	713	4,783
Hills	166	36·7	633	3,640
Highland	152	157	424	3,776
Average for the three regions .	140·7	124·2	590	4,066

Dr. Lupu found, during an inquiry he made in 1906 into the diet of 40 families, consisting of 180 individuals, that the individual averages did not exceed 73·40 grammes of albuminoids and 27·84 grammes of fats, hydrocarbons alone being up to normal. 'It appears from these inquiries', concludes Dr. Banu, who cites the results of those conducted after the War, 'that during the last ten years the nourishment of the peasants has improved, which is very probably due to the better material situation resulting from the land reform.'[1] One may note especially the considerable rise in the proportion of fats, which shows that besides eating more wheat the peasants have begun to consume a reasonable proportion of the animal food-stuffs produced on their holdings.

These welcome improvements can, of course, not be taken to mean that the food of the peasants is now adequate in quantity and quality, or that they could uproot in the present generations the ill effects of a lifetime of starvation. Dr. A. Urbeanu records in his recent book that one bad harvest suffices to cause a recurrence of pellagra. From this he draws two conclusions: first, that physiologically the power of resistance of the peasant is still very low—which, unfortunately, is only what could be expected; and, secondly, that pellagra is not due to the eating of mouldy maize, but is the consequence of the peasant's organic exhaustion. Pellagra, in his opinion, will reappear as long as the peasant's nourishment is biologically inadequate, independent of the fact of whether he eats bad or good maize, or whether he replaces maize with wheat. Dr. Urbeanu justly added that this

[1] Dr. G. Banu, *Biologia Satelor*, pp. 97–9.

problem had a cultural side to it. The life which the peasants have led for several generations—man, wife, and children all working excessively in the fields, and coming home to a larder which besides maize-flour contained only a few vegetables and perhaps a little cheese—through such a life the peasants have lost the art of cooking. To help them to a knowledge of how to choose and prepare their food, therefore, is biologically as important a part of the problem as to have enabled them to acquire larger supplies.

(*c*) *Alcoholism.* Indigence is apt to cause those who suffer from it to forget, or to become indifferent to, the healthy purposes to which greater affluence, should it come their way, might be usefully applied. One effect of the increase in the means at the disposal of the peasants has been a corresponding growth in the consumption of alcohol.

The alcoholization of the Rumanian villages, which had already gone far before the War, began in the eighteenth century. The Porte having prohibited the export of corn from the Rumanian provinces, the surplus which it did not claim itself began to be used by the landlords in distilleries which gave both fodder for the fattening of animals, and spirits. For these spirits an outlet had to be found. The landlords opened public-houses in the villages on their estates, obliging the villagers to buy a certain quantity each year; sometimes labourers were paid in spirits. Later, the export of corn was left free, but drinking had become a habit, and potatoes replaced corn as raw-material for distilling, while the making of spirits from plums at home spread among the peasants. The State made no effort to check the growing production and consumption of alcohol. After the rising of 1907 the number of public-houses in rural areas was legally restricted to one for each 100 inhabitants, i.e. roughly, for each 20 households. After the War a law allowed peasants in the wine-growing districts to prepare and retain for personal consumption wine and spirits from their own vineyards and orchards. The opening of new public-houses was, if anything, encouraged. M. Vintilă Brătianu stated in the *exposé de motifs* to his law for the restriction of the manufacture of alcohol that 'in every one of the provinces the number of public-houses has been allowed to grow under pretext of increasing trade conducted by Rumanians, overlooking the great harm that was being caused to the Rumanian people in facilitating the spread of alcoholism. To sum up, we could say that since the end of the War we have pursued, rather, a policy for the alcoholization of the country'.[1] The number of factories and distilleries for the production of alcohol reached 159 by 1926; most of them were enjoying the protection of the law for the encouragement of national industry, which enabled them, e.g. to import their machinery duty free, &c. A law passed in 1924 limited the number of factories to those already in existence, but it did not limit the quantity or the strength of their output. This state of things was boldly curbed by M. Vintilă Brătianu, in 1926, though as Minister of Finance he risked losing resources thereby just when the Treasury was badly depleted. The 1927 law obliged all alcohol factories to organize themselves into a syndicate, which was to control production and sales. The Ministries of Finance, of Industry, and of Agriculture were each to have one delegate on the board of directors. No new distillery could thereafter be established; total production was fixed for periods of three years, being reduced from period to period, and each factory was allotted a maximum yearly production. All sales had to be made through the Syndicate; alcohol could not be sold of a strength exceeding 30°, and it had to be aromatized before being put on the market. Finally, the duty on such alcohol was raised from 11 to 104 lei per kg., while fruit or wine alcohol only paid 10 lei per kg. As a result of the law the yearly production of industrial alcohol fell from 5,400 wagons to 580, in 1928. Another result has been the spreading of clandestine manufacture of alcohol, and of the discovery during 1928 that wood alcohol was making its appearance in the drinks put on the market.

The growing consumption of alcohol is in a way evident from the increase in the revenue derived from it:

	1924	1925	1926	1927	1928
			(in million lei)		
Estimate . . .	1,390	1,420	1,560	2,695	2,972
Revenue . . .	1,693	1,744	2,000	1,789[2]	—

[1] Article by Dr. Alex. Manolescu in *Adevĕrul*, August 11, 1925.

[2] First nine months.

The consumption of alcohol in degree-decalitres was:

1920–1	190,422,722
1921–2	300,478,152
1922–3	300,080,400
1924	472,746,600

The consumption per head of population, reduced to 100° alcohol, was as follows:

Year	Old Kingdom	Transylvania	Bessarabia	Bucovina	Average
1920–1	13·5	9·6	4·3	21·3	11·3
1921–2	16·6	21·1	7·9	37·8	17·9
1923	13·1	26·7	5·9	40·5	17·9
1924	18·6	43·1	10·4	71·8	28·2 [1]

The effect of M. Brătianu's Act on the production and consumption of industrial alcohol is shown in the table below:

Year	(in litres)		
	Production	Consumption	Consumption per head of population
1925	76,954,722	51,799,871	3·0
1926	60,643,663	40,167,834	2·3
1927	38,191,388	24,173,525	1·5 [2]

The consumption, per head of population, of the several kinds of beverages varied as follows:

Year	(in litres)		
	Spirits	Beer	Wine
1922–3	6·4	4·5	13·5
1923	5·4	4·5	14·4
1924	4·9	4·4	16·7
1925	2·1	4·4	15·3
1926	5·0	4·08	15·0
1927	2·5	5·1	12·4

These figures do not include the many other varieties of spirits put on the market.

The peasants' favourite drink is the *ţuica*, a brandy distilled from plums, with usually a strength of 30°. Its consumption varies with the yearly output of the plum orchards, but it is

[1] *Buletinul Statistic*, 1925, No. 4. [2] *Ibid.*, 1928, No. 3.

beyond doubt larger than that recorded in the official figures, as it is impossible adequately to control all the private stills, of which there are said to be nearly 50,000. The consumption of that spirit in the villages far exceeds that of wine and beer, and is the more harmful as most of these home-brewed brandies, according to the result of official tests, contain dangerous chemical impurities. The number of licensed houses was as follows:

In 1926

Province	Public houses		Out-licenses	Total
	Rural boroughs	Urban boroughs		
Old Kingdom	14,301	10,085	1,833	26,219
Transylvania	13,693	4,708	213	18,614
Bessarabia	3,839	1,332	349	5,520
Bucovina	1,275	1,094	24	2,393
Total	33,108	17,219	2,419	52,746 [1]

The total figures work out at one licensed house to about 350 inhabitants. But the actual number of public houses is certainly much greater, many of them being carried on without a license but with the connivance of the local police. The inquiry conducted by the Ministry of Labour, early in 1929, into the sale and consumption of alcohol in the industrial centres of the Jiu valley, in Transylvania, found, e.g., 35 unlicensed houses in one place, at Petroșani, and about 30 at Vulcan. At the anti-alcoholic congress held at Bucarest in 1926, it was stated that there were 168,000 public houses in the country, i.e. roughly one to each 100 inhabitants, which is about the proportion that one meets with in most places.

The drink problem is grave in Rumania. It means that some 100,000 wagons of corn are daily transformed into alcohol and dispensed in that damaging form to a population which succumbs to it the more easily as it is undernourished and weakened by disease. According to Dr. Gerota, the yearly consumption per head of population was 28·2 litres of alcohol and 12·47 kg. of meat (35·6 kg. in the towns and 2·77 kg. on the land); 11,627 million lei were spent in 1924 on alcohol and 6,470 millions on

[1] *Buletinul Statistic*, 1928, No. 3.

meat. Drunkenness is appallingly frequent. The *exposé de motifs* to M. Brătianu's bill mentioned that peasants bought alcohol of 96° and watered it down to half strength, for personal consumption, and also that spirits of 96° were being consumed. In 1925 Dr. Obreja stated that, divided by the number of adult inhabitants, the average yearly consumption of alcoholic drinks amounted to 150 litres per individual. Since the War, the peasant women have abandoned their former restraint and can be seen openly drinking in the licensed houses of villages and market-towns. The problem can, of course, not be solved by fiscal measures alone; the whole social and cultural basis of peasant life will have to progress to a happier stage before the hold which drink has gained on the village could be shaken. Official policy can contribute in many ways to that end, by curbing all sorts of abuses which the drink trade has hitherto been able to perpetrate unmolested. On inquiry into the causes of labour unrest among miners in the Jiu valley, in the winter of 1928–9, the Labour Ministry found, e.g., that debts contracted by the miners at public houses were retained from their wages, which in consequence dwindled each week. The same inquiry established the existence of a large number of unlicensed public houses. The Minister of Labour in the new Peasantist Government, M. I. Răducanu, besides taking steps for ending such abuses, attacked the problem directly by establishing in the various centres, with the aid of the mining companies, tea-houses in which cups of tea are sold at about $\frac{1}{4}d.$ each. During the first four days the six tea-houses sold 3,000 cups of tea. An article in the *Dreptatea* of February 15, 1929, affirmed that the unrest in the valley, with its constant threat of strikes, faded away when tea began to take the place of alcohol. The gravity of the problem, apart from vital issues, may be gathered from the statement of the Directorate of the Gendarmerie that 24 per cent. of the crimes committed in rural districts during 1918–23 were attributable directly to alcoholism. In 1929 the new Peasantist Government decreed that public houses should be closed on Sundays and holidays; it is instructive to find that this wise measure was welcomed by the peasants themselves.

Section 3

Health and Vital Statistics

What has been said hitherto concerning the peasants' housing, food and drink is sufficient to suggest that one cannot expect their health to be flourishing. The effect of the reform in that respect has not so far been very marked, but there are many reasons why progress should have been slow during the few years which have elapsed since the enactment of the new land laws.

Rumania is far from being confronted as yet with the problem of over-population. She only has about fifty-seven inhabitants to the square kilometre and her agricultural production is capable of expanding considerably. The flight from the land keeps within moderate limits; there is not sufficient industry to give employment to a large number of people and the peasant's attachment to the soil is very strong. The yearly increase of the rural population still exceeds the number of those who leave the village. Emigration, on the part of the rural elements, was only 5,625 men and 431 women in 1926, and these belonged mainly to the national minorities. What effect the new agrarian legislation will have on the natural increase of the rural population can only be vaguely forecasted on the basis of social experience. The changes in the conditions of bequeathing peasant land, described in Chapter VI, will probably leave younger sons little prospect of becoming themselves owners of holdings, however small, and with the disappearance of large properties agriculture will be able to give employment only to a restricted number of labourers. On the other hand, it has been established in France and elsewhere that smallholders tend to limit the size of their families. Agricultural labourers show no such restraint, being hopeful of deriving in old age aid from the labour of their children, but the peasant who acquires a piece of land seems anxious to avoid the splitting of his property among several sons; apart from the restictive influence which a rise in the standard of living generally exercises on the number of children. The fact of the land reform having been delayed till now has left free play to the natural increase of the rural population, and it will be interesting to see how far and how quickly that increase will be checked by the resettlement of the peasants. The new distribution of property, as resulting from the reform, is bound to lead to the gradual intensification of Rumanian agriculture, and as the smallness of the holdings limits the use of machines, more intensive agriculture will in the first place mean an increase in the use of hand labour. The peasant households, as we have seen, still have a good deal of labour to spare, enough to balance the requirements of intensive cultivation; so that altogether it seems reasonable to assume that the land will be able to maintain the present population in a higher standard of living without the need of emigration from the villages, but also without much room for an increase in the number of cultivators.

The number of marriages showed a continuous and high proportional increase:

1900	13·4 per 1,000 inhabitants
1905	16·0 ,, ,, ,,
1910	18·4 ,, ,, ,,
1915	14·3 ,, ,, ,,
1918	17·6 ,, ,, ,,
1919	22·0 ,, ,, ,,
1920	25·0 ,, ,, ,,

In 1920 the country registered an exceptional increase in marriages in rural districts, followed by a steady decline:

	1920	1921	1922
Rural communes	173,848	165,348	139,885
Urban communes	32,628	32,650	29,912

This phenomenon appears still more clearly from the figures below relating to the Old Kingdom alone, and leave no doubt that the increase was directly attributable to the land reform:

Rural Marriages in the Old Kingdom

1913	1914	1915	1918	1919	Average	1920
56,256	53,243	44,164	48,123	58,144	52,186	79,943

Observations have established that generally the number of marriages is directly proportional to the output of agriculture. The number of divorces increased in the same measure as that of marriages:

In the Old Kingdom

1900	1,282
1905	1,722
1910	2,847
1915	3,226
1919	2,781
1920	4,409

The total number of divorces in the whole of the country was 7,716 in 1920. Divorces were much less frequent in the country:

	Rural communes	Urban communes
1920	7·6 (per 10,000 inhabitants)	20·0
1921	10·8 ,, ,, ,,	21·5
1922	8·8 ,, ,, ,,	17·0

The number of births, per 1,000 inhabitants, showed the following variation:

1901–5	39·4
1906–10	40·4
1911–15	42·0
1918	15·8
1919	36·0
1920	33·7

And in absolute figures for the whole of Rumania:

	Rural communes		Urban communes	
		Per cent.		Per cent.
1920	471,283	87·4	68,076	12·6
1921	541,686	88·0	78,771	12·0
1922	535,094	87·2	78,632	12·8

The average yearly increase in the rural population, in the years before the War, was 20 per thousand inhabitants, and in the urban population only 7·4 per thousand. Rumania had the second highest percentage of births in Europe, 36·6 per thousand inhabitants, coming after Jugoslavia which had 38·6 per thousand. This high proportion of births was offset by an equally high proportion of deaths:

1901–5	25·4 (per 1,000 inhabitants)
1906–10	25·8 ,, ,, ,,
1911–15	24·4 ,, ,, ,,
1918	45·7 ,, ,, ,,
1919	33·7 ,, ,, ,,
1920	26·0 ,, ,, ,,

Still-births and infant mortality showed a high degree of frequency:

Still-births (per 100 births)

	Whole country	Rural communes	Urban communes
1900	1·6	1·1	4·2
1905	2·31	1·84	5·18
1910	2·72	2·36	5·18
1915	2·76	2·39	5·26
1918	3·40	2·61	7·11
1919	2·51	3·99	5·50
1920	3·04	2·38	5·36

Infant mortality was as follows, per 1,000 inhabitants:

	Up to 1 year	Up to 18 years
1901	30·4	56·7
1905	30·6	53·3
1910	31·5	56·1
1915	32·8	57·1
1918	10·4	42·7
1919	22·4	46·1
1920	30·9	56·33

General mortality of the rural population in the post-war years was as follows:

Years	0–1	1–5	5–20	20–40	above 40
1921 . . .	34·7	13·1	11·0	9·8	31·4
1922 . . .	35·4	12·5	9·4	10·0	32·7
1923 . . .	35·5	14·1	9·0	9·8	31·6
1924 . . .	34·2	14·3	8·0	10·1	33·4[1]

In villages infant mortality showed the following proportions per 1,000 inhabitants:

	0–1 year	0–5 years	5–15/18 years	Total
1901 . . .	31·4	17·7	9·3	58·4
1905 . . .	31·6	15·9	7·9	55·4
1910 . . .	33·8	17·8	7·9	59·5
1915 . . .	34·3	17·8	7·4	59·5
1919 . . .	23·3	18·7	11·6	53·6
1920 . . .	30·9	12·6	12·8	56·3

At the beginning of the century infant mortality in the first month after birth was higher by 6 per cent. on the land than in the towns, and in the first year after birth by 2 per cent.

[1] After Dr. G. Banu.

Compared with the mortality of children in the first year after birth in the neighbouring peasant countries, the Rumanian figure referring to the period 1895–9 was much higher:

Rumania, 21·5 per cent.
Bulgaria, 15·0 per cent.
Serbia, 16·1 per cent.

Very significant in the light of the history of the land problem is the following table, cited by M. Bibicescu on p. 21 of his book:

MORTALITY OF CHILDREN IN THE FIRST YEAR AFTER BIRTH

	Rural communes	Urban communes
	Per cent.	Per cent.
1870–4	16·0	28·3
1875–9	17·9	26·4
1880–4	18·0	24·0
1885–9	17·7	23·5
1890–4	21·6	24·0

From the time of the peasants' emancipation to the end of the century the mortality of children under one year therefore as steadily increased in the villages as it decreased in the towns.

According to Dr. Glicsman the high frequency of infant mortality on the land was due largely to the peasant woman's excessive labour during the period of pregnancy and feeding, to her insufficient nourishment, to the lack of proper medical attendance, and to social prejudices. Many babies were born prematurely and were under weight. There were many cases of asphyxiation through whooping-cough, the babies being left at home alone while the mothers worked in the fields. A local inquiry into the causes of rural mortality conducted just before the War and referring to a district with 50,000 inhabitants (the land belonging to five owners) confirmed the excessive mortality among children of one and two years of age, but it also came to the conclusion that most of them could have been saved. The inquiry extended over three years between 1910–13. During that period there were 2,604 births; 455 babies died in their first year, that is, 17 per cent., and mortality of children up to the age of two years accounted for one-third of all the deaths. The causes of 1,000 deaths among children up to three years of age were:

344 diseases of the breathing organs;
154 gastro-enteritis; 212 innate debility;
164 tetanus; 122 epidemics.

The great number of child deaths through tetanus, a phenomenon unknown in more developed countries, is a proof of the degree of misery and ignorance which prevailed on the land. Frequently the woman gave birth lying on the ground on some straw, or in the fields. During birth the woman was considered 'unclean' and for three days she and her child were approached by no one save the midwife, when there was one, or some old woman of the family. Mother and child were then washed with holy water, after which they were considered to be clean again. Dr. Glicsman states that he made an effort to act preventively in two communes, by vaccinating newly-born babies against tetanus, but that he was forced to abandon the attempt because it was strongly opposed by the local priest. He added that there was one single midwife for all the 50,000 inhabitants, which meant that pregnant women had little chance of getting expert help even when they wished it.

The effect of the absence of medical attendance may be seen from the following figures, relating to the rate of mortality among the rural population of the Old Kingdom and to patients up to the age of nineteen years:

	Scarlet fever Per cent.	Diphtheria Per cent.
1905	18·89	18·84
1910	20·32	15·72
1915	18·21	19·23
1919	16·55	14·76
1920	17·02	11·76
1921	17·22	19·34
1922	17·95	31·76

Another index to the health of the population may be found in the table below, giving the number of suicides between 1919 and 1925:

	Rural inhabitants	Total number
1919	73	357
1920	66	331
1921	67	339
1922	109	408
1923	105	434
1924	90	460
1925	104	550

Most of the suicides on the land were attributed to drink and pellagra.[1]

We shall have again to refer to the State's health policy, which is no doubt partly responsible for the high rate of mortality. Preventive action is needed above all, and action of a social nature more than of a medical kind. We cannot here argue the point at length, though it is a point which peculiarly concerns our study, but will merely refer in brief to pellagra, which is undoubtedly a social malady; and to certain facts established by the army authorities, which offer remarkable evidence in support of the view that social rather than biological conditions are responsible for the bad health of the peasant population.

Pellagra, which among Transylvanian peasants is known as 'poor man's sickness', was unknown in the Rumanian provinces, notwithstanding frequent famines, as long as millet, wheat, and barley formed the basis of the people's food. The disease appeared with the cultivation of maize and spread with it. Maize was introduced in the Rumanian provinces about 1700, and it became rapidly popular with the peasants just because as an autumn crop it withstood the climate better than the frequently-failing summer crops. But it must be remembered that the change from millet to maize coincided with the lowering of the peasants' status and well being.[2]

[1] *Buletinul Statistic*, 1927, No. 2.

[2] The social origin of the disease is brought out by Dr. Creighton in his *History of Epidemics in Britain* (1891, pp. 107–13), by way of an interesting comparison with medieval leprosy. 'The varying types of diseases, or their existence at one time and absence at another, are a reflex of the variations in the life of the people—in food and drink, wages, domestic comfort, town life or country life, and the like.' The appearance and disappearance of leprosy in medieval Europe was a case in point. 'The pellagra of the North Italian peasantry (and of Rumania, Gascony and some other limited areas) is the nearest affinity to leprosy among the species of disease. . . The two diseases are closely allied in the insiduous approach of their symptoms, in their implicating the tissue-nutrition through the nerves, or the nervous functions through the nutrition, in their cumulating and incurable character, and in their transmissibility by inheritance.' Symptoms and effects being alike, there must also be a relation between the causes. 'The most general expression for leprosy is a semi-putrid or toxic character of animal food, just as for the allied pellagra, it is a semi-putrid or toxic character of the [maize] bread or porridge,' consumed steadily from day to day, and aided and abetted by other conditions. 'These aiding things are for the most part the usual concomitants of poverty and hardships . . . [Leprosy] was a *morbus miseriae* of the Middle Ages, but on the whole not a very common one; and it was easily shaken off by the national life when the conditions changed ever so little.' That is happening in Rumania with pellagra, after the land reform.

The first cases of pellagra were not recorded till about 1830 in Moldavia and 1850 in Muntenia.[1] After that the disease spread both in absolute numbers and in proportion to the size of the population:

	Number of patients	per 1,000 inhabitants
1880	10,626	—
1898	21,282	—
1899	32,271	—
1900	40,786	—
1901	34,776	6·8
1902	36,239	7·0
1903	40,660	7·7
1904	43,676	8·1
1905	54,689	10·5[2]

According to Professor Babeş there were 30,000 patients in 1911, 40,000 in 1913, and about 100,000 just before the War. The average mortality from pellagra was 5 per cent. There is no doubt that the incidence of the disease has diminished considerably after the War, and Professor Babeş has pointed out, in bringing the following table to the attention of the Rumanian Academy, that the largest number of patients was found in the counties in which maize was more extensively cultivated.

	1912	1914	1916	1918	1919	1920	1921	1922
Dolj	539	432	286	127	210	184	147	160
Gorj	347	322	195	79	141	176	168	146
Mehedinţi . .	473	404	268	162	143	119	108	118
Romanaţi . .	481	411	251	107	128	96	94	72
Vâlcea	463	363	226	137	140	143	105	106
Botoşani	416	1,301	79	945	225	682	633	?
Fălciu	871	1,099	680	631	681	601	490	?
Iaşi	583	536	394	320	355	403	323	?
Putna	1,143	1,035	632	612	575	564	568	?
Suceava	1,210	1,552	1,039	?	438	1,753	629	?
Tutova	1,242	1,190	640	278	1,088	303	207	?
Vaslui	1,165	1,296	1,167	137	140	143	105	?

It is interesting to note the statement of Professor Babeş, that

[1] According to Professor V. Babeş, the disease is contracted during the winter, but makes its appearance, with its characteristic rash, with the spring sun. Professor Babeş made an experiment with white mice which were fed on mildewed maize; the animals which were kept in the shade recovered, but those which were exposed to the sun succumbed. Pellagra affects the patient in three ways: it attacks the skin, the digestive organs, and the nervous system, causing a burning sensation in the throat, pains in the stomach and an abnormal nervous state.

[2] After Bibicescu, p. 22.

pellagra is prevalent among the colonies of Rumanian emigrants in the south-western regions of the United States.

The social origins of the population's ill-health is also confirmed by the serious extension of tuberculosis. We will mention only the evidence obtained a few years ago during an inquiry undertaken among the schoolchildren of Cluj, by Professor Iuliu Haţiegan. He found that among the 291 boarders of a girl's secondary school only 137 displayed no evident signs of tuberculosis; but 92·5 per cent. of the girls reacted positively to the Pirquet test. Almost all the pupils belonged to country districts. In a University Hostel for women students, to which admission was granted after a medical examination, only 47·89 per cent. of the 119 boarders displayed no evident signs of tuberculosis; and from 364 men and women chosen at random from the students of Cluj University, only 147 were free from all symptoms. In 1923 deaths from tuberculosis reached in Rumania 37 per 10,000 inhabitants, as against 30 in Hungary and 9·9 in the United States.

On another occasion Professor Babeş reported that the average percentage of young men found unfit for military service increased as follows:

1890–92, 5·6 per cent.
1893–96, 7·0 per cent.
1897, 8·3 per cent.

It is possible, of course, that the larger number of rejections may have been due to stricter standards. But certain conclusive evidence was obtained from the examination of 57,958 recruits, 20–1 years of age, born in 1893. An anthropometric study[1] made by M. Nicolae Tabacovici, sometime chief of the statistical service in the Rumanian War Office, found that the average height of those examined was 165 cm., which was equal to the average European height, and that this size was sufficiently well represented with 8·02 per cent. of the measured recruits. But the evidence assumed an altogether different aspect when the height was related to the chest measurements of these men, a relation which forms a reliable index to the physical state of a

[1] *Arhiva pentru Stiinţa şi Reforma Socială*, vol. i, No. 1, 1919, pp. 217–23.

population. The men were divided into three groups: those whose chest measurement was less than half their height, those in which it was equal to it, and those in which it was above it. A comparison with corresponding Bulgarian figures gave the following results:

	Group 1	Group 2	Group 3
	Per cent.	Per cent.	Per cent.
Rumania . . .	16·28	6·35	77·37
Bulgaria . . .	3·41	18·88	77·71

The number of those which fell below standard, therefore, was considerably greater in Rumania. M. Tabacovici did not press the point that Bulgaria was a country of smallholders, but he brought out further significant facts; of the examined recruits 88·88 per cent. came from rural districts. According to their height the rural recruits gave a much greater percentage in the lower groups. 'These inquiries entitle us to conclude that from the point of view of height our rural population is in a state of considerable inferiority as compared with the urban population.' This result was the more serious as the rural inhabitants formed more than four-fifths of the country's population. M. Tabacovici, therefore, made an attempt to trace the cause of that physical deficiency, excluding for obvious reasons all possibility of innate physiological differences between the urban and rural groups or between the various geographical regions. Working on the hypothesis that the distribution of property among the inhabitants of a country is the best index to their economic well-being, he grouped for each county: (*a*) the recruits whose height remained below 165 cm. and (*b*) the rural inhabitants owning less than 2 ha. of land each. The resulting figures established a remarkably close correlation between the two factors, as is shown by the accompanying diagram. 'The parallelism of the corresponding curves is statistically evident and in consequence the reason for the unsatisfactory height of our rural population must be sought in the unsatisfactory distribution of land property.' M. Tabacovici privately informed us of a further experiment on similar lines made at the War Office in 1916, the material of which unfortunately was lost, with many other documents, during the retreat. In the second case the urban recruits were eliminated from the test and the remainder were examined with a view to establishing, according to counties, three sets of factors: (1) Physical deficiency as resulting from the unsatisfactory relation between height and chest measurement; (2) the prevalence of tubercular arthritis; and (3) insufficient peasant property (below 2 ha.). The resulting curves, according to M. Tabacovici, displayed an even closer parallelism than the diagram here re-

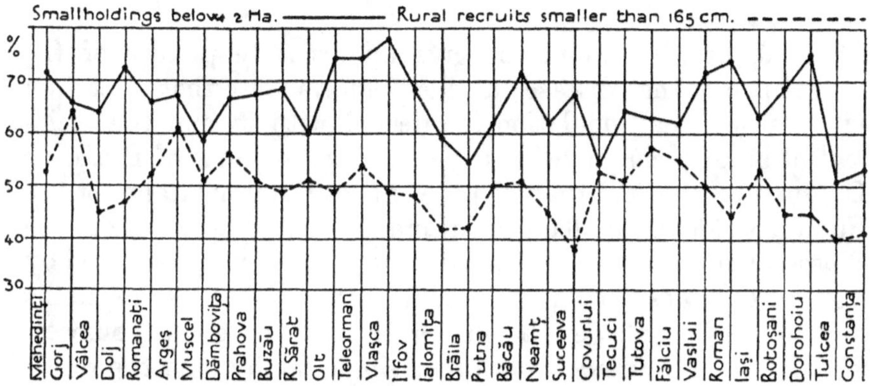

produced; and in the case of one county, apparently Băcău, all those rejected on medical grounds from one particular class of recruits belonged to that section of the peasantry which owned less than 2 ha. land.

The average length of a peasant's life in most civilized countries, according to Dr. Babeş, was 50 years before the War, whereas in Rumania it was only 30 years. 'As there exists a relation between mortality and morbidity, we are entitled to state that the Rumanian peasant is much more unhealthy than the peasant of any other civilized country.'[1] M. Mihalache has asserted that during the War the Rumanian soldiers more easily fell victims to epidemics than the Rumanians of Bessarabia and Transylvania.[2] After the War the peasant nourished himself better, especially quantitatively, and there is an incipient improvement in housing. But war exhaustion and the spread of alcoholism have not yet allowed the improvement in material conditions to bear palpable results, and to this must be added

[1] Cited by I. Bibicescu, p. 26.

[2] Speech in the Chamber, 1921, p. 36.

the appalling increase in sexual diseases. The number of rural inhabitants receiving medical treatment for syphilis was:

1912	10,027
1914	9,867
1916	8,614
1918	11,652
1919	17,634
1920	23,464
1921	26,818
1924	59,258
1925	77,264

These figures by no means give an accurate picture of the prevalence of that disease; medical officers state that in certain villages half the population is infected with it, and the sociological inquiry conducted last year by the seminar of Professor D. Gusti established that 75 per cent. of the inhabitants of the Fundul Moldovei, a village of Bucovina, reacted positively to the Wassermann test. A very painstaking inquiry was conducted by a group of Transylvanian doctors into the prevalence of syphilis in about fifty villages, belonging to four different counties. Blood tests were made from every single family, and the result was that in certain villages the percentage of those infected was as high as 25, while the general average was 10 per cent.[1] Dr. Leontin Munteanu, after keeping under observation over a period of 26 months 8,700 soldiers quartered in Oradea-Mare (Transylvania) and making 9,000 Wassermann tests, proved that syphilis infection increased with the length of military service. He found among first-year soldiers 3·437 per cent. infected with syphilis, and among second-year soldiers 5·131 per cent., while among those who had signed on again and had served more than two years the percentage was 8·722. In the second place, he found that the total number of those infected rose from year to year:

1923–24, 2·5 per cent.
1924–25, 3·5 per cent.
1925–26, 4·15 per cent.[2]

The provision which the State made for the prevention and curing of disease never had been anything but trifling, yet even

[1] Cited by Dr. G. Banu, pp. 109–10, from the Annals of the Ministry of Health.
[2] *ib.*, p. 113.

that was reduced after the War. The proportion which expenditure for public health represented in the total budget of the State was 4 per cent. in 1908, 3·4 in 1911, 2·5 in 1914, 1·3 in 1920, 2·5 in 1924, 2·7 in 1925, 2·5 in 1926, 2·7 in 1927 and 2·9 in 1928. The percentage of the last few years was in reality still smaller because separate accounts had been established for the railway budget. One must also remember that this decrease coincided with the limitation of the activities of hospitals kept up by private endowment. These functioned almost exclusively in the towns, while public expenditure was devoted mainly to health services in the country districts. It is therefore important to note that expenditure on public health, besides being reduced, was also redistributed in a way which still more deprived the rural districts of medical and sanitary assistance. Before the War, and during the first post-war years, the public health services were centralized in the General Sanitary Directorate, but in 1924 they were transferred to a newly-constituted Ministry of Health. The effect of that change on the distribution of the Department's budget may be seen in the table below: [1]

Year	Central administration	Inspecting services	External services
1913–14 [2]	27,322,200	5,670,000	504,199,005
1916–17 [2]	29,789,100	5,767,000	557,221,275
1921–2	7,982,150	—	258,955,661
1925	69,080,616	—	651,932,802

As long as the old organization lasted, therefore, the central administration used up 1/19, 1/19 and 1/36 of the total budget, but with the creation of the Ministry about 1/9 was spent on central administration, leaving the active services, which had in their care a population twice as large as before, correspondingly depleted of resources. It is not surprising that, as officially admitted in 1923, of the 870 rural health divisions only 409 had permanent medical heads, while 164 of the positions were vacant, 83 were occupied by medical students, and 217 had provisional incumbents. While in the towns the profession was seriously overcrowded, the authorities could not find applicants even for

[1] Dr. Al. P. Ilie, article in *Aurora*, January 31, 1926.
[2] The original gold lei sums multiplied by 45.

the rump services maintained in the country districts. One of the reasons for this was disclosed by the *rapporteur* of the 1929 budget of the Ministry of Health. He told the Chamber that at the end of 1928 the Ministry had unpaid estimates amounting to 190,000,000 lei left over from previous budgets; these included transport allowances to rural medical officers, some of them dating back to 1926; allowances for the maintenance of patients; for fuel, &c. The 1929 budget could only provide 4,950,000 lei for the payment of those arrears, this being an amount for which judgement had been obtained in the Courts against the Ministry.[1] Therefore not even the poor nominal sums inscribed in the budget had in fact been forthcoming. When the central administration enjoyed such favourite treatment it was not unnatural that candidates for the neglected field posts should be scarce; especially as those who would have liked to devote themselves to the urgent work that is to be done in the villages found their usefulness cramped by an excess of duties and an insufficiency of means.

The many legal provisions concerning public health are rendered senseless by the practical impossibility of carrying out their intention; medical officers are obliged to establish and isolate contagious cases but hospitals or infirmaries for their isolation do not exist. Moreover, neglect from the centre engendered indifference in the branches, to judge from the circular which the Ministry of Health had to issue in June 1927. The Ministry's inspectors, it complained, had found that many of the rural health officers were not living at their official place of residence but congregated in the towns, and some of them even in the capital. The Ministry ordered them to settle forthwith at the places to which they were appointed. What a rural medical officer is expected to perform appears from the figures which Dr. Glicsman related in 1920 concerning the county of Mehedinţi. For a population of 271,000 there were 8 district hospitals, but two of them were closed; there were 4 hospital doctors, and one district doctor, who had also periodically to inspect the communes; further, 50 so-called sanitary agents for 171 communes, and 21 midwives. The county had no infirmaries. The total

[1] *Adevĕrul*, January 9, 1929.

number of trained midwives was 2,749 in the towns and 1,397 in rural districts, or one to 6·2 rural communes—a commune generally including several villages. In 1924 the number of births was 81,715 in the towns and 522,721 in the villages, which meant that one midwife had, theoretically, to attend yearly to 395 child-births spread over a large area.[1] The situation would seem to be better in Transylvania where, according to M. Enescu, there was in 1920 one doctor to 4,824 inhabitants, one midwife to 1,425, one chemist to 9,467 and one hospital to 53,894 inhabitants. Infant mortality has remained at the same high level for the last 20 years or more. 'For purposes of hygiene', says Dr. Banu, 'we have neither organization nor any systematic policy.' And the curative services on the land work under such limitations of personnel and equipment that the peasant has only a narrow chance of recovering from the maladies which he is not helped to avoid.

Section 4

Cultural and Psychological Changes

(*a*) *Education.* An intense activity in all fields of education has been manifest in Rumania since the War. The country-side is displaying extraordinary keenness for more and better schooling, and public authorities have made praiseworthy efforts to open the doors of education to the masses who have acquired land and political power; they have relatively done more for education than for any other field of social policy.

Such an effort was more than called for by the high percentage of illiterates which the country still had. The census of 1899, the first to be more carefully conducted, found that 78 per cent. of the inhabitants above 7 years of age could neither read nor write, the number of women illiterates reaching 90 per cent. The 1912 census established that, of the inhabitants above 8 years of age, 39·3 per cent. could read and write and 60·7 were illiterates. The proportion for the rural districts alone was worse,

[1] Dr. G. Banu citing a report of Dr. T. Ionescu, of the Ministry of Health, 1926.

the total number of illiterates reaching 67·4 per cent. and among women 84·2 per cent. Of the several provinces Oltenia had the worst figure with 71·2 per cent. rural illiterates while Dorodgea, though economically backward, had the best figures with 75·2 per cent. illiterates in 1899 and 54·8 per cent. in 1912 (60·8 per cent. of the rural population in 1912). In his booklet on the activity of the Ministry of Education Dr. C. Angelescu stated that on the eve of the War the number of illiterates was 43 per cent. in the Old Kingdom, 40 per cent. in Transylvania, 60 per cent. in Bucovina, and 94 per cent. in Bessarabia. As the figure for the Old Kingdom is much below that established by the census of 1912, one can regard the above percentages as only approximately accurate, that for Bessarabia[1] seemingly applying merely to the Rumanian inhabitants of that province; Hungarian statistics gave in 1910 the percentage of illiterates in that Kingdom as 33·8, and Austrian statistics for the same year gave the percentage for Bucovina as 53·9. Speaking generally, more than half the population was illiterate when the new State was formed after the War.

Elementary Education. That circumstance is especially surprising in view of the fact that elementary education was made free and compulsory in Rumania as early as 1864, these principles being inscribed into the Constitution of 1866. But Cuza, the author of that law, could do no more than proclaim the principle; as buildings and teachers could not be created by decree, his law very naturally added that attendance at schools was obligatory 'wherever schools exist'. Since then schools have not come into being in sufficient numbers to make of that Constitutional principle a reality. In 1884 the country had 1,968 rural schools with 61,977 pupils (61,504 boys and 473 girls) and 1,988 teachers. The number of urban schools was 165 with

[1] The zemstvo statistics of 1905 and 1907, cited in the *exposé de motifs* to the new law for elementary education (1925), gave the following percentages of those being able to read and write among the various nationalities of Bessarabia:

Nationality	Men	Women	Nationality	Men	Women
Germans . .	65·6	62·9	Bulgarians . .	31·4	6·4
Poles . .	55·6	52·9	Turcs . .	21·1	2·4
Jews . . .	49·6	24·1	Ukrainians . .	15·3	3·1
Russians (Little)	42·3	11·5	Moldavians .	10·5	1·7
Russians (Great)	39·9	21·1	Gipsies . .	0·9	0·3

23,260 pupils and 516 teachers. On the eve of the War, in 1912–13, the situation of elementary education was as follows:

Communes	Number of schools	Number of teachers	Number of registered pupils	Average number of pupils per school	Average number of pupils per teacher
Rural . .	4,686	6,826	531,634	113	77
Urban .	370	1,414	84,936	229	60

There were in addition 256 private elementary schools in the towns functioning with the permission of the Minister of Education. These schools, notwithstanding their overcrowding, were far from being able to accommodate all the children of school age, as may be seen from the following figures, giving the position in 1922:

	Children of school age	Registered pupils	Children unable to gain admittance
Bessarabia	398,695	136,172	262,523
Bucovina	108,498	104,301	4,197
Transylvania	679,240	457,567	221,873
Old Kingdom	1,355,031	834,472	520,559
Total . . .	2,541,464	1,532,312	1,009,152

Many of the registered pupils, however, did not attend school either because their parents found a pretext for keeping them at home or because the conditions of the buildings, &c., prevented the regular functioning of the school, so that it is safe to say that in 1922 about half of all the children of school age were receiving no tuition at all. Only about half the schools had buildings of their own, while the remainder were carried on in rented houses which were unfit for the purpose. The bulk of the schools had only one room, and during the winter many could not hold their classes because of lack of fuel. The number of children who were unable to gain admittance increased, therefore, in proportion to the number of those registered, especially in the rural districts.

The number of pupils who finished their elementary schooling was in fact much smaller than those registered. Between the census of 1899 and that of 1912, 281,303 pupils took their

School year	Children of school age	Registered pupils	Regular attendance	Percentage of those registered attending regularly
				Per cent.
1893–4	634,342	186,403	143,570	77·0
1897–8	663,536	228,476	138,604	60·7
1900–1	704,039	282,225	190,674	67·6
1905–6	765,754	428,887	246,107	65·7
1908–9	910,203	481,074	300,276	62·4
1914–15	910,203	550,010	415,673	73·0
1919–20	1,015,347	681,272	435,943	61·3
1920–1	1,202,677	722,588	464,240	64·5

certificate in rural schools and 129,851 in urban elementary schools. The increase in the total number of pupils registered may be seen from the table below:

	Population	Pupils registered in elementary schools
		Per cent.
1864–5	4,424,961	85,237 or 1·9
1873–4	4,356,000	82,617 ,, 1·9
1885–6	5,173,452	140,000 ,, 2·5
1891–2	5,400,000	211,000 ,, 4·0
1896–7	5,700,000	287,110 ,, 5·0
1899–90	6,000,000	336,300 ,, 5·7
1910–11	7,000,000	611,300 ,, 8·8
1914–15	7,500,000	620,565 ,, 8·3
1919–20	7,500,000	748,765 ,, 10·0
1923–4	17,000,000	1,759,885 ,, 10·5

The state of elementary education in 1923–4, before the passing of the new law, was as follows, in the Old Kingdom:

Number of schools, 7,415;
Number of teachers, 16,415;
Number of children of school age, 1,329,381;
Number of registered pupils, 949,314;
Number of pupils who took the certificate, 72,307.

A new law on elementary education was promulgated on 24th July 1924. It changed the age for compulsory school attendance from seven to five, the first two years having to be spent in a kindergarten, and it extended the duration of attendance at elementaryschools from four to seven years. During the last three years general teaching was to be combined with training in handicrafts, for which purpose 2,000 workshops of various kinds were set up all over the country. For the carrying out of this program an average of 2,000 new teaching posts were created yearly from 1922 until 1926, the year Dr. Angelescu wrote his report, while the number of school buildings was almost doubled during that time:

	Number of elementary schools		Number of teachers	
	1921–2	1925–6	1921–2	1925–6
Bessarabia . . .	1,564	4,320	2,938	6,736
Bucovina . . .	512	580	2,056	2,475
Transylvania . . .	3,424	6,157	5,022	8,735
Old Kingdom . . .	6,508	12,000	13,676	18,730
			Handicraft teachers	1,091
Total . . .	12,008	23,057	23,692	37,767

Dr. Angelescu was hopeful that if this effort were kept up, in five or six years room would have been found for every child of school age, and illiteracy would then speedily disappear. The budget which the State placed in 1900–1 at the disposal of the Ministry of Education represented 10 per cent. of total expenditure (the Army receiving 19·4 per cent.); between 1901–8 it fell to 9·5 per cent., and in 1909–10 to 7 per cent., remaining at this level up to the War. In 1923 the Ministry received 7·3 per cent. of the budget, and in 1926 it reached 2,643,114,240 lei or 10·2 per cent. of the total budget.[1] In its turn the Ministry allotted 59 per cent. for elementary education in 1900, 56 per cent. in 1916, and about 44 per cent. in 1926. Again, from the amount allotted for elementary education 69 per cent. was spent on urban schools and 31 per cent. on rural schools in 1865; in 1895–6 the rural schools received 52 per cent. and the urban 48 per cent.; in 1910 the proportion was 65 per cent. and 35 per cent. and in 1924 roughly 74 per cent. and 26 per cent. In proportion to the size of the population and to the number of pupils the yearly sums the State spent on elementary education amounted in the period

[1] In reality that percentage was substantially lower, as in 1926 the railway budget, which showed a heavy deficit, was isolated from the general budget.

1914–15 to 2·5 gold lei per inhabitant and 22 gold lei per pupil, and in 1924, 46·25 lei per inhabitant and 460 lei per registered pupil, the average expenditure having thus been reduced by roughly one-third. In 1910 the United States were spending 96·65 lei per pupil, England 91·60, Germany 52·20, Bulgaria 26·40, Spain alone with 12·50 lei having a lower average of expenditure per pupil than Rumania.

The most striking aspect of that post-war development is the share which the peasants had in making it possible. The State was not in a position to supply the funds for the carrying out of the construction program described above, and the Minister of Education appealed therefore for private assistance—not to the wealthy people of the country, but to the mass of the peasantry. The response was such that within four years all the schools were repaired, and 7,500 schools were newly built or radically repaired, together with several thousand houses for the head-teachers. Many villages, and sometimes individual peasants, offered land for the school; they contributed ready money, or part of their produce; most of the villagers concerned helped in the actual construction—making bricks, cutting timber, carting sand, and building the walls.

> 'The State contributed free timber from its forests, for a value of almost two milliards, for the construction of these schools. The Ministry of Education, the county and communal authorities helped the poorer communes with several hundred million lei. The larger part of the money, however,' says Dr. Angelescu, 'was collected from the peasants in the form of voluntary gifts of money—which up to the present have reached over one and a half milliard lei—gifts in kind and in labour, with their own hands and with their carts, contributed free to the communes for the building of the schools.' [1]

This movement, begun by the peasants of the Old Kingdom, spread to the new provinces in 1923, and to the towns. There is no doubt that the initiative came from the peasants themselves, who soon after the distribution of the land began to send in requests for the creation of schools, offering to supply the land and to put up the building at their own cost if the Ministry would only provide them with a teacher. In the county of Constanța, e.g., the Ministry contributed merely 600,000 lei for the building

[1] *Activitatea Ministerului Instrucțiunii*, p. 10.

of elementary schools, the County Council 1,000,000 lei, and the peasants themselves 12,000,000 lei. Or, take the suggestive case of the commune of Șomartin (county of Făgăraş, Transylvania), with a mixed Rumanian and Saxon population of some 300 families. As, for some reason, it was not found possible to build a new school, the parish meeting unanimously decided, on the proposal of the Saxon priest, in October 1928, to close the communal public house and to use the building for school purposes.

Professional Education. Side by side with the building of elementary schools the Ministry approved the construction of 54 training schools for teachers, 48 secondary schools, 37 trade schools, and 20 commercial schools. In their case also private initiative harnessed itself to a task which the public authorities had no means of fulfilling, the funds being collected by special local committees, who in almost every case, moreover, supervised the actual building, so as to save the cost of contractors and other intermediaries.

Besides elementary schooling, the rural population needs above all training in agricultural methods, and in handicrafts for the winter months. The more wideawake peasants would seem to understand this very well. Those of eight villages in the county of Gorj, led by their mayors and priests, petitioned in 1915 for the creation in one of their communes of an ambulant handicrafts school:

> 'We think the time has come for the sons of peasants to be given both handicraft and agricultural tuition, for many of them have good holdings of land, and should they till it rationally, knowing also some trade which they might carry on in winter and when the weather is bad, they could earn good money to make their life easier. As the secondary schools are of no use to the peasants, it would be a good thing to abolish them and use the money allotted to them in the budget for the establishment in their place of as many ambulant handicraft schools . . . wherever a diligent population has the desire to learn these handicrafts, knowing full well that that is the only means of bringing about a more peaceful existence and one without want.' [1]

Steps for the satisfaction of such wise desires were only taken recently, with the establishment of the 2,000 handicraft schools to which reference has been made. Agricultural teaching as such

[1] Cited in O. A. Anastasiu, *op. cit.*, p. 101.

is only just beginning to receive attention, as may be seen from the following table giving the agricultural schools of all kinds:

Schools	1900	1910	1919	1924
Higher agricultural schools	—	1	3	3
Secondary agricultural schools	1	—	2	2
„ viticultural schools	—	—	1	1
Lower agricultural schools	4	8	26	34
„ viticultural schools	—	1	4	6
„ horticultural schools	—	—	2	6
„ apicultural schools	—	—	—	1
Elementary agricultural schools	2	15	15	11
„ viticultural schools	—	2	2	—
„ horticultural schools	—	—	2	1
„ handicrafts schools	—	—	—	1
Winter schools	—	—	—	1
Agricultural apprenticeship schools	—	5	11	18
Girls' domestic science schools	—	6	6	16
	7	38	74	101

According to standards these schools were grouped into:

3 high schools, of University standard;
3 secondary schools;
47 lower schools;
13 elementary schools;
1 winter school;
18 apprenticeship schools.

Nine lower and two elementary schools were maintained by private institutions, the remainder were State supported. The two points worth noting are, first, the absence of agricultural teaching till 1910,[1] in keeping with the 'Raubwirtschaft' which had been the rule in Rumanian farming; and, secondly, the efforts made in recent years to widen agricultural education. Of the 47 lower schools, 11 were established in 1923–4. The

[1] Agricultural teaching had been decreed as early as 1864, yet it was neglected, in keeping with the State's general policy, in favour of the trade schools, established much later:

	Agricultural Schools		*Trade Schools*	
	Number of schools	Number of pupils	Number of schools	Number of pupils
1919	68	1,645	122	1,972
1928	97	3,690	314	28,730

At the end of 1928, four different Ministries shared in the control of the Agricultural Schools.

higher schools act in effect as training schools for the teaching personnel of the others. The secondary schools are supposed to supply experts for experimental stations and teachers for the elementary schools; the course involves three years of school work and one year of practical work. The lower schools prepare their pupils for the management of small farms or for subordinate positions on larger undertakings; the course lasts three years, followed by one and a half years of practical work. The schools of Transylvania and Bucovina had, till 1923, a course of only two years. The elementary schools, with a curriculum extending over two years, were introduced in 1901 with a view to training sons of peasants in rational farming. This type of school appears to be on the decline, as most of the peasants send their boys to school in the hope of their obtaining afterwards some post, and they therefore prefer to enter them into the higher grade schools. Now, after the extension of elementary education over seven years, the last three of which are to include elementary training in agriculture, the specialized elementary schools will have no useful function to perform. Finally, the schools for apprentices, attached to the State's model farms, are based mainly on practical work, theoretical teaching being limited and sporadic. The course extends over three years.

Altogether, the organization and extension of agricultural teaching bore no relation to the technical needs of the industry. The figures given later on show that the whole system hardly touched the practising peasant farmer; there was only one winter school, of the kind so successfully active elsewhere, and ambulant teaching by lecture and demonstration was only begun in 1924. The total teaching staff, including that of the private schools, numbered 407 in 1924; of these only 290 were permanently provided for in the budget, the remainder being extraordinary appointments, mainly at the lower grade schools, which could be confirmed or terminated. The number of pupils was 3,213 during the session of 1922–3—or 39·4 per school and 13 per form —and 354 who had passed on to practical work; that was 2,267, or 190 per cent., more than in 1913–14 (in the Old Kingdom) and 1,943 or 40 per cent. more than in 1920–1. A good test for the practical usefulness of these schools is supplied by the survey which M. Ernest Grințescu, Director of Agricultural Education, has prepared, to show the careers taken up by those who finished the various courses. The table below only refers to those schools which kept such a record:[1]

Schools	Agricultural State officials	Agricultural private functionaries	Cultivators for their own account	Changed to other occupations	Continuing their studies	Unknown	Dead
	%	%	%	%	%	%	%
1. Higher Agr. School of Bucarest (since 1862) .	35	23	11	8	2	9	12
2. Secondary Agr. School of Roman (since 1885) .	44	6	8	10	—	22	10
3. Secondary National Agr. School of Chișinău (since 1921)	22	2	4	3	18	49	1
4. Lower Agr. Schools:							
(*a*) Old Kingdom . .	38	3	9	7	7	26	9
(*b*) Transylvania . .	4	8	14	10	—	71[2]	3
(*c*) Bessarabia . . .	39	8	7	9	10	16	10
5. Elementary Agr. schools .	24	5	29	13	4	12	13
6. Schools of domestic science	10	5	50	20	10	2	3

By far the largest proportion, therefore, of the graduates from the higher and even from the lower Rumanian schools have become officials. That tendency has not ceased to operate, to judge from the figures relating to the newly-established Bessarabian schools. Hence the activity of the High Schools seems to lead mainly to an expansion of the various administrative organs, the number of teaching posts being very limited; moreover, according to Dr. G. Antipa, about 100 highly-trained Transylvanian agronoms, who before the reform had been in charge of various undertakings, are now without employment. Most of them probably belong to national minorities and are, therefore, less favoured with official positions. The percentage of graduates from the higher and lower Rumanian schools who have settled down as practical farmers on their own account is less than 10 per cent., and is equalled by the number of graduates who have altogether abandoned the agricultural profession.

[1] *Școalele de Agricultură din România*, 1925, p. 32. Percentages as in the original.

[2] The greatest number of the diploma holders had remained in Hungary, so that their present occupation could not be traced.

The Transylvanian figures show a more satisfactory distribution and M. Grinţescu suggests that the percentage of those who became practical farmers would have been found to be much higher if more of the Transylvanian graduates could have been traced. The figures relating to elementary schools are relatively even less satisfactory; their purpose was to train sons of peasants for the rational cultivation of their own smallholdings, yet fully one quarter of the graduates have become officials and 13 per cent. have abandoned agriculture.

'One of the greatest needs felt by most schools in the Old Kingdom', says M. Grinţescu, 'results from the absence of teaching material. Almost every school shows in this respect a lack of material such as maps and diagrams, collections of seed samples, simple physical and chemical apparatus, &c. There are several elementary schools and schools of domestic science which literally have none of the usual teaching material. In that respect the schools of the new provinces have little to complain of, especially the agricultural academy of Cluj and the secondary viticultural school of Chişinău, which are provided with laboratories and everything that is needed for facilitating teaching and the activities of the staff. All the lower schools of the new provinces are equally well provided.' [1]

As a result of the land reform the agricultural schools depending on the Ministry of Agriculture were provided with 6,000 ha. land, that is 73 per cent. of the total area of 8,247 ha. at their disposal. The central authorities contribute the cost of the staff and the maintenance of the students, but for the rest the schools would seem to be self-supporting. In 1923 the total working budget of all the schools was as follows:

Revenue, 9,054,432;
Expenditure, 7,087,031;
Net profit, 1,967,401.

The net income could have been higher but for the fact that part of the revenue was spent on stock and building repairs, and that part of the produce was sold below cost to the canteens of the students and personnel. 10 per cent. of the gross revenue was set aside as a reserve fund; of the net profit 10–20 per cent. was destined as a bonus to the heads of the schools, while 80–90 per cent. were distributed among the students. The insufficiency of

[1] *Şcoalele de Agricultură din România*, p. 42.

the endowment received by these schools is seen from the fact that in 1923 their working capital only amounted on an average to 966 lei or about £1 4*s*. 0*d*. per ha.

The law on agricultural schools, passed in July 1929, represents the tendency of the new Government to lay the emphasis on practical rather than on theoretical education. It places agricultural schools under the control of the Ministry of Agriculture. Pupils are not to be admitted till they are sixteen years old, when it is assumed they will have made up their minds to take up farming as a profession. The law also expresses the decentralizing policy of the Government, as it allows each school to adapt its work to the needs and character of the district. It will be seen, therefore, that the difference between Liberals and Peasantists goes as deep in educational matters as in economic policy.

Adult Education. Adult Education in Rumania had until recently implied only the attempts made, in the Army and elsewhere, to teach grown-ups how to read and write. After the War, however, adult education has extended both in scope and activity. Public authorities and private organizations are endeavouring to cover the country with a system which shall spread some knowledge of the sciences, arts, and crafts among the masses. The Ministries of Education, of Agriculture, of Labour, and of War all have a program for adult education. The law of 1924 which reorganized elementary teaching also contained provisions for the establishment of schools and courses for adults and made attendance at them compulsory. The Ministry of Education put the number of those who attended these schools in their first year of activity at 730,000; private estimates mention only 100,000. The Ministry of Agriculture has organized ambulant adult schools which supply courses in agricultural subjects consisting of 6–22 lectures, followed by practical demonstrations. The courses are held in the larger rural centres and facilities for board and lodging are provided for those who attend them.

Of the private organizations devoting themselves to popular education the oldest and most important is the Transylvanian 'Astra', founded in 1861, the full name of which is 'The Association for Rumanian Literature and Rumanian Culture'. After the

War the 'Astra' extended its activities to the other provinces of the country. In addition to special courses for illiterates the 'Astra' works mainly by means of individual lectures; it encourages and assists in the organization of study circles of which there were 608 in 1926, the number of lectures reaching 772. In the Old Kingdom the 'League for the Cultural Union of all Rumanians', founded in 1891, has a similar activity to that of the 'Astra', though on a more moderate scale. Of the post-war organizations, those which are active on the land are the so-called National Clubs and the 'Prince Carol Cultural Endowment'. The Endowment has been specially concerned with the establishment of centres in the villages from which all cultural and social activities should be carried on. In principle these centres are intended to contain lecture- and reading-rooms, and to be provided with libraries, first-aid stations, and even with land for agricultural experiments. Until now, 600 such centres have been established, though the bulk are inevitably in a rudimentary stage. A monthly publication is intended to assist them in their work and to keep them in touch with each other. Their libraries now contain about 100,000 books, and the yearly number of lectures and social evenings is about 3,000. One should also mention the successful University extension work carried on mainly in Transylvania by the Professors of Cluj University. Finally, the State obliges every village teacher to organize a certain number of lectures, song and theatrical evenings, &c., in the year; and village priests to hold every three weeks regional meetings in alternate centres, with a joint service, sermons, and other edifying proceedings.

The Official Bulletin of the Ministry of Education gave in its first issue for 1925 the report in which the district administrator of Alexandria, in the county of Teleorman, described what steps he had taken to apply the new law on adult education. The document is valuable both because it confirms the keenness of the villagers and because it throws a sidelight on the country's administrative methods. The administrator was determined that the new legal provisions should be a success in his district. Therefore he obliged every peasant to give a hundredth part of his produce, and other contributions were imposed upon publicans, millers, and upon any one who had business with the authorities. The result was that, on November 1, 1924, 29 adult schools, with 3,446 pupils, began to work in the district. Courses were held daily, between 5–7 p.m., by 64 village teachers to classes of at least 50 adults. The attendance was so strong that in certain places duplicate classes had to be arranged. Moreover, 'many peasants thought that two hours' teaching was not enough, and asked permission from the teachers to be allowed to come in the day-time also, with the children, for "it being winter they have anyhow nothing to do at home" '. This keenness of the peasants to improve their minds is undoubtedly the most striking, as well as the most solid, element in the new educational movement. It is a reproach to the former system which had left them without educational opportunities, though there is no doubt that the stimulus has come largely from the new conditions on the land; and it puts in its proper place the argument with which some people opposed the land reform, as being beyond the level of the peasants' education. The peasants are better off, and they have more freedom in the use of their time. Their spirits have been roused by the War, and their resettlement as landowners has filled them with a new sense of personal dignity. The extension of the franchise has helped to draw the village into public life. Their minds are astir, and, as we have seen, great efforts are being made by public authorities and private institutions to live up to that call for education. Some of these efforts, however, have met with criticism as to ways and means, and have roused doubts as to their fitness to the end in view.

The activity of the administrator from Alexandria, however praiseworthy in intent and achievement, shows that, for the carrying out of the new educational provisions, local officials are allowed to devise and impose an additional system of taxation not discussed or approved by Parliament. This school tax, moreover, is levied probably solely upon the peasants, and there is reason to believe that it is but one in a whole set of impositions dictated by local officials on their own authority. The case of Alexandria is by no means isolated. Further, it is widely affirmed that many of the new school buildings have been left unfinished and are rotting away, and that others are but flimsy structures destined to a short life. In the second place, the artificial basis of the movement is said to appear in the poor quality of the teaching personnel; in the feverish rush to create new schools many unfit elements have been engaged as teachers. Expansion without regard to available means has caused the Government to pay salaries which were driving the better elements away. A public appeal issued by the General Association of Teachers in 1925 stated that village teachers were receiving monthly salaries of 1,600 lei, and those with 25 years service 2,600 lei—or about £1 12*s*. 0*d*. and £2 12*s*. 0*d*. a month. 'We can say, without exaggerating, that in the measure in which the number of school buildings rose the quality of the teaching personnel fell, and bad remuneration has undermined the will to work of the good elements among the teachers.'[1]

The new Minister of Education, M. I. Costăchescu, stated in the Chamber in January 1929, that many schools had been found to work irregularly; the teachers left their posts without leave of absence, taking trips to the towns and neglecting their classes.

In the general haste to make up arrears, says Dr. G. Strat, the erection of school buildings has been confused with education. As usual, elaborate legal provisions have been devised in the abstract without relation to the possibility of giving them substance. Some critics doubt whether the peasants in their poverty —which the State has done little to mitigate—could afford to dispense for seven years with the help of their children. The severest strictures have been directed against the setting up of secondary schools in country districts. The peasants had seen how functions were continuously created for people who secured a certificate or some academic degree, and so, in their anxiety to give their children a better chance in life, they do all they can to send their sons to a secondary school. These are unable to accommodate all the applicants, while technical and commercial schools are abandoned to the minorities or to the more practical Transylvanians. The educational authorities have pandered to that misguided attitude by creating more such secondary schools, modelled on the French classical *lycée*. The consequence is that, 'especially in the absence of technical schools, the sons of

[1] N. Daşcovici, article in *Societatea de Măine*, March 1926.

peasants, that is all the coming life-blood of our nation, fall over each other in their anxiety to get into the secondary schools. The majority, of course, are left on the way and fail to reach the University. What becomes of these young men? Do they return to the village? No. Do they take up some trade? Never. They come to the town to swell the pitiful mass of applicants for official posts and of the most barren proletariat.' For, the pupil of a technical school, if he discovers in himself a gift for study, will pass to a theoretical school, but the unsuccessful pupil of a *lycée* 'will abandon it, but never enter a technical school. He becomes an official.'[1] It is the same in the Universities. The faculties of arts and law are overcrowded to a degree which leaves no room for serious teaching and learning. All these throngs of students merely crave a diploma which should open the door to some official post. In pleasant contrast are the instances to be noted among sons of the few old landed families, which form the real élite of the country and are now ruined by the expropriation; they are devoting themselves to study for the sake of scholarship, aspiring as it were to regain through intellectual prowess the leadership which their families formerly held through unearned privileges.

All improvements in agriculture, said Ion Ionescu seventy years ago, in his heartfelt patriarchal style, must begin with the improvement of the tiller himself—'of him whose function it is to receive the rays of light and to guide them towards the soil. If we first cultivate and improve the man, there is nothing which we may not thereafter improve and cultivate in our country'. Because those who ruled the national State took no notice of such sagacious advice they had in the end to hand over the nation's main patrimony, the land, to a mass of peasants who had in no way been prepared for getting the best out of their charge. Since the War, an attempt has been made to atone for that neglect, but the effort would have been more convincing if it had not been linked to a general policy which socially and economically was unfavourable to the peasants. Professor Virgil Bărbat has pointed out how differently Spiru Haret had acted. He realized that schooling could not be effective among people

[1] Professor C. Rădulescu-Motru, in the *Adevĕrul*, April 16, 1927.

who were too depressed to make use of it; his first action as Minister of Education, therefore, was to encourage village teachers to take an interest in the spreading of co-operative organizations, for the improvement of the peasants' existence. If nevertheless the educational effort of the last few years rests on solid foundations, though some of the upper-structure may be flimsy, it is because it has been demanded and supported by the people themselves. Many of the peasants' elementary material wants are still unsatisfied, yet they have readily sacrificed money and leisure on the altar of knowledge. None of the social effects of the land reform has justified that measure so immediately, or promised so well for the future, as that keen desire which the peasants displayed, as soon as their chains were loosened, to attend to the improvement of their minds.

(*b*) *Crime and Conflict.* It is a matter of endless surprise to the visitor from the civilized West to find how honesty is taken for granted among the peasants of the ill-reputed Balkans. Ion Ionescu mentions in his monographs that it was a widespread habit among landowners to store their corn in barns and lofts built out in the fields, and no one ever touched it. We have ourselves come across villages in which a number of newly resettled peasants were using a big barn, taken from a large owner, as a common store-room. Each peasant's heap of corn was piled next to another's, and there was no suggestion that this promiscuity might lead to loss through pilferage. The peasants' honesty was ingrained; not even revolutions affected it. In a series of articles on the revolution of 1848, probably written by Ion Ionescu, it was affirmed that 'during the three months of revolutionary régime in our country not a single obligation of the peasants towards landlords remained unfulfilled'.[1] Later, in the measure in which the conditions under which the peasants lived were depressed, the number of criminal acts committed by them increased. 'Criminality', said Liszt, 'is, after all, but one aspect of social life.'

With the advent of the exploiting tenants and usurers, the peasant, it would seem, adopted two moral codes: one which he applied to dealings with his fellow-villagers, the other in his relations with the large tenants, with officials and other masters.

[1] *Ţăranul Român,* November 19, 1861.

With the latter he was sly and tricky as often as possible. Because, as Professor I. Simionescu says, 'The Rumanian peasant —accused of being lazy, sly, and given to drink—is at bottom a primitive being who led a latent life of his own imagining. . . . Many of the vices imputed to him were in fact his only means of self-defence and preservation.'[1] The main causes of violence among the peasants themselves were alcoholism and, above all, land disputes. The Directorate of Gendarmerie, as we have seen, attributed 24 per cent. of the crimes committed in rural districts during 1918–23 directly to alcoholism. Before the War land disputes caused most of the quarrels and crimes. It is only natural, therefore, to find that rural crimes have decreased after the reform, and that improvement would no doubt have been still more marked if abuses in the application of the reform had not left so much bad blood behind in many villages. The general increase in criminal offences, a universal post-war phenomenon, is shown in the following table referring to the Old Kingdom and Bessarabia:

Province	Population	1914		1919		1920	
		Number	%	Number	%	Number	%
Moldavia	2,200,200	250	1·1	332	1·4	644	2·9
Muntenia .	3,200,000	256	0·7	354	1·1	589	1·8
Oltenia .	1,450,000	138	0·9	225	0·7	223	1·7
Dobrogea	650,000	39	0·5	49	0·7	111	1·7
Bessarabia	2,500,000	—	—	192	0·7	643	2·5

The next table divides the sentenced individuals according to their nationality, religion, occupation, and degree of culture:

Year		Nationality			Religion				Culture			Occupation				
	Number sentenced	Rumanians	Foreigners	Unknown	Greek Orth.	Roman Catholics	Jews	Others	Higher schooling	Literates	Illiterates	Agriculturists	Traders	Workers	Clerks and officials	Without occupation
1914	309	266	34	9	298	6	5	—	3	100	206	277	7	20	5	—
1919	404	361	42	1	385	2	16	1	1	193	210	283	9	66	18	28
1920	538	494	42	2	517	8	10	3	—	256	282	451	16	37	14	20
Increase, 1914–20												% 65	% 128	% 85	% 180	

[1] 'Townsman and Peasant' in *Economia Naţională*, vol. xviii, No. 6, 1927.

A more recent table of those sentenced for criminal offences in the Old Kingdom and Bessarabia shows them to have been distributed according to their occupation as follows:

Year	Agriculturists	Traders	Workers	Without occupation
1919	372	11	70	31
1920	656	21	57	25
1921	1,006	31	99	323
1922	1,037	42	139	120
1923	883	21	79	153 [1]

One will note the terrible increase of crime among individuals without occupation. Agriculturists made 79 per cent. of those sentenced in 1923 as against 89 per cent. in 1914; in Bessarabia the percentage decreased from 92 in 1919 to 78 in 1923. The fall is explained by the fewer number of conflicts concerning land property after the reform. And the years to which the above figures refer were those during which the land reform was being applied. The report of the Directorate of Gendarmerie on the period 1918–23 says on this point: 'Among the main causes which determine the commission of criminal offences by our rural population is the existence of a state of transition which must prevail until the completion of the land reform.'

A large number of lawsuits, besides these criminal offences, are due to the insecure state of the right of property. The absence of a ground book, of surveys and plans, or, frequently, of any other documentary evidence, makes it difficult to establish the title of ownership; hence the importance of possession in Rumanian law. Proof of possession, and of location of boundaries, depends mostly on the evidence of witnesses; and to the assertion of one set of witnesses that of another set can be indefinitely opposed. 'That state of insecurity is the cause of most of the rural lawsuits. There are lawsuits concerning possession, claims, divisions, boundaries—and their consequences: quarrels, insults, material damage, physical violence, murder.'[2] There have been at first few lawsuits between the peasants resettled in 1864, but they became more numerous with the splitting-up

[1] *Buletinul Statistic*, No. 2, 1925.

[2] Andrei Rădulescu, *Viața Juridică și Administrativă a Satelor*, pp. 474–8.

of their properties. There will be more than enough between the peasants now resettled, unless more reliable means for the establishing of property rights are introduced. M. Rădulescu, a judge of the supreme court, believes that a peasant likes nothing better than a quiet life, but that he has an ingrained sense of justice. His right is with him not merely a question of wealth, but of dignity. 'He resents not so much the material loss, as the attack on his rights and dignity.' The townsman is apt to look at everything in terms of money and to work out the possible profit and loss before he engages in a lawsuit. But to the peasant his property is a part of his being. If he thinks he is in the right, he will accept no compromise; he will rather get himself into debt than give in. 'This predisposition undoubtedly has its drawbacks, for sometimes it causes him to go too far, wasting days and weeks, spending money, poisoning his relations with other villagers. Yet it has also great value, because that state of mind has contributed to the safeguarding of property.' One might hope, therefore, to see crime and conflict diminish still further on the land when property rights are safeguarded by proper administrative instruments, and not merely by the stubbornness of the peasant owners.

(*c*) *The Changing Psychology, as seen in Religious and Social Outlook.* The anxiety of the peasants to emerge out of their slough of ignorance is part of a comprehensive psychological evolution, stimulated by the reform. They are becoming, so to speak, mentally enfranchized. They are developing a personality, as individuals and as a class; or rather their personality has been released from physical and social bondage and is claiming its freedom, especially in things of the spirit, not without some of the exaggerations which are frequently the first-fruits of suppression. 'The old leaders, who in a certain measure enjoyed the confidence of the peasantry, are most of them brushed aside to-day. The priests have lost a good deal of their former prestige and authority, the teachers almost as much. Public administration is weak and compromised by the politicians.' The peasants are not so simple as to have failed to see that most of these supposed leaders, spiritual and social, were working for their own selfish interests. Mistrust in the character of these individuals has engendered doubts in the doctrines they were expounding. 'During the last decade the minds and hearts of our peasants have undergone a far more radical transformation than in a century.'[1]

Religion. As everywhere else, the War has set free much spiritual questioning in Rumania, affecting men's attitude to religion and to the established Church. That effect is not the same in village and town. For the ruling classes in the several Balkan States the Church had been primarily an instrument in the national struggle against the Turks, and afterwards amongst themselves. The Church was a State institution, and the State was nationalist and oligarchic. Religious fervour, which in the first place would have meant communion with the sister Churches of the Greek-Orthodox creed, was impossible in such conditions. The Church had no inner life of its own, and religious devotion was exhausted with formal observances. Now however, for the first time in centuries, a religious revival is noticeable in the towns, partly due to the psychological upheaval caused by the War, partly no doubt to the fact that the States are more consolidated, and the national struggle, therefore, no longer emotionally so all-absorbing. This new current, initiated by a few of the clergy, rests on what are as yet small but high-minded groups in the younger generation, who are seeking to retrieve from under the ashes of lifeless externals the mystic fire of the Eastern creed. While in the towns, therefore, that revival of religious sentiment finds expression in a closer communion with the established Church, the revival in the villages is leading rather away from it. The Rumanian peasant has never been truly religious. His piety sprang from a 'fear of sin' and its consequences, which induced him strictly to perform all the formal prescriptions of the Church; in fact he performed many other rites, rooted in old pagan superstitions. He probably feared more than he loved the Church and its servants, because of its authority to register and to punish worldly sins, and he showed nothing but gratification when the lands of the monasteries were secularized. The ease with which that complete confiscation was carried out dealt a heavy blow to the prestige

[1] Onisifor Ghibu, article in *Societatea de Mâine,* vol. ii, No. 35, 1925.

of the Church, and the growth of the political State relegated it to a subordinate position. Internal weakness hastened that decline. After the secularization many churches were left without priests; the priesthood was neglected by the State and the quality of the incumbents continuously fell, especially as the organization of the national State was at the same time opening many new attractive fields of activity. Most village priests were crudely ignorant, many were almost as illiterate as their flock, and some, as M. Radu Rosetti, the novelist, averred, were outspoken atheists. Those to whom poverty was irksome coupled more lucrative occupations with their ministry; like the priest who set Ion Ghica musing 'on the compatibility between the calling of priest and the trade of publican'. They encouraged and defended the peasants' superstitions, which gave them an easy hold on their flock, against the intrusion of more progressive views. As a body, the rural clergy were unfit for spiritual or moral leadership. The Church was merely anchored in the pagan superstitions and beliefs which are ever present with simple people whose life is conditioned almost as much by the accidents of nature as by their own work. These superficial formal ties are apparently giving way now under the mental and spiritual searchings which are disturbing the traditional passivity of the villages. That questioning of transmitted views and beliefs has reached even the religious life of the villages, says M. Octavian Goga, 'shaking the supreme moral factor of the multitude, sometimes through cloudy mystic tendencies, at times through impatient rationalism. A hard struggle is going on around our village churches, their old walls seem to weaken, the ancient Christian ideology is suffering hasty revisions. An extraordinary crop of religious sects, with tens and hundreds of thousands of followers, has emerged all over the country during the last few years out of that moral storm, helped by the organic weakening of the Church.'[1] Leadership of the villages has passed to the abler peasants themselves, who are often preaching on behalf of one or the other sect. State and Church are inclined to regard this phenomenon as a danger to themselves, yet repression of these sects, without reform of the established Church, is likely

[1] Article on 'The Religious Problem', in *Ţara Noastră*, vol. iii, No. 11, 1922.

to prove as little effective as it did in pre-war Russia. At any rate, nothing reveals the new temper of the peasantry more strikingly than this craving for spiritual adventure, after the hollow discipline of their past religious life.

Social Psychology and Class Division. The dominant characteristic of the Rumanian village had been its unvarying stability of outlook and habits. Generations followed each other without making any change in their ways of living. The peasant's guiding line was traced by the conduct of his fathers and forefathers in similar circumstances. If he was called upon to face any fresh problem a peasant never took a decision before he saw 'what the others were doing'. All things had first to be talked over with the whole village. From this social conservatism sprang his gregarious habit of life. The peasant preferred to live in the village, at a great sacrifice of time and labour, rather than settle on his holding, a few miles away. That conservatism was in no way the result of deadness; at the worst it was a symptom of stagnation. As Professor Rădulescu-Motru finely said: 'Like the wheat in the country's neglected furrows, the soul of the peasant cannot ripen into fruit. It lies, a tired reserve, below the surface of the soil.'[1] But that dormant seed is beginning to germinate. 'A spirit of independence is running through the peasant masses,' says M. Goga, in the article already cited. 'One notices an unaccustomed buoyancy in their ways, a livelier gleam in their eyes, a ripple of boldness, and, above all, a critical temper which knocks at every gate. That nervous fluid has overflowed into the farthest hamlet, arousing many questionings. . . . Who are we? How many are we? What is it we are receiving? Who gives it us? What is our right?' That awakening began during the War. Fearful suffering, due in a large measure to bad management, coupled with the obvious dislike of facing risks shown by many members of the urban upper-class, made the peasants bitterly critical of the rulers. When some officer belonging to that class happened to fall in action, the comment of the peasant-soldiers was: 'From them one is fallen, from us one is left.' Those among them who were taken westwards as prisoners of war, and those who took part in the march on Budapest, saw towns and villages

[1] 'The Psychology of the Villages,' in *Societatea de Mâine*, vol. iv, No. 3, 1927.

which gave them a vision of a world better than that in which they lived. All this made the peasants more critical of their surroundings and of their superiors, and the grant of land and of the right to vote gave them more confidence in themselves. They are beginning to feel that at last they count for something in their country. They may still be hesitant, being doubtful of their new position, but a more independent temper is asserting itself in their attitude towards other classes. Formerly, any townsman was invariably greeted hat in hand, with 'I kiss your hands'. Now, strangers are simply passed by, people they know may get a 'good day'; while the local landlord is merely a 'sir', and no longer the *coconaș* (young boiar) of only a few years ago.

The peasants, in brief, are becoming class conscious. They are beginning to take things into their own hands. 'This is the explanation of the phenomenon, which up to a point one may welcome, that in those provinces with a better culture the part which used to be played by the intellectuals has passed to the more wideawake peasants.'[1] The general run of village teachers and priests, recruited as they were from inferior material, could command authority only as long as the State and the Church they represented imposed implicit obedience upon the peasants. On the other hand, flight from the land takes in Rumania the form of a flight not of poor labourers, but of the sons of the better-off peasants and of the village intelligentsia. Prolonged suppression, first by alien rulers and then by a nationalist oligarchy, has caused the livelier young men to aspire to escape from the plough and the village, and to become 'boiars' too, that is to join the ruling class. As official appointments had been made conditional on the possession of a degree, the possession of a degree was assumed to entitle its owners to an official appointment. Degrees in the faculties of law and arts were easy and cheap, and the political parties found it simplest to recruit armies of partisans by creating official jobs for all those degree holders. 'The Rumanian', once said the wise and witty M. Carp, 'is born a bursar, lives an official, and dies a pensioner.' The addition of the new provinces, in which the Rumanian population

[1] P. Nemoianu, article on 'The Leaders of the Village' in *Țara Noastră*, November 1925.

had been kept away from office, brought fresh water to that current after the War. One aspect of it is the top-heavy growth of the towns; the other aspect is the abandonment of the village to the care of the least qualified among those with some kind of schooling. Professionally and culturally the village is bound to suffer thereby, and socially the tendency might be dangerous, were it not for the uncommonly sound sense and balanced temper of the Rumanian peasant.

The land reform has initiated a recasting of social strata, both on the land, and as between land and town. The process is only beginning on the land, and its outcome could hardly be forecasted, as many factors might intervene to divert it from the path on which it has been started by the reform itself. The landlord class, of course, has disappeared without chance of return; if the peasants could ever be induced or forced to relinquish their stubborn clinging to their holdings it would only be in favour of some Socialistic form of ownership. Social stratification in the village, therefore, could only mean some differentiation within the mass of the peasant class. The reform itself has had a levelling effect. It has raised many landless peasants into the ranks of small owners, and it has left little arable land from which well-to-do peasants might increase their holdings. Whether that effect will last depends on many technical and social factors. No peasant would give up his holding, however small it might be, if improvement in the methods of cultivation or the development of village industries, or both, should enable him to eke out a living on his own piece of land. Without such developments, and in the absence of opportunities for work on the large estates, it is probable that some of the smallest holders will drop out, especially if towns and industry offer them an opening.[1]

[1] The Communist theoreticians, chained to their traditional Marxian premises, have no doubts whatever on the social evolution of the village. In a private letter, written in June 1927, M. S. Timov, of the International Agricultural Institute in Moscow, told a Rumanian correspondent what he thought would happen in Rumania: 'It seems to me that the agrarian reform must create an unimportant stratum of rich and well-to-do peasants, and contribute at the same time to the proletarization of the great mass of the rural population, thus facilitating the existence of the large agricultural undertakings. . . .' That process 'will bring about before long a class differentiation in the village. . . .' One cannot see how the reform is supposed to have created a class of rich peasants. It may possibly turn some peasants into proletarians, because it has done away with the large estates which helped the very small holders

So far, though bad blood has been caused by abuses in the application of the reform, no conflict divides the villagers into those with land and those without any. Those peasants, who for one reason or another have not received land under the recent reform, are confident that their turn will come in good time. Moreover, what remains of the large estates, as well as the State's model and experimental farms, still offers some outlet for labour. As long as these reserves remain, the landless peasants will fasten their hopes and claims upon them. 'The State farm serves as a lightning-conductor for local discontent,' says M. Henri Stahl in his notes on Rușeți. The only thing the peasants can see in it is that it withholds from them land they need, for purposes which they regard as trivial; for, said one of them, 'the horse has oats, while man has no wheat.' It is curious to note that the powers which ruled in Rumania during the last decade imposed upon the village a social stratification which exactly corresponds to the Communist standpoint. The constitution of the Agricultural Chambers, which were conceived as purely professional bodies, accorded right of membership only to persons owning more than 3 ha. land. The smaller peasants were thrown pell-mell into the Labour Chambers with industrial workers and artisans, thus being more or less officially classified as proletarians. The arrangement smacks of an intention to divide the village professionally and politically. The new Agrarian League is attempting just the contrary.

If the disappearance of the landed class has left the village a more compact unit than it was before, it has conversely taken away with it the main link between town and village. The old social separation between the two has been deepened by a clear-cut difference of professional interests. Town and village were not only two separate worlds, but two worlds almost alien to each other. Except the language they had hardly anything in common in their lives. Within the last decade the relations between

to keep going. But if the disappearance of the large estates were to have such an effect on some peasant families, in what way could that help the large estates —which are gone—to exist? The only way of making any sense of that statement is to regard it as a 'hopeful wish'. What M. Timov probably had in mind was that when the peasants had become proletarians the turn would come for the large-scale Socialistic farms, cultivated by 'armies of labourers', in the words of the Communist Manifesto. This, of course, would be in accordance with sound Marxian dialectics.

town and village have been thoroughly transformed. The widening of political life has brought the town politician to the village, and the new economic organization has sent the peasant to town. After the reform, the trading activities of the peasants have greatly expanded, both in the way of buying and selling. These more intense trading relations have not so far produced a better understanding between the two sections of the population. The peasant is apt to consider the townsman merely a consumer, not a producer, as a parasitic factor which lives from extortionate prices and profits; especially as during the period under review the better organized traders and bankers also received preferential treatment in the State.

Nevertheless, the peasant has no sense for social revolution. If he revolts, it is solely against abuses. He never thinks of changing his status except through legal means, and he is anxious to make due compensation for any boon he may receive. We have seen how on every occasion when the peasants had an opportunity of stating their claims to the land, they invariably offered more than fair compensation for any privileges which the landlords might have had to forgo, although those privileges were built on flagrant abuses. And even in the revolutionary atmosphere of the post-war years—to quote from our own experience, in the whole of the Balkans, not only in Rumania—we have never once heard peasants suggest that they ought to get the land without having to pay for it. In the same way they had no thoughts for political revolution. Their quarrel was with their direct tormentors, landlords and local officials, while they looked upon King and Central Government rather as a court of appeal, as it had been during the rule of the old Princes; though the peasants' chances of being heard had sorely lessened from the time when Mihaiu Șutzu set aside Monday of each week as a day on which he received only peasant petitioners. Modern democratic methods require the peasants to send in stamped petitions for any claim or complaint, which are more often lost than solved in the endless compartments of the bureaucratic hierarchy. The new reforms having brought the peasants into more direct contact with the machinery of State, it is also probable that discontent will vent itself more directly against the central authorities.

Various circumstances have in fact contributed during the last decade to harden the peasants' old political mistrust: the miscarriages in the application of the land reform; the great extension of the corps of gendarmerie, and its repeated use for the frustration of the peasants' new right to vote; the demagogic courting of the village by politicians violently abusing their opponents and outbidding each other in promises which seldom come to fruition.

It cannot be long, however, before the peasant masses will be in a position to see clearly what they want, and to insist on getting it. Their general standpoint was outlined by M. Mihalache in his great speech on the reform, in 1921. To him all reforms could only have one goal: to raise the peasants out of their morass of physical and moral misery—a misery more complete in reality than that illustrated by the few facts given in this chapter. He refused to consider the land reform primarily from the angle of production. 'We cannot look merely at the surplus in the balance of exports, but must also take into account the deficit it causes within the country. For if the surplus of several thousand wagons which large property gave for export was obtained at the price of ruining the health of the mass of peasant labourers, of their being kept in ignorance and poverty—then, gentlemen, we prefer a thousand times the health and vigour of the people to the surplus wagons for export.'[1] Nor were they willing any longer to trust the accomplishment of such improvements to the fairness and public spirit of other classes. The only hope lay in the redistribution of political power, as the ablest Peasantist thinkers had realized long ago. After the rising of 1907, M. Sebastian Moruzzi admitted that 'our society had been constituted in such a way that the few who governed had always interests differing from those of the many who were governed'. As a consequence the various legislative reforms, pushed through by a few generous spirits, had in practice remained a dead letter. M. Moruzzi, therefore, appealed to all the parties to join hands for the purpose of saving the peasant. In reviewing the pamphlet M. C. Stere replied that nothing was to be expected from an agreement among groups who in one way or another were all interested in keeping

[1] Speech in the Chamber, 1921, p. 35.

the peasant under. A more logical deduction from the premises established by M. Moruzzi himself was that nothing could offer a serious guarantee for genuine social reform except 'a reform of the franchise, which would shift the centre of political gravity towards the "many who are governed", and destroy the very basis of the actual political system'.[1]

Section 5

The Political Effects of the Reform

A victorious war usually gave the people and the class who conducted it a stronger hold on power than they had before. Authority, and those who represented it, acquired a fresh reputation for strength and ability. But in eastern Europe that habitual trend was reversed after the last War. In that region power passed not to the bureaucratic-nationalist upper class, which had ruled till then, but to the mass of peasantry. The semi-autocratic feudal structure of the eastern States proved economically and socially incapable of standing the strain of a modern war. The mighty Tsarist empire crumpled up like an empty egg-shell. Rumania might have suffered the same fate had not two-thirds of the country fallen speedily into the hands of the enemy and the other third been arrested in passive self-defence; and had the rulers, above all, not hastened of their own accord to surrender land and power to the peasants. The dictatorships of the Right or of the Left, open or veiled, which have since then pullulated in that region, disclose how unwillingly those voluntary reforms were granted.

'Land and Liberty!' In the minds of the eastern peasants the two ideas have at all times been inseparable. The peasants seem intuitively to remember that in the not very distant past possession of land had been the condition for the enjoyment of personal liberty, the loss of the first bringing with it the loss of the second. The two notions have become identified in their sentiments, an association which goes a long way to explain why peasants are filled with an overpowering desire to possess a strip of land of their own. Nor does that desire lead them astray in

[1] *Viața Românească*, October 1907.

the walks of our modern social organization. Land without liberty or liberty without land would indeed be no more than half a freedom. The Rumanian ruling-class could afford to put off the clamour of the peasants with occasional grants of land, as long as it kept control of power. The liberal provisions of the Constitution and the many reform laws may have been well meant by the few idealists who proposed them, but they always worked crookedly in the hands of those who applied them. Occasionally a flash of repentance would cause one or the other of the rulers to confess that their professions were being denied by their practice, but failings admitted at the confessional have seldom been known to reform history. In reality the Rumanian peasants have been kept in serfdom right up to the Great War. There was no possibility of reforming that state of things by means of Constitutional action, for the State was ruled by the interested landed class. Nor could it be ended by revolutionary action, because the sufferers themselves were morally and physically exhausted; and also because Rumania was wedged in between two reactionary empires who would not have tolerated a revolutionary movement at their frontiers. Hints in that sense—not an ultimatum, as has been suggested—may have been partly responsible for the frantic repression of the rising of 1907. That geographical situation goes a long way to explain why the peasants submitted to ill-treatment before the War, as well as the resigned surrender of the ruling class in 1917. The only hope for a change, in the conditions prevailing before the War, lay in opposing the new plutocracy to the old landed class. That consideration induced the small but able Socialist group to join the Liberal Party, early in the century; and no doubt it was due to their influence that many of the well-sounding reform laws were passed before the War. But what advance could a few administrative texts achieve when the whole State and its attendant offices, like the economy of a feudal manor, were maintained from the tithes and corvées of the peasants?

All the conditions on which that structure rested were swept away by the War. The rulers lost their prestige, and the ruled their patience. Revolution swallowed up autocracy and feudalism in the neighbouring empires. Rumania itself was enlarged with three new provinces, not conquered by feats of arms, but detached from Russia and Austria-Hungary by their own revolutionary action. In all three of them the peasantry was more class-conscious than in the Old Kingdom, and formed the backbone of the dominant parties. As the Old Kingdom is the centre of political action, however, it was inevitable that there also the peasantry should be actively organized, so that it could defend and consolidate the gains of War and Revolution. For, under the new conditions, the struggle is not a mere issue between parties disagreeing as to whether certain ends, which they all accept as desirable or unavoidable, should be approached warily or in haste, by conservative concessions or by a radical sweeping of the boards. In Rumania the conflict reaches down to the bedrock of social outlook and economic interest. The peasants are divided from the other groups by the many discordant traits of outlook and aim which separate a class of frugal manual workers from the more self-indulgent middle and upper layers of our type of society; by all that is mutually jarring in the ways of countryman and townsman; and also—the peasants now controlling agriculture—by the clash of policy between agrarianism and mercantilism. Almost the only political sentiment which these classes share in common is their determined belief in the institution of private property. But, even there, private property as conceived by the small peasant holder is a vastly different thing from the private property to which the banker and manufacturer aspire. The peasant believes that land should be distributed to all, according to the personal needs of each household; the capitalist believes that wealth is there to be illimitably accumulated by each man, according to his ability to prevent others from getting an equal share of it. From which it is clear that in creating several million of new private owners the recent agrarian reforms have not generated a multitude of petty capitalists, as the Marxian doctrinaires chose to maintain. By extending the system of smallholdings the Russian Revolution and its corollary movements have, in fact, like the French Revolution, checked the progress of large-scale capitalist property and production on the land.

The main point lies in the different political background of the two movements. The French Revolution gave the peasants a great deal, but it did not give them political influence. Power remained in the hands of the urban middle class, who proceeded to fashion a society to its liking. In that respect the prospect is now radically different in eastern Europe. The real centre of political power is the village; and, moreover, the village knows something, and is rapidly learning more, of what it wants and can do. The break-up of feudalism in the West opened the way for the coming of the industrial-capitalist 'acquisitive' society. What will be the nature of the society which the break-up of feudalism, under conditions which deny the succession to capitalism, will bring forth in the eastern half of Europe?

The Time and Temper of the Nationalist Middle Class. The rise of the peasants from serfdom to mastery, without any violent destruction of their opponents, can only be explained by a peculiarity in Rumania's social structure. We have described in the introductory part how the Rumanian people entered upon political independence without a national middle class. There was at one end the mass of the servile peasants, and at the other end, far removed, the small class of privileged landlords, with a sprinkling of men in liberal professions. The country's economic organization was still rudimentary; articles of everyday need were manufactured by artisans, and the very limited trade —the peasants being well-nigh self-sufficient—was in the hands of foreign merchants. That patriarchal way of life, together with the intrusion of an alien administration over a period of several centuries, had left no room for the growth of a national middle class.

With the creation of the national State the former obstacles to economic and social evolution were removed, and the country appeared to make good use of its new opportunities. Public services were organized with a will and endowed in regard to personnel on a generous scale. The towns expanded rapidly, and the demand for manufactured goods rose in proportion. Nevertheless, social evolution retained some of its former stultified traits. The change was so sudden, and the ambition to emulate the organization of the western States so keen, that every one who had any schooling whatsoever was quickly absorbed in the machinery of the new State. Together with the traditional repugnance of landed people for mercantile occupations, and the absence of a money and labour market, that circumstance caused the supply of manufactures to be left to foreign producers, and the role of intermediaries to be abandoned as before to the country's foreign inhabitants. Public offices and liberal professions, on the other hand, were made the closed preserve of the few educated Rumanian elements, after the style of the old boiar offices. In the early stages the continuation of a system of privileges was no doubt inspired by jealous anxiety for the unhampered progress of the new State. It was only natural that after a protracted struggle for national independence the leaders should be suspicious and circumspect, especially as the neighbouring autocratic empires did not even then cease their disintegrating intrigues. But after a short time, the privilege to serve was turned into a privilege to enjoy. The people devoting themselves to a public career became, under the guise of an exclusive patriotism, a political brotherhood—neither more principled nor more zealous, and certainly not more indulgent towards the peasants than the monastic brotherhoods of the Middle Ages. As this bureaucratic middle class produced nothing and consumed everything that entered the public purse, the working of the system depended on their retaining undivided control of the political machine. Civic rights were denied to the foreign section of the population, which was the only one to perform the economic function of a middle class, while the mass of the real people was deprived of social freedom. In that way all competition for power in the State was effectively dammed up. The few elements which were able gradually through good luck or keen wits to break through their social disabilities and rise from below, were initiated into the mysteries of the bureaucratic brotherhood and quickly absorbed into it. In the words of Mr. and Mrs. Hammond, 'Selection and assimilation, as de Tocqueville saw, and not exclusion, are the true means of preserving a class monopoly of power.'

The national middle class which was thus evolving generation by generation bore a very different character from that of its Western counterpart. In both regions the middle class was the expression of the urban section of the population, in the East if anything more so than in the West, because the break between the towns and the feudally ordered country-side was so very abrupt. The eastern class, too, represented a popular reaction against the former privileged rule of the few great landlords. But here the similarity with the West ended. The Rumanian middle-class movement was a purely political expression. It did not rise up with an industrial revolution, or even with a capitalist reorganization of agriculture—not, in short, on a wave of changed economic and social outlook and requirements, but as a nationalist opposition to foreign domination. The only interest of that middle class, once the independence of the country was assured, was to curtail the political privileges of the boiars, so that power should be vested in themselves. They were bent upon organizing not an economic system but a political one, not an industry but a State; and what in consequence they required was not free labour and a free market, but an easy flow of revenue. This need harmonized with the anxiety of that new class to ward off political competition from the wider masses of the people, and they showed therefore no haste to reform the social organization of the country. The *sui generis* servile system instituted by the laws on agricultural contracts was devised and flourished under the nominal reign of the new Constitutional liberties: 'That period of slavery, during which the peasants were utterly exhausted, was just the golden epoch, the time of glory for the Liberal Party'—the party of the nationalist middle class. 'That was the time', said M. Mihalache, 'when the new bourgeois class came into being, with the motto: "Get rich".' 'Get rich' from the public purse, that is. For the way of the Liberals was to cause 'the middle class, instead of trying to work and prosper, to throw itself upon the powers of the State, in order to govern.'[1]

Moreover, besides having its roots in political rather than in economic furrows, the new middle class was not socially an unalloyed layer of urban society. Though its members spent their life and their income in the towns, the great majority of the political leaders derived their incomes from the land; and

[1] Mihail Eminescu, *Scrieri Politice*, reprinted 1914, p. 543.

the atavistic attraction of the land was so powerful that it was every Rumanian's ambition to own a *moşie*, an estate, if only of a hundred acres or so. These estates, large or small, were invariably worked by the peasants under the system of agricultural contracts, which, as we have seen, brought in a safe return without requiring either capital investments or personal management from the landowners. Even that small section of the middle class which later began to take an interest in industrial undertakings probably, in most cases, invested in them a surplus obtained from neo-servile agriculture. Hence there was no clear-cut division between the landed class and, at least, the upper layer of the new middle class. One could have counted on one's fingers the public men who did not belong, or who were not related, to one of the landed families. The two classes overlapped extensively in their membership, and therefore in their outlook and interests. Directly or indirectly the private material interests of the middle-class leaders were intertwined with their interests as landowners; and, in the second place, their interests as private individuals coincided with their interests as public men. Because, in the absence of any other sources of production, the State, of which they were the privileged beneficiaries, was supported almost solely from the contributions of the peasants. Nor was that burden shifted in any appreciable measure from the shoulders of the peasants when later on the middle class began to foster the growth of industry. That policy was likewise inspired by nationalist rather than by economic motives, and, in consequence, the State distributed more favours to the new industries than it received benefits from them. It will be remembered that even after the War one of the arguments used by those who opposed the land reform was that agriculture alone, run on the old basis, was in a position to give a surplus for the maintenance of the State. So that, to sum up, a whole constellation of circumstances mitigated against the release of the peasants from their bonds. As the middle class developed from political roots, having no interest in the release of a labour supply or in the creation of a greater demand for goods, but every interest in avoiding the setting free of fresh political forces, and as materially they were, both as private individuals and as public men, still dependent on extracting from the land an excess of revenue without capital expenditure or personal effort—what, in such conditions, could have induced the new middle class to free the peasants from the galleys of the nationalist-bourgeois State?

Having begun by cold-shouldering the emancipation of the peasants in 1864, the nationalist leaders only gave a thought to land reform when the nationalist State of their creation was passing through some crisis, due to dangers of war or revolution. Social reform, so to speak, was an emergency item in the policy of the nationalist middle class. The proof of that lies in the fate of the sporadic reforms, enacted under the pressure of some crisis, the advantages of which were subsequently and in almost every case cancelled or reduced by subversive laws or desultory application. The latest reforms, the unmistakable children of the War and of the Russian Revolution, have proved no exception to that rule. We have in Chapter X given a sketch of the many obstacles which were placed in the way of agriculture during the difficult period of its transformation, after the War, and we have seen how, especially, all the favours of the State were one-sidedly diverted towards industry and finance. For the latest reform at last forced the nationalist class to seek an economic foundation for its existence in the traditional middle-class fields of activity. Yet, characteristically enough, it did not attempt to gain a footing in those fields by private initiative and open competition, under the aegis of the classical *laissez-faire* principles, but, on the contrary, through the old methods of State intervention, carried to extreme lengths. 'We have a privileged leading class', said a Peasantist writer, 'which no longer draws its revenue from land properties, as under the old régime, but from the modernization of economic life, from capital investments, credit, and international exchange. But this it does through the State and with the help of the State, just as under the old régime. . . . The land monopoly has changed into a monopoly of credit and of political power: the rest is a mere parody. Far from representing the western Liberal conceptions, Rumanian Liberalism . . . is actually obliged to impede the growth of a bourgeoisie, in the real sense of the term.'

As soon as the land had to be abandoned to the peasants,

indeed, the interference of the State in the country's economic life increased to lengths reached nowhere hitherto except in Soviet Russia. Prices, trade, imports and exports, foreign credits, and, to a certain extent, banking, were all placed under State control. And when in 1923 that policy culminated in the nationalization of the country's mineral wealth, one had the strange spectacle of a nationalist and not otherwise progressive middle class rushing headlong into the sphere of Socialism. True to its character and tradition, that class was not concerned with opening new sources of production, but rather with securing new sources of revenue for the State which it controlled and exploited. It did not really matter to the ruling-class if they lost control of agriculture as long as they could gain control of some other pivotal source of revenue. While the system thus evolved into a vast enterprise of bureaucratic parasitism, it is probable that not all of those interested looked upon it merely as a means for personal or party profit. There was at least candid conviction, if not wisdom, behind the nepotism of the best among the nationalist leaders. They firmly believed that political independence could only rest securely on industrial and financial self-sufficiency. Any attempt to give the State a truly modern organization, wrote M. Corteanu, by allowing free play to private initiative and opening the doors to the civilizing influence of foreign capital and enterprise, 'is in the eyes of the Liberals an attempt against the sacred national traditions.'[1] Before the foreign capitalist could be safely let in, the country had to possess a national moneyed class of its own; and as the nationalists were impatient to display a neatly-trimmed and well-armed State, no mere accumulation of hard-won savings year by year would have answered their purpose. The nationalist class, the guardians of the nationalist State, had somehow to be enriched quickly. Whence could such wealth be extracted? In an apology for nationalist economics M. Zeletin wrote a few years ago that it was easy for the English and the French, who had all their colonies to exploit, to produce a rich middle class. But Rumania had no colonies, and therefore the ruling class was driven to exploit its own people. The people had to be bled for the good

[1] *Tendinţele de evoluţie ale clasei Ţărăneşti*, 1926.

of the State. The toll taken from the peasants' social freedom was a levy for nationalist independence.

The Peasant Revival. Politically a middle class acts as a buffer between the two social extremes, softening the asperity of their assaults upon each other. The absence of a genuine middle class in Rumania had enabled the landowners to continue enslaving the peasants even under the régime of Constitutional liberties. But for the same reason the landed class found itself defenceless when the War threw the peasant masses into ferment. Faced with the spectre of revolution, the ruling class had to surrender everything, just as before it had grasped everything. In the words of Marx, 'the expropriators are expropriated.' Because of its ambition to rule, and to rule alone, that class had neither taken the trouble to organize itself into a manufacturing class, nor had it allowed the producing urban middle class, mostly of foreign origin, to acquire political status; it made no effort to organize agriculture itself, nor did it encourage the growth of a rural middle class from among the peasants. When the crisis came, therefore, there was between them and the mass of poor peasants no reserve having an interest in maintaining the exisiting social order, or the influence to do so. The *débâcle* of the State and of those who ruled it left a political vacuum, through which the only other social class, the peasantry, rose automatically, so to speak, to the surface of political life again. The very backwardness of economic and social policy in the nationalist State had indeed helped to conserve the peasantry as a class. The neo-servile agrarian system had sorely exploited the working peasant, but a more progressive capitalist system might have destroyed him altogether. As it was, the crisis in the nation's history found the Rumanian peasant in the condition in which Mr. and Mrs. Hammond have described the English peasant to have been before the enclosures: 'Standing in rags, but standing on his feet.'

The warped progress of the Rumanian nation had held up the rise of a middle class until our own times, when Constitutional rights could no longer be withheld from the mass of the people. But the letter of the Constitution never materialized into political liberty and social equality. It was the misfortune of the peasants to secure their emancipation just when the nationalist middle class had freshly come into power, and could therefore feel no inclination towards allowing any competitor to grow up against it. The coming of government by the people, therefore, in no way meant government for the people. After 1864 the burdens of the peasants increased; the screen of law and order only seemed to leave them more helplessly isolated from the seat of power; the establishment of a systematic administration only seemed to increase the numbers of exploiters. As Mihail Eminescu, Rumania's greatest poet, once said, 'the boiars at least had the merit of being few—about one to 30,000 inhabitants—and limited in their needs. But there is no limit to the numbers and voracity of the "Liberal" bourgeoisie.' The same circumstances have produced the same effects after the latest reform. The two great reforms, contends M. Mihail Șerban, have in truth, on both occasions, only changed the peasants' masters. In 1864 they escaped from the feudal landlords only to fall into the hands of the new landowners and tenants; in 1918 they escaped that agrarian-social dictatorship only to fall victim to exploitation by finance and the whole mercantile system. In fact, they now paid two tithes instead of one: one they paid to the bankers in the form of usurious interests, the other to the State in the form of export and other taxes. In 1864 their land was freed and their labour conscripted; after 1918 their labour was freed but their produce conscripted. 'It is true', admits M. Mihalache, 'that to-day the peasants are no longer the serfs of the great proprietors; that is the only evident progress, and a merely formal one at that. For serfdom has not disappeared, it has merely changed its aspect: now it is serfdom to the banking trust, which dictates the conditions on which the produce is sold. Free labour, but taxed and coerced trading—that is the modern method of serfdom.'[1] Never, perhaps, have the peasants been so enraged by the rule of the landlords, under whom, after all, generations of them had been accustomed to labour, as they appear to be now under the economic pressure of urban trade and finance. The one was an old familiar evil, bound up with the land; the other is a new evil sprung upon them

[1] *Noul Regim Agrar*, 1925, p. 4.

from the mistrusted town, just when they had rid themselves of the first. Even in the sober review published by members of the Cluj University, popular indignation bursts through, in verses which at the end snap with the writer's pent-up feelings:

> 'The scurvy and craven *ciocoiu*[1]
> Has moved from the manor into a bank with steel doors. . . .
>
> . . . .
>
> 'Whatever the plough draws from the rich soil,
> The *ciocoiu* seizes with his knavish tricks. . . .
>
> . . . .
>
> 'One fine day he will rob you even of roads and sky:
> The *ciocoiu*,
> The usurer,
> The banker. . . .[2]

There was, however, a vast difference between 1864 and 1918. If the nationalist ruling groups were now driven to acquire the economic attributes of a class, it was because the land reform and the War, together with the coming of reinforcements from the new provinces, had shaped the peasants also, into a conscious social class. 'The thundering of the guns', said M. Mihalache, 'was too fierce not to shatter the film which covered the eyes of the many thousands who had been in the trenches, and to let them see light in its own true clarity.'[3] The attempt to repeat the feat of 1864—to nullify the effects of a measure which it was not possible to refuse—had now to face grim and organized opposition. The peasants who had the land in their hands, and who had seen pomp and power collapse in brittle ineffectiveness all around them, were no longer in a mood to lie low under abuse. Lenin himself, in announcing the new economic policy, admitted that 'to use force towards the peasant middle class would be the worst thing we could do. A class which contains so many million people must be treated with consideration.' And a Rumanian writer, echoing that sentiment, declared that 'it was possible to stamp one's foot

[1] The Rumanian name for a voracious species of ravens. Applied in earlier times to the servants who collected taxes on behalf of landlords, &c.; and, later, as a nick-name, to the large tenants, to officials, and in general to the upstart members of the landowning and ruling class.

[2] A. Cotruş, in *Societatea de Măine*, November 1, 1928.

[3] Speech in the Chamber, 1920, p. 36.

when it was a question of two or three thousand large owners, but the several million of new small owners must be faced hat in hand'.

When even the Communist dictatorship has to acknowledge the latent political strength of the peasants, they must be counted as a correspondingly strong but active political force in Rumania, where together with the land they have acquired universal suffrage. Throughout the country's recent history the position of the peasants with regard to land rights was closely parallelled by their position as regards political rights. During the long period of Turkish domination political life, if it could be called thus, was limited to the intrigues of the great boiars for or against the ephemeral occupants of the throne. Settled political rights could not exist, for either lord or villein, in the absence of an established system of government. Only with the return to national autonomy were the two Rumanian provinces endowed with a body of constitutional law; and the Organic Statutes, which consecrated the boiars as lords of the land, also set them up as the sole holders of political power. Nor did the Paris convention do more than dilute that monopoly of privilege, linking up political rights with landed property. 'Land property', said Ion Ionescu, 'was everything, and man, work, and intelligence nothing.' The first national Parliament was elected by 3,796 owners of land. For a moment, after his *coup d'état*, Cuza transferred power to the mass of the people, by means of universal suffrage. But though the first Rumanian Constitution (1866) recognized political rights to all sections of the population, the electoral system turned that good principle into a queer practice. The detailed working of the system was described by Ion Ionescu in his monograph on the effects of the agrarian reform in Mehedinți.[1] In that county the electors were divided into four colleges: 1. The first college included the large landowners, with more than 300 ducats yearly revenue; it consisted of 31 electors, paying together 18,397 lei in annual taxes, who elected one deputy. 2. The second college consisted of 69 electors with less than 300 ducats yearly revenue each, paying together 9,192 lei taxes. They, too, elected one deputy. 3. The third college included the

[1] *Județul Mehedinți*, pp. 190–2.

qualified urban electors, 243 in all, who elected two deputies. 4. Finally, the fourth college consisted of 467 electors, delegated by the mass of the urban and rural population (one delegate for fifty families), who elected one deputy. The taxes paid by this college amounted to one million lei annually. In 1884 the franchise qualifications were altered and the number of colleges was reduced to three, but the change only benefited the mass of the urban population. Moreover, even within these limits the system was never allowed to work fairly. In reality the peasant masses were virtually disfranchised, for they were never able to secure the election of deputies from among themselves, as they had in the divans *ad hoc*. The absence from Parliament of even one single peasant deputy was bewailed by Ion Ionescu in connexion with the first Constitutional Parliament of 1866; and we have seen that this was still the same in 1917, when the great Constitutional land and franchise reforms were debated at Jassy.[1]

[1] Professor N. Basilescu gave on pp. 46–8 of his book the following outline of the principles and working of the old electoral system:

The Rumanian electors were divided for the *Chamber* into three classes or colleges according to the fiscal census.

The first college included those who had a yearly income of at least 1,200 lei from real estate; the second college, those who resided in the towns and paid at least 20 lei yearly in direct taxes; and the third college those who did not come within the scope of the first two.

The electors of the third college did not, however, vote all of them directly. Those who had a yearly income of 300 lei from real estate or paid a rental of 1,000 lei yearly, as well as teachers and priests, voted directly. The others, the mass, voted indirectly, each fifty of them electing one delegate who voted with the direct voters.

In the second place, the first college elected 77 deputies, the second 72, while the third only had 40. The official statistics relating to the elections of 1911 gave the number of first college electors on the register as 15,301, of the second college as 33,270 and of the third college as 52,758 direct and 976,638 indirect voters. Consequently, 1,029,406 electors were represented by 40 deputies and the 48,571 electors of the first and second colleges by 149 deputies.

The million peasant citizens had no voice at all in the election of the *Senate*, which was elected by the large landowners and by the towns. The *Senate* had equal powers with the *Chamber*, the right to veto any bill, and to upset any Government which did not enjoy its confidence. 'It would be idle to believe that the 40 deputies of the third college really represented the peasants. All the parties were alike in having never allowed that to happen.' Pressure was used to prevent teachers and priests, who were public officials, from voting, if they were suspected of independent views. 'Moreover, the election of delegates almost never takes place in fact. The village mayor appoints some of his minions, or party agents indicated by the prefect: on the day of the poll these men will vote as ordered by the Government. To make still more sure of the result, the Government confiscates the cards of suspected electors and makes its own agents vote with them. All the electoral operations are nothing but a fraud from beginning to end. Most of the 40 deputies supposed to have been chosen by the peasants are in reality elected by the Government.'

Elsewhere the introduction of representative government and the gradual widening of the franchise have included in their scope urban and rural citizens alike. The limiting qualifications were based on various individual and social criteria, but nowhere on a sectional differentiation between town and country except in Rumania (and now, to the same effect, in Soviet Russia). After the second Balkan War, when the Liberal Party raised the question of land reform, they also advocated a reform of the franchise. The intention apparently was to give the vote to all who could read and write, voting together in a single college. The arrangement would have meant a great improvement on the one in force at the time, but it would still have left the rural population at a disadvantage. In any case franchise reform, like land reform, was allowed to be dormant, till the outbreak of the Russian Revolution induced the passing of a hasty radical measure. Together with the expropriation of the land, in 1917 the Iassy Parliament voted the expropriation, so to speak, of the political privileges which the upper class, and to some extent the urban population, had enjoyed till then. An amendment to the Constitution established the principle of universal franchise, with proportional representation, as part of the country's fundamental law.[1]

Like the land reform, the electoral reform passed from one extreme to another, from a retrograde three-class franchise, on the Prussian model, to full universal suffrage. And, in its case also, the general principle alone was adopted at Iassy in 1917, the detailed provisions for its application remaining to be enacted by means of a special law, after the end of the War. The subsequent evolution of the franchise reform has served the critics as a pointed commentary on the spirit which lay behind the Iassy decisions. It was taken for granted that the mass of the peasants, who were determined to get the right to land, which they had already possessed, had much less understanding for the importance of the right to vote, which they had never exercised. So convinced were the ruling-class that once the peasant got the first he would not trouble about the second, that when the final

[1] The so-called Labour group advocated its extension to women also, but found no support for this view.

land law was passed, in 1921, M. Argentoianu exclaimed: 'And now the Peasant Party may rest in peace!' Unlike the land reform, therefore, which was carried through immediately after the Armistice, the final suffrage arrangements were delayed for nearly five years. The new unified Constitution was passed in 1923, and its provisions bore distinct traces of the change of temper which had intervened during that delay. In 1917 the Iassy Assembly had voted for 'universal, direct, equal, and compulsory suffrage, with proportional representation'. 'Proportional' became in 1923 'minority representation' (Art. 64), without any further elucidation of what the latter meant. The subsequent electoral law provided in fact that a party securing 40 per cent. of all the votes cast in a general election should fill two-thirds of the seats in the Chamber. Moreover, the political influence of the new universal voter was side-tracked by the provisions laid down concerning the composition of the second Chamber. Article 67 of the 1923 Constitution maintained in full the legislative powers of the Senate, equal in every respect with those of the Chamber, but it took the election of Senators to a large extent out of the hands of the electorate. Part of the Senators were to be elected, the others were to sit by right. Of the first, some were to be elected by the voters of forty years of age and above, and the others by various professional bodies. The qualifications of the members by right were such that, at least for a generation or two, the bulk of these Senators will inevitably be derived from among the groups who had ruled the country hitherto. Because of these and other provisions none of the Opposition Parties took part in the discussion or voting of the measure; holding also that the Parliament of 1923 had no Constitutional powers. They left the Liberal Party to bear the whole responsibility for the new Constitution.[1]

Hence the curtailment of the rights formally granted to the peasants in 1917 was pursued in the political field more brazenly than in the economic field, where it had to be done in a roundabout way, through the oppression of agriculture. Moreover, during the first ten years after the War the country was in fact

[1] For a detailed criticism of the 1923 Constitution see my article in *The Journal of Comparative Legislation and International Law*, vol. vi, Part I, February 1924.

ruled by a veiled dictatorship of the Liberal Party. The electoral machine was tampered with to a degree which went far even for Rumania; administrative interference with the exercise of political rights kept pace, that is, with the legislative extension of those rights. Professor Iorga's remark that between 1866 and 1917 Rumania was really without a Constitution applies with equal truth until the end of 1928. One might say, by way of comparison, that England had a working Constitution which was not written, and Rumania a written Constitution which was not working. The experience of those ten years proved that notwithstanding War and Revolution what M. Stere predicted in 1907 still held good—that no reform would work until the political system was itself reformed and until its spirit conformed to the letter of the country's democratic laws. Far from having made the Peasant Party superfluous, they showed that the peasant revival would not bear palpable fruit until the peasants were organized for picking it. The land reform had not solved the social problem. It had put into the peasants' hands a powerful instrument, but they could not use it to advantage without the concurrence of benevolent government. 'What matters from the standpoint of the Peasant Party', exclaimed a Peasant deputy, 'is not a little more land, or other such benefits, for the peasants, but the assurance of liberty, of culture, of justice which would follow from the governing of the country by those who labour with their hands and with their brains.'[1]

The Peasant Movement. The Peasant Party was founded immediately after the War by a group of village teachers and priests, together with a few progressive intellectuals from the towns. The appearance of such a Party was not unnatural when the whole political life was in the melting-pot, and the land and franchise reforms seemed to give the peasants a dominant role in the State. But while the organization of the Party was spontaneous, the doctrine on which it rested was of old standing. It has been represented for many years by the Populist[2] current led by M. C. Stere, and grouped round the able review *Viața Românescă*, of Iassy.

[1] M. V. V. Haneș in the Chamber. *Aurora*, December 10, 1926.

[2] In Rumania *Poporanist*, from *popor* = people.

The Populist current, of which the Rumanian Poporanist current was a section, was in the agrarian countries of eastern Europe the counterpart of the Labour movement in the West. The Industrial Revolution had raised economic and social issues which had thrown the masses into ferment everywhere. 'While the democratic virus was naturally most active among the compact industrial masses, the slower moving country-side could not remain altogether untouched by it. Modern education and means of intercourse were bringing the great public controversies into the village.' The loss of land the peasants suffered, through the enclosures in England and at the time of their emancipation in other European countries, together with the ruinous agricultural crisis in the second half of the century, 'made the village ripe for a stimulus that would push it into line among the contending popular forces.' In the great struggle against feudalism, burghers and peasants had been accustomed to fight side by side, and in 1848 the towns generally found the peasants ready to help them. More and more, however, the growing division of labour was differentiating between the interests of town and land. In the measure in which the several fields of production were becoming specialized, the agrarian countries of eastern Europe appeared to follow a different social trend from the industrial West; and, in the latter, the urban working-class appeared bound towards a goal other than that of which the village population dreamt. For that reason the early leaders of the Labour movement made no effort to win understanding and support from the peasant. They feared that he would be a hindrance rather than a help in their revolutionary march; and they were convinced that large-scale production would before long exterminate the peasant as it had destroyed the artisan. Their program therefore stood uncompromisingly for the nationalization on the largest possible scale of property and production, in agriculture as much as in industry. The peasant's cardinal ambition was thereby scorned and flouted, and the natural community of interests between handworkers in factory and farm was confused from the very beginning of the political mass movement.

This circumstance had the effect in western Europe of driving the peasants into the Conservative camp; and, as a consequence, in those countries, which like France and Germany, had a considerable peasant population, strong sections of the Labour movement demanded and effected a revision of Socialism's agrarian program. If the impossibility of that program was obvious in Germany and France, 'it was bound to be glaring in the peasant countries of eastern Europe. As one travelled eastwards across the Continent factories became scarcer while farms multiplied, until agriculture spread itself out almost without rival on the vast plains of Russia. The revolt against Marxism followed exactly that variation in economic structure. It travelled eastwards in ever widening circles until it struck the shores of Russia; there it suffered intense local influences, and the returning ripple came back, transformed out of all recognition, as Populism, and in that guise overflowed into the neighbouring agrarian regions.'[1] The social reformers of eastern Europe discovered that the Marxian brand of Socialism offered them no guidance for the solution of their specific problems. The social problem they had to face was in the nature of a peasant question and not of a proletarian question, and they were not willing to subordinate the most intense aspirations of a people 'to the claims of a formula'. The practical consequences of that standpoint were summed up by M. Stere in the series of articles published by him in 1907 under the title 'Social-Democracy or Populism'. In a country like Rumania, in which the peasants formed 94 per cent. of the taxpayers, political progress had no meaning unless it tended towards a rural democracy; and that ideal meant in turn that 'our economic evolution, as the whole structure of our State, will necessarily have to retain its specific peasant character. Hence economic progress must tend above all to organize the nation's economic life on peasant foundations: a vigorous peasantry, owning the land it tills and uniting through a comprehensive co-operative system all the virtues of smallholding with all the technical advantages which to-day are accessible only to large farmers.'

This profession of faith, written more than twenty years ago,

[1] This and the other quotations are from my essay 'Marx v. the Peasant', which forms an introduction to the history and philosophy of the Peasant Movement.

would no doubt be accepted to-day by the leaders of the Peasant Party as a correct and comprehensive statement of their own political creed. M. Mihalache only gave a more sentimental turn to the same idea when he said, in 1920, that 'the creation of the national State on democratic foundations was bound to lead to the recognition of that fundamental element on which the State rests: the Peasant. . . . Any wise policy must make of the peasant—who is the producer of wealth, the soldier, the tax-payer, the life-spring of the people's leaders—the kernel of our national prosperity.' That standpoint determines every aspect of the political and economic program of Peasantism, as well as its attitude towards other Parties and social groups.

The central pillars of the Party's creed are, first, an unmitigated belief in the virtues of representative government, and, secondly, an equally strong conviction that the way to the greatest happiness of the largest number of Rumanians lies through co-operative peasant agriculture. With such beliefs as its guiding tenets it was inevitable that the movement should come into conflict with the groups which have ruled the country hitherto. Political contest, indeed, is only just beginning in Rumania. The recent reforms have broken through the walls which defended the bureaucratic-manorial existence of the upper class. The expropriation of large property, especially, has destroyed the social basis of the old Conservative Party. It lingered impotently for a short time after the War, soon, however, to break up, its scattered remnants attaching themselves to various other Parties. An attempt made by General Averescu to create a new political grouping of the Right failed mainly on account of the absence of any social reason for its existence, and also because of the inefficiency of the chief protagonists. This although the attempt was fostered by the Liberals, either because they wished to have on their right a grouping with which they might alternate in power, as they did in 1926–7, and thus put off the advent of popular government, or because they hoped by that means to divert from themselves some of the shafts which the new Peasant movement was launching. In any case, the disappearance of the Conservative Party and the birth of a political mass movement have pushed the Liberal Party to the extreme Right. In that position it finds itself altogether on the defensive. It has to protect from assault its old political power and privileges, as well as the urban-mercantilist organization of society in which it has its roots. And as its power was obtained until now not through the exercise of some crucial social function, but only through the exploitation of political privileges, the Liberal Party could withstand that assault only by working up from political licence to a practical dictatorship.

During the past decade the Liberal supremacy has been more exclusive than it ever was before the War, when Liberals had to alternate in power with the Conservatives, at the will of the King. It was only because the Crown supported it that a one-party dictatorship could assert itself after the enactment of the land reform and of universal franchise. The emergence of a popular movement clamouring for strict Constitutional government struck the Court as a prelude to revolution. And so, in a sense, it was bound to be. For in a country in which the Crown, notwithstanding the existence of a written Constitution, had been accustomed to exercise quasi-autocratic powers, and the ruling groups to do with the country as they pleased, a popular demand for 'law and order' claimed nothing less than a revolution in the customary methods of government. That explains the attraction which the Peasant Movement had for the more idealistic of the younger intellectuals, and the Liberal Party's failure to establish itself as the recognized exponent of the urban and industrial population. It also explains why the movement gained the new provinces, which were chafing under the strain of an excessive and incompetent centralism.[1] That the Bessarabian Peasant Party, with its Radical temper, should unite with the Peasant Party in the Old Kingdom, was in the nature of things. But subsequently (1927) the group was also joined by the National (Transylvanian) Party, though its leaders were recruited from the province's urban intelligentsia, with a bourgeois rather than Peasantist outlook. Finally, it explains why the Socialist industrial workers joined hands with the National-Peasant Party

[1] In the general election of November 1928—the first really free election after the War—neither the Liberal nor the People's (Averescu) Parties were able to secure a seat in Transylvania and Bessarabia.

in opposition to the mercantilist Liberals. The possible limits of the Socialist movement are in Rumania even narrower than the limits of industry. The bulk of the workers engaged in the chief industries—mining, forestry, &c.—are peasants who own land, and who never hesitate to return to the plough if they can make a living from it. M. Madgearu is convinced that 'in agrarian countries it is much easier for peasants and workers to co-operate politically, seeing that the industrial proletariat has only recently emerged from the peasantry'.[1] In earlier years, Rumanian Populists and Socialists had joined in the fiery doctrinal battle provoked by Marxism between handworkers in town and on the land. Now the Socialists have not been unwilling to find shelter under the wings of the Peasant Party, because it has become clear that none of the urgent social problems could be solved until the Constitutional issue was settled. However much they may have been divided by class doctrines or sectional idiosyncrasies, all these groups felt united by their faith in the practice of representative government. Therefore they naturally rallied round the newly released peasant force, because through it alone could government by the people be peacefully secured and securely held.

To gain some insight into the fundamentals of the Peasant Movement, one might seek to extract from its social structure and from its economic tenets an answer to two generic questions. What place does the movement take within the line of our customary political divisions? And, secondly, is the movement likely to further or to traverse the typical organization of modern society evolved in the West? The Peasant Party's program makes it possible to give a direct answer to the first question, though not a simple answer. That program does not fit neatly into any of the traditional political doctrines. It is an eclectic program, having appropriated from all the existing doctrines such of their traits as best suited the aspirations of the new movement. In its attachment to the principle of private property the Peasant movement is probably more orthodox than the staunchest Conservative parties. It shares with Liberal ideology an unswerving devotion to the practice of representative government on a fully democratic basis, which is natural enough with a mass

[1] Article on 'Peasantism and the Town Workers', in *Aurora*, July 19, 1923.

movement; as well as a dislike of excessive State interference with the life of groups and individuals. The attitude of the Peasant Party to protection, which it admits only for those industries which stand a chance of surviving on such materials and labour and markets as the country itself can offer, comes very near the essential meaning of Free Trade. But most of these similarities are held with a characteristic difference. The Peasants' idea of private property is widely different from that of Conservatives in the western 'acquisitive Society'; and their dislike of State interference is far from a belief in *laissez-faire*. In regard to land, especially, the peasants believed empirically in a 'property of use' long before that idea was worked out theoretically in some of the post-war projects and laws. As long as the peasants hold such views on the ownership of land it is unlikely that they would approve the unrestricted accumulation of other forms of property. A cardinal aim of the Peasant program is, in fact, increasingly to limit the power and function of 'capitalist' middlemen—financiers, traders, and others of their kind; but it intends achieving this through co-operation, the axis on which their whole economic system revolves, and not through the Socialist method of nationalization. State ownership and control is approved, however, for essential public services, and even for such large-scale industries as mining, &c., which merely collect the country's natural resources. Co-operation is preferred, in other words, where the individual is a more or less complete unit of production and where individual effort contributes more than machinery to the finished product; but where the individual worker is merely a cog in a vast machine, requiring for its effective working concentration of property and production, then it is considered just that property and control should be vested in the nation. One must add to this the detail that the Peasant program definitely favours the organization and protection of labour, and the provision of equal opportunities for all. Hence, while that program is at variance with the standpoint of the Marxist Socialist Parties of the Continent, its economic, industrial, and social traits bring the movement on the whole closer to English Labour than to either Conservatives or Liberals. In brief, the Peasant program represents economically and socially a select combination of the instinct of Liberalism with the ideal of Socialism; a combination that may become even in the industrial west the program of a progressive movement indifferent to doctrine but intent upon the early realization of social justice, through applied evolution.

One finds a similar unorthodox blending of views in the national outlook of the Peasant movement. No other section of the people is so firmly attached to transmitted national characteristics, and less cosmopolitan in its customs and habits. But apart from the Socialists, no other section of a people is less eaten up with the ambitions of political nationalism. Every pronouncement and action of the eastern Peasant movement has shown that it places as much faith in international co-operation, as it does in economic co-operation at home.

Any attempt to answer the second question, by constructing out of these programmatic points a picture of the society they are likely to produce, must inevitably touch upon the borderland of speculation. One or two premises may be stated definitely enough. The new movement seems determined to encourage the country's agrarian development, and to base it on small peasant farming. Such a line of evolution would break away sharply from the path upon which the industrial West is irretrievably set. Instead of mammoth works and companies and trusts—fated, as Marx has predicted, to outgrow the safe limits of private control and to run straight into the arms of Socialism—small units of ownership and enterprise, made efficient by co-operation; instead of the blind struggle of each for himself, a large measure of adjustment and mutual aid; instead of cold-blooded control by unknown masters, constant personal contact between the agents of production; instead of the soulless machine and the hated factory, the living and beloved land; instead of the indefinite subdivision of stereotyped labour, an infinite variety of work and the satisfaction of creative achievement; instead of the restless and bitter spirit caused by the spectre of unemployment, the daily lesson in patience and perseverance brought home by the contest with the forces of nature. These basic differences between one civilization and the other open up such a wide vista of speculation as to their likely effects, as cannot possibly be compassed within a few pages. What will be their effect on the life of the people and on the growth of population? What the effect on the temper of the nation and on the psychology of its several classes? Or on the growth of cities and on the structure of government?

In the measure in which the Peasant Movement has an opportunity of asserting itself, the immediate effect will no doubt be a reaction from the excesses of the recent past. Some form of devolution will come to replace the extreme centralization of government in force hitherto. The utter neglect of the village may be followed by some check on the superficial but expensive brilliance of the capital town and of the host of State dignitaries. In Rumania the mercantilist excesses of the last ten years have sharpened the antagonism between town and country which has been felt throughout central and eastern Europe after the War. 'The conflict between peasantry and bourgeoisie in the field of the circulation of goods', says a Rumanian writer, 'is as acute as that between bourgeoisie and proletariat in the field of distribution.' And M. N. Lupu-Kostaki considers that 'in a country where, as in Rumania, the interests of agriculturists, who constitute the producing class *par excellence*, are constantly crossed and injured, the machinery of State becomes an abusive institution and resistance to it a real act of civic bravery'.[1] Such a demand as that put forward by the Serbian Peasant Party a few years ago, that as much of the public money should be spent on the village as the village contributed in taxes and rates, would greatly reduce the resources available for urban development, be it good or bad.[2] And the diversion of the State's

[1] Article in *Ţara Noastră*, 1925, p. 1019.

[2] The *exposé de motifs* to the new Administrative Law (1929) illustrated in figures the extreme poverty of the rural administrative units. The average yearly revenue of the rural communes was:

In the Old Kingdom	133,411 lei
,, Bucovina	260,195 ,,
,, Transylvania	259,556 ,,
,, Bessarabia	345,352 ,,

In Bessarabia and the Old Kingdom, however, the communes were artificial administrative units generally composed of several villages, whereas in the former Austro-Hungarian provinces each village was a unit in itself. If one divided the communal revenue by the number of villages, therefore, one obtained the following more accurate comparison:

benevolence towards agriculture may force some painful readjustments in the life of the urban upper and middle-classes. But this need not necessarily be a bad thing for the mass of the urban inhabitants. Enough has been said throughout this book, and especially in the other sections of this chapter, to show that under the old régime the arbitrary control of the State profited only a narrow caste of the urban hierarchy, while the mass of the town-dwellers derived little advantage from it either in the way of individual freedom or of communal services. Now workers and lower middle-class are bound to get their share from any measure of democracy which the Peasant movement may enforce. What the towns may lose in splendour the town population should gain in comfort. Be that as it may, the real point is that in a country in which eighty-five per cent. of the population lives on the land, any check on the towns which helps to bring more light and health to the village must ultimately balance out for the nation's good.

One could not presume to define the limits which that redressing of the social balance in favour of the peasants would reach. As the peasants have only just been truly emancipated, and agriculture placed in a position to adopt modern methods of farming, it is too early to analyse the social structure which those changes will produce. A general bias in favour of agrarianism is certain, but less so the strength of that bias. Peasant policy will receive its emphasis from the evolution of the peasantry as a class; and seeing that the peasant holdings, which already occupy almost the whole of the arable area, average merely 3½ ha. each, the standard of living within the peasant class will clearly be set by the development of farming technique and of co-operation, rather than by social doctrine.

So much seems sure, that neither of the two political extremes can hope to find recruits among the peasants. As long as Socialism, to say nothing of Communism, remains intent upon

Old Kingdom	51,095 lei
Bessarabia	140,482 ,,
Bucovina	229,680 ,,
Transylvania	251,635 ,,

The revenue of villages in the Old Kingdom was far below the minimum that would have been needed merely for the payment of such officials as the law required.

experimenting with Marxian tenets in agriculture, it can expect nothing but hostility from these inveterate aspirants to private holdings. On the other hand, the idea of Professor Rădulescu-Motru that a Conservative policy might be based upon 'the religious, traditional and Conservative peasantry', would seem to spring from a confusion between the peasants' psychological traits and their political needs.[1] A class which has still to gain most of the good things of life and the right to control its own destiny, cannot but be an army of advanced reformers. M. Garoflid came closer to the realities of modern society when he contended that 'in agrarian countries, democracy can only be real if it rests on medium-sized property. . . . Without such property, any political régime, be it Conservative, Liberal or Peasantist, becomes dictatorial. The creation of such a property requires only one condition, that which the bourgeoisie enjoyed everywhere—liberty'.[2]

This is a theme which, for a number of reasons, M. Garoflid always espoused. Together with most Rumanian agricultural experts he was convinced that better farming could be initiated only through a class of well-to-do peasants, owners of good-sized holdings. The creation of such a class, moreover, was desired by the more far-sighted landowners in their own interests. We have seen that some Conservative writers advocated it because the medium-sized farms would have acted as a buffer between large owners and the mass of the peasants, and because in that way many of those left without land would have given the large estates a regular supply of labour. As M. Garoflid well knew, 'liberty', i. e. *laissez-faire*, would have turned many peasants into landless labourers as surely as it turned the artisans into factory hands. Finally, the above quotation suggests that with a rural middle-class M. Garoflid hoped to check the incipient mercantilist domination of the Liberal bourgeoisie, as well as to forestall a possible mass movement from the Left.

[1] See *Ţărănismul, Un Suflet şi o Politică*. Professor Rădulescu-Motru is one of Rumania's most distinguished and progressive thinkers. The pamphlet, written in 1922, apparently represented a revulsion against the spurious Liberalism then in power. It is characteristic of the man and of the movement that Professor Motru has since joined the Peasant Party.

[2] Lecture on 'The Social Role of the Medium-sized Property'.

A peasant middle-class would still be good for agriculture, and the hard-pressed remnant of large property needs it more than ever as a buffer. But politically the hour of its usefulness is past. For the peasant mass movement which M. Garoflid dreaded has come; and while this will undoubtedly check the attempt to establish a mercantilist supremacy, it is unlikely to favour the rise of an agrarian capitalism. It is significant that when the Agrarian League was founded recently, under M. Garoflid's presidency, with the declared aim of furthering the professional interests of all cultivators, its advent was but half-heartedly welcomed in Peasantist quarters. A leader in the Party's official organ[1] expressed the suspicion that 'certain people seem to look upon the League as an attempt to revive Conservative-feudal agrarianism', and warned them against any attempt to short-circuit the work of the Peasant Party. The kind of Conservative agrarianism so skilfully pursued hitherto by the German *Landbund* is indeed out of place in the countries of the Peasant movement. Even in Czechoslovakia, where agriculture had reached an advanced stage of capitalist development, the agrarian movement, started with just those middle-class views and elements which M. Garoflid favoured, had, after the War and the relatively moderate agrarian reform, to be tuned up to the more Radical temper of the smallholders.

To sum up, the elementary conditions which govern the Peasant movement in eastern Europe can be stated very simply. In Rumania, as in most of the neighbouring countries, the large owners are as a class reduced to insignificance. Nor is the remaining land reserve sufficient to feed the growth of a weighty rural middle-class. Hence, agriculture and any agrarian movement will, as far as one can see, be dominated by the mass of small peasants. On the other hand, there is for many years to come no possibility of industrial expansion on a scale big enough to create a numerous industrial proletariat, holding Socialistic views on property and production. Until this happens there is no room for the growth of a powerful political organization to the Left of the Peasant mass. From which it follows that through its own inner structure, as well as through the place it occupies

[1] *Dreptatea*, January 15, 1929.

in relation to other social groups, the Peasant movement must necessarily lean towards the Left wing of political alignment in Rumania.

Nevertheless, the emergence of this movement is likely to give stability to the politics of a country situated in a disturbed region and which enters upon its adolescence in such restless times as ours. Even the Party which has done least for the Peasants sees and admits that. 'The good sense, patriotism, intelligence, and moderation of the peasantry', said a Liberal leader, 'will bring into the life of the political parties, and hence into the life of the country, an element of steady national development.'[1] A mass movement with such a strong emphasis on the individual should be capable of rising to a democratic conception of social justice without having altogether to rely for its fulfilment on the heavy-handed discipline of the State. It should be able to promote equality without the devastation of liberty. It should instil into public life both the patience which is inborn in the yeoman tiller of the soil and also his dogged resistance to violence and abuse. Dictatorships, Red or White, should find it hard to flourish where the Peasant movement is established. Any advance resting on it may make but slow progress, yet the progress should be steady. For if the peasants are more difficult to organize than industrial workers, their powers of resistance are infinitely greater. They can bide their time. And to that elemental strength great idealism is now allied. The movement is astir with that crusading spirit which inspired the early Labour movement in England. Like the latter, it fills many of the best among intellectuals, old and young, with a passionate zeal to serve the people; a zeal akin in eagerness to that 'going-to-the-people' which perhaps was the most moving episode in Russia's revolutionary history, but a zeal applied in our case to practical politics, and with the odds of power on its side.

So much of this picture has had to be sketched with the brush of imagination that no one could expect to see it come out unaltered from the workshop of Time. Many of the details are

[1] M. N. Chirculescu, *Report on the New Constitution submitted to the Chamber on 5th March 1923*, p. 3.

bound to vary a little, and some of them may in the long run vary a great deal. Or, not impossibly, some unforeseen discovery or a fresh jerk to the course of history may change the picture out of recognition. What historian of a revolutionary change could think of predicting finality for its results? But at least one general conclusion emerging from this study should remain unaffected by age: that the more far-reaching effects of this great reform, which a political upheaval has called forth, will blossom and bear fruit in the social life of the people, rather than in the field of agrarian economics.

CONCLUSION

THE Rumanian reform may be likened to an historical *carrefour*, a meeting-place from which many paths radiate in all directions—paths of inquiry and of speculation. The law itself only marks the stage where past and future trends in the country's historical development cross each other. A description of its texts would not suffice for the elucidation of that story. For this one must go farther and deeper, drawing light from history and economics, from sociology and politics. The highway of our work was mapped out as a sociological study of the evolution of the peasantry as a class; but at many points we were forced, or perhaps merely tempted, to make excursions into the by-ways of relevant sciences. If that was needed as a means of giving relief to the main story, it had the subsidiary advantage of bringing out certain aspects in the social progress of eastern Europe which may be new to the Western reader. It has, for instance, revealed some peculiar features in the economics of a backward agrarian country, and it has shown what a poor alloy such a backward economic structure makes with the forms of advanced government. It has furnished for political science some striking material on the nature and habits of nationalism; and it has incidentally raised the question whether we should not revise our estimate of the main forces and events which have moulded the history of eastern Europe.

The customary historical verdict has been satisfied with charging to the Turkish invasion all the evils which the peoples of south-eastern Europe have suffered. The coming of the Turks and their prolonged domination, certainly contributed to arrest and warp the political development of the Balkan peoples. Yet for the mass of the populations concerned the consequence was not altogether bad, for the same circumstances likewise delayed the social hardening which goes with the growth of the modern nationalist-militant State. It is characteristic that the first Prince to lower the status of the peasantry was Mihaiu the Brave, who was also the first to equip a professional army and to use it for expansion and not merely for defence. And it was only with the emergence of a nationalist ruling class, and the erection of an elaborate political and military structure, that the people were called upon to make contributions of a size which they could not give without prostration and which therefore could not be secured without coercion. In fact, the peasant never touched, during the times of quasi-feudal serfdom, the depths of misery which was his lot during 'neo-serfdom', under Constitutional government. During the several centuries of foreign domination the peasants' ancient land rights were sporadically abused, but with the coming of national government they were quickly and systematically suppressed. And if in the lawlessness of the first period the peasant's larder was occasionally raided and emptied of its good things, in the period of national independence the exactions of State and upper class seldom left him with anything to put in it. The history of the Rumanian people reveals a close and plain correlation between the rise of national Government and the social depression of the peasantry.

Very rare were the voices to condemn such maltreatment of the mass of the people, and no heed was paid to them at all. The interest and ambition of the ruling class were absorbed in a frenzied pursuit after the trimmings and trappings of civilization. In 1906 the State celebrated with great pomp its golden jubilee. In 1907 it was visited by a desperate peasant rising, put down with wholesale massacre. This brought home to most people how rotten were the foundations on which the State rested, and caused many of them to preach and press for reform. The need for reform was conceded by the rulers in 1913, when the country expected at any moment to be dragged into the Balkan wars. But the danger passed and nothing was done, until the Great War imperatively forced the question to the fore again. Further hesitation and tergiversation were only ended by the Russian Revolution. Between the two countries there was a great similarity of conditions. In neither of them was the organization of the State and the morale of the people capable of standing the strain of prolonged effort, and it was inevitable that, in the case of a breakdown, the masses should be tempted to break their chains. In Russia the war with Japan was followed by the abortive rising of 1905, while the Great War ended for her in the Revolution of 1917. The lesson was too fearful and too near to be lost upon Rumania's rulers. What years of argument had failed to achieve even on a moderate scale was accepted in 1917 overnight and in a sweeping measure. The inertia from which the good intentions expressed before the War suffered, and the conspicuous ill-will shown to the beneficiaries of the reform after the War, leave no room for doubt that War and reform stand in close causal connexion to each other.

This conclusion finds further support in the manner in which land was distributed under the reform. An agrarian reform applied in normal conditions would have based the selection of the beneficiaries on certain economic standards. Under the Rumanian reform, however, land was given not to those who could make the best use of it, but to those who suffered most in the War. Even the Peasant leader, M. Mihalache, had to admit 'that a different criterion would cause a moral revolt in the villages'. On the same grounds the size of the holdings was fixed with a view not to sound farming but to the satisfaction of the largest possible number of claimants. In every other aspect the reform bears the imprint of hasty work carried out in the throes of an emergency. The whole reform, therefore, amounted to a redistribution of land property, with little account taken of the economics of the agricultural industry. The new law thus continued the tradition of all the reforms from 1864 onwards, but with two important differences. Whereas the ultimate result of the reform of 1864, when the peasants were emancipated, had been to enlarge the area and, especially, to strengthen the status of large property, the latest reform has almost annihilated it. And while in 1864 the balance of compensation weighed heavily in favour of the landlords, they have now received no more than a nominal compensation for the land that has been taken away from them. These two variations would justify the description of the new reform as a revolution—a social revolution carried through by peaceful means, but a revolution none the less when measured by the sudden and sweeping change it has caused in the relative position of the two rural classes.

It is instructive to consider in retrospect the effects which were expected to follow from such a reform, as forecasted by one of the country's leading politicians. In a pamphlet on 'The Advantages of Expropriation', written early in 1914, M. Vintilă Brătianu attempted a comprehensive list of them, as summarized below:

For the Landowner:

1. He will secure peace on the land, and a strengthened defence of property.
2. With the money received as compensation he will pay his debts, improve his farm, or invest in industry and commerce.
3. The value of the land left to him will be enhanced.
4. What remains of large property will be better farmed, as the peasants will have more and better animals, while the landless peasants will derive advantage from more intensive cultivation.
5. The development of national industry will draw workers into the towns, thus reducing hunger for land.
6. Food and health will improve, and 'one million well-fed children will mean one million healthy and spirited soldiers for the defence of the country, and one million vigorous workers for agriculture and industry'.

For Industry and Trade:

1. The landowners' purchasing power will not fall; the liquid money which they will have at their disposal, or else improved cultivation, will help trade and industry.
2. The peasants' purchasing power will increase greatly. 'Traders and manufacturers have a greater interest than any other class in the application of an agrarian reform.'

For the Peasant:

1. He will no longer have to pay rent.
2. He will apply his labour more economically, and
3. also that of his animals.
4. He will be more independent, and will be able to till the land at the proper time, thus obtaining larger and better crops.
5. He will be able to introduce a greater variety in his crops, which will make him more secure against bad harvests and bad prices.

General:

1. National wealth will increase with the all-round improvement in farming.
2. Agriculture will improve, because only small cultivators can give labour, manure, &c., for intensive farming. 'The gold that comes

into the land at present should not mislead us. *It represents in part the draining of the richness of our soil. . . . To direct agriculture towards small intensive farming would be a measure of foresight and progress for our whole national economy.*'

3. All this would strengthen the financial position of the State.

Many of the good results predicted by M. Brătianu were self-evident. That national wealth would gain from the development of the country's chief industry was clear; and there could be no doubt that trade and manufacture stood to benefit from an increase in the purchasing power of four-fifths of the population. If M. Brătianu was less successful in telling the fortune of the landowners it is because he and his Party intended taking less land and paying more for it than they were able to do through letting things wait till 1917. Of special interest for our study is M. Brătianu's unhesitating forecast that production would improve, for in reality it has to some extent fallen off, and this has been made the chief plank in all criticisms of the reform. His view was based on the undeniable fact that nothing could be worse than the system of large-scale agriculture practised in Rumania before the reform; and on the assumption, now generally held, that only the smallholder can profitably apply himself to highly intensive farming. In Rumania the passing of yeoman farming was not brought about by the pressure of economic requirements, and the result was not an economic advance. On the contrary, both the methods and the results of farming steadily deteriorated with the increase in the size of farms. Exports were the outgrowth of extreme *Raubwirtschaft*, inflicted upon the people as well as upon the soil. It is doubtful whether that system could have survived even a slight redress in the social balance of power, which was inevitable after the War, or the fresh crisis into which European growers have been plunged by the fabulous development of corn-growing in North America. Professor Gustav Cassel has stated in an interview that with the new machines which cut and thresh at the same time American farmers can perform these operations at a cost of forty cents an acre, and that under fair conditions the cost of the whole process of cultivation, from ploughing to threshing, amounts to eighty-four cents per acre. The cost of corn production has therefore fallen to one-fourth within the last three years.[1] The large Rumanian owners and tenants who had no equipment and little capital of their own could not have kept pace with these changes.

Professor Cassel further pointed out that relatively the consumption of corn is decreasing, because (*a*) mechanical transport is replacing animal traction, (*b*) many countries are reducing the manufacture and consumption of spirits, (*c*) human diet is showing a marked change. Cereals and meat are being increasingly replaced by dairy produce, vegetables, and fruit. But these are just the articles in the production of which the peasant smallholder easily outstrips the larger farmer. In agriculture production on a large scale is not always practicable or profitable. The small farmer, and especially the peasant, owning his holding and working it himself with the aid of his family, is usually the more successful producer of those crops and food-stuffs which require intensive hand labour and constant care. For that reason agriculture, unlike industry, has not seen a continuous concentration of the units of production, but rather a variation in the size of farms related to the nature of the crops grown on them. Speaking of the slow change in the distribution of land-property in France, H. Passy, in a book published about the middle of last century, suggested that 'it is the nature of agricultural work itself, conditioned as it is by local circumstances, at one time to further the break-up of land properties, and at another time to arrest it and to provoke a movement of concentration'. When Europe needed bread for its rapidly expanding industrial cities the way was opened to large-scale farming through the emancipation of the peasants and other measures. Now that the overseas countries supply more corn than is demanded, European agriculture is tending to concentrate on the production of animal and dairy products and on market-gardening, and in every country large farms are being broken up into smallholdings.

The Rumanian land reform merely let loose with a rush,

[1] *Argus*, October 25, 1928. On the other hand a writer in the *Revue d'Économie Politique*, Jan.–Feb. 1929, argued that the United States were likely to become importers of meat.

through the breach the War made in the wall of social and political restrictions, a current which was already coursing throughout rural Europe. The change was easy in Rumania because the large farms were generally cultivated by the peasants with their own animals and implements, the system being like the domestic piece weaving at the beginning of the Lancashire cotton industry rather than like the modern capitalist production on a large scale; and the change was urgent not only to meet the altered requirements of the market, but also technically, in order to introduce a more rational crop-rotation which should spare the strength of the soil. Finally, the opinion has been expressed that the reform has facilitated the task of reconstruction. In his report to the International Agricultural Congress which met in Rome, in June 1927, M. Ionescu-Sisești declared that 'it is a general opinion in our country that, leaving aside its social importance, the agrarian reform has helped to make good in agriculture the damage caused by the War. Without the intervention of the reform the resumption of cultivation on the large latifundiary estates and the renewal of live-stock would have been much more difficult and greatly delayed'. The reform must indeed have been fully justified economically seeing that it has been commended even by M. S. Timov, the spokesman of the International Agrarian Institute in Moscow. He wrote to M. Ionescu-Sisești saying 'I agree with you that the agrarian reform has done no harm whatever, and that, on the contrary, it has contributed to the progress of rural economy and even of national economy'.[1]

These predictions and expectations have not been realized so far. Production has been unable to keep either absolutely or relatively even to the low pre-war level. The argument that without the reform things would have been still worse, though advanced with undisputable authority, is hypothetical and does not tell us how that fall in output is to be explained. The retrogression is the more puzzling as with the huge transfer of property went only a very limited change in the agents and methods and means of production. Hence the inevitable falling off—to be expected while the industry adapted itself to the new

[1] Letter to M. Ionescu-Sisești dated June 20, 1927, kindly communicated by the recipient.

conditions—should have been slight and of short duration. More serious was the damage caused by the War. The renewal of live stock, in the rearing of which the peasants excel, may have been exceptionally rapid; but ten years should have sufficed to make good the damage suffered by corn-growing.

We will not here repeat, even in summary, the detailed discussion of this problem contained in Chapters VIII–XI. One need merely state the relevant conclusion that the reform failed to make good economically because during a period of double reconstruction agriculture did not dispose of the credits and other facilities required to that end. And still more because agriculture, while being starved of support, was burdened with one-sided restrictions and impositions in the shape of export taxes, controlled prices and so on. The psychological effect of that policy must have been at least as deterring as its material consequences. A manufacturer or a capitalist farmer has to keep his undertaking going even in the face of adverse conditions, in order to cover as far as possible his overhead running expenses. But an unfriendly policy may bring the frugal and almost self-sufficing peasant to the point of troubling no longer to grow supplies for the market, as happened in Russia after the Bolshevik revolution. In Rumania the hostile interference of the State did not go far enough to dry up production, but it certainly took away all stimulus from it. After having been given very cheaply the land they coveted, the peasants were made to pay its real value several times over by means of indirect contributions. They escaped the selfish exploitation of the large owners only to fall under the stepmotherly tutelage of a mercantilist State. The injustice of such a policy was officially recognized in theory. The *exposé de motifs* to the fiscal law of 1923[1] emphatically declared that 'A State cannot be democratic if, at the moment when a large rural property disappears, it permits a few people to accumulate fortunes from trade and industry, while leaving the mass in the state of the serfs of yesterday, who were unable to share in the benefits of our general prosperity'. In practice, however, this was the very thing that happened. The income-tax returns for the period

[1] *Monitorul Oficial*, January 16, 1923, p. 278.

1923–7 showed that 6 per cent. of the private traders and manufacturers secured 47 per cent. of the total private revenue. 'The situation which before the War existed on the land', says Dr. Creangă, 'where a number of latifundiary owners retained the greater part of the agricultural revenue, has now passed into the domain of trade and industry.'[1] As a peasant tersely summed up the result: 'He's pushed the bowl nearer, but he's given me a shorter spoon.'

The latest land reform has thus given the ruling class merely another opportunity to prove their lack of interest in the fate of the peasantry. Circumstances had forced them to hand the land over. But instead of turning that necessity to national advantage, by endowing the new owners with the means of making the best of their valuable possession, all the resources of the State were thrown into the service of a policy of industrial expansion. To achieve this aim reasonably one of two possible ways might have been followed. The interested class might have taken a long view of it and gradually prepared the required conditions—capital, labour, and a market—by developing to the utmost the existing productive organization, that is agriculture. Or it might have appealed for speedy achievement to the help of foreign capital and enterprise. Being a privileged ruling class they would not allow the first, and being a nationalist class they could not admit the second. The ambition to create a national industry was coupled with a policy of making agriculture pay for it. The result was that instead of enlarging economic life with the help of foreign capital, it choked economic life by extracting tribute for the creation of a national capital. On the occasion of a private visit to Rumania, Professor Werner Sombart exposed in an interview the obvious folly of that course. 'In the present phase of Rumania's economic evolution', he is reported to have said, 'the export of agricultural products forms the principal element of capitalist progress. The ambition to make the country economically self-sufficient, by creating all the means of production needed for modern social life, is altogether fallacious. The possibility of reaching economic autonomy is determined by natural conditions, and not by the will of the rulers to organize

[1] 'Veniturile și Averile României Mari', in the *Bul. Inst. Ec. Rom.*, January 1927.

it.'[1] In fact, the neglect of agriculture has made the road of Rumanian industry more difficult. It has delayed the growth of the home market, while exports are out of question; and it has prevented the improvement of the supply of food needed for the maintenance of an industrial population.

The question whether that attempt to foster a national industry was right or wrong, and of the motives which inspired it, is however not directly relevant to our argument. The point is that in the manner and circumstances in which it was made that attempt acted as a serious deterrent to agriculture. Hence the issue resolved itself not into a technical problem, which agriculture might have solved for itself, but into a political issue which agriculture had to fight out against the forces of mercantilism. Some years ago a Liberal Minister said in the Chamber that, unlike industry, agriculture deserved no special encouragement, as its progress depended on natural elements, such as soil and rain. The opposite point of view was put in Parliament by M. Iuliu Maniu, the present Prime Minister, in a speech delivered on 4th August 1927. He insisted that as Rumania was an agrarian country, all the other branches of economic activity depended on agriculture. 'After decades of an unnatural economic policy, directed towards a forced industrialism, which demanded enormous sacrifices from State and consumers, the time has come to inaugurate an agrarian policy. . . . Such a new orientation of our economic policy is the more necessary as the reform has put four-fifths of the country's soil in the hands of small cultivators.' No other line could indeed be pursued after such a reform, added M. Mihalache. It is not possible to contemplate a systematic organization of production 'carried out *against* the interested class—the peasants; it can be done only with *its participation and under its control*, both in devising legislation and in applying it'.[2] Lenin admitted as much when he put forward his new economic policy in 1921: 'We must now endeavour to develop a national economy based upon the real psychology of the well-to-do peasant, whose motives and sentiments we have been unable to change during these three years.' That was equally true of Rumania. The economic and financial policy of

[1] *Adevĕrul*, January 4, 1929.

[2] *Noul Regim Agrar*, 1925, p. 5.

the past ten years has only succeeded in vitiating the national effect of the reform. Instead of setting free a new spirit of professional hope and emulation, it has created an atmosphere of bitter political conflict. In a negative way, at any rate, there is in this further proof that the chief obstacle to the successful maturing of the reform has been political rather than technical. The technical problem, indeed, was hardly put at all. It would be naïve to believe, said M. Mihalache, that the reform is finished, and that all that remains is to revise abuses in its application. The revision of abuses is merely a moral necessity, so as to restore confidence in law and government. But 'everything is still to be done so far as the work of agrarian development is concerned'.[1]

The technical improvement of farming, and the systematic organization of agricultural economics, is of vital importance for Rumania, because almost everything in the progress of State and nation depends on that. Without it the population problem must before long become acute. Hitherto agrarian unrest has been periodically placated by grants of land. But when in twenty years' time the problem crops up again, 'in a more threatening manner', writes a Professor at the Cluj Agricultural College, 'having no longer at our disposal the latifundiary estates, we shall then be unable to have recourse to the facile solution applied at present. . . . The agrarian-social crisis which simmers to-day, but which to-morrow will boil over, has its roots in the technique of agriculture, to which until now we have given little thought or none at all.'[2] The random distribution of land at the last reform has rendered the technical problem still more difficult. When after two or three generations the holding of 5 ha. has been reduced to 2–3 ha., peasant farming will contribute little towards the needs of the towns unless in the meantime agriculture is brought technically and commercially to a high degree of efficiency. And as industrial expansion is impossible without first raising the purchasing power of the peasants considerably, failure to develop farming intensively would mean that agriculture will be left to maintain an increasing rural population on its present poor output. In such conditions the population

[1] *Noul Regim Agrar,* 1925 p. 27.

[2] Dr. M. Chiriţescu–Aron, article in *Societatea de Măine,* vol. i, No. 6, 1924.

problem would soon become serious on the land. To favour large-scale farming and corn-growing, as some experts who regret the former exports advocate, would only render the problem more acute. For if they are to compete with oversea corn, Rumanian growers would have to rely increasingly upon machines, and that would leave correspondingly less room for agricultural labour. The pressure of population is not as strong in Rumania as it is in the West, and it would not be felt at all if farming were more productive. Ultimately, however, the issue will resolve itself, for the Rumanian statesman also, into a population problem, but as far as one can foresee into a rural and not an urban problem.

The general tendency towards a return to smallholdings reflects something more significant than a mere change in the diet of the civilized countries. The hot-house growth of industry during the War and in the first years after it has in many parts gone too far, and as a consequence the need is now felt to redistribute economic activity. This finds expression above all in programs and policies supporting a return to the land. A century ago, when the new industries were drawing a large number of workers from the village, that flight from the land favoured a concentration of farming and the laying down of the land under corn and grass. Now that industrial centres suffer from an over-supply of labour the process is being reversed: the tendency is to break up estates and large farms into smallholdings, and to replace corn-growing with farming activities that will allow agriculture to absorb as large a proportion as possible of the surplus population. That reversal of policy, made necessary by the artificial growth of industry under the stimulus of tariffs, &c., is rendered possible by the fact that we are as yet far from having exhausted the possibilities of agricultural production. The relation between agricultural production and the problem of population was discussed in a valuable paper which Professor Friedrich Aereboe read before the Viennese *Gesellschaft für Sozialpolitik* in 1926.[1] Briefly, his argument was that through the improvement of agricultural technique an ever smaller extent of the original area suffices for the feeding of a community. Production, in fact, would be

[1] 'Die Wachstum-Möglichkeiten der Landbevölkerung.' Summarized in the *Buletinul Agriculturii*, October–December 1926.

excessive but for certain obstacles. The main obstacles are lack of capital and instability of prices. One could say that an increase in population never causes a crisis in food-supplies if such obstacles do not prevent the best use being made of an area, or its extension, for the production of food-stuffs. The relative overcrowding of China, e.g., is due not to the density of the population, but to its inability to use to the full the possibilities of agricultural production. Speaking generally, the available areas suffice for the existence of an increasing population. But in the West the drawback resulting from a wrong distribution of land will have to be removed. The institution of property cannot remain bound in iron hoops if it cramps the forces of production. The maintenance of extensive land property causes an excessive fall in agricultural wages, and this results in a flight from the land, with all the attending social evils. The institution of entail becomes an obstacle to economic development if the latifundiary owner can no longer insure the full and rational exploitation of the land. Admitting, therefore, that the soil offers an indefinite prospect for the increase of production, the removal of the obstacles which impede that progress—bad distribution of property, lack of capital, instability of prices—would bring with it the possibility of paying better wages, which in its turn would curtail the flight from the land. So that any improvement of economic conditions in agriculture would absorb some of the unemployed industrial workers, and result in a parallel improvement of urban conditions.

If we survey the Rumanian land reform in the light of this argument, we see that it has carried through the most difficult of the changes which Professor Aereboe laid down as essential if the population problem is to be solved through agriculture. The burthen of most criticism is that the measure went too far in the right direction, propelled as it was by the momentum of a revolutionary wave. But that is a drawback which could be remedied by leaving a certain latitude to natural selection;[1] and it is outweighed by the advantages derived from that political

[1] The National-Peasant Government has opened the way for such a process through the law, passed in 1929, which removes the prohibition to sell and to mortgage holdings distributed under the land reform.

unrest. That alone made it possible to carry through the reform without any serious opposition or disturbance, and without any cost to the State. In that respect, therefore, Rumania starts with a considerable advantage over the Western countries in the reform of the agrarian system. But everything else, as M. Mihalache said, remains to be done, including the education of the peasants into capable farmers. The hopes based on the economic achievement of the reform can have no fair chance of being fulfilled except in the measure in which the reform is economically and technically completed. Yet even under the present conditions the results cannot be relatively anything but good. For the change, let it be repeated, was not from rational farming on a large scale to small and therefore inevitably less rational farming, but from latifundiary *Raubwirtschaft* to peasant cultivation, always more careful in the use of the soil. The supreme proof which the wife of a Jersey smallholder offered, in telling me of the many qualities of her husband, was that 'he manures his land as well as any man on the island'.

With a reform like the Rumanian, moreover, which was the outcome of a social claim, any judgement on results must lay as much or more weight on distribution as on production. Even if the present drop in production could be laid without doubt to the door of the reform, the falling away of exports should weigh lightly in the balance against the lifting of several million people out of grievous misery. The Socialist critic is apt to think with the orthodox economist that the large farmer produces mainly for the market and the peasant mainly for himself. This view, once held as an axiom, has been disproved by the trading activities of well-organized peasant communities like those of Denmark, Switzerland and others. But in any case it was a narrow view, inspired by the anxiety of people concerned mainly with industrial problems for the needs of the urban consumers. How the change from peasant to capitalist farming reacted on the rural poor was ignored. Yet, almost without exception, the expansion of large-scale farming, organized for supplying the market, left those whose labour created the produce without an adequate share of it. Mr. and Mrs. Hammond have described in *The Village Labourer* how after the enclosures the poor villagers could get neither butter nor unskimmed milk for their food, while many farmers gave the surplus milk to their pigs. In Rumania, as we have seen, after the development of corn-growing on a large scale, village children grew up without knowing how cow's milk tasted. The social reformer cannot, therefore, appraise economic organization solely for its mechanical efficiency, or for the impetus it gives to the flow of trade. The decisive test for his verdict lies in the measure in which each system favours a fair all-round distribution of the produce of labour. In Rumania, at any rate, it was generally true that 'where there is large property there is also great poverty and great ignorance', as M. Mihalache affirmed; 'the poorest villages I have met with in our country were on the large domains. I found in them neither school nor church nor village hall.'[1] And M. Maniu confirmed that in Transylvania, too, one could observe a marked difference in well-being between the regions of large and small property. In agrarian countries like Rumania, where eighty-five per cent. of the inhabitants live on the land, the disappearance of corn exports, and even a decrease in the supplies which reach the towns, may nevertheless mark a social improvement if that result means that the rural population no longer suffers from chronic starvation. From general evidence that would seem to be one of the main causes for the change in the supply of Rumanian corn. Hence the essential effect of these land laws, adopted for the purpose of bettering the existence of the people, cannot be described otherwise than as a great social advance.

Rumania's political rulers could claim great legislative merit for these laws. And the willingness with which so many landowners resigned themselves to the transfer of their land to the peasants deserves recognition. The event might have remained a monument to wise and generous statesmanship, had it marked a resolute break with an inglorious past. Unfortunately, what followed afterwards stamped it rather as an interlude. The policy adopted during the ten years which followed the reform created a feeling that the rulers were less repentant for their erstwhile selfishness than for their war-time liberality. Perhaps no social class can be expected to part lightly with the power

[1] Speech in the Chamber, 1921, p. 36.

and privileges it has been accustomed to enjoy; or to break suddenly with the methods through which that favoured position was secured. But in that case the makers of the reform builded better than they knew or willed. Like Goethe's 'Sorcerer's Apprentice', they have conjured up a force which they can no longer subdue or control. The attempt having been made—only to give rise to bitter political conflict—perhaps a wiser outlook will prevent its renewal. It is not too late to make of this reform, so great in spirit and in scope, the starting-point of a new era of social peace and co-operation, for the good of a people whose many trials in truth entitle it to a happier future.

APPENDIX I

TRADITIONAL FORMS OF JOINT LANDHOLDING IN RUMANIA

RUMANIA still contains a number of villages whose inhabitants have a common title to the land, in accordance with old-established if varying rules.

The typical old Rumanian village rested on a threefold unity: (*a*) A unity of blood-relationship, the village having been generally established by an ancestor of the actual inhabitants, called *moşneni* in Muntenia and *răzeşi* in Moldavia.[1] It carried his name, with the suffix *eşti* or *eni* (Olăn-eşti, Topolov-eni, &c.). (*b*) A unity of property, the title to the land being held in common. The arable strips were in private possession and remained by inheritance in the same family. But the possessor's private rights were not absolute. He could sell only with the consent of his relatives and neighbours, who had the first refusal. A deed of sale always stated that 'This sale has been made with the knowledge of my kinsmen and neighbours . . .'; otherwise the sale could be attacked. (*c*) A fiscal unity. The main tax, the *bir*, was paid on the *cislă* system (Serb. *cislo*, number). The tax was imposed upon the village as a whole, and the village elders then apportioned it among the villagers, according to each family's capacity to pay.

The unfree villages, inhabited by *rumâni* or *vecini*, had only the fiscal unity, but no unity of either blood-relationship or of ownership. Therefore such villages had a less stable existence. M. Radu Rosetti estimates that 20,000 villages have disappeared in Moldavia alone from the time of the foundation of the Principalities till the coming of the Organic Statutes.

In the yeoman villages grazing (*islaz*) and woodland (*branişte*) were common land; each villager had a right of use, not determined, and when a stranger was admitted to share in that right, the whole village took part in the action. Arable land (*ţarină*) was in the private possession of each family. But the village community had a title to it in so far as the land of a family which died out returned to the common patrimony, and as the holdings could be sold only to members of the village clan. In spite of that restriction, yeoman land increasingly passed into the hands of the boiars. Whole villages sold their freedom away to the local boiar, when they were in distress, and the boiar found means to penetrate even into the other yeoman villages. They could not buy *răzeş* land, but their way was to get one of the villagers to make them a gift of his holding, thus preventing the donor's family from making use of their right of *protimesis*[2]; and having thus become themselves

[1] *Moşnean* from *moş*=ancestor; *răzeş* from the Magyar *részes*=partner.

[2] The term is of Byzantine origin and is commonly used by Rumanian writers to indicate this traditional restriction on an individual's right to dispose of his land, but M. Fotino, e.g., disputes that this Rumanian custom is of Byzantine origin.

members of the village, they were afterwards entitled to purchase land from the other villagers.

In an attempt to stop such abuses Al. Mavrocordat decreed in October 1785 that gifts of land were permissible only 'between relations and people of the same standing, or by the wealthy, if they so liked, to the poor, and by any one to the holy monasteries', but not by the poor to the rich. The decree was to have retroactive force. (Poni, *Statistica Răzeşilor*, p. 76.) Later Mihaiu Sturdza so interpreted the decree as virtually to cancel it, being himself interested in a transaction of that kind.

Declarations made before the Popular Assembly of 1817 show that the great majority of the inhabitants owning land could not produce documents to prove that they had received land through gifts from the Princes, and these were precisely the small owners, the răzeşi. (Philippide, *Incercări asupra stărei sociale . . .*, pp. 53–4). Even in the boiar villages a small class of men was found who by good luck or obstinate resistance had been able to keep their yeoman land. They were called *călăraşi* (*cal* = horse), because in return for being exempted from taxation they served in time of war as cavalry men, with their own horses. Cantemir, in his *History*, speaks of them, but he deals only with Moldavia, and it is difficult to know whether they also existed in Muntenia. (Philippide, *op. cit.*, p. 81). The persistance of such yeomen in the midst of boiar villages must be taken as a proof that the other villagers had once enjoyed the same status. (The tradition has continued to the present day in the regiments of călăraşi; they are recruited from men who bring their own horses and supply fodder for them at short periods of training, over a number of years, and in return are exempted from the continuous compulsory military service, which in the cavalry lasts three years.)

The dispossession of the *răzeşi* reached its climax between 1830 and 1850, after the Principalities had regained national autonomy. Article 391, Ch. viii of the Organic Statutes, was meant to check that abuse, but it remained a dead letter. In 1857 the *răzeşi* deputies in the Moldavian divan *ad hoc* submitted a proposal that the divan should appoint from among its rural members a commission of eighteen, which should inquire into the whole question, with power to cancel arrangements made in violation of the Organic Statute and other ordinances. But the majority of the divan rejected the proposal. (Poni, *op. cit.*, p. 94.)

In the yeoman villages which have continued to exist the system of land tenure is essentially the same as of old. In most of them land is held under two forms: common ownership (*devălmăşie*), in the case of woodland and grazing, and *răzăşie*, which is merely a common title, in the case of arable land, clearings, gardens, vineyards and farm-yards, that is, all the land which human labour has improved and kept in condition. The basis of the first is always the same, that of the second varies in details from place to place. The transition from *devălmăşie* to *răzăşie* may happen in a number of ways. Individual villagers may clear a piece of forest or of waste. Or, secondly, the villagers may deliberately divide among themselves the grazing land, in the same way as was done with arable land; this is usual when pasture is changed into meadow. All the taxpayers are entitled to a share. After haymaking the land is opened indiscriminately to all the village cattle, so that in fact there is a seasonal return to common ownership and use. Finally, the transition may take place as a result of a gift of an individual holding, from the common land, to a villager who has rendered the village some special service or who is poor.

Răzăşie is a form of landholding half-way between common and individual property. The villagers form together a *ceată*, i.e. a group or clan, and every *cetaş* (also *obştean* or *răzeş*) holds a distinct strip of land, called *delniţe* (also *opcine* or *obştine*). The title of ownership is vested in the family,[1] not in one of its individual members, and the proof of blood-relationship is valid in law in support of a claim to a share of the village land. The extent of the share is measured after the number of descendants. Originally a group of *răzeşi* inhabited the same village, and the title of co-ownership rested on the possession of house and yard within the village boundaries, in the *vatra satului* (the village hearth). In Bucovina and Maramureş there are still districts where the sale of house and yard entails the sale of the title to the village estate.

M. Garoflid estimated in 1908 the area covered by *răzeş* property at 1,500,000 ha., i.e. 37 per cent. of the properties below 100 ha. But the bulk of *răzeş* land is forest and mountain pastures, not arable. The poor mountain slopes, so difficult of access, did not exercise the same attraction upon the boiars; and perhaps the populations concerned also received more protection because they inhabited frontier regions. At any rate, they have remained sole owners of the mountains.

This is the case, e.g., in the highland district of Vrancea.[2] The forest is joint property, and every villager has an equal right to the timber, (though on rare occasions there are villagers with several 'rights'), but exploitation is individual. This applies also to grazing. As neither demand any preliminary individual effort and care, there is no inducement for private possession. In these cases there is no right of family inheritance, nor much weight laid on blood relationship. The sale of any part of the common forest requires the decision of the village meeting, which appoints special delegates with power to conclude the transaction. Taxes were distibuted among the villagers by the same method of proportional assessment as of old; and in 1808, when some of the mountains had to be

[1] M. Georges Fotino points out that in Rumania this family property may be inherited by women also; this is essentially different from the Slav institution, from which the Rumanian is generally assumed to derive. (*Contribution à l'Étude* . . . p. 117.)

[2] The information relating to Vrancea is from notes prepared by M. Henri Stahl during a sociological inquiry in the summer of 1927, under the guidance of Professor Dimitrie Gusti, who kindly allowed me to see the MS.

bought back from the boiar Roznovanu, the money was collected on the *cislă* system, the mountains being afterwards partitioned among the several villages in proportion to their contribution.

Until 1818 that system of common ownership extended in Vrancea to a whole district, the villages forming a union. From that date the villages proceeded to a voluntary division of their joint property, and only the common properties of individual villages survive.

One remnant of that district commonalty is the joint ownership of the salt mines which are to be found all over Vrancea, and from which salt may be extracted by any inhabitant of the district without any payment whatever. When a decision of some importance is to be taken, each village assembly elects one delegate; the delegates meet and decide the issue, the decision being then submitted by each delegate to his own village.

New villages grow in time round the old settlements, the surplus population emigrating in search of arable land, or setting up new communities on the edge of the common forest and in the midst of the pastures. Sometimes neighbouring villages stretch out until they touch each other, in which case they collect the taxes together and hold the mountains jointly.

If the highland *răzeşi* escaped the covetousness of the boiars, they have in our time to withstand the intrusion of capitalist companies interested in the exploitation of timber. The well-to-do peasants cannot resist the temptation offered by these companies, and since the beginning of the century, therefore, the process which led to the division between villages of their joint property has developed into a tendency to divide up the property of a village among its individual members, with the more astute villagers securing the lion's share. Most village commons have in that way been split up. It would seem that this was done under the impulse of some momentary trend, as many villages have since endeavoured as well as they could to re-create a common grazing. But in these cases each peasant's contribution in land remains nominally his private property, with at any time a theoretical right to enclose it. It is a common formed by mutual agreement, not inherited as such by the village as a whole.

Throughout the process of enclosure, the *răzeşi* have fought for their rights with an obstinacy which has become legendary. In earlier times, disputes within the village group were rare and were settled by the village meeting or elders; the villagers were united by a sense of kinship and by the fact that each used only what he needed. But when outsiders began to intrude and deprive the *răzeşi* of their ancient rights, there began 'almost everywhere the long series of *răzeşi* lawsuits, which in many parts form the history of large property'. (Rădulescu, *Viața Juridică*... p. 33.) Some of the lawsuits lasted tens of years, impoverishing the villagers, but they refused to give in. One could see groups of villagers, with their food in a bag, spending weeks on the road and about the Courts in the defence of their property—'the *răzeşi* with a scrap of land and a sack-full of documents', as the popular saying goes; and many of the lawsuits are not ended.

That the obliteration of *răzeş* rights went hand in hand with the development of corn-growing in Rumania, appears clearly from the fact that the process was most active after the opening of the Black Sea to international shipping (1829), as well as from the distribution of the present *răzeşi*. According to the census of 1912, there were 1,272,519 families in the rural communes of the Old Kingdom (Dobrogea excluded); of the heads of families, 334,234 or 26·2 per cent. were *răzeşi*. About 60,000 of them no longer had any land at all. The *răzeşi* were distributed as follows: 19·33 per cent. in the highland districts, 63·34 per cent. in the hills and 17·33 per cent. in the plain, their proportion decreasing with the altitude of the land, almost none being left in the corn-lands along the Danube. (Poni, *op. cit.*, p. 12.)

APPENDIX II

'MÉTAYAGE' IN RUMANIAN AGRICULTURE

THE development of corn-growing can be dated in Rumania from the opening of the Black Sea to foreign shipping, in 1829, though its main impetus came from the circumstances connected with the Crimean War. Till the middle of the last century Rumanian agriculture was predominantly pastoral, and cattle-rearing was almost completely in the hands of the peasants. In the second place, the beginnings of economic development coincided with the beginnings of political independence; the small landed upper-class was absorbed in public life, and could devote neither time nor money to agricultural pursuits. They wished merely to derive from their estates a revenue with as little personal trouble as possible, and this, together with the fact that the peasants possessed most of the working animals and implements, caused the expansion in corn-growing to be based largely on the métayage system. There was in that respect, however, a marked difference between Rumania's several provinces. In Moldavia, for historical reasons, the peasants were poorer and less well provided with animals and implements, and the same reasons had attracted to that province a more numerous moneyed element of foreign origin. Therefore corn-growing was generally carried on in Moldavia by tenants of very large estates on their own account, the peasants' wages as well as the rent owed by them being calculated on a money basis. In Muntenia and Oltenia the peasants had a better supply of live and dead stock, and in those two provinces métayage was in consequence widespread before the War.

Métayage (*dijmă*) was practised in Rumania under two forms. One of them was the customary form of a sharing of the produce (*dijmă deavalma*), the landlord giving the land and the peasant raising the crop; a second more peculiar form rested on a division of the surface to be cultivated (*dijmă la tarla*), the peasant receving a piece of land for his own use in return for an obligation to cultivate another piece of land for the owner. Less frequent was a combination of these two forms, a division of the produce and, in addition, certain labour obligations for the peasant (*ruşfet*). In general, métayage was resorted to by the landlords only on the poorer soil and for the raising of the less valuable cereal crops, especially maize, which required more labour and cleaned the ground in preparation for wheat. Even in Moldavia that part of the estate which was yearly rented by the peasants had to be laid under maize. During the last twenty years large-scale farming had spread in Muntenia and Oltenia also, but it never covered as much as half of the estate. Large property occupied half of the arable land, but it owned merely one-tenth of the live and dead stock.

M. Garoflid wrote in 1908 that since 1880 the landlord's share had risen in most districts, but especially in the rich cornlands, from 1/5 to 1/2; there were occasional variations, either in the share of the landlord or of the peasant, in accordance with the density of the population and the local demand for land. But the métayage agreements did not rest on a simple sharing of the land or of the harvest; they always contained many other points relating to money loans and their repayment, to grazing, to payments in kind with produce other than that raised in métayage, &c., &c. The more complicated the contract, the wider was the possibility of abuse, especially as few landlords kept proper books. Frequently the division of the harvest was delayed and did not take place until the peasants had entered into an agreement for the following year.

The legislation of 1908 endeavoured to check such abuses by means of various provisions concerning the measurement of the land and the division of the harvest, the nature of the contracts and the manner of their registration, &c. Above all, it prohibited altogether métayage on the *tarla* system, that is on the basis of a division of the land. The landlords were accused of keeping for themselves the best and giving to the peasants the worst land, and also of forcing the peasants to work during the most propitious time on the landlord's part. There is no doubt that this prohibition was often circumvented by means of a double contract, one providing for the letting of land to the peasants and the other for the tilling of the landlord's share. The system thus approached closely to that customary in Moldavia, where the peasants were let a portion of land in return for an obligation to labour on the estate.

M. Garoflid (*Chestia Agrară*, pp. 100–18) considered that the *tarla* system really favoured the peasant. The landlord was forced to let him labour whether the crop made it worth while or not, and at the end of the year the peasant was thus in any circumstances acquitted of his obligations. Under the *deavalma* system the landlord or tenant was unable to introduce a sound crop-rotation, to use expensive seed or manures; and if the harvest failed, the peasant risked being left in the end with a debt for seed, &c. The more general view, however, was that in the second system the chances and risks were shared equally by both parties, and that it tended to make the landlords take some interest in the kind of seed the peasants used and in the way they tilled the soil.

M. Garoflid's conclusion was that the restrictions of the 1908 law would have the effect merely of causing only the least fertile soil to be cultivated in métayage, and only those crops which either demanded much labour or fetched a low price in the market. Otherwise landlords and tenants would find it more profitable to enter into money contracts with the peasants. The main advantage which landowners and their tenants had derived from the métayage system, besides the fact that it required no capital investments and no personal efforts, was that under it all the loss resulting from farming at a great distance fell upon the peasants. Many landowners, therefore, were willing to see the system extended, by means of legal provisions, and thereby secure their revenue while canalizing the peasants' land hunger. After the rising of 1907 a number of schemes were put forward by influential landowners, aiming in essence at the same result. They proposed that a limit should be set by law—some of them put it at 250 ha.—beyond which the land of every estate should have to be let to the peasants in métayage. (Garoflid, *op. cit.*, pp. 42, 65.) In effect that would have signified a return to the feudal tithe-system, with the difference that the peasants would have had to give one-half or more of the produce instead of one-tenth.

The post-war reforms have instead partitioned the great estates in a large measure among the peasants. But métayage has not disappeared; on the contrary, it is now found even in the provinces in which it had been almost unknown. Both the landowners and the peasants have an interest in its continuation. In many cases what is left of an estate can no longer support a family of intellectuals or of *rentiers*; they follow other occupations and therefore tend to let their land be worked in métayage. The uncertainty of the labour-supply has strengthened that tendency. Even some of the holdings distributed at the reform, to officials, &c., are worked in that way. On the other hand, the reform has been unable to give land to all the peasants. Some of them have received none at all, while others not sufficient to employ all their labour, and these men are so anxious to cultivate as much land as they possibly can that they are greatly dissatisfied if an estate which formerly had been worked in métayage is now farmed by the owner himself. Further, the depreciation and fluctuations of the currency have caused the peasants often to prefer a métayage arrangement to money contracts.

As a consequence métayage has spread to Moldavia, and even to Transylvania, where before it was rarely met with. A second change is that métayage arrangements are now made not only between large owners and peasants, but also between the peasants themselves. (G. Ionescu-Siseşti, *Structure Agraire et Production Agricole*, pp. 22–9.) As land was distributed on the basis of social considerations, some peasants have received more than they have means to cultivate, while others have been left with more dead and live stock than they can employ on their own holdings. Not being able to rent additional land from the reduced large estates, they are willing to work in métayage the holdings of small owners who, for one reason or another, cannot farm themselves.

The Ministry of Agriculture carried out in 1922 an inquiry which established that in the Old Kingdom métayage was to be found everywhere; in Transylvania it was altogether absent from only one county, and in Bessarabia and Bucovina from only two counties. The following figures were obtained in a somewhat rough and ready fashion, and were considered by M. Ionescu-Siseşti (*op. cit.*, p. 24) to be below the actual state of things:

	Area cultivated in métayage Hectares	Percentage of total cultivated area per cent.
Moldavia	76,742	5·9
Muntenia	478,932	20·6
Oltenia	54,439	6·0
Dobrogea	37,070	4·7
Bessarabia	88,580	3·5
Bucovina	28,697	12.3
Transylvania	382,707	16·4
Total	1,147,167	11·0

In Moldavia métayage was hardly ever resorted to before the reform. In Dobrogea medium-sized property predominates. In Oltenia there has been a great decrease after the reform, peasants and large owners having in that province been more active in adapting their farming to the new conditions. The high percentages in Bucovina and Transylvania were no doubt due to the fact that the application of the reform was not yet terminated. In Bessarabia the reform had been more radical, so that each landowner knew from the outset that he would not be allowed to retain more than 100 ha. M. Siseşti considered that in 1922 an area equal to about one-half of the above total was worked on the basis of money contracts, with the equipment of the peasants, which were closely allied to métayage. The remainder of the land, covered mainly by small and medium-sized holdings, was worked by its owners themselves.

Most of the métayage arrangements involved in the above table were based on a partition of the harvest. In ten departments the inquiry found a revival of the *tarla* system, which could not be prohibited, the 1908 laws having become inoperative after the reform; M. Ionescu-Siseşti estimated that it did not represent more than 1/11 of the area worked in métayage.

In the *deavalma* métayage agreements are made yearly, one to three months before the beginning of the work. The surface cultivated in métayage changes from year to year, to suit the crop rotation of the large owner or tenant. The harvest is usually divided on a half-and-half basis, the seed being retained by the party which supplied it. Conditions vary with population; and in parts which have suffered much through the War, the agreement is more favourable to the peasant. In Dobrogea there were parts where the landowner received only one-third or even one-fourth of the harvest, the *métayer* being obliged to clear the land which had remained uncultivated during the War; the landowner received one-half only if he supplied the working animals and implements. Likewise in Bucovina the landowner takes one-third or one-half if he first has the land ploughed and cleared on his own account. In the other provinces the landowner takes one-third, two-fifths or one-half of the harvest if he gives merely the land; but one-half, three-fifths or two-thirds if he gives the land already ploughed or ploughed and sown. Cases have been recorded, however, when the landowner gave merely the land and received three-fifths of the harvest or two-thirds, where the demand for land was great. The conditions of sharing vary also with the nature of the crop and the amount of labour it requires.

Generally the value of those labours represents the rental value of the area worked in métayage; but frequently the labours are worth more than the money rental (Ionescu-Siseşti, *op. cit.*, p. 27). Métayage, therefore, is in favour of the landowner when he gets one-half of the harvest, and even more so if he gets a larger share; which explains why the system is still in such great favour. In certain parts most of the métayage agreements were between smallholders, one party being unable to work the holding it had received. The peasants only sell their land in extreme circumstances.

Before the reform the bulk of the peasants depended on the large owners and tenants for land, and they had to accept such conditions as were offered them. Socially, métayage represented the transition from serfdom to yeoman farming; and technically, it was a half-way house to farming on a large scale by the owner or an entrepreneur. The large owner or tenant had an interest in seeing that cultivation was carried on under the best possible conditions that could be obtained without capital investments and expert direction. After the reform, the peasants have generally been in a better position for bargaining, and the tendency among them is to develop intensive methods and crops, which will leave less scope for métayage. That is also the line which large cultivation is likely to follow; the relatively small areas left to the former large owners no longer lend themselves to haphazard extensive cultivation. Only intensive farming, with paid labourers, under the active direction of the owner or tenant, is likely to be profitable on what is left of the large estates.

Appendix III

FINANCIAL ACCOUNTS OF THE REFORM ON JANUARY 14, 1929

A. Expropriated Estates for which Payment has been Claimed

Province	Number of estates	Expropriated area Hectares	Value of mortgages with the Rural Credit Institute taken over by State Lei	Total payments assumed by State Lei	Average payment per ha. Lei
1. Old Kingdom					
First expropriation	4,441	1,944,971	461,997,258	5,342,719,216	2,747
Second ,,	1,448	147,751	9,325,999	381,083,944	2,580
2. Bessarabia . .	2,984	1,073,539	73,598,442[1]	973,680,587	908
3. Bucovina . .	370	54,990	—	345,796,863	6,288
4. Transylvania .	2,259	785,429	—	814,918,362	1,038
Total . .	11,502	4,006,680	544,921,699	7,858,198,972	1,961

[1] Mortgages with Russian Banks

B. Bonds issued until June 28, 1929

	Lei
5% Redeemable bonds . .	5,591,183,500
5% Perpetual bonds . .	585,526,700
Total	6,176,710,200

C. Amounts Debited to the Peasants

Province	Area Hectares	Number of recipients	Surveying expenses Lei	Cost of Holdings Lei	Total price Lei	Average price per ha. Lei
1. Old Kingdom:						
First expropriation	1,427,265	422,220	121,133,755	2,348,121,047	2,469,254,802	1,730
Second ,,	359,857	—	79,530,956	484,074,059	563,605,015	1,566
2. Bessarabia . .	997,893	435,145	76,810,228	800,799,085	877,609,313	879
3. Bucovina . .	45,165	18,435	5,466,842	38,519,982	43,986,824	985
4. Transylvania .	61,318	39,752	9,711,529	129,937,591	139,649,120	2,278
Total . .	2,891,498	915,552	292,653,310	3,801,451,764	4,094,105,074	1,415

D. Payments made by the Peasants during 1922–8

Budget Year	Estimates	Budgetary Revenue [1] (over 18 months)	Surplus Revenue [2]	Total
1922	—	230,031,426	—	230,031,426
1923	187,500,000	493,779,592	—	493,779,592
1924	290,000,000	495,536,952	—	495,536,952
1925	240,000,000	180,356,536	106,233,875	286,590,411
1926	240,000,000	166,207,785	210,773,241	376,981,026
1927	240,000,000	168,677,526	142,625,288	311,302,814
1928	250,000,000	132,943,604 [3]	69,697,211 [4]	202,640,815
Total	1,447,500,000	1,867,533,421	529,329,615	2,396,863,036

[1] Payment of amounts actually debited to the peasants.

[2] Payments in advance, beyond the amounts debited till then. This surplus was handed over to the National Bank, for account of the Treasury; it is now being deposited with the Savings Office (*Casa de Depuneri şi Economii*).

[3] Receipts from January 1, 1928 to April 30, 1929.

[4] Receipts from January 1, 1929 to April 30, 1929.

The first three tables are compiled from figures supplied by the Directorate of Land Mortgage in the Ministry of Agriculture. No explanation was given for the relatively high compensation apparently paid to landowners in Bucovina, nor for the fact that the Transylvanian peasants, according to these figures, were made to pay for the land they received twice as much as the State paid to the expropriated owners. In Bessarabia, it seems, the peasants bear nearly the whole cost of compensation.

The fourth table was obtained from the Ministry of Finance. The peasants have paid off from their debt four hundred millions more than the State expected, and in addition have made advance payments of over half-a-milliard lei; altogether they have paid nearly one milliard lei more than was budgeted. Payments fell off sharply after 1924, as conditions in agriculture became worse.

BIBLIOGRAPHY

BIBLIOGRAPHY TO PART I

ALEXANDRESCO, MIRCEA. La propriété paysanne roumaine et l'égalité du partage successoral. pp. 204. (Thesis.) Paris, 1913.

AMIABLE, L. La Question des Paysans en Roumanie. (Extr. 'Journal des Économistes'.) Paris, 1861.

ANGELESCU, I. N. Origina răzeşilor şi moşnenilor. pp. 116. Bucarest, 1909.

ANTIM, ST. Chestia Ţărănească, pp. 45. Craiova, 1919.

——. La Question Sociale en Roumanie. Paris, 1912.

ARGETOYANU, I. C. Marea noastră proprietate şi exproprierea. Contribuţiuni la Studiul Reformelor agrare. Bucarest, 1913.

ARION, C. La Situation Économique et Sociale du Paysan en Roumanie. Paris, 1895.

ARION, DINOU C. Le Nomos Georgikos et le régime de la terre dans l'ancien droit roumain jusqu'à la réforme de Constantin Mavrocordat. pp. 210. Paris, 1929.

——, SCARLAT C. Starea Ţăranului Român. Bucarest, 1914.

ARICESCU, C. D. Chestiunea Proprietăţii, Desbătută de Proprietari si Plugari la 1848. Bucarest, 1862.

ARSAKE, SP. La Question de la Propriété devant les Assemblées Législatives. pp. 56. Bucarest, 1860.

BIBICESCU, I. G. In Cestiunea Agrară. pp. 137. Bucarest, 1907.

BOGDAN DUICĂ, G. Viaţa şi Opera Intâiului Ţărănist Român, Ion Ionescu dela Brad (1818–91). pp. 149. Craiova, 1922.

BOGDAN, I. Despre cnejii români. Annals of the Rumanian Academy, Bucarest. II series, XXVI, 1903.

BRAESKO, C. Le Paysan Roumain et la Question Paysanne en Roumanie. pp. 364. (Thesis.) Paris, 1906.

BUSUIOCESCU, DR. D. Sistemele de Arendare Aplicate pe Moşiile Academiei. Studiu Economic. Bucarest, 1911.

CARAGIALE, I. L. 1907: din primăvară pînă'n toamnă. pp. 32. Bucarest, 1907.

CARP, P. P. Auswärtige Politik u. Agrarreform. (Reden und Zeitungsartikel.) pp. 78. Bucarest, 1917.

CHEBAP, GR. GH. Regimul Legii Agrare dela 1864 în Efectele lui Asupra Economiei Rurale şi Stării Generale a Ţăranului. pp. 79. Bucarest, 1902.

Colecţie de Toate Instrucţiile şi Deslegările ce s'au dat în Aplicaţia Nouei Legi Rurale. pp. 69. Bucarest, 1864.

Colecţie de Aşezăminte Făcute Inaintea Regulamentului Organic, pentru Statornicirea Drepturilor şi Indatoririlor Reciproce ale Proprietarilor de Moşii şi ale Locuitorilor Săteni din Moldova.

CORNATZEANO, VISORIE I. Du Maximum et du Minimum comme Principe de la Loi Agraire du 23 Déc. 1907 et de l'Opportunité d'une Caisse Rurale en Roumanie. pp. 142. (Thesis.) Paris, 1908.

COSĂCESCU, N. Date şi Observaţii în Chestia Rurală. pp. 30. Bucarest, 1907.

CREANGĂ, DR. G. D. Proprietatea Rurală în România. Bucarest, 1906.

——. Grundbesitzverteilung und Bauernfrage in Rumänien. Berlin, 1914.

——. Der Bauernstand in Rumänien. Brünn. 1901.

——. Creşterea Arenzii Pământului în Bani şi în Dijmă, a Păşunatului şi a Preţurilor Muncilor Agricole dela 1870–1906. pp. xiv, xv, 102. (Min. Interne.) Bucarest, 1908.

Decretele Guvernului Provizoriu şi Manifeste (1848). 'Ţăranul Român.' Vol. ii. No. 44–50. Bucarest, 1863.

Dobrogeanu-Gherea, C. pp. 498. 'Neoiobăgia.' Bucarest, 1908.

Dragu, T. La Politique Roumaine après les Troubles Agraires de 1907. L'Affaire Kogălniceanu. Paris, 1908.

Duca, I. N. Ion C. Brătianu şi Chestia Agrară. pp. 24. Bucarest, 1921.

Filitti, I. C. Clasele Sociale in Trecutul Românesc. pp. 23. Bucarest, 1925.

Fotino, Dr. Georges. Contribution à l'Étude des Origines de l'Ancien Droit Coutumier Roumain. Un Chapitre de l'Histoire de la Propriété au Moyen Age. pp. 460. Paris, 1925.

Garoflid, C. Chestia Agrară în România. pp. 428. Bucarest, 1920. (2nd edition.)

Georgescu, Constant C. Rezultatele Reformelor Agrare din 1907. Bucarest, 1913.

——. C. La Réforme Agraire en Roumanie. (Thesis.) Paris, 1908.

Giurescu, C. Vechimea rumâniei în Ţara Românească şi legătura lui Mihaiu Viteazul. Annals of the Rumanian Academy, Bucarest. II series, xxxvii, 1914, pp. 479–543.

——. Despre rumâni. Annals of the Rumanian Academy. II series, xxxviii, 1915.

——. Organizarea Socială în Ţara Românească. (Lectures delivered 1915–16; stencilled MS. in the Library of the Rumanian Academy.) pp. 213.

——. Despre boieri. pp. 128. Bucarest, 1920.

Golesco, A. G. De l'Abolition du Servage dans les Principautés Danubiennes. pp. 156. Paris, 1856.

Grama, Eftimie. Chestiunea Agrară şi Chestiunea Ţărănească din Muntenia şi Moldova. pp. 50.

Grigoresco, C. Sur la Politique Agraire en Roumanie. pp. 213. (Thesis.) Paris, 1912.

Haret Spiru, C. Chestia Ţărănească. pp. 80. Bucarest, 1905.

Ioaniţescu, D. R. Istoricul Legislaţiei Muncii în România. Bucarest, 1919.

Ionescu, Dr. D. B. Die Agrarverfassung Rumäniens, ihre Geschichte und ihre Reform. pp. 132. Leipzig, 1909.

——, dela Brad. Agricultura Română în judeţul Dorohoi. Bucarest, 1866.

——, Agricultura Romāna in judeţul Mehedinţi. Bucarest, 1868.

——. Agricultura Romāna in judeţul Putna. Bucarest, 1870.

——, Ion. 'Ţăranul Român.' (Periodical, 1860, &c.), pp. 861–3.

Ionescu-Siseşti, G. Politica Agrară cu Privire Specială la România. pp. 206. Bucarest, 1910.

Iorga, N. Constatări istorice cu privire la viaţa agrară a Românilor. pp. 91. Bucarest, 1908.

——. Situaţia Agrară, Economică şi Socială a Olteniei în Epoca lui Tudor Vladimirescu. Documente Contimporane. Bucarest, 1915.

——. Développement de la question rurale en Roumanie. pp. 58. Jassy, 1917.

——. Évolution de la question rurale en Roumanie jusqu'à la réforme agraire. (Paper read before the International Agricultural Congress, Bucarest 1929.) pp. 24.

Kogălniceanu, Mihail. Autobiografie. pp. 84. Bucarest, 1916.

——. Imbunătăţirea Soartei Ţăranilor. Cuvântu Rostit în Adunarea Generală a României, şedinţa din 25 Maiu 1862. pp. 88. Bucarest, 1862.

——, V. M. Chestiunea Ţărănească. pp. 109. Bucarest, 1906.

——, V. M. Improprietăririle Ţărăneşti. pp. 32. Bucarest, 1906.

Leon, N. George. Istoria Economiei Publice la Români. pp. 142. Bucarest, 1924.

Leonescu, N. V. Anul 1907. Răscoala Ţăranilor. pp. 52. Iaşi, 1924.

LOVINESCU, E. Critice: I. Istoria Mişcării 'Semănătorului'. pp. 208. Bucarest, 1925.
——. Istoria Civilizaţiei Române. 3rd vol. Bucarest, 1924–5.
LUPUS, JUSTIN. Măna Moartă şi Răscumpărarea Averilor ei. Bucarest, 1911.
MALTEZIANU, DR. CONST. Die Rumänische Agrarbewegung. Berlin, 1913.
MĂNDREA, R. Politica Agrară Conservatoare. pp. 17. Bucarest, 1906.
MANOLESCO-MLADIAN, S. L'Évolution de la Classe Moyenne en Roumanie. (Thesis.) Paris, 1909.
MITRANY, D. Rumania. (In 'The Balkans'.) Oxford, 1915.
MORUZI, SEBASTIAN. Câte-va vederi în chestia ţărănească. pp. 30. Fălticeni, 1907,
MOTEANU, C. Consideraţiuni asupra pulverizării proprietăţii ţărăneşti. 'Analele Statistice şi Economice', No. 5, 1918.
NEGULESCO, P. Étude sur le protimis dans l'ancien droit roumain. 'Nouvelle Revue Historique', xxiii, 1899, pp. 213–24.
OPRESCU, ANTON. Din Trecutul Agriculturei Noastre. 'Viaţa Agricolă', January–August 1921.
PANGAL, LÉON. La Question Agraire. pp. 30–7. Paris, 1914.
PANU, GHEORGHE. Cercetări Asupra Stărei Ţăranilor în Veacurile Trecute. Bucarest, 1910. 1 vol. (2 pts.) pp. 652.
PĂUCESCU, GR. Imbunătăţirea Soartei Ţăranilor (1888). (Political Fragments.) pp. 69.
PHILIPPIDE, A. S. Incercări Asupra Stărei Sociale a Poporului Român în Trecut. (Pt. II. Proprietatea După Actele Publice). pp. 83. Iaşi, 1896.
POPESCU, DR. CONSTANTIN. De ce Trebue Imbunătăţită Starea Sătenilor. pp. 138. Bucarest, 1914.
Proecte de Legi Rurale Prezentate Corpurilor Legiuitoare în Sesiunea 1907–8. Bucarest, 1907–8.
Programul Democrat-Radical. pp. 49. Bucarest 1890.
RACOVSKY, C. Chestia agrară. Articles in 'Viitorul Social', Bucarest, Aug.–Sept., 1907.
RĂDULESCU, ANDREI. Dreptul de protimis la arendare. 'Dreptul', xxxvi. No. 40, 1907.
RĂDULESCU-MOTRU, C. Cultura Română şi Politicianismul. pp. 254. Bucarest. (without date).
RĂŞCANU, DR. TEODOR. Problema Pământului în România. pp. 282. Bucarest, 1922.
RATEŞ, STEFAN. Viaţa Agrară a Românilor din Ardeal. Documente Contimporane. Vol. i. 1508–1820. pp. 3021. Bucarest, 1922.
Regulament Privitor la modul de întrebuinţare a Islazurilor Communale. (Ministerul Agr. şi Domeniilor.) pp. 14. Bucarest, 1910.
ROSETTI, C. A. Discursul Asupra Tocmelelor Agricole Rostit în Şedinţa Camerei dela 12 Feb. 1882. pp. 44. Bucarest, 1882.
——, RADU. Acte şi Legiuiri Privitoare la Chestia Ţărănească. (2nd Series.) Ploeşti–Bucarest, 1907–8.
——. Pământul, Ţăranii şi Stăpânii în Moldova. pp. 555. Bucarest, 1905.
——. Pentru ce s'au Răsculat Ţăranii. pp. 699. Bucarest, 1908.
ROSETTI, GENERAL R. L'influence du régime de la propriété foncière sur l'organisation, la tactique et la stratégie des armées roumaines au XVe siècle. (Paper read before the sixth Historical Congress, Oslo 1928.) pp. 8. Bucarest.
SANIELEVICI, H. Poporanismul Reacţionar. pp. 440. Bucarest, 1921.
ŞĂRBESCU-LOPĂTARI, D. Propietatea Moşnenească în Indiviziune. Studiu Economico-Social. pp. 75. Buzău, 1906.

ŞERBAN, MIHAIL. Agrarpolitische Studien über Rumänien. pp. 68. (Thesis.) Halle-Wittenberg, 1913.

SIMU, SIMEON. Individualismul şi Pământul la Românii şi Barbarii Daciei sau Procesul de Individualizare Socială a Barbarilor Daciei în Primele opt Veacuri. pp. 43. Lugoj, 1924.

SIRIANU, ION RUSU. Iobăgia.

SLĂVESCU, VICTOR. Die Agrarfrage in Rumänien. pp. 169. (Thesis.) Halle-Wittenberg, 1914.

STERE, C. Articles in the 'Viaţa Românească', 1905, &c. Jassy.

STURDZA, DIMITRIE, A. Memoriu Aspura Legilor Agrare din România. pp. 148. Bucarest, 1914.

STURDZA-SCHEIANU, D. C. Acte şi Legiuiri. Bucarest, 1906 &c.

TANOVICEANU, I. Formarea proprietăţei funciare în Moldova. (In 'Prinos lui D. A. Sturza'.) pp. 413–33. Bucarest, 1903.

TAŞCĂ, G. Les Nouvelles Réformes Agraires en Roumanie. pp. 212. (Thesis.) Paris, 1910.

VLĂDESCU-OLT, M. Improprietăririle din Dobrogea. pp. 107. Bucarest, 1905.

XENOPOL, A. D. Istoricul chestiunei agrare. 'Le Mouvement Économique.' Bucarest, June 1910.

——. Proprietatea mare şi cea mică. 'Viaţa Românească.' Iaşi, viii, 1913.

——. Proprietatea mare si cea Mică în Trecutul Ţărilor Române. Iaşi, 1913.

'Ziarist.' Factorul Nou (Ţărănimea). Colecţie de Articole scrise înainte şi după Răscoala din 1907.

BIBLIOGRAPHY TO PART II

Agrarian Reform in Eastern Europe. 'The Economist', London, 19–26 Aug. 1922.

Agrarreform. pp. 48. Sibiu, 1921.

ALEXANDRESCU, TRAIAN. Drepturile Proprietarilor şi Creditorilor Ipotecari în raport cu Legislaţiunea Agrară. Studiu Teoretic şi Practic, Urmat de Instrucţiunile şi Formularele necesare Pentru Incasarea Titlurilor de Rentă. pp. 64. Bucarest, 1923.

ALVAREZ, ALEJANDRO, &c. Agrarian Reform in Roumania and the case of the Hungarian optants in Transylvania before the League of Nations. (Opinions by legal authorities.) pp. 320. Paris, 1927.

ANTONESCU, MIHAIL. Al II-lea Memoriu Aprobat de Adunarea Generală a Proprietarilor şi Arendaşilor. (Uniunea Generală a Sindicatelor Agricole.) pp. 8. Bucarest, 1918.

——. Regimul Agrar Român şi Chestiunea Optanţilor. Bucarest, 1928.

ARONESCU, G. Ce Trebuie să Ştie şi să Facă un Propietar Faţă de Noua Lege a Expropierei. pp. 35. Câmpulung, 1921.

ARSENESCU, VALERIU. Condiţia Juridică a Subsolului Faţă cu Reforma Agrară din Vechiul Regat. pp. 39. Bucarest, 1925.

AVERESCU, AL. Sărbătorirea Reformei Agrare. pp. 16. Bucarest, 1921.

Balkanisation of Agriculture. Budapest, 1921.

BANESCU, EUGENIU P. Fixarea Preţului Pământului Expropiat Pentru Cauză de Utilitate Naţională în V. Regat. pp. 54. Craiova, 1921.

BASILESCO, N. La Réforme Agraire en Roumanie. pp. 260. Paris, 1919.

Bergheanu, V. Păreri Asupra Reformei Agrare. Exproprierea. pp. 70. Tighinea, 1919.

Bogdan, F. I. Desnaţionalizarea Românilor Ardeleni, sau Politica Agrară Ungurească. pp. 64. Bucarest, 1916.

Bonjug-Alibeichioi, N. I. Făţărnicia Partidului Naţional-Liberal Faţă de Improprietărirea Ţăranilor dela 1858–1918. pp. 19. Tulcea, 1922.

Borchard, Edwin M. Opinion on the Roumanian-Hungarian Dispute. pp. 67.

Bosiano, C. La Politique Paysanne en Roumanie depuis la Guerre. pp. 96. (Thesis.) Paris, 1920.

Bontescu, Victor. Reforma Agrară din Ardeal sub Raportul Financiar. 'Voinţa.' Cluj, April 15, 1921.

Brăileanu, Const. Chestia Agrară cum Trebue Privită, cum Trebue Deslegată. pp. 48. Bucarest, 1919.

Brătăşanu, Pavel. Improprietărirea Ţăranilor şi Reformele Democratice. pp. v +33. Bucarest, 1919.

Braesco, J. et Sescioreano, G. La réforme agraire en Roumanie. 'Bulletin de la Société des Législations comparées,' Paris. July–Sept., 1925.

Buescu, Vasile P. Terenurile din Zona de Inundaţie a Dunărei Faţă cu Exproprierea. Constituţia Modificată de Legea Agrară. pp. 44. Bucarest, 1921.

Burbur, W. E. de. Despre Reforma Agrară. 'Viaţa Agricolă.' Bucarest, 1 Aug. 1922.

Cantacuzino, Mihail. Discurs rostit la Cameră. pp. 33. Bucarest, 1919.

Căpăţineanu, D. G. Discurs la Reforma Agrară. pp. 52. Bucarest, 1921.

Cardas, Dr. Agricola. Reforma Agrară în Basarabia. 'Moldova Agricolă.' Jan. 1922. An. 1, Nos. 6–7.

Carp, P. Partidele Politice şi Reformele Constituţionale. pp. 92. Iaşi, 1917.

Cecropid, N. La Loi Agraire en Roumanie et ses Conséquences Économiques. pp. 111. (Thesis.) Paris, 1924.

Chiriţă Paraschiv, B. Prădarea Moşnenilor. pp. 69. Râmnicu-Vălcea. 1921.

Chiţoiu, D. G. In apărarea reformei agrare din România. 'Viaţa Agricolă,' an. xiv, No. 20, 1923.

——. Reforma agrară din România. 'Buletinul Agriculturii,' an. V, No. 10–12, 1924.

Comitetul Agrar. Direcţiunea Islazurilor, Instrucţiuni relative la Aplicarea Legii pentru Infiinţarea Păşunelor Comunale. pp. 45 'Mon. Oficial', 24 Sept. 1920.

Conacher, H. M. Agrarian Reform in Eastern Europe. 'International Review of Agricultural Economics.' Rome, Jan.–March 1923.

Confiscarea Pădurilor Particulare din Basarabia. pp. 16. Bucarest, 1922.

Connert, Fritz. Zur Frage der Agrarreform in Siebenbürgen. 'Nation und Staat.' Vol. i, No. 4. Vienna, 1928.

Constantinescu, Alex. Declaraţiile cu Privire la Discuţia Reformei Agrare la Senat. pp. 89. Bucarest, 1921.

——, Mitiţă. L'Évolution de la Propriété Rurale et la Réforme Agraire en Roumanie. pp. 479. Bucarest, 1925.

Coruţiu, Petre. Reforma Agrară. Studiu Parţial asupra celui de al doilea proect al dlui Garoflid. 'Voinţa,' Cluj. an. I, No, 110–53.

Creţu, T. Cuvântare Asupra Legii Agrare. pp. 51. Bucarest, 1921.

Cristodorescu, Const., and Ştefănescu-Priboi, Dem. C. Codul Legislaţiunei de Expropiere Pentru Utilitate Naţională, Pentru Uzul Organelor de Aplicare şi Părţilor Interesate la Expropiere. pp. 285. Bucarest, 1922.

Criza 'Comunităţii de Avere' cu un Mănuchi de Dovezi şi o Muncă Depusă Intru Apărarea Intereselor Graniţei. pp. 64. Caransebeş, 1922.

Currey, Muriel. The Hungaro-Rumanian Dispute. pp. 34. London, 1929.

Danielopol, G. D., &c. La Loi Agraire. L'Expropriation des Sujets Étrangers en Roumanie. pp. 30. Bucarest, 1921.

——. Reforma Agrară şi Politica. pp. 7. Iaşi, 1917.

Daşcovici, N. La réforme agraire en Roumanie et les optants hongrois de Transylvanie devant la Société des Nations. Paris, 1924.

Deák, Francis. The Hungarian-Rumanian Land Dispute. pp. ix+272. New York, 1928.

Decret Lege Relativ la Reforma Agrară Votată de Sfatul-Ţării din Basarabia. 'Monitorul Oficial,' Nr. 220. pp. 16. Bucarest, 1918.

Diaconescu, Ilie S. Chestiunea Ţărănească în România. pp. 366. Bucarest, 1928.

Dobos, Filaret. Reforma Agrară o Necesitate Economică, Socială şi Naţională. pp. 16. Cernăuţi, 1921.

Dobrescu, Aurel. Reforma Agrară. pp. 36. Bucarest, 1921.

——, Dem. I. Cuvântare la Senat. pp. 112. 1921.

Dragu, Toma. Reforma Agrară. pp. 22. Bucarest, 1921.

Dron, Pr. C. Legea pentru Reforma Agrară. Păreri şi Reflexiuni după Votare. 'Viaţa Românească.' Iaşi, May 1922.

Duca, I. G. Ion C. Brătianu şi Chestia Agrară. pp. 24. Bucarest, 1921.

Dugdale, Mrs. Edgar. The Hungaro-Rumanian Dispute. pp. 78. London, 1928.

Dumitrescu-Bumbeşti, G. Împroprietărirea Ţăranilor. Discurs la Senat. pp. 45. 1921.

Emandi, Teodor G. Pământ Obştiilor, nu Ţăranilor. pp. 16. Bucarest, 1919.

Evans, I. L. The Agrarian Revolution in Rumania. pp. 197. Cambridge, 1924.

Fenner, Gerhard & Loesch. Die Neuen Agrargesetze der Ost und Südosteuropäischen Staaten. Vol. i, pp. 258. Berlin, 1923.

Filipciuc, Dr. Vasile. Problema Maramureşană. pp. 12.

——. Reforma Agrară în Maramureş. pp. 16. Turda.

Filitti, I. C. Preţul Pământului Expropiat. pp. 15. Bucarest, 1920.

Florescu, Const. Discurs la Senat. pp. 15. Bucarest, 1921.

Florescu, M. P. Împroprietărirea şi sacrificiul naţional al marei propietăţi. 'Pagini Agrare şi Sociale,' Bucarest. an. II. No. 6–7. 1925.

——, Radu A. Pentru Dreptul Românesc în Cauza Ţărănească. pp. 50. Bărlad, 1919.

Garoflid, Const. Memoriul Agricultorilor Mari cu Privire la Expropiere. pp. 45. Bucarest, 1920.

Georgescu, Ghinaru Leonida. Indrumări Noi Ţărănimei. pp. 24. Craiova, 1922.

——, Const. Pământ Pentru Ţărănime şi Vot Obştesc. Discurs la Cameră. pp. 19. 1919.

——, Gr. Discurs la Senat. pp. 62. 1921.

Ghilezan, Dr. Liviu T. Chestiunea Băncii Agrare din Cluj. pp. 15. Bucarest, 1921.

——. Contribuţiuni la Noua Reformă Agrară Pentru Transilvania. pp. 132. 1921.

Ghica, G. Chestiunea Proprietăţii în Dobrogea. Bucarest, 1880–1.

Gonnard, René. La Réforme Agraire dans les Pays de l'Europe Centrale. 'Revue Politique et Parlementaire,' 10 June 1921.

Grigorescu, Petre. Memoriu Asupra Exproprierilor în Dobrogea. pp. 15. Constanţa, 1921.

IOACHIMOVICI, EMIL. Bolşevismul sau Atentatul contra Proprietăţii. pp. 135. Bucarest, 1922.

IOAN, NICOLAE. Reformele Agrare şi Administrative. pp. 26. Călăraşi, 1920.

IARCA, DEM. C. Preţul Pământului Expropiat. pp. 16. Bucarest, 1921.

Instrucţiuni Asupra Constituirei Obştiilor Prin Care Sătenii Intră din Primăvara Anului 1919 in Folosinţa Terenurilor expropriate din Marile Proprietăţi. Circulara 625. pp. 16. Caracal, 1919.

IONESCU-SISEŞTI, G. Reforma Agrară în diferite Ţinuturi ale României. 'Arhiva p. S. s. R. S.', Jan. 1920.

——. The Agrarian Reform in Rumania. 'Manchester Guardian,' Reconstruction Supplements. Vol. vi. 1923.

——. Evoluţia propietăţilor rurale. 'Argus,' an. XV, No. 3255, 1924.

IORGANDA, GEORGE SP. Legătura de Brazdă. Expropierea Embaticului. pp. 70. R.-Sărat, 1922.

ISVORANU, I. T. Discurs Asupra Reformei Agrare la Senat. pp. 21. 1921.

JORNESCU, C. Improprietărirea Ţăranilor. pp. 24. Discurs la Cameră. Bucarest, 1921.

JOUSSE, PIERRE. Les Tendances des Réformes Agraires dans l'Europe Centrale, l'Europe Orientale et l'Europe Méridionale (1918–24). pp. viii+238. Paris, 1925.

KOSITCH, DR. MIRKO. Agrarne Reforme u Rumuniji (Serbian) 'Letopis', Novisad. November 1924. pp. 36–46.

La Réforme Agraire en Roumanie et les Optants Hongrois de Transylvanie devant la Société des Nations, Mars–Juillet 1923. Paris, 1924.

LAHOVARY, EM. N. Chestia Reformelor. pp. 36. Bucarest, 1919.

LATEA, G. I. Asupra Problemei Agrare. 'Viaţa Agricolă.' Bucarest, 15 Feb. 1921.

LAZAR, L. La Mise en Œuvre de la Réforme Agraire en Roumanie. pp. 127. (Thesis.) Paris, 1924.

Legiferările Partidului Poporului. Improprietărirea sătenilor. Legea Agrară Scrisă pe Inţelesul Tuturor. pp. 105. Bucarest, 1921.

LEON, DR. GEORGE. Dreptul de Expropriere. pp. 45. Iaşi, 1918.

LEONARDESCU, GR. Câte-va Observaţiuni Asupra unor Articole din Legea de Reformă Pentru Basarabia. pp. 14. Chişinău, 1922.

Ligue pour la Protection des Minorités Nationales de la Roumanie. Mémoire au Sujet de la Réforme Agraire de Roumanie. Budapest, 1921.

L'importance Économique et Sociale de la Réforme Agraire en Roumanie. 'L'Europe Nouvelle.' Paris, 1925.

LUPU, DR. N. Pătura Conducătoare Faţă de Ţărănime. Discurs la Cameră. pp. 61. Iaşi, 1917.

MĂGURĂ, I. Discurs la Senat. pp. 36. Bucarest, 1921.

MANTOU, GEORGES. L'Expropriation de la Grande Propriété Rurale et la Question Agraire en Roumanie. 'Journal des Économistes.' Paris, 1920.

MARGHILOMAN, AL. Criza Economică şi Soluţiuni. pp. 38. Bucarest, 1921.

MASLOV, SERGĚJ S. Princip Soukromého Vlastnictví v Pozemkových Reformách Poválečné Evropy. (Czech.) pp. 45. Prague, 1927.

Memoriu Agricultorilor Avănd peste 25 ha. Pământ din Basarabia. pp. 35. Chişinău, 1921.

Memoriu in Chestiunea Exproprierii Pădurilor Grănicereşti din Fostul District al Nasăudului. pp. 20. Bucarest, 1923.

Memoriul Marilor Proprietari Agricultori din Moldova Asupra Reformei Agrare. pp. 24. Iaşi, 1920.

MIHAIL, MARGARETA D. La Roumanie Agricole après la Guerre Mondiale. pp. 102. (Thesis.) Paris, 1921.

MIHALACHE, ION. Proectul Legei de Improprietărirea Ţăranilor. pp. 63. Bucarest, 1920.

——. Dreptul Ţăranilor la Pământ, Islazuri şi Păduri. pp. 176. Bucarest, 1922.

Ministère de l'Agriculture et des Domaines. La Réforme Agraire en Roumanie (Ancien Royaume et Bessarabie). pp. 76. Bucarest, 1919.

MIRONESCU, C. M. Discurs la Senat. pp. 39. Bucarest, 1921.

MITRANY, D. The Transylvanian Land Dispute. 'Foreign Affairs.' London, April 1928.

MOLDOVAN, LEONTE. Reforma Agrară. Păreri Asupra Anteproectului Partidului Naţional Liberal şi Asupra Decretului Lege de Expropriere. pp. 31. Brăila, 1919.

MOTEANU, C. Legea arendărilor obligatorii şi rezolvarea problemei agrare. 'Analele St. şi Ec.' No. 9, 1918.

Muncile Agricole în 1917. Referate, Decrete, Regulamente, Instrucţiuni. pp. 204. Iaşi, 1918.

NASTA, Al. La réforme agraire en Roumanie. (Report submitted to the International Agricultural Congress, Bucarest, 1929). pp. 30.

NEDELCEANU, I. Legi şi Regulamente Privitoare La Proprietatea Imobiliară din Dobrogea. Bucarest, 1919.

New Agrarian Legislation in Central Europe. 'International Labour Review.' Geneva, Sept. 1922.

NICULESCU, IOAN LUCA P. Răvaş către săteni. pp. 89. Bucarest, 1922.

ODOBEŞTIANU, R. La Propriété Agraire en Roumanie. pp. 120. (Thesis.) Montpellier, 1925.

OPRIŞANU, GEORGE. Exproprierea Rurală. pp. 46. Bărlad, 1924.

OSVADĂ, V. C. Reforma Agrară în Transilvania, Banat, Crişana şi Maramureş. pp. 60. 1921.

PACATIAN, TEODOR. Legea Despre Islazurile Comune neimpărţite. pp. vii+45. Sibiu, 1913.

PARTIDUL ŢĂRĂNESC. Proectul de Program al Partidului Ţărănesc din România. pp. 59. Bucarest, 1921.

PĂRVULESCU, GEORGE D. Domnia Legilor sau Anarhie ? Constituţiunea, Exproprierea şi păşunile comunale. pp. 72. Tg. Jiu, 1921.

PICHA, ZDENKO. Bodenreform in Rumänien. Prague, 1920.

POPESCU, LAZAR and RĂDULESCU TOMA. Legea Pentru Infiinţarea Păşunelor Comunale cu Discursurile la Senat a Dlor . . . pp. 63. 1920.

——, PION. Réclamation des Optants Hongrois de Transylvanie contre la Réforme Agraire en Roumanie. Bucarest, 1927.

POPOVICI-LUPA, N. O. Preţuirea Pământului Expropriat. 'Viaţa Agricolă.' Bucarest, 1st March 1921.

POPOVICI, I. MAX. Chestiunea Colonizărilor ce se fac Totodată cu Reforma Agrară. 'Viitorul.' 8 and 23 February 1921.

PORUTIU, PETRE. Reforma Agrară. Studiu Parţial Asupra celui de al Doilea Proect al Dlui Garoflid. 'Voinţa,' Cluj. January–March 1921.

POTÂRCĂ, VIRGIL. Legea Agrară. Legea cu Comentarii Juridice. pp. 295. Craiova, 1921.

Proect de Lege de Reformă Agrară Elaborat de Comisia Agrară din Sfatul Ţării. pp. 13. Chişinău, 1918.

PRECUP, VICTOR. Pădurile şi Reforma Agrară din Transilvania. Revista Pădurilor. Bucarest, 1922.

RADUŢĂ, N. Scrisoare Către Ţărani. pp. 48. Târgovişte, 1921.

Reforma Agrară. Cercul de Studii al Partidului Naţional-Liberal. 1. Legea Exproprierei. 2. Legea Improprietărirei. 3. Legea de Organizare a Casei Ţărăneşti. pp. 40. Iaşi, 1918.

Reichsministerium für Ernährung und Landwirtschaft. Berichte über Landwirtschaft. New Series, No. 1. (Contains a Study on Agrarian Reform in Rumania.) Berlin, 1923.

ROMANESCU, NICOLAE P. Reforma Agrară. pp. 95. Bucarest, 1921.

ROSETTI, RADU. O Schiţă de Reformă Agrară. April 1919. (Unpublished proofs in the Library of the Rumanian Academy.)

RUMER, WILLY. Die Agrarreformen der Donau-Staaten. pp. 170. Innsbruck, 1927.

SACHELARIDE, C. Câteva Reflecţiuni Asupra Reformei Agrare în Vechiul Regat. 'Viaţa Agricolă,' 15 August 1921. Bucarest.

SAVIN, G. I. Ridicarea Individualităţii Pământurilor Rurale. 'Viaţa Agricolă.' Bucarest, 15 April 1921.

SCHIAVI, AL. La Reforma Agraria in Rumenia e i suoi primi effetti. 'La Riforma Sociale.' Turin, November–December, 1927.

SCHIFF, PROF. WALTER. Die Grossen Agrarreformen seit dem Kriege. pp. 37. Vienna, 1926.

SEBESS, DENES VON. Landownership Policy of New Rumania in Transylvania. Budapest, 1921.

SERING, PROF. MAX. Die Umwälzung der Osteuropäischen Agrarverfassung. pp. 20. Berlin, 1921.

SIMU, S. Criza 'Comunităţii de Avere'. pp. 64. Caransebeş, 1922.

STACCA, G. La réforme agraire en Roumanie. 'Bulletin du Bureau International Agraire,' Prague, vol. 5–6, 1925.

Statutele Obştei de Arendare Constituită în Vederea Improprietărirei. Ed. II. pp. 27. Bucarest, 1919.

STINGHE, V. N. Pădurile şi Legea Păşunilor Comunale. (Bibl. Economică Forestieră, No. 15.) pp. 15. Bucarest, 1921.

STURZA, R. V. Neconstituţionalitatea Legii Agrare. pp. 14. Bucarest, 1921.

SZÁSZ, ZSOMBOR DE. The Minorities in Roumanian Transylvania. pp. 414. London, 1927.

TĂTĂRĂSCU, GH. Pe Drumul Anarhiei. pp. 20. Bucarest, 1920.

The Expropriation of Land in Rumania. 'Nation,' vol. 110, 14 February. New York, 1920.

TIMOC, IOAN. Stabilirea Valorii Pădurilor Expropiate pe Baza Legii Agrare 'Revista Pădurilor'. Bucarest, 1922.

TONESCU, VASILE. Exproprierea. Art. 19. din Constituţie. Decretele Legi. Preţurile Regionale de Arendare. Comisiunile Judeţene. Casa Centrală. Adnotate şi Comentate. Ed. 2–a. pp. 128. Bucarest, 1919.

TRANCU-IAŞI, GR. Reforma Agrară şi Electorală. Discurs. pp. 99. Iaşi, 1919.

TUTUC, I. Reforma Agrară, Plata Pământului. 'Independenţa Economică.' Bucarest, January–March 1921.

VĂLCOVICI, VICTOR. Reforma Agrară, Exproprierea. pp. 31. Iaşi, 1918.

VELICAN, V. I. Cum se Interpretează Art. 68. din Ref. Legii Agrare de Către Comisiunile de Ocoale din jud. Gorj. 'Viaţa Agricolă,' 15 September. Bucarest, 1922.

VELICAN, V. I. Ogorul şi Plugarul. pp. 31. Târgu-Jiu, 1921.
VRĂBIESCU, IULIAN C. Reforma Agrară şi Electorală. pp. 31. Craiova, 1917.
WAUTERS, ARTHUR. La réforme agraire en Europe. p. 292. Bruxelles, 1928.

BIBLIOGRAPHY TO PART III

AGRONOMILOR, SOCIETATEA. Sporirea Producţiunii Agricole a Ţării. pp. 296. Bucarest, 1921.
ALIMĂNIŞTEANU, AL. Organizarea Muncei Rurale. pp. 24. Bucarest, 1927.
ANASTASIU, A. ORESTE. Contribuţiunile Directe ale Societăţilor Anonime. 'Buletinul Institutului Economic Românesc,' November–December 1928.
—— Industriile Săteşti. pp. 205. Bucarest, 1928.
ANGELESCU, DR. C. Activitatea Ministerului Instrucţiunii, (1922–6.) pp. 87. Bucarest, 1926.
——. Lege Pentru Invăţământul Primar al Statului. pp. 95. Bucarest, 1925.
ANTIPA, DR. G. L'Occupation Ennemie de la Roumanie et ses Conséquences Économiques et Sociales. pp. 176. Paris, 1929.
——. Paralizia Generală Progresivă a Economiei Naţionale şi Mondiale. pp. 56. Bucarest, 1923.
——. Problemele Evoluţiei Poporului Român. pp. 406. Bucarest, 1919.
ASOCIAŢIA ECONOMIŞTILOR. Criza de Aprovizionare în România. 'Comunicări,' February–March, 1919. Bucarest, 1919.
BĂICOIANU, C. I. Exproprierea şi Lărgirea Colegiilor Electorale. pp. 16. Bucarest, 1914.
BĂLUŢĂ, CRUCERU C. Norme Pentru Parcelarea Terenurilor Expropriate în 1919 pentru Interes de Utilitate Naţională. Studii şi Propuneri cu Privire la Reforma Agrară. pp. iv+76. Bucarest, 1920.
BAMBERGER, F. Agrarian Reform in Rumania. (In 'Agrarnaya Revolyutsiya v Evrope'. Prof. Max Sering, Editor.) (in Russian). pp. 25–70. Berlin, 1925.
BANU, DR. G. Biologia Satelor. 'Arhiva p. S. s. R. S.', vol. iv, Nos. 1–2.
BECKMANN, DR. FRITZ. Die Internationale Agrare Arbeitsteilung Europas. pp. 24. Jena, 1926.
BELDIE, V. C. Cronica Agrară. Situaţia Noului Improprietărit. Câţi Ţărani Rămân Fără Pământ. 'Viaţa Românească.' Iaşi, 1921.
BERCARU, VALERIU. La Réforme Agraire en Roumanie. pp. 91. Paris, 1928.
BEZA, M. The Rumanian Church (in Papers on the Rumanian People and Literature). London, 1920.
BLANK, ARISTIDE. Contribuţiuni la Rezolvarea Crizei Economice. 'Arhiva p. S. s. R. S.', vol. iv, No. 1.
BRANCOVICI, E. M. Importanţa Agriculturii şi în Special a Cerealelor din România-Mare (Faţă de Industrie). 'Lupta Economică,' 10 Jan. 1922. Bucarest.
BRĂTIANU, VINTILĂ I. Burghezia de Eri şi de Mâine. 'Democraţia,' Jan.–Feb. 1922. Bucarest.
—— Foloasele Eproprierei. pp. 14. Bucarest, 1914.
BUNGEŢEANU, D. Câteva Cuvinte Despre Viaţa Casnică şi Agricolă a Populaţiei Rurale. pp. 25. Severin, 1885.
CAMARACHESCO, JEAN. L'Agriculture Roumaine est fonction des conditions naturelles du climat et du sol. (Paper read before the International Agricultural Congress, Bucarest, 1929.) pp. 18.

CARABELLA, A. D. Grâul Romănesc din Recolta 1923. pp. 15. Bucarest, 1923.

CARDAS, DR. AGRICOLA. Aspecte din Reforma Agrară Basarabeană. pp. 53–131+pl. Chişinău, 1924.

CAZACU, AL. Producţia şi consumaţiunea băuturilor alcoholice în România'. 'Buletinul Statistic,' July–Sept. 1928.

Ce Impozite Plătesc Ţăranii Către Stat. pp. 8. Bucarest, 1921.

CHRISTODORESCU, GH. Problema Agrară. Agricultura Ţărănească, Starea ei de Inferioritate Faţă de Agricultura de pe Marile Exploataţiuni. Bucarest, 1914.

CIOMAC, ION L. Privire Asupra Diviziunii, Specialităţii şi Organizării Muncii în Agricultură. 'Viaţa Agricolă,' 1 April 1922. Bucarest.

CIORICEANU, GEORGES D. La Roumanie Économique et ses Rapports avec l'Étranger de 1860 à 1915. pp. 443. Paris, 1927.

CIPĂIANU, C. Exploataţiunile Agricole Faţă cu Noile Impozite. pp. 12. Bucarest, 1921.

——. Inrăurirea Reformei Agrare Asupra Producţiunei Agricole în România Intregită. pp. 20. Bucarest, 1921.

CIPĂIANU, G. Desvoltarea Agriculturii în ultima sută de ani (1829–1929). 'Bul Inst. Ec. Rom.', May–June, 1929.

CIUPAGEA, GH. Trei Gospodării Ţărăneşti. 'Viaţa Agricolă,' 1 July 1921. Bucarest.

COLESCU, DR. L. Comerţul Exterior al României Inainte şi După Războiul Mondial. 'Buletinul Institutului Economic Românesc,' November–December, 1928. Bucarest.

CONNERT, FRITZ. Zur Frage der Agrarreform in Siebenbürgen. 'Nation und Staat', vol. i, No. 4. Vienna, 1928.

CONSTANTINESCU, MARIUS and POPESCU, TEODOR. Monografia Comunei Petroşani (Vlaşca). pp. 40. Bucarest, 1922.

——, G. N. Reforma Alimentară şi Creşterea Animalelor Sub Noul Regim Rural. 'Independenţa Economică,' January–March 1921. Bucarest.

CORNĂŢEANU, DR. N. D. Die Bedeutung und Rentabilität der Viehzucht in Rumänien nach Der Agrarreform. pp. 114. Bucarest, 1928.

CREANGĂ, DR. G. D. Veniturile şi Averea României Mari 'Buletinul Institutului Economic Românesc'. January 1927.

D. B. The Agrarian Situation in Rumania. (In Russian.) 'Na Agrarnom Fronte,' No. 4. Moscow, 1925.

Débats du Parlement National Paysan réuni à Bucarest (Juillet 26–7, 1928). pp. 64. Bucarest, 1928.

Directoratul Jandarmilor Rurali. Mersul şi cauzele delictelor şi crimelor săvărşite pe teritoriul rural al ţării româneşti dela 1918–23. Bucarest, 1924.

DOBROGEA. Cincizeci de Ani de Viaţă Românească, 1878–1928. pp. 793. Bucarest, 1928.

DUŞIAN, I. Chestiunea Agrară în Basarabia. 'Dreptatea Socială,' Bucarest, an I, No. 5, 1923.

ENESCU, I. I. Ardealul, Banatul, Crişana şi Maramureş din Punct de Vedere Agricol. pp. 254. Bucarest, 1920.

ENGLISCH, R. Die Agrarreform im Nösnergau. (In 'Die Sächsische Landwirtschaft im Nösnergau,') Part II, pp. 56–9. Bistriţa, 1924.

Exposé général de l'état sanitaire en Roumanie à l'occasion du centenaire Pasteur. Bucarest, 1923.

FILIP, NICOLA. Mica Proprietate şi Creşterea Vitelor. 'Viitorul,' 6 January 1921. Bucarest.

Frâncu, St. Agricultorii şi Impozitele Dlui Titulescu. 'Viaţa Agricolă,' 1 July and 1 August 1922. Bucarest.

Garoflid, Const. Rolul Social al Proprietăţii Mijlocii, p. 19. Bucarest, 1926.

——. Un Program Agrar. pp. 40. Bucarest, 1923.

——. Problema Monetară şi Agricultura. pp. 27. Bucarest, 1924.

——. Păreri Economice şi Financiare. pp. 229. Bucarest, 1926.

Georgescu, D. Contribuţiuni la Studiul Boalelor Sociale în Mediul Rural. (Thesis.) Bucarest, 1926.

Georgianu, Ilie I. România sub Ocupaţia Duşmană. Part II. Exploatarea Economică a Ţării. pp. 176. Bucarest, 1920.

Giurgea, Eugeniu. Lucrările de Expropriere Făcute în Basarabia până la 1 Febr. 1923. 'Buletinul Agriculturii,' an. IV. No. 1–3, 1923.

Glavan, D. I. Consideraţiuni Asupra Stării Sanitare a Ţăranului. pp. 12. Bucarest, 1920.

Glicsman, Dr. L. [Dr. Ion Fulga.] Mortalitatea Rurală in România; Cauza-remediile. pp. 22. Bucarest, 1916.

Goga, Octavian. Coşbuc. pp. 52. Bucarest, 1923.

Graf, Dr. Oscar. Die Industriepolitik Alt-Rumäniens und die Grundlagen der Industrialisierung Gross-Rumäniens. pp. 198. Bucarest, 1927.

Grigorescu, Elvir V. Hrana Ţăranului Român din Punctul de vedere al Igienei Alimentare şi Sociale. (Thesis.) Bucarest, 1926.

Grigorescu, P. Exproprierile în Dobrogea. Cazul de la Tariverde. pp. 25. Constanţa, 1923.

Grinţescu, Ernest. Les Écoles d'Agriculture de Roumanie. pp. 58. Bucarest, 1925.

Griscov, Mihail N. Câteva Observaţiuni Asupra Culturei Pământului şi a Producţiunei Basarabiei de înainte şi după Răsboi. 'Basarabia Economică,' An. III. Nr. 2. Chişinău, 2 August 1922.

Halaceanu, Const. Păreri Asupra Expropierei. Metoda Ştiinţifică Pentru Determinarea Cotei de Reducere. pp. 48. Iaşi, 1917.

Hallunga, Alexandre. L'Évolution et la Revision Récente du Tarif Douanier. Paris, 1927.

Iacob, Dr. Ioan. Chestia Agrară în Ardeal. Oradia Mare, 1924.

——. Organizarea proprietăţilor. Comasările în Ardeal. 'Ţara Noastră', an. VII, Nos. 45–7.

Iarca, G. Viticultura Faţă cu Nouile Impozite. 'Revista de Viticultură,' July–August–September 1921. Piteşti.

——, Traian I. Contribuţii la Influenţa Reformei Agrare Asupra Creşterei Vitelor. pp. 47. Orăştie, 1921.

——. Urmările Economice ale Reformei Agrare Pentru Ardeal. 'Revista Economică,' June–August 1921. Sibiu.

Ioaniţiu, Ing, G. V. The Peasant Industry in Rumania. Bucarest, 1926.

Ionescu-Brăila, G. Industria şi Comerţul Mondial de Carne şi Participarea României la Acestea. 'Buletinul Institutului Economic Românesc,' March–April 1928.

Ionescu-Siseşti, G. L'Agriculture de la Roumanie pendant la Guerre. pp. 135. Paris, 1929.

——. La Mezzadria in Rumenia. Bulletin of Social and Economic Institutions, Institute of Agriculture. No. 2. Rome, 1923.

——. Participarea României la Producţiunea şi Comerţul Mondial de Cereale 'Buletinul Institutului Economic,' May–June 1928.

Ionescu-Sisești, G. Repartiția Proprietăților Agricole și a Impozitelor pe Aceste Proprietăți în Perioada 1923–6. 'Buletinul Institutului Economic Românesc,' pp. 165–87, March–April 1927. Bucarest.

——. Structure Agraire et Production Agricole de la Roumanie. pp. 61. Bucarest, 1924.

——. Reforma Agrară și Producțiunea. Un Program Pentru Ridicarea Agriculturei. pp. 48. Bucarest, 1925.

——. Reforma Agrară in România. pp. 42. Bucarest, 1920.

——. Agricultura și Capitalismul. 'Arhiva p. S. s. R. S.', vol. vi, Nos. 3–7.

——. La Culture des Céréales au Point de Vue Économique et Social. Paper read before the Thirteenth Agricultural Congress. Rome, 1927.

——. Rumäniens Bäuerliche Landwirtschaft. Bucarest, 1912.

Iorga, N. La Réforme Agraire en Roumanie. 'L'Information Roumaine,' Nos. 4–7. April and July 1923.

Jinga, V. Cerealele Românești în Producția și Comerțul Mondial și Finanțarea Exportului lor. 'Independența Economică,' an. XII, No. 1. 1929.

Kirițescu, Const. Anuarul Invățământului Secundar în România. 1924–5. Bucarest, 1925.

Latiève, H. Étude Agricole sur la Roumanie. pp. 20. Extr. du 'Bulletin Mensuel,' March 1903. Imprimerie Nationale, 1903.

Lazar, L. La Mise en Œuvre de la Réforme Agraire en Roumanie. pp. 127. (Thesis.) Paris, 1924.

Lazeano, C. La Nouvelle Réforme Fiscale en Roumanie. pp. 144. (Thesis.) Paris, 1921.

Legea Societăților Civile de Credit Funciar Rural. 'Monitorul Oficial,' 31 May 1923. No. 67.

Lupu, Dr. N. Crushing the Rumanian Peasant. 'Nation,' New York, vol. 115, 27 December 1922.

Madgearu, Virgil. 'Țărănismul,' pp. 68. Bucarest, 1921.

——. Doctrina Țărănista, pp. 23. Bucarest, 1923.

——. Revoluția Agrară și Evoluția Clasei Țărănești. 'Arhiva p. S. s. R. S.', vol. iv, No. 3.

——. Rumania's New Economic Policy. pp. 64. London, 1930.

Magniet, Léopold. Les Systèmes de Culture en Roumanie, etc., 'Le Moniteur Économique,' February–September, 1911. Bucarest.

Măgură, Ioan. Problema Agrară se Pune Iarăș Pentru Ridicarea Clasei Țărănești. pp. 78. Galați, 1923.

Maior, Dr. George. România Agricolă. Studiu Economic. pp. 102. Bucarest, 1895.

'Manchester Guardian'. Rumanian Supplement. November 28, 1929.

Mann, Fritz Karl. Kriegswirtschaft in Rumänien. pp. 64. Bucarest, 1918.

Manoilescu, M. Regimul Impozitelor Față cu Producția Națională și în Special Față cu Producția Industrială și Minieră. pp. 42. Bucarest, 1921.

——. Țărănism și Democrație. pp. 56. Bucarest, 1922.

Mantu, George. Proprietatea Mică și Exproprierea Față de Producție. pp. 37. Bucarest, 1920.

Metes, Stefan. Viața Agrară-Economică a Românilor din Ardeal și Ungaria. Documente Contemporane, vol. i (1508–1820), pp. xv+303. Bucarest, 1921.

Mihaescu, N. I. Mica Improprietărire sau Minimum de Existență Pentru Muncitorii Agricoli, pp. 7. Bucarest, 1921.

MIHAILESCU, C. Rolul Marei Proprietăţi în Evoluţiunea Tehnică a Agriculturii Româneşti. 'Arhiva p. S. s. R. S.', vol. iv, No. 1.

——. Intervenţia de Stat în Indrumarea Tehnică a Agriculturii Noastre. 'Arhiva p. S. s. R. S.', vol. iv, Nos. 4–5.

MIHALACHE, I. Noul Regim Agrar. pp. 31. Bucarest, 1925.

——. Partidul Ţărănesc în Politica Ţării. p. 48. Bucarest, 1925.

MINCA, NICOLAE. Constatări, Conclusiuni şi Propuneri Agrare. pp. 50. Brăila, 1923.

MITRANY, D. The New Rumanian Constitution. London, 'The Journal of Comparative Legislation and International Law.' 1 February 1924.

——. The Bibliography of Peasantist Reconstruction. 'Manchester Guardian Reconstruction Supplements,' No. 12. 1923.

——. Marx v. The Peasant. (In 'London Essays in Economics in Honour of Edwin Cannan'.) pp. 319–76. London, 1927.

MOGHILIANSCHI, I. The Production and Distribution of Cereals in Bessarabia. (In Russian.) Chişinău, 1916.

MUGUR, GH. D., &c. Căminul Cultural. Indreptare Pentru Conducătorii Culturii la Sate. pp. 118. Bucarest.

NASTA, AL. Satul Model. (Arhiva, p. S. s. R. S.), vol. vii, Nos. 1–2.

——. Reforma Agrară şi Problemele Agricole ale Viitorului. pp. 17. Bucarest, 1926.

——. Problema Colonizării. pp. 14. Bucarest, 1925.

——. Imbunătăţirile Funciare în Noua Fază a Organizărei Agrare. (Arhiva p. S. s. R. S.), July–October 1919. Bucarest.

NEGRUŢI, VASILE. Lotul de 5 Ha. 'Viaţa Agricolă.' Bucarest, 1 April 1922.

NIKOLITS, DANIEL. Reflecţiuni la executarea reformei agrare. 'Glasul Minorităţilor'. an. II, No. 2. 1924.

NIŢESCU, DR. CONSTANT. Monografia Regiunii Sud-estice a Judeţului Vlaşca. pp. 225. Bucarest, 1928.

N. P. D. Situaţia Lucrărilor de Expropriere şi Improprietărire. 'Economia Naţională.' Bucarest, August 1927.

OPRESCU, G. Arta Ţărănească la Români. pp. 74 and 58 plates. Bucarest, 1922.

——. Peasant Art in Rumania. pp. 182. London, 1929.

PAMFILE, TUDOR. Pământul După Credinţele Poporului Român. pp. 59. Bucarest, 1924.

PASVOLSKY, LEO. Economic Nationalism of the Danubian States. (Washington Institute of Economics.) pp. 609. New York, 1928.

PONI, P. Statistica Răzeşilor. pp. 220. Bucarest, 1921.

POPA, ISTRATI. Transilvania, Banatul, Crişana şi Maramureşul. Bucarest, 1915.

POPESCU, AURELIU ION. Variaţiile Sezonale ale Leului. pp. 32. Bucarest, 1927.

POPOVICI-LUNGA, N. Elemente de Economie Rurală. Bucarest, 1912.

POPOVICI, MAXIMILIAN. Un Testament de Politică Agrară. (Arhiva p. S. s. R. S.). Bucarest, April 1919.

PRAGER, DR. Rumäniens Landwirtschaftliche Klimatographie. Halle a/S. 1909.

RĂDULESCU, ANDREI. Viaţa Juridică şi Administrativă a Satelor. (Arhiva p. S. s. R. S.), vol. vii, Nos. 1–2.

——, A. A. Monografia Comunei Ghiojdeanca. pp. 64, Bucarest, 1924.

——, CODIN C. Comuna Corbii din Muscel şi Locuitorii săi. pp. 112. Piteşti, 1921.

RĂDULESCU-MOTRU, C. Ţărănismul: Un Suflet şi o Politică. pp. 29. Bucarest, 1924.

RĂDULESCU, SAVEL. La Politique Financière de la Roumanie, de 1914 à 1922. Paris, 1925.

ROTARU, I. Improprietărirea Ţăranilor şi Sporirea Producţiunei Agricole. Exemple din Basarabia. 'Călăuza Agricultorilor şi Viticultorilor Moldoveni.' Part I. Nos. 4–5. Cernăuţi, 1922.

Roumanie Agricole, La, pp. 45. Bucarest, 1929.

Rumania. An economic handbook prepared in the Eastern, European and Levantine Division, Bureau of Foreign and Domestic Commerce. Washington: Gov. Pr. Off. 1924. VII, 166 S. (Department of Commerce.) Special Agents Series. Nr. 222.

SADOVEANU, MIHAIL. Poezia Populară. Bucarest, 1923.

SANIELEVICI, M. and PRAPORGESCU, N. Mortalitatea Populaţiei Generale a României în Pragul Veacului XXlea. (Arhiva p. S. s. R. S.). April–October 1920.

SCHAFFNIT, DR. ERNST. Die Landwirtschaftlichen Verhältnisse Rumäniens. pp. 149. Hannover, 1921.

SCUTARU, ION. Reforma Agrară în Bucovina. 'Viaţa Agricolă,' No. 2, 1926.

ŞERBAN, M. Problemele Noastre Social-Agrare. pp. 208. Bucarest, 1924.

——. Tendinţa de evoluţie ale clasei ţărăneşti. pp. 80. Braşov, 1926.

SERGIU, DR. Raport general asupra pelagrei. pp. 18. Bucarest, 1888.

SFINŢESCU, EUG. CINCINAT I. Edilitatea Oraşelor şi Satelor din România Mare. pp. 65. Bucarest, 1921.

SIGERUS, FRED. Grossrumänien, ein Neues Wirtschaftsgebiet. Berlin, 1920.

Sindicatul Agricol al Jud. Ialomiţa. Câteva Observaţiuni cu Privire la Reforma Agrară, Reforma Financiară Precum şi la Producţiunea Agricolă şi Comerţului de Cereale. pp. 31. Bucarest, 1922.

SION, DR. V. Importanţa Socială şi Igienică a Commasării Satelor. pp. 12. Iaşi, 1918.

SLĂVESCU, VICTOR. Organizaţia de Credit a României. Bucarest, 1926.

——. Organizaţia Creditului Agricol în România. 'Buletinul Institutului Economic Românesc.' January 1925.

STAMATION, PETRE. Monografia Comunei Novaci (Târgu Jiu). pp. 18. 1921.

STANCULESCU, FL. Gospodăria Românească. pp. 30. Bucarest, 1927.

ŞTEFAN, RADU. Soc. de Construcţiuni şi Problema Locuinţelor Economice la Sate. pp. 176. Bucarest.

ŞTEFĂNESCU-GUNA, G. Mémoire relatant à l'Organisation et l'Activité du Cadastre en Roumanie. 'Revista Cadastrală,' August–October. Bucarest, 1928.

STERE, C. Partidul Ţărănesc din Basarabia si Acţiunea lui Politică. pp. 31. Jassy, 1922.

STERE, C. ROMAN. Despre Comasare şi alte Măsuri Destinate a Inlătura şi a Impedica Fărămeţirea Proprietăţilor Ţărăneşti. pp. 63–73. 'Buletinul Agriculturii,' October–December 1927. Bucarest.

STERE, R. C. Particularităţile Economice ale Micilor Gospodării Ţărăneşti. pp. 32–64. 'Buletinul Agriculturii,' Bucarest, April and June 1927.

SYNADINO, P. V. Insemnătatea Reformei Agrare în Basarabia. 'Buletinul Camerilor de Comerţ şi Industrie.' Chişinău, May–June 1926.

SZÁSZ, ZSOMBOR DE. The Minorities in Roumanian Transylvania. pp. 414. London, 1927.

SZABO, L. Agrarreform in Siebenbürgen. Berlin, 1928.

TABACOVICI, N. Talia şi perimetrul tinerilor clasei 1893. 'Arhiva p. S. şi R. Socială', vol. i, No. 1.

TATARANO, C. Les Nouvelles Tendances Économiques de la Roumanie d'après la littérature économique et les discussions parlementaires. pp. 144. (Thesis.) Paris, 1922.

TEXTOR, LUCY M. (Paper on Nationalist Tendencies in the Agrarian Reforms, read before the Congress of English and American Historians.) Richmond, Virginia. January 1925.

The Agrarian Reform in Roumania and its consequences. (Official pamphlet.) pp. 23. Bucarest, 1925.

TIMOV, S. The Agrarian Problem in Roumania. (In Russian.) 'Na Agrarnom Fronte,' No. 9. Moscow, 1925.

——. The Rumanian Agrarian Reform. (In Russian.) 'Agrarna Probleme,' No. 2. Moscow, 1928.

Transilvania, Banatul, Crişana, Maramureşul, 1918–28. 2 vol. pp. 1337. Bucarest, 1929.

URBEANU, DR. A. Probleme Sociale. Bucarest, 1927.

VASILIU, EM. D. B. Situaţia Demografică a României. pp. 112. Cluj, 1923.

VOLKWART, DR. H. Wirtschaftliche Typen: Grossgrundbesitz und Bauernschaft. 'Siebenbürgisch-Deutsches Tageblatt', 1 March 1922. Sibiu.

ZABIRANIC, M. Fondul Crizei. 'Furnica.' Chişinău, December 1921.

ZAHAREANU, DR. M. Consideraţiuni asupra Impozitelor ce vor fi Aplicate Producţiunei Agricole. 'Agrarul' Bucarest, 9, 11, 12, 13, and 21 May 1921.

ZAHARESCUL, P. Spre Reforma Şcolii Săteanului. pp. 40. Bucarest, 1916.

ZELETIN, DR. STEFAN. Revoluţia Burgheză în România. (Arhiva p. S. s. R. S.), vol. iii, Nos. 2–3, vol. iv, Nos. 1, 2. July–October 1921.

——. Furt Juridic şi Furt economic. 'Independenţa Economică. Bucarest. December, 1922.

——. Burghezia Română. Origina şi Rolul ei Istoric. pp. 256. Bucarest, 1925.

——. Neoliberalismul. pp. 278. Bucarest, 1927.

BIBLIOGRAPHY TO PART III (CO-OPERATION)

Anuarul Cooperaţiei Agricole pe 1925. pp. 102. Bucarest, 1926.

Agricultural Associations in Rumania. Bull. of Ec. Soc. Institutions. Rome. Vol. i, No. 2, October–November 1910. pp. 197–215.

Analele Băncilor. August–September, vol. ix, Nos. 8–9. pp. 382–9. Bucarest, 1927.

ARGHIR, NICOLAE. Din Istoricul Cooperaţiei Săteşti şi Rolul Caselor Naţionale Alături de Cooperaţie. Ed. II, pp. 32. Bucarest, 1921.

AXENTE, TITUS. Le Mouvement Coopératif de Crédit en Roumanie à la fin de l'année 1925. Bucarest, 1926.

——. La structure et la situation du mouvement coopératif en Roumanie. (Paper read before the International Agricultural Congress, Bucarest, 1929.) pp. 12.

'Buletinul Secţiei de Studii Cooperative.' Institutul Social Român. No. 1, 1927.

Cassa Centrală a Cooperaţiei şi Improprietăririi Sătenilor. Mişcarea Cooperativă la Sate in România. pp. 32. Bucarest, 1921.

Caisse Centrale de la Coopération Centrale des Banques Populaires Rurales et de leurs Fédérales dans l'Ancien Territoire de la Roumanie, pour l'année 1919. pp. xvi+272. Bucarest.

Centrale des Banques Populaires et Sociétés Coopératives en Roumanie. pp. 140. Bucarest, 1924.

Centrale des Coopératives. Le Mouvement Coopératif de Production et de Consommation en Roumanie. pp. 49. Bucarest, 1925.

Codreanu, Emil G. Pentru Ridicarea Clasei Ţărăneşti: Sindicatele Agricole. pp. 19+2 pl. Galaţi, 1922.

Coffey, Diarmid. Agricultural Co-operation in Jugoslavia, Rumania and Italy. pp. 95. Oxford, 1922.

Cooperaţia de Peste Prut. 'Furnica.' Chişinău, 2 Oct. 1921.

Cooperative Movement in Rumania. 'International Review of Agricultural Economics.' pp. 569–94. Rome, October–December 1924.

Damian, C. Les Banques Populaires en Roumanie. pp. 112. (Thesis.) Paris, 1922.

Decret Lege relativ la înfiinţarea Casei Centrale a Cooperaţiei şi Improprietăririi Sătenilor. 'Monitorul Oficial', No. 225, 3 January 1919. Bucarest, 1919.

Demetrescu, Eugen. Sensul Nouei reforme a cooperaţiei Româneşti. 'Independenţa Economică', vol. xii, No. 1. 1929.

Dron, Pr. C. Tovărăşiile Ţărăneşti. Botoşani, 1921.

——. Situaţia Actuală a Cooperaţiei în România Intregită. 'Viaţa Românească.' Annul. XIII, No. 7. Iaşi, 1921.

Enescu, I. Cooperativele Săteşti. Sibiu, 1912.

Eremie, N. Les Associations Paysannes de Production en Roumanie. pp. 119. (Thesis.) Paris, 1912.

Filipescu, C. Bilanţul Economic al Obştiilor de Arendare. pp. 9. Bucarest.

Galan, A. G. Développement et caractère du mouvement coopératif en Roumanie. (Paper read before the International Agricultural Congress, Bucarest, 1929.) pp. 36.

Ghenzul, V. Cincizeci de Ani ai Cooperaţiei de Credit din Basarabia. Chişinău, 1924.

Ghiulea, N. Asociaţiile Ţărăneşti. pp. 208. Bucarest, 1927.

——. La Coopération en Roumanie. 'Revue des Études Coopératives.' Paris, 1923.

Juvara, D. Le Problème du Crédit Agricole en Roumanie. pp. 280. Paris, 1928.

La Cooperazione Rurale. Riv. Int. delle Inst. Ec. Soc. Int. Institute Agriculture, No. 4. Rome, 1923.

La Situation des Coopératives de Crédit de Village dans l'Ancien Royaume au 3 Décembre 1920. 'Buletinul de Studii si Informaţiuni Economice-Financiare,' vol. iii, January–August, No. 1. Bucarest, 1922.

Lege pentru Unificarea Cooperaţiei. 'Monitorul Oficial', 20 February 1923. No. 269.

Les Banques Populaires et les Associations Coopératives de Paysans en 1909. 'Le Moniteur Économique,' August–September 1911. Bucarest.

Macavei, M. Le Crédit Agricole et le Crédit Foncier en Roumanie. pp. 199. (Thesis.) Paris, 1912.

Madgearu, Virgil, and Mladenatz, Gr. Reforma Cooperaţiei. Bucarest, 1923.

Mihaescu, N. I. Viitorul Cooperaţiei în România Nouă. 'Curierul Cooperaţiei Săteşti din România,' October–November, 1921. Bucarest.

Ministerul Agriculturii. Regulament pentru punerea în aplicare a Legii Casei Centrale a Cooperaţiei şi Improprietăririi Sătenilor. 'Monitorul Oficial,' 12 May 1922.

Ministerul Agriculturei. Cooperaţia în România Intregită. Bucarest, 1920.

Mladenatz, Gr. Die Geonossenschaftsbewegung in Rumänian. Bucarest, 1929.

Moisil, I. Prima Pagină din Istoricul Băncilor Populare din Vechiul Regat. pp. 29. Bucarest, 1922.

PARTHENIU, CESAR. Le droit et la coopération. (Paper read before the International Agricultural Congress, Bucarest, 1929). pp. 22.

POPESCU, DEM. G. Le Mouvement Coopératif en Roumanie. 'Buletinul Agriculturii.' Bucarest, October–December 1925.

POPOVICI, D. Despre Cooperaţia Basarabeană. 'Furnica.' Chişinău, 9–16 October 1921.

RĂDUCANU, DR. I. La Coopération Rurale en Roumanie. 'Le Moniteur Économique.' Bucarest, December 1910.

RĂDULESCU-LIVEZI, GH. GH. Contribuţiuni pentru Refacerea Agriculturei. Statul şi Cooperaţia. 'Lamura Agricolă.' Bucarest, April 1921.

ŞERBAN, M. Le rôle de la coopération dans l'organisation de la petite propriété et dans l'augmentation de sa production. (Paper read before the International Agricultural Congress, Bucarest, 1929.) pp. 11.

SLĂVESCU, VICTOR. Istoricul şi Dezvoltarea Băncilor Populare. 'Bul. Inst. Ec. Ro.', March 1925.

TURNEANU, G. Cooperaţia în Ardeal. 'Economia Naţională,' Anul XIII, No. 11. Bucarest, November 1921.

VASILIU, AMILCAR. Chestia Sindicatelor Ţărăneşti. 'Cooperaţia Nouă.' Bucarest, 15 February 1922.

Statistics

Ministère de l'Agriculture. L'Agriculture de la Roumanie: Album Statistique. pp. 90. Bucarest, 1929.

Ministry of Agriculture, Yearly Statistical Bulletins on Extent of Cultivation, Production and Animals.

SIGERUS, FRED. Wirtschafts-Statistisches Jahrbuch Rumäniens. Sibiu, 1928.

PERIODICAL PUBLICATIONS AND JOURNALS

'Monitorul Oficial.' (For legislation and Parliamentary Debates.)

'L'Indépendance Roumaine.' For the events of the period 1916–18.

'Neamul Românesc.' For the events of the period 1916–18.

'Buletinul Agriculturii.' (Six issues yearly, published by the Ministry of Agriculture.)

'Buletinul Statistic al României.' (Quarterly, published by the Institute of General Public Statistics.)

'Analele Statistice şi Economice.' (Six issues yearly.)

'Buletinul Institutului Economic Românesc.' (Monthly.)

'Arhiva pentru Stiinţa şi Reforma Socială.' (Published by the Rumanian Social Institute.)

'L'Économiste Roumain.' (Weekly, published by the Ministry of Industry and Commerce.)

'Argus.' (Daily, for economic information.)

'Adevĕrul.' (Daily, for political information.)

INDEX

Absentee landowners, definition in reform laws, 176–8; treatment at expropriation, 124, 130–1.
Absenteeism, 70, 552.
Ackermann, Convention of, 22, 26.
Adam Smith, 252.
'Adevĕrul', on agrarian reforms, 102 n.
Administrative Law (1929), 561 n.
Adrianople, Treaty of, 25, 26; freedom of trade in virtue of, 339.
Adult education: deficient organization, 522–3; in Alexandria, co. Teleorman, 521–2; provided for in the law of 1924, 520.
Aereboe, Prof. Fr., 579; 'Die Wachstum-Möglichkeiten der Bevölkerung', 578.
Agrarian Bank: and compensation of landowners in Transylvania, 154.
Agrarian capitalism: opposed by Peasant movement, 564.
Agrarian Committee: attitude to land reforms, 113, 135; its procedure, 198–9; number of appeals to, 200.
Agrarian Cultural Association, 457.
Agrarian economists: congress of, 116.
Agrarian League, 457; attempts at social consolidation in the villages, 534; its attitude to self-help and State aid, 458; combining peasants and landowners, 462; its comprehensive character, 458; resolution at inaugural meeting, 458.
Agrarian legislation: application of expropriation to individual estates, 106; Bessarabia, 167; Bucovina, 168; decree of 16/xii/1918, 111; condemned by Garoflid as unscientific, 107; decrees of 22/xii/1918, 112; definition of arable in the law of 17/vii/1921, 131; frequency of, 95; failure of 1907 legislation, 89; forest rights of peasants ignored in 1919 and 1921, 145; Garoflid's law of 17/vii/1921, 116; Garoflid's programme of reform, 109; indefiniteness of laws, 106; King Ferdinand's proclamation, 111; law of 17/vii/1921 and communal grazings, 131; law of 17/vii/1921 and foreigners' estates, 130–1; memorandum of Moldavian cultivators owning more than 25 ha., 168; original drafts of land laws, 104; peasantists' attitude, 113; Prof. Basilescu's view of draft laws, 105; provision of the law of 17/vii/1921 concerning expropriation, 129–30; reaction under Marghiloman, 108; reform laws re-enacted, 110; Transylvania, 166; variations in the several laws, 168 ff.
Agrarian movement: in Czechoslovakia, 564.
Agrarian policy: need of effective, after the war, 415.
Agrarian problem: its economic aspect insufficiently considered, 185.
Agrarian program: of National Peasant Party, 457; of the Peasant Party, 456.
Agrarian reforms: appeals to Agrarian committee, 200; approximate nature of statistics of, 185; change in distribution of land in Bessarabia, 204; changes in land distribution in Bucovina, 206; change in land distribution in Transylvania, 210–11; chief obstacles political, 577; complaints against the application of, 199; criticism of their application, 194, 200; and Economic and Financial policy, 414; effects in Bessarabia, 200; effects in Bucovina, 204–6; effects in Transylvania, 206–11; effects of, 200–19; effects upon agriculture, 199; effects upon agricultural industries, 371–3; effects upon co-operative movement, 373; effects upon peasantry summarized, 480–1; effects upon production, 306; effects upon prices and exports, 338–56; effects upon production in Transylvania, 325–8; effects upon rural economy, 357; effects upon rural industries, 365; excessive haste, 195; influence on cultivated area, 308–12; levelling effects on peasantry, 533; Marxian view of their effects not justified, 539; no grounds for pessimism as to effects, 283; organs of, 197–8; political effects, 537; public attitude to, 198; reduction in leases, 244; surveying difficulties, 195–6.
Agrarian statistics: changes in distribution of land after 1907, 187; changes in the distribution of land after the expropriation, 189; distribution of land in 1905, 186; for Bessarabia, 201–4; results of first expropriation (1918), 188; results of second expropriation (1921), 188; unsatisfactory state of, 194 n.
Agrarianism, *vide* Peasant movement.
'Agrarna Probleme', 239.
Agricultural Chambers: restriction of membership, 534.
Agricultural contracts: contravention of law of, 82 n.; enforcement by military abolished, 77; food for workers, 87; Gherea on, 77; inquiry by Iassy District Council (1862), 78; joint responsibility by law of 1872, 75; jurisdiction on, 77; law of March 1866, 73; enforcement of, by law of 1872, 74; law of 1882, 73; law of May 1882, 75; law of 23 Dec. 1907, 87; law of 1907 fails to ensure the quality of food for labourers, 487; law of 1907 on labour leases, 240; legislation of

Barbu Stirbey, 77; Liberal policy on, 79; Moruzi on maladministration of 81; penalizing clauses in, 82–3; Radu Rosetti on effects of the law of 1882, 82; Senate resolution of 1865, 73; terms of labour contract, 74.
Agricultural co-operation: *see* Co-operative movement.
Agricultural credit: attitude of National Bank to, 424; effects of Government policy upon, 423; for large owners, 424; for peasants, 425–6; law of 1929 on, 426; law for 1929, 455; supply of, 421–6.
Agricultural Credit Institution: Law establishing the, 426.
Agricultural education: absence of teaching material in schools, 519; budget of agricultural schools in 1923, 519; laws on, 455; no relation to needs of the agricultural industry, 517; on choice of careers by those educated in agricultural schools, 518; practical tendency of the Agricultural Schools Law of 1929, 520.
Agricultural equipment: imports of, 279; Rumanian production of, 279.
Agricultural implements: high prices of, 435.
Agricultural industries: effects of reforms upon, 371.
Agricultural insurance: bill on, 455.
Agricultural labour: employment bureau, 88.
Agricultural labourers: increase in, 236; their rations controlled by law on agricutural contracts of 1907, 487.
Agricultural machinery: customs policy in respect of, 451–2.
Agricultural personnel: high standards of, 454.
Agricultural policy: in Rumania, 454; of the National Peasant Government, 455; planned by the Congress of Agriculturists in 1920, 456.
Agricultural prices, artificially kept down, 448; local differences in, due to transport conditions, 442; movement of, 348 ff, 450.
Agricultural production: decline after the reforms, 573; effects of hostile policy, 574; influence of reforms on, 312–37.
Agricultural reconstruction: lack of inducements to, 451.
Agricultural Society: failure of, 457.
Agricultural statistics: large and small cultivation in Rumanian, 270; Rumanian, 228–9.
Agricultural Syndicates: Congress in 1923, 456.
Agricultural technique: influence of agrarian reforms on, 280; vital necessity of improvement, 577.
Agricultural wages: District Commissions on, 87; inquiry by Ministry of Home Affairs, 82; stationary (1862–1905), 82.
Agriculture: after Cuza's reform, 65; and the agrarian reforms, 199; burdens of, 427–44; changes in management, 69; change in means of production, 252; changes in scale of farming, 232; corn prices and extension of arable, 25; effects of agrarian reforms on, 228; excessive concentration of farming units, 233–4; extent of farming by owners, 235; identified with peasant farming, 477; labourers in, 236; large estates and large-scale farming, 232; leases, 234; not substantially altered by the post-war reforms, 250–1; recovery after rising of 1907, 91; resources of, 416–26; signs of reorganization of, 294; State aid to, 452.
Agronomical Research, Institute of, 453.
Alba Iulia: revolutionary demand for land reform 1 Dec. 1918, 166; National Assembly of 1848 in, 207.
'Albina' Bank, 376.
Alcohol: statistics of production and consumption, 491–2.
Alcoholism, 490–4; effect on crimes, 494; encouraged among miners, 494; estimates of average consumption, 493–4; Minister of Labour inquiry in 1929, 493; growing consumption of alcohol, 491; restriction by law of 1924, 491; number of public houses restricted in 1907, 490; origins in Rumania, 490; grave problem, 493; as a cause of suicides, 501.
Alimentation: worse after peasant emancipation, 487; adverse effects of the war, 487; effect of land reform on, 488–9; law of 1907 fails to improve labourers' diet, 487; of peasants, 486–90.
Allied victory: influence on reform, 107.
America: peasant emigrants to, and reform, 176; pellagra among Rumanian immigrants, 503.
Anastasiu, O. A., 368, 369, 370 n., 430, 515 n.
Andrews, U.S. Minister to Rumania, 100.
Angelescu, Dr. C.: on illiteracy before the war, 510; report on schools, 513–14.
Antim, St., 13 n.
Antipa, Gr., 308 n.; on protection of domestic industries, 366; on unemployment among agronoms, 518.
Arable: distribution among large and small properties, 222; distribution of, 187; per cent. of leased, 248.
Argentoianu, C., 552; against agrarian reform, 104; on exhaustion of land reserve, 223.
Army recruits: relation of holdings to health of, 503–5.
Associations of Resettlement: administra-

tion of, 159, 398; system dropped, 159, 409.
'Astra', 520–1.
Aurelian, P. S.: early attempts at co-operative organization, 375.
'Aurora' Bank, 376.
Averescu, General, 86 n.; fails to create new Conservative group, 556; coming to power, 115; promises to peasants, 118, 456; protest against political inefficiency, 110.

Babeş, Prof. V.: on fitness of army recruits, 503; on extension of pellagra, 502; on peasants' length of life, 505.
Balasian, Dr., on bad quality of food given to labourers, 487.
Balkan wars: effect on agrarian policy, 91.
Balta Liman: Treaty of, 35.
Bankers: peasant antagonism to, 547–8.
Banu, Dr. G., 498 n.; 506 n.; on absence of sanitary policy and organization, 509; inquiry in peasants' diet after the reforms, 489; on national characteristics of homesteads, 482 n.
Bărbat, Prof. V., on Spiru Haret's educational policy, 525.
Barley: production 320; and export, 1921–5, in various countries, 338.
Basilescu, Prof. N.: on reform laws, 103; on electoral system, 550 n.; demand for wider expropriation, 190; on the revolutionary nature of the agrarian laws, 185.
Bessarabia: classification of land properties, 201; distribution of landed property before reform, 202; after reform, 203–4; effects of reform on production in, 328; features of land reform, 113–14, 126; re-union with Rumania, 111; revolutionary origin of land reform, 165, 167–8; Zemstvo statistics of literacy, 510 n.
Bibescu, George (1843–8): restricts peasants' freedom of movement, 33.
Bibicescu, I. G., on attitude of upper-class to agrarian reform, 95, 470, 505 n.; figures of child mortality, 499; on peasants' services to the State, 470; on abuses of land settlement, 67 n.
Bir, 583.
Births: 497.
Bivolari: report on disturbances (1907) at, 85.
Boerescu, B.: on peasants' status before emancipation, 47–8 n.
Boiars, origin, 12; character, 21 n.; position, 27; judged by Kisselev, 34.
Bolsheviks, 99–101. *See also* Russian Revolution.
Bordeie, 484.
Bornemann, Dr. Felix: on labour requirements of peasant farms, 286.
Brăncoveanu, Constantin: on abuses of tax-collectors, 16.
Branişte, 583.
Brătianu, Ion C., 456.
Brătianu, Ionel: on agrarian reform, 97; bargain with Take Ionescu, 103; hasty execution of agrarian reform by, 195; on land reform, 91.
Brătianu, Vintilă, 456; on advantages of expropriation, 570; stops free import of agricultural machines, 452; anti-alcoholic policy, 491; demands limited expropriation (1917), 97; denounces expropriation of mortmain estates (1913), 97; on insufficiency of pre-war agrarian reforms, 86; on deliberate flouting of 1907 laws, 89; on peasants' lack of milch-cows, 487; on encouragement of alcoholism since the war, 490–1; on superior production of smallholders, 277.
Bucarest, Peace of (1913), 92.
Bucovina: changes in the distribution of land in, 206; distribution of landed property before reform, 204; effects of reform on production in, 329–30; expropriation in, 205.
Buildings: statistics of agricultural, 269.
Bulgaria: health of recruits, 504; State assistance to farming in, 453.
Buteniev: Kisselev's letter to, 34.
Buzdugan, Ion: 100.

Cadets, Russian: similarity between (1917) agrarian program and Rumanian reform, 121.
Călăraşi, 584.
Calimach, Gavril (Primate of Moldavia): demands greater labour servitudes, 19.
Camarăşescu, Jean: on climate and vegetation, 313 n.; and reform in southern Dobrogea, 179.
Cantacuzino, Grigore: his Conservative Government on expropriation of mortmain, 96.
Cantacuzino, M.: acceptance of expropriation principle, 97; influence of the law regarding oilfields, 105.
Cantemir, Prince D.: 'History', 584.
Căpităneanu, N. R.: on size of large estates, 70 n.; inquiry concerning extent of large estates in 1905, 186; on rents, 258.
Capital, invested in large industries, 424.
Capital, national: cannot be created by co-operation, 408; must extract tribute from the people, 545–6.
Capital in agriculture, 264–7; as affected by reforms, 416–21.
Capitalism: effect of French and Russian Revolutions on, 539–40; in Rumanian agriculture, 52–4, 69, 80, 239, 255, 257, 262–3, 282, 474.

Capitalist farms: decrease in number of, with increase in area, 262.
Carada: code of, 20.
Caranfil, Dr. G.: on landowners' losses, 463.
Carol, King: 63, 90; neglect of peasant question, 97 n.; reform manifesto (1907), 86.
Carp, M.: on extent of leasehold farms, 234.
Carp, Peter: attempt to organize landowners, 457; on the side of large property, 107; on Rumanian preference for official jobs, 532.
Carra, 25.
Casa Rurală, established 1907, 38.
Cassel, Gustav, Prof.: on effects of new agricultural machinery in U.S.A., 571; on relative decrease in corn consumption, 572.
Caziot: on minimum size of holdings, 194.
Ceată, 585.
Central of Co-operative Societies for Production and Distribution, 393, 395.
Central of the Popular Banks, *see* Co-operative Banks.
Central Resettlement Office, 132, 134, 135; colonizing work of, 147.
Cercul de Studii Economice: on increased consumption of wheat after reform, 300, 488.
Cereals: internal consumption of, 298–300, 488; diminishing proportion of crops, 292–3.
Cernăuţi, Regional Bank of, 133.
Cetaş (obştean, răzeş), 585.
Chambers of Agriculture: turned into party instruments, 454.
Chebap, Gr. Gh.: on the periods of peasants' history, 472 n.; on the policy of extending the peasants' social rights and reducing their rights to land, 473.
Child mortality: causes of, 499.
Children: employment of, 87.
China: relative overcrowding, 579.
Chirculescu, N.: on choice between Constitution or Revolution, 102 n.; report on new Constitution, 565 n.
Chiriţescu-Aron, Dr. M.: on future population problem on the land, 577.
Chuprov, Prof. A. I., 481 n.
Church: demands heavier peasant servitudes, 19; effect of establishment on, 530; character and position, 529; attitude to religious sects, 530; and nationalism, 529; shaken by reform, 530; and women in child-birth, 500.
Ciocoiu, 548.
Cioriceanu, Dr. I.: on replacement of foreign currencies, 418 n.; estimate of value of expropriation bonds, 419 n.
Cislă (system of taxation), 583, 586.
Clacă, 18, 375.
Clăcaşi, 6 n., 51.
Class consciousness: among peasants, 532.
Clergy, attitude to peasants, 19; unsatisfactory personnel and conditions, 530.
Cluj, Agrarian Bank of, 133.
Cluj University: extension work, 521; members of on peasants' position, 548.
Cnezi, 5 n.; relations with Princes, 10–11.
Coandă, General: Transitional Government of (1918), 110.
Coconaş, 532.
Codaş, 28.
Coffey, Diarmid: on the relations of state and co-operative movement in Rumania, 407.
Cojoc, 369.
Colescu, Dr., 237; on leases, 248; on value of corn exports, 348 n.
Colonies: flourishing condition of foreign peasant, 275.
Colonization: in Bessarabia, 148; law of 1921 in the Old Kingdom, 147.
Commercial Code, 1887: on co-operation, 378.
Commercial policy: in favour of corn-growing, 64.
Commission on Land Reform, 98.
Committees for the improvement of agricultural production, 455.
Communist manifesto, 253.
Communists: propaganda by, 100; misread growth of co-operative landholding societies, 401 n.; on growth of peasant farming in Rumania, 239; approve Rumanian reform, 573; on differentiation among peasants as result of reform, 533–4 n.
Compensation: according to the Constitution, 148; basis of, 150 ff; differences between the several laws, 173–4; in Bessarabia, 153; in Bucovina, 153; V. Brătianu's plan in 1914, 148; below value of expropriated land, 416; discussion of, 148–51; Take Ionescu's plan, 1914, 148; Ionescu's appeal to landowners in 1917 on, 151; memorandum of Moldavian landowners on, 151; modified by law of 1921, 154–6; and the tax on capital, 150; in Transylvania, 154; proposal of Peasant Party, 152 n.; real value of, 416–17; bonds issued, 419–20, 592–3; privileged treatment of French and British landowners, 178–9, 419 n.
Compulsory cultivation: differentiation between landowners and peasants in 1917, 99.
Compulsory labour: revival after emancipation, 73 ff.
Conacher, H. M.: on different agrarian ideas of peasants and townspeople, 460.
Concentration: of farming, 257; tendency of farming towards, 262.
Congress of Agriculturists in 1920, 456.
Connert, Fritz: on abuses in application of

Transylvanian reform, 217 n.; on Agrarian League, 458.
Conservative Party: social basis destroyed by expropriation, 556.
Conservatives: and large estates, 95; attitude on agrarian reform, 49, 79, 95; attitude on expropriation, 96–7, 104–5, 108.
Consolidation: in Bessarabia, 161; Garoflid's Law on, 161; Mihalache on, 160; of peasant holdings, 455; attitude of Transylvanian peasants on, 160; in Transylvania (after 1848), 207.
Constantinescu, Al.: his law of 1/iv/14 on expropriation in Dobrogea, 179; on non-Rumanians resettled in Transylvania, 218; restriction of rights of Hungarian optants by, 178; return to power in Liberal Government, 180.
Constituent Assembly of 1913, 98; of 1917, 102; dissolved by Marghiloman, 108.
Constitution of 1866: on inalienability of holdings, 76; drafted and passed in one day, 53; makes elementary education free and compulsory, 510.
Constitution: written constitution, incompatible with serfdom, 50.
Constitutional Government: peasants' depression under, 568.
Consumers' Societies, *see* Co-operative movement.
Consumption: of cereals in Rumania, 298; declines with increase of large-scale farming, 486–7; after reform, 488; of animal products, 486, 489.
Contingentation system, 436.
Co-operation, achievement of peasants on Bordeiu Verde estate, 276; spirit and tradition of, among Rumanian peasants, 374–5.
Co-operation and education, 525.
Co-operative associations: for execution of land reform *vide* Associations of Resettlement.
Co-operative landholding: Communist view of effect on village, 401 n.
Co-operative Banks: changes in capital distribution, 386; criticisms concerning peasants' participation in, 385; distribution of capital shares in, 385; distribution of deposits, 386; distribution of loans in, 387–8; law of 15/iii/1905 on federations of, 383–4; loans raised by, 388; types of popular banks, 379; weakness of credit co-operation, 392.
Co-operative code of 1929 (summary of provisions), 409–11.
Co-operative movement in Bessarabia, 403–4; in Bucovina, 405–6; in Transylvania, 376, 404–5; effects of land reform upon the, 373; statistics for 1921–8, 412–13; Co-operative Code of 12/vii/1928, 381; Decree-law of 3/i/1919 on co-operation and resettlement, 380; government control of the, 381; helped by policy of Minister of Education, 1900–2, 378; history of the, 373–8; law of 28/iii/1903, 378; obstacles to its development in the past, 374; State tutelage of, 406–13; neglect by National Bank, 407; and Capitalism, 408–9; and Nationalism, 408; Peasantist ideal, 374, 409.
Co-operative production, *see* Co-operative movement.
Co-operative Societies: legal status, 378–81; for credit, 381–93; for consumption and production, 393–6; for production, 396–8; for leasing and purchasing land, 399–402; Agricultural, 402–3.
Corn surplus: changes in Rumanian, 341; Bessarabian, 341.
Corteanu, Al.: reactionary nationalism of Liberals, 545.
Cost of living in Rumania, 463.
Costăchescu, I.: Minister of Education, 523.
Costaforu, Al., 101.
Cotruş, A., 548 n.
Creangă, Dr. G. D., 422 n.; on transfer of wealth from landowners to trading classes, 575; statistics of landed property, 186.
Credits for agriculture, 424–5.
Creighton, Dr. Charles: on social causes of pellagra, 501 n.
Crime and conflict, 525–8.
Crimean War: stimulates corn-growing in Rumania, 25, 339, 588.
Crimes: decrease on the land since the reform, 526; distribution of criminals according to occupation, &c., 526–7; due to drink, 526.
Crops: area of cereal crops, 1920–7, 336; average production of the chief cereals in 1920–7, 337; effect of reform on the distribution of, 284–95; distribution of chief cereal crops, 1910–15, 332; distribution of chief cereal crops in 1920–7, 333–334; variation on large and small farms, 287; increase in grass, 319; intensive crops in large and small farms, 288; relative decline in cereal, 292–3; variations in yield, 271; total production of cereal crops in 1920–7, 335; yield compared with other countries, 317; yield on large and peasant farms, 271–2.
Cskonics, Count: his estate in Jimbolia, 326.
Cultivation: changes in proportion of land under, 290; distribution of area between large and small farms, 310; recovery in the area under, 310.
Currency: rapid depreciation of, 466; effect of fall of currency on agricultural prices and exports, 352, 448, 590; and on village industries, 371.
Customs duties: effect on agricultural

exports and prices, 352, 360, 363–4; effect on village industries, 371; receipts from, 438; change in receipts from, 440.
Cuza, Prince Alexander: imposes land reform, 48; forced abdication, 49, 103; educational reforms of, 510; introduces universal suffrage, 549.
Czechoslovakia: State policy for agriculture in, 454.

'Dacia Felix', 3.
Danube: law for the improvement of the Danube regions liable to flooding, 455; European Commission, 64; opening delta to navigation, 44, 64.
Daşcovici, N., 523 n.
Daszinska-Golinska, G., 307 n.
David, Dr. E., xxv.
Death-rate, high death-rate offsets high birth-rate, 497.
Deavalma: system of share cultivation, 591; compared with *tarla* system by Garoflid, 589.
Deflationist policy, 420.
Delniţe (*opcine*, *obştine*) 585.
Democratic Unionist Party, 168.
Depreciation: of currency affects working and salaried classes, 466; partly due to decline in agricultural production, 466.
Descălecare, 9.
Dettweiler, on labour norms on farms, 286.
Devălmăşie, 584.
Devolution, 561.
Dictatorship, opposed by Peasant movement, 565.
Dijmă (*métayage*), 588.
Dissescu, Senator C., 103.
Distribution of land, 186–7.
District Agronoms, according to law of 23/xii/1907, 87, 88.
District Commissions, law of 23/xii/1907, 87.
Divans, election of rulers by, 26.
Divans ad hoc: convened in 1857, 46; discussion of agrarian question in Moldavian, 46–7, 115.
Divorces, increase in, 496–7; in villages and towns, 497.
Dobrogea: Averescu reform law (1921) for, 179; colonization in, 212–13; land laws and Rumanian reforms, 179–80.
Dobrogeanu-Gherea, C.: 66 n.; on consequences of 1864 reform, 62.
Domestic industries: their importance, 366; effects of reforms on, 365; general considerations concerning, 371; nature and distribution, 368–70; State indifference towards, 365–6; Transylvanian statistics of, 368.
Domni, 5.
Draught animals: Garoflid on cost of teams, 261; on large and peasant farms, 260.
Drink problems, *see* Alcoholism.
Duca, Prince Constantin: resists enslavement of peasants, 16.
Duca, I.: reform decree (1918), 100, 112, 197; supporter of co-operative movement, 407.
Duguit, on property, xxiv.
Dumitra co-operative bank, 377.

Economic aspects of land reform, 231.
Economic doctrine of nationalists, 408, 544–6; of Peasant Party, 120–1, 457, 559–60, 576–7.
Economic motive, absent at peasant emancipation, 53–5, 61–2; absent in post-war reform, 185, 228.
Economic policy: post-war mercantilist bias, 230, 444 ff., 544–6; its cost paid by agriculture, 448.
Education: 509 ff.; development after the war, 509; agricultural schools, 516–20; elementary, 510 ff.; *see also* Adult education, Elementary education, Professional and Agricultural education; proportion of illiterates, 509; excessive striving after degrees, 523.
Electoral system: Prof. Basilescu on crooked working of, 550 n.; tampered with by Liberals, 553.
Elementary education: assisted by peasants' own contributions, 514–15; made free and compulsory by Constitution of 1866, 510; position in 1912–13, 511; statistics, 511–12.
Emancipation of the peasants: background of Cuza's reform, 52; central commission, 47; compensation of landlords, 50; Al. Cuza's policy, 49; decree of 14/viii/1864, 50; effects on agriculture, 53; Moldavian *divan* (1857) and agrarian question, 46; servitudes inconsistent with constitution, 50; motives of, 61; practical effects of, 57; relation to political reform, 53; nationalist elements hamper it, 48–9; carried by *coup d'état*, 49; settlement of landless peasants on State domains, 52; Talleyrand's policy, 47; lack of economic motive, 53–4; imposed by Great Powers (Treaty of Paris, 1856), 45–6.
Emigration, 465–6 n., 495.
Eminescu, Mihail, 22 n., 24; on greed of the Liberal bourgeoisie, 547; 'Scrieri Politice', 542 n.
Enăşescu, admission of frauds in the application of reform, 191.
Enclosures (in England), effect on peasants' food, 581.
Endowments: Bucarest, of civilian hospitals, 468–9; effects of expropriation upon cultural and charitable, 466–9; St. Spiridon (Iassy) Hospitals and Almshouses, 467–8.

Enescu, on medical service in Transylvania, 509.
Enescu, Fotin, role in preparing draft of decree of 16/xii/1918, 111.
Englisch, M., on Saxon losses in Transylvania, 218.
Estates, large, seldom cultivated by owners, 282.
Export duties, protest by Central Union of Agricultural Syndicates, 436.
Export taxes, 432, 435; borne mainly by agriculture, 438–9; effect on agriculture, 441; effect of frequent changes in, 439; effect on peasants, 440; rates of, 437–8; reduction of, 440.
Exports: 'outgrowth of Raubwirtschaft', 571; post-war corn, 340; prohibition of wheat, 300; quantities 1919–26, 356; large surplus for, 339; value of, 1919–26, 355.
Expropriation: absentee owners' estates, 176; and housing in Transylvania, 128; and transfer of land, 172; attitude of Parties to, 117; authorities charged with, 173; basis extended in 1921, 130; Bessarabia, 122; in Bucovina, 122, 204–5; compensation for, 148, 173, 416–17; cost of expropriated land, 419; decree laws concerning, 122; dissatisfaction with results of, 189; distribution of expropriated land, 220; Max Weber on ease of French, 282; easy in Rumania because landlords not farmers, 282; economic criteria in second, 129; economic considerations taken into account, 123; extent of proposed, 104; extension by law of 17/vii/1921, 116; exemption limits, 125, 171–2; features of, in Transylvania, 126–7; first, 122; gradual, and according to the law of 1921, 131; land subject to, 124; large owners' attitude to, 114; leases of unexpropriated area, 174; levelling tendency in Bessarabia, 126; mechanism of, 131–6; nationalist tendencies in, 175; of forests by law of 1924, 173; of land belonging to foreigners, 130, 178; of leased land, 170–1; of municipal lands, 466; Old Kingdom, 122; Old Kingdom and Bucovina, 125–6; mechanism of, in Bessarabia, 132; mechanism of, in Bucovina, 132–3; mechanism of, in the Old Kingdom, 132; mechanism of, in Transylvania, 133; price of, 151–6; of Hungarian optants, 177; provisions for benefit of production, 123–4; provisions concerning existing tenants and crops, 156; scope of, 168–70; second (Garoflid's Law, 1921), 129; statistics for Bessarabia, 202–3; criticism of application, 194–200; and national minorities, 211–19; total area expropriated, 220; fails to solve agrarian problem, 222–6; in Transylvania, 122–3, 209–10; varies in the several provinces, 168–72, 175–82; in southern Dobrogea, 179–80.
Expropriation bonds, amount and value of, 417, 419–20, 592–3.

Fallow, proportion of land under, 289–90.
Family *Wirtschaft*, 461.
Famine: in Russia (1833) stimulates Rumanian corn-growing, 25; of 1928 in Bessarabia, 300; in Russia increases with corn exports, 83.
Farm buildings, improved since the reform, 485.
Farm equipment: distribution between large and peasant farms, 260; distribution of, 268; in Bessarabia, better on peasant farms, 269; difficulty of replacement after war, 278–9.
Farming capital: ownership of, 267; on large estates, 264–5; on peasant farms, 266.
Farming trusts: in Rumania, 263.
Farms: classification by size, 241; proportion of freehold to leasehold, 242.
Ferdinand, King, 92; proclaims reform, 100–1; claims initiative of reform, 101; second proclamation, 110–11; attitude to Peasant Party, 115, 557.
Feudalism, in genuine form unknown in old Rumania, 6–7, 11, 21 n., 474.
Finances, effects of nationalist policy upon, 447.
Financial policy, unfavourable to agriculture, 451.
First Rural Credit Society, 424.
Fiscal Law of 1923, exposé de motifs, 574.
'Fish-days', 486.
Flight of peasants in eighteenth century, 16; after Organic Statutes, 38; drastically punished, 41.
Foreign capital: nationalist attitude towards, 408, 447–8, 545.
Foreigners: expropriation of 130, 178.
Forests: communal woodlands in Transylvania, 146; communal ownership restricted by Moruzi in 1792, 144; compensation for expropriated, 156; Emancipation Law of 1864 on, 145; expropriation in Bessarabia, 145; expropriation by constitution of 1923, 146–7; expropriation in Transylvania, 145; Forestry Code, 145; in the Moldavian Organic Statute, 144; in the Muntenian Organic Statute, 144; law of Barbu Stirbey, 1851, 145; law of Mihaiu Sturdza, 1844, 145; Organic Statutes of 1829 on, 144; Peasant Party and communal woodlands, 145; rights of peasants ignored in laws of 1919 and 1921, 145; loss of peasant rights, effect on housing, 484.
Fotino, Georges, 583, 585 n.
Franchise: Constitution of 1923, 552; hasty

reform in 1917, 551; after Independence, 549–50; widening of, in 1884, 550; peasants denied, 550.
French Revolution: comparison with Rumanian reforms, 185, 282, 539–40; with Russian Revolution xxviii f.
Frontier regiments: joint properties of, 170, 215–17.
Fruntaş, 28.

Garoflid, C., 166, 186, 233 n., 234, 590; and administration of Associations of Resettlement, 159; as Minister of Agriculture, 115; attitude to agrarian reforms, 113; attitude to Iassy reforms changed, 115–16; attitude to the Saxon University, 216–17; 'Chestia Agrară', 257, 474 n.; condemnation of latifundia by, 267; criticism of resettlement by, 192; establishment of the Institute of Agronomic Research, 453; estimate of *răzeş*, property, 585; exemption of agricultural machinery from import duties by, 451; his estimate of peasants' annuities, 476; his estimate of peasants' incomes before and after the reform, 477–8; his expropriation law, 129; his view of *métayage* systems, 589; mistaken views of large-scale farming, 295; 'Memorandum of Large Cultivators', 256, 309; on agrarian democracy, 563–4; on concentration of farming, 257; on development of cattle-breeding, 365; on feudalism and free trade, 64; on cessation of wheat exports, 303; on intensive large farms in Transylvania, 325; on landlord's share in *métayage*, 589; on landowners' dissatisfaction with expropriation, 189; on large farming, 255; description of latifundiary farming, 257 ff.; on oligarchy, against middle-class, 79; on minimum holdings, 72; on peasants' earnings after the reform, 478; on reform of 1864, 55; on social revolution effected by reform, 181; on the restriction of agricultural prices, 448; on wheat prices, 352–3; on the 'sweating' of the soil, 263; passage of his law of 17/vii/1921, 117; President of Agrarian League, 458; *re* Hungarian optants, 177.
Gendarmerie, reports of the Directorate of Gendarmerie, 526–7.
General Association of Teachers, 523.
General Sanitary Directorate, 507.
Georgianu, I. I., value of Rumanian production according to, 348 n.
German call to revolution, 98.
Gerota, Dr., estimate of consumption of alcohol, 493.
Gesellschaft für Sozialpolitik (Vienna), 578.
Gheorghiu, D., on change in wheat prices, 450.
Ghibu, Onisifor, on changes in peasant outlook, 529.
Ghica, Alexander, on relations between peasants and landowners, 40.
Ghica, Grigorie, 19, 35.
Ghica, Ion, on peasants, 470; on clergy, 530.
Ghiulea, N., 'Asociaţiile Ţărăneşti' 396 n.
Gipsies: as slaves, 50 n.; resettled, 225.
Gîrbacea, Dr. I., 'Şomajul în Ardeal şi Emigrarea', 466 n.
Giurgea, E., 201 n.; on farm equipment in Bessarabia, 269 n.
Glăvan, on misery of peasants in 1919, 487.
Glicsman, Dr., on causes of rural child mortality, 499–500; on rural medical service, 508.
Goga, Octavian: on changes in religious outlook, 530; on nationalist scope of reform, 181; on new spirit among the peasants, 531.
Golesco, Al.: number of *scutelnici*, 31; on extent of monastic property, 52 n.; on number of gipsy slaves, 50 n. 1; on number of serfs, 50 n. 2; on peasant emigration, 38.
Government: advanced forms conflict with backward economic organization, 567.
Grain: average weights of, 322; impurities in, 321.
Grain exports, 64; change in the nature of, 345; changes in various countries, 1909–13, 1921–5, 343–4; effects of the breaking-up of large estates on, 347; effects of reform on, 338; effects of restrictive policy, 344; influence of policy and taxation on, 352; Prof. Iorga on exports at the expense of peasants' diet, 471–2.
Graniţă, 7 n.
Grass, increase in grass crops, 319.
Grazing: communal grazing lands, 87; extension by law of 27/vii/1921, 143; extension of artificial, 292; Garoflid's law of 27/ix/1920, 142; in Bessarabia, 143; in Transylvania and Bucovina, 143–4; land expropriated for communal, 131; law of 27/ix/1920 criticized by Mihalache, 143; omitted in decree law of Dec. 1918, 141; rapid decrease in nineteenth century, 26; distribution among large and small owners, 359; shortage of, 59.
Great War: losses suffered by agriculture, 277–80; political effects in Eastern Europe, 537.
Grinţescu, Dr. Ernst, on agricultural education, 518–19.
Ground book, bill concerning survey and, 455.
Gusti, Prof. D., 220 n., 506, 585 n.

Hagemeister, Jules de, on corn prices (1835), 25.
Hammond, Mr. and Mrs.: on class mono-

poly of power, 23, 541; on English peasants before the enclosures, 546; *The Village Labourer*, 580.
Handicraft schools: peasant petition for, 515; in villages, 515.
Haneş, V. V., 553 n.
Haret, Spiru: on peasants on State lands, 82; and co-operation, 378.
Hasnaş, Dr. N., on Dobrogea land question, 180.
Haţiegan, Prof. Iuliu, inquiry into tuberculosis among school children, 503.
Health: effects of the reform upon health and vital statistics, 495–509; relation of holdings to, 503–5.
Hetairia, 24, 42.
Hilferding, Dr., xxiii f.
Holdings: consolidation of, 160; bill concerning consolidation, 455; economic size, 193–4; for newly-married couples, 66, 67; inferior quality of peasants', 272; inheritance of, 161; injurious effects of scattering, 314; law of February 1887, 67; minima fixed by Peasant Party, 194; pulverization of, 193; relation of health to, 503–5; sale and mortgage of, 162, 579 n.; size fixed without regard to farming efficiency, 569; sizes, 193; sizes of resettlement allotments, 173; splitting-up of, 60.
Horses: Rumanian horse-breeding in the past, 357.
Hotar, 7.
Housing: assisted by the Central Resettlement Office, 485–6; of peasants, 481–6; of peasants in the Old Kingdom, 483; inquiry of 1906, 484; quality of, 484–5; decline with loss of timber rights, 484; effect of reform on, 483, 485.

Iacob, Ioan: on abuses with reform in Transylvania, 218; on strengthening of minorities through the land reform, 218–19.
Iassy: proclamation by the King, 92; treaty of, 22; Bolshevik agitation in, 100.
Ilie, Dr. Al. P., 507 n.
Illiteracy: high proportion, 509–10; among various nationalities in Bessarabia, 510.
Import duties, 433; on agricultural implements, 433–4.
Indemnities (for emancipation), effect of, 61.
Independence, proclamation of, 66; effect of, on social policy, 49.
Index numbers: of retail prices not indicative of conditions, 464; Rumanian, 450.
Industrial expansion, at the expense of the peasants, 575.
Industrialization, criticism by Sombart, 575.
Industries, dependence on agriculture, 372–3.
Industry: capital invested, 445; cost of industrialization borne by agriculture, 448; encouragement of, 443, 445–6; financial policy favourable to, 447; growth of limited companies, 446; influence of agrarian reform on, 445–6; law for encouragement of, 372, 446; losses of, 447; materials used, 445; proposed remedies against unemployment, 447; attitude of Peasant Party towards, 457, 559–60, 576.
Infant mortality, 498.
Inheritance, of boiar estates, 24; of peasant holdings, 161.
Intensive cultivation and increase in manual labour, 285–6.
International Agricultural Institute in Moscow, 239.
International Institute of Agriculture (Rome), on the agricultural crisis, 307.
Ioncu, Teofil, 'Creditul in Basarabia', 426 n.
Ionescu, Ion, 57 n.; on agricultural education, 524; on better equipment of peasants, 268; on effects of emancipation, 57; on electoral system, 549; on peasant housing, 484; on impoverishment of peasants, 61; on relations between peasants and landlords, 59; on peasants' honesty, 525; on the absence of peasants in the first Parliament, 550; on the social predominance of property, 549; on the quality of land allotted to peasants, 272.
Ionescu-Siseşti, G., 308 n.; Director of the Institute of Agronomic Research, 281; on costs of wheat cultivation, 303; on effects of reform, 306; on export duties, 432; on manures, 324; on relation of money rents to labour cost, 592; on statistics of *métayage*, 591; on application of the land reform, 194; on the diminution in corn exports, 344; on the yield of the land tax, 430–1; report to the International Agricultural Congress in 1927, 573; *Structure Agraire et Production Agricole*, 290 n., 590.
Ionescu, Take: expropriation of foreign properties, 178; on expropriation, 97; opposition to communal grazing lands, 87.
Ionescu, Dr. T., 509 n.
Iorga, Prof. N., 7 n., 102 n.; on injustice of the old agrarian system, 471; on abuse of Constitution, 553; on the origin of the Royal Proclamation, 100.
Islaz, 583.
Istrati, Dr. C., 378.

Jasny, N., 315; on production in Transylvania and Bucovina, 329–30.
Jireabie, 7.
Joint land holding: old forms, 7–8, 583 ff.; co-operative, 158–9, 399–400; in Transylvania, 215–16.

Joseph II, measures to mitigate serfdom, 206.
Județ, 5; privileges of, 5–6; rights to land, 10.
Județie, 5.
Jugar, meaning of term, 127 n.
Jus valachorum, 7.
Juzi, 5 n.

Kisselev, Count Paul, 26, 30, 33, 34 n.; on size of holdings, 28 n.; on labour servitudes, 29; on tyranny of native boiars, 34.
Kogălniceanu, Mihail: carries through peasant emancipation, 78–9; on levelling up society, 55; on immobility of land property, 56; plea for peasants, 49.
Kossinsky, Prof. V. A., 481 n.
Krafft, on number of ploughing teams, 258.
Kutshuk Kainardji, treaty of, 22.

Labour, compulsory: reintroduced through agricultural contracts, 76–7; during the War, 108.
Labour, proportion to extent of farms, 265.
Labour Chambers, and smaller peasants, 534.
Labour contracts: enforcement of, 76; laws on, 73 ff.
'Labour Group', origin of, 103.
Labours, agricultural: in latifundiary cultivation, 258 ff.; character in Rumania, 273; effect of reform upon, 280–1, 283.
Laissez faire, on the land, 563.
Land, distribution according to use, 290–1.
Landbund, conservative agrarianism in Germany, 564.
Land disputes, as causes of crimes and violence, 526.
Land distribution, principle of, 569.
Land hunger: attitude of ruling class, 67; begins with return to national autonomy, 33; law of April 1889, 70.
'Land and Liberty': inseparable for eastern peasants, 537; motto of Russian revolutionaries, 81.
Land policy after 1864: first period, 66–9; second period, 69–72.
Land prices: dependence on density of population, 480; rise in, 83.
Land reserve, abandoned to landlords, 52, 66.
Land tax: assessments increased in 1927, 440; low yield of, 428.
Land tenure: original forms of, 7–8, 9–10, 21; transition to landlordism, 13, 22–3, 28, 32–3, 36; establishment of private property, 56; Hungarian and Byzantine influence, 10; joint, old system of, 7–8, 583 ff.; co-operative, 158–9, 399–400.
Landowners: attitude of Peasantists to, 113; character of Rumanian landlordism, 474; Garoflid's view of their role, 113; losses estimated by Dr. Caranfil, 463; opposition to reforms by, 114; return to the land of, 281; wiped out as a class, 461–2; Unions of Syndicates of Moldavian, 114.
Large estates: before reform, in Old Kingdom, 186; in Bessarabia, 202, in Bucovina, 204; in Transylvania, 208–9; extent expropriated, 220; in Old Kingdom, 188; in Bessarabia, 202; in Bucovina, 205; in Transylvania, 209; extent after reform, 221–2, 227.
Large farmers, existed by exploiting peasants, 265.
Large farming: conditions of, in Rumania, 257–8; no enduring superiority, 256; poor equipment of, 265.
Large landowners: and expropriation, 114; reduced to insignificance, 564.
Large v. small farming, 252–4, 580.
Latifundia: condemnation of, by Conservatives and Radicals, 267; differ from big industry, 262; farming on Rumanian, 257–60; origins of, in Rumania, 261.
Laur, Prof. Ernest, on production of small farms, 254.
Law: *jus valachorum*, 7, 138; first written, 9; *Pravile* of Vasile Lupu and Mateiu Bassarab, 9; Code of Carada, 20; *urbariu* of Calimach, 21; *see also* Constitution, Organic Statutes.
Lawsuits: their causes, 527; peasants and, 528.
Lazar, Livius, 204 n., 209 n.
'League for the Cultural Union of all Rumanians', 521.
Leases: extent of, 234; extent of, in 1907, 248; extent of, in 1927, 249; inquiry of 1911, 237–8; per cent. of arable leased according to sizes of properties, 248; restriction of area by antitrust law, 89; Synadino on, after the reform, 249–50; variations according to locality, 243; growth of money, 239.
Lenin: on strength of peasant middle class, 548; on the economic role of medium sized peasantry, 576.
Leon, G. N., 6 n., 21 n.; on compensation of landowners, 149 n.
Leroy-Beaulieu, on Russia, xxviii.
Levasseur, on agricultural statistics, 228.
Liberalism, 43; characteristics in Rumania, 50, 53.
Liberal Party: pushed to extreme right, 556; supremacy due to Court support, 557; land policy, 68.
Liberals: against expropriation, 96; manifesto of 1888 ignores agrarian problem, 96; mercantilist tendency of, 96, 444 ff., 544–5, 575–6; opposed to small cultivation, 277; Spiru Haret against Peasant

proprietorship, 96; sudden conversion to land reform, 97 n.
Liszt, on crime, 525.
Live stock: changes in the quantity of, 362; effects of reforms on, 357–65; exports of, 345–6, 363–5; decline in, 1860–1911, 358; in Rumania compared with other countries, 358; measures for the improvement of, 363; peasants position improved by the reforms, 359; distribution between small and large farms, 269, 359; total numbers of, 361; war losses in, 361; yearly increase in, 361.
Loria, on servile labour, 81.
Losses, on industrial investments, 447, 463.
Lupaş, Dr., 482 n.
Lupu, Dr. N., 102 n.; inquiry into peasants' diet, 489; on conditions in Maramureş county, 331; on achievements of peasant co-operation, 275–6.
Lupu-Kostaki, M. N., on State v. agriculture, 561.

Mackensen, Marshal, 98.
Madgearu, V., on co-operation between peasants and workers, 558.
Madgearu and Mladenatza, Reform Cooperaţiei, 385 n., 403.
Maior, Dr.: on decline in wheat yield, 274; on working conditions on latifundia, 260.
Maiorescu, Titu, 97 n.
Maize: conditions of cultivation, 303; decrease in consumption by peasants, 299–300; diagram of prices, 351; increase after reform, 294; importance of, in peasants' diet, 304; origins of its cultivation in Rumania, 304; pellagra due to consumption of, 305; production and export, 1921–5, in various countries, 339; varied use of, 304.
Mămăliga (polenta), 300, 304, 488.
Mandrea, R., defends large farming, 256.
Măndru, T., 'Cooperativele Agricole', 406 n.
Maniu, Iuliu: on agrarianist policy, 576; on relation of poverty to size of property, 581.
Manolescu, Dr. Al., 491 n.
Manoilescu, Mihail: estimate of peasants' incomes, 478; on real wages and salaries, 464.
Manure: defficient use, 323; use of artificial, 323–4.
Mantu, G., 238.
Maria Theresa, enactments on the treatment of serfs, 206.
Marian, Emil, 296.
Marmorosch, Blank & Co., industry largely controlled by, 424.
Marriages: increase in, 496; numbers proportional to agricultural output, 496.
Marţian, D., 235.
'Marx v. the Peasant', 555 n.
Marxism: agrarian revolt against, 555; unacceptable to peasants, 563.
Marghiloman, Alexander: against expropriation, 97; and compensation, 157; in favour of large states, 107; reform proposals (1918), 108.
Mathiez, A., xxix.
Mavrocordat, Al.: on gifts of land, 584; preparing emancipation, 473.
Medical attendance, absence of, 500.
Medical service, inadequacy in rural areas, 508–9.
Memorandum of Bessarabian Cultivators, 168.
Memorandum of Large Cultivators, 286, 309.
Memorandum of Rumanians in Transylvania (1892), 207.
Memorandum on the Balance of External Trade, League of Nations, on Rumanian agricultural exports, 346–7.
Mercantilism v. Agrarianism, 444 ff., 544–5, 547–8, 555, 558–9, 576–7.
Métayage, 588–92; decrease in, 244; extension planned by landowners, 590; extension among peasants, 590; inquiry by the Minister of Agriculture (1922), 590–1; in the landowners' agrarian program, 474; survival after reform, 590; law of 23/xii/1907 on, 87; restrictions in law of 1908, 589; spread to Moldavia and Transylvania, 590.
Mezincescu, Prof., inquiry into peasants' diet, 488–9.
Michael, L., *Agricultural Survey of Europe*, 340 n.
Middle Ages: foundation of Principalities, 8; land tenure, 7–8; rise of upper class, 11–13; rulers, 5; social organization, 5; status of peasants, 6.
Middle class: pressed by reform into economic activities, 544; created by nationalism, 542; difference from western, 53, 541–2; bureaucratic character of, 541; monopolistic tendencies of, 544; rural interests of, 542–3; absent in Rumania at political emancipation, 54, 540; not clearly divided from landed interests, 543; slow growth in Rumania, 540–1.
Mihăilescu, Crum, on yield of peasant crops, 318.
Mihaiu the Brave, 15; first to lower peasant status, 472, 567.
Mihalache, I., 114, 165; agrarian measures in Dobrogea, 180; estimate of land left to large owners, 189; reform bill, 112; commission on reclamation, 312; proposal for expropriation, 113; doctrine of land property, 120–1; more effective agricultural policy under, 455; on agrarian demagogy, 117; on co-operation, 374; on co-operative struggle

against capital, 409; on decrees of Dec. 1918, 112; on educating peasant farmers, 580; on philanthropic endowments as landlords, 469; on historical grounds for abolition of large farming, 268; on reform of 1917, 104; on Liberal agrarian legislation, 452; on greed of nationalist middle-class, 542; on Peasantist ideals, 113; on peasants' control of production, 576; on peasant poverty on large domains, 581; on peasants' serfdom to banking, 547; agrarian programme of Peasant Party, 457; on Rumanian soldiers, 505; on war and the peasants' outlook, 548; on fundamental role of peasantry, 556; on principle of land distribution, 569; on road tax, 431; 'Noul Regim Agrar', 547 n.; on scope of reform, 536.
Mijlocaş, 28.
Millo, A. V., on holdings, 55.
Ministry of Agriculture, budget of, 452–3.
Ministry of Health, budget and activity, 507–8.
Minorities: illiteracy among Bessarabian, 510; churches affected by reforms, 169; distribution of land in Transylvania among, 214; criticism of reform by national, 166, 212, 216–18, 327; strengthened by reform, 218–19; resettled peasants belonging to, in Bessarabia, 203; in Bucovina, 205; in Transylvania, 210, 219; treatment of Rumanian, in Transylvania, 207, 214–15; treatment of, by land reform, 175–82, 327; criticism of co-operative régime, 407; welcome new co-operative code, 411.
Minorities problem: and land reform, 211 ff; partly removed by reform, 219.
Mitrany, David, 552 n., 555 n.
Mixed commission, on Agrarian Reform, in Muntenia (1848), 43.
Moldavian cultivators, memorandum of, 168.
Monasteries, exploitation of peasants by, 18, 24, 39.
Monastic estates: increase in revenue, 33; secularization of, 51 n., 529; sale of, 67.
Money, shortage of, 426.
Morărescu, Stan, on State and co-operative movement, 408.
Mortality: among rural population, 498; incidence for various diseases, 500; child, causes of, 499–500.
Mortgages: on new holdings, 162; wiped out, 428–9; bonds for, 592.
Moruzi, Alexander, restriction of peasants' grazing rights (1803), 22, 471.
Moruzi, Sebastian, appeal on labour contracts for peasants, 81–2, 536.
Moşie, 543.
Moşneni, 6 n., 583.
Moţi, 368.
Munteanu, Dr. L., 506.
Muntenian Revolutionary Government, 43.

Nart, 19, 29, 32.
Naspuken, 305.
Nasta, Prof. A., on cost of expropriated land, 419.
National Agricultural Society, seed-selecting stations, 323.
National Assembly (1918, Alba Iulia), 166.
National Clubs, 521.
National Moldavian Party (Bessarabia), 167.
National Peasant Government: agrarian policy, 223 n., 455; co-operative policy, 409 ff.; in power, 457; supported by Socialists, 558.
National Peasant Party: formation (1927), 457; economic program, 457, 559–60, 576–7.
National Radical Party (Transylvania), 166.
National (Transylvanian) Party, 166, 557; agrarian program, 166.
Nationalism: agrarian policy in Dobrogea, 212–13; agrarian policy in Transylvania, 213–19; and peasant oppression, 471, 545, 568, 584; and social reforms, 181, 568; discrimination in the application of reform, 212; complaints by minorities in Transylvania, 217; costs of economic, 447; in agrarian reform, xxv, 175, 180–1; limited influence on land reform, 182; minorities problem and agrarian reform, 211; of Rumanian Liberals, 541–2, 544–5; nature revealed by Rumanian evolution, 567; and Church, 529; reform during post-war wave of, 219.
Nationalist character of reform inevitable, 214.
Nationalization: of mineral wealth, 545; of industry, 447–8, 545; influence upon State finances, 447.
Negură, M., on the work of the Agrarian Committee, 200.
Nemoianu, P., 532.
Neniţescu, S., estimates of pre-war mortgages, 422 n.
Neniţescu, D., on interest charged on loans to peasants, 84.
'Neo-serfdom', 66, 80, 462.
Netta, Dr. Xenofon, 418 n.
Nistor, M. I., 168.

Oats, production and export, 1921–5, in various countries, 339.
Obcine, obştine, 585.
Obreja, Dr., on consumption of alcohol, 494.
Obştean, 585.
Obştii, 158.
Oltenia, illiteracy worst, 510.
Optants, treatment of Hungarian, 177–8.

Organic Statutes, 25–4; reduction of holdings through, 473; attempt to protect *răzeşi*, 584; effect on conditions of peasants, 29–30; Moldavia, 27; Moldavia labour dues, 29; Muntenia, 31; Muntenia labour dues, 32; Muntenia right to land, 31–2; report of European Commission, 44; revision, 34; taxation in Moldavia, 30; vesting on boiars both land and power, 549; 'voluntary servants' in Moldavia, 30; first modern national laws, 26–7, 34; impositions mitigated by Kisselev, 33–4; practical effects summed up by Rosetti, 33.
Oromolu, on the credit policy of the National Bank, 424.
'Our Office' (Casa Noastră), 132, 134, 135.

Paris, Treaty of, 44.
Paris Congress, and national autonomy, 44.
Parliamentary commissions, 103.
Passy, H., on scale of farming, 572.
Patraşcanu, D., on peasant conditions in Rumanian provinces, 471.
Peasant Bank (in Bessarabia), 200.
Peasant movement, 553; attracts intellectuals, 557; benefits workers and lower middle-class, 562; hostility to Socialism and Communism, 563; crusading spirit of, 565; left leanings of, 564–5; prospects of, 561; resists abuses, 561; opposed to dictatorship, 565; opposition to agrarian capitalism, 564; origin and doctrine, 555–6; 'Populists' and 'Poporanists', 553–4; stabilizing effects in politics, 565.
Peasant Party: agrarianism of, 113, 560; favours international co-operation, 560; cardinal aim of, 559; blends Liberalism with Socialism, 560; views on private property, 120–1, 559; attitude to protection, 559; eclectic program, 558; and expropriation of land, 104; in Bessarabia, 557.
Peasant property, law on sale of, 455.
Peasant question in Eastern Europe, 460.
Peasant revival, 546.
Peasantism: program of, 556, 559; representative government and co-operative farming essential to, 556; social ideal of, 560.
Peasants: Alexander Moruzi's decree on grazing rights, 22; Alexander Ypsilanti's decree on service in Muntenia, 20; area of land received or bought, 72; average length of life, 505; and industrial workers, 558; Barbu Stirbey's laws in Muntenia, 35–7; causes of their disabilities, 475; chronic poverty after 1864, 476; Code of Carada on labour dues, 20; complaints of oppression, 40–1; condition on the eve of national independence, 38–41; conflict with bourgeoisie, 561; chronic underfeeding among, 486; control Rumanian farming, 282; financial exploitation, 547; difference in position after reforms of 1864 and 1917–21, 476; effect of corn-growing on land rights of, 25–6; effects of reform upon, 480–1; emancipation in Bessarabia, 45; emancipation in Bucovina, 45; emancipation in Rumanian provinces, 45–62; emancipation movement, 44; emancipation in Transylvania, 45; emancipation reform of 1864, 47 ff.; emancipation beginning by Constantin Mavrocordat, 16; Eminescu on oppression of, 78 n.; estimates of their incomes before and after reform, 477–8; extension of liberty accompanied by restriction of right to land, 473; fall in agricultural output, 314; Gherea on their position, 80; Grigore Ghica's Acts in Moldavia, 35; Grigorie Ghica's decree on services, 19; discontent with central government, 535–6; fiscal exploitation by mercantilist policy, 574; housing, 482–3; holdings after emancipation, 51 n.; increase in servitudes, 17–20; inheritance of holdings, 161; inquiry into their diet, 488; interests charged for loans to, 84; connexion between culture and farming of, 281; kept in serfdom until the Great War, 538; labour servitudes, 39; land hunger, 38; land law of Grigorie Calimach on peasants' right to land, 21; landlords' policy of tying peasants to the village, 474; limitation of rights to land, 21–2; material conditions after reform, 481–90; migration to Bessarabia and Transylvania, 38; and religion, 529; not revolutionary, 535; periods in status of, 472; political parties and rural middle class, 79; political rights of, 37, 550, 584; position after 1864 and 1918 compared, 547; position better in subjected provinces, 471; position on State lands, 82; relations with towns, 534–5; poverty related to large property, 581; restriction of rights to land by Ioniţa Sturdza, 22; position under laws on agricultural contracts, 76; protective laws not applied, 81–2; right to land under Organic Statute in Moldavia, 27–8; inquiry into wages of, 82; risings 24, 38, 49, 69, 85–6; incited by Germany, 98; Rosetti on oppression of, 78; safeguarding their property, 160–4; Sering on extension of peasant ownership, 251; Sturdza's estimate of service days, 20; taxation of, 16, 84, 427–9, 431–2; burdens of, 81; drink among, 492–3; honesty of, 525–6; political power of, 549; present economic situation, 475–81; services to State, 470; continued demand for land, 222–4; position under *urbarial* system, 16; winter contracts, 40.

Peasant risings, Russian influence on, 85.
Pellagra: practical disappearance after reform, 488; causes, 486; as a cause of suicides, 501; a social malady, 501; appearance and extension in Rumania, 502.
Pencov, on expropriation in Dobrogea, 212.
People's Party, agrarian policy, 456.
Phillipide, M., 15, 584.
Philippovich, Prof. E. V., on small v. large farming, 253.
Physiocrats, on large v. small farming, 252.
Poinsard, Léon, 89 n.
Political parties attitude to reforms, 118.
Political rights, Constitution of 1866, 549.
Political situation, influence of geographical position, 538.
Politicians, demagogy in villages, 536.
Poni, P., 584, 587.
Poporanist, 553 n.
Popovici-Lupa, V. O., 416 n.
Popular Assembly of 1817, 584.
Popular Banks, *see* Co-operative Banks.
Population: effects of reform on increase of, 495; problem discussed by Aereboe, 578.
Population problem, and land tenure, 577–8.
Populism: Mihalache's view of, 556; Stere's exposition of principles, 555.
Populists: parallel with Western Labour movement, 554; also Poporanists, 554.
Postnikov, A. S., 481 n.
Prăjini, 32.
Pre-emption, State's right of, 163, 174, 223 n., 224.
Pretor, 327.
Prices, effects of reform on, 338.
Primogeniture, 24.
'Prince Carol Cultural Endowment', 521.
Prisoase, cost of leasing, 58.
Production: value of agricultural, 348, 349; compared with pre-war, 315–16; decline in, 312; of large and small farms, 281–2.
Prohaska, I., 426.
Property: right of, 33; restriction of large, 164; change in public attitude to, 148–9; Peasantist view of, 120–1; Russian Cadets (C.D.) programme, 121; affinity of new and old conception of property in Rumania, 121; changed conception of, xxiv, 120; German Constitution on, 120; statistics distribution of (1927), 227; effect of reform on distribution of, 220–7; summary changes in distribution of, 221; medium-sized property at basis of agrarian democracy, 563; peasantist view of, 539, 559.
Protection, Peasantist attitude to, 559.
Protimis (*protimesis*), 164, 583.
Proudhon, xxiv.
Psychological changes, 528–37.
Public health: expenditure for, 507; legal provisions ineffective, 508.
Public houses: numbers of, 493; increase encouraged after War, 491; to close on Sundays and holidays (1929), 494; kept by landlords, 490.
Public lands, sale of, 89 ff.
Pulverization of holdings, 193–4.

Radishchev, on Russian peasants, 79.
Răducanu, Ion, 409 n., 494.
Rădulescu, Andrei, 527 n., 528, 586.
Rădulescu-Motru, Prof. C., 524 n.; on peasant stagnation, 531; on peasant conservatism, 563.
Railways and agriculture, 442–4.
Rakowsky, C., 100.
Rates and taxes, increased by reform, 466.
Rau, 253.
Raubwirtschaft, 580; and exports, 571; on latifundia, 274.
Răzeşi, 6 n., 583; dispossession of, 23, 584.
Răzăşie, 584.
Religion: revival in towns, 529; pagan character among peasants, 529; nonconformity spreading, 530.
Religious sects, spread after reform, 530.
Rents: fixed by District Commissions, 87; inquiry into, 70; nature of agricultural, 263; on large and small properties, 258; rise in peasant, 83–4.
Representative government, in Peasantist program, 556.
Requisitions, of corn, 449.
Resettlement: associations of, 158–9; in Bessarabia, 136–7; Bucovina, 137; communal grazings and woodlands, 141; in Transylvania, 157; criticism of execution, 192; cultivation of new holdings, 160; disposal of buildings, 140; distribution of holdings, 136–40; groups having preferential claims, 191; mechanism of, 140–1; number of registered claimants, 191; in Old Kingdom, 138–40; peasants' payments in Bessarabia, 156; peasants' payments in Bucovina, 157; peasants' payments in the Old Kingdom, 157; prices of holdings, 156–8; payment of annuities, 593; sizes of holdings, 173, 192; statistics of, 197; summary results on 1/ix/1927, 221; in Transylvania, 137–8; cost to State, 418.
Revolution, difficult in Rumania, 538–9.
Revolutionary tendencies among peasants, 535.
Risings, *see* Peasant risings.
Road tax, 431.
Roads, compulsory labour for upkeep, 431.
Roman, Visarion, 376.
Roscher, W., on economic size of holdings, 194.

Rosetti, Radu, 13 n., 33, 82, 583; on oppression of peasants, 78.
Rosetti, Radu (General), on origins of serfdom, 14–15.
Roznovanu, 586.
Ruggiero, on economic rationalism, 62.
Rumâni, 6 n., 583.
Rumanian Academy, effects of reform on, 467, 469–70.
Rural economy, effects of reforms on, 357.
Rural middle class, use of, 563.
Rural office, 88–90; experience of re-settlement, 158.
Rural poor, and transition to capitalist farming, 580.
Ruşfet, 60, 588.
Russian famines, increase with corn exports, 83.
Russian Protectorate, 22.
Russian Occupation, 26 ff.
Russian Revolution: compared with French Revolution, xxviii ff., 539; influence on reform, 92, 99–101, 102 n., 181, 568–9.
Russian risings, influence on Rumanian peasants, 85.
Rye, production and exports, 1921–5, in various countries, 338.

Salaries of State employees, 465.
Sămânţa, 323.
Sate domneşti, 11.
Saxon University, expropriation of, 216.
Say, I. B., 252.
Schäffer, Dr. ('Augenzeuge'), 91 n.
Scutelnici, 31.
Secondary schools, 523–4.
Sects, religious spreading after reform, 530.
Secularization of monastic estates, *see* Monasteries.
Seignobos, Prof., on peasantry, xxxii.
Senate, system of election, 552.
Serbian Peasant Party, fiscal program, 561.
Şerban, M., 193 n., 234 n., 240 n.; peasant inquiry into leases, 237–8; inquiry into organization of large farms, 263–5; of peasant farms, 266–7; estimates value of dead stock, 269; on labour outlay for wheat and maize, 286; on value of buildings, 269; on the peasants' working days, 273; statistics of leases, 239.
Serfdom: Constantin Duca's rebuke to boiars, 16; Constantin Mavrocordat's decree of 1746; declaration of mixed commission (1848), 207; local differences in evolution, 15; origin of, 13.
Sering, Prof. Max, on Rumanian reforms, xxvi, 251, 283.
Şezători, 375.
Sfatul Ţării, 122.
Share cultivation, *see* Métayage.
Simionescu, Prof. I., 526.
Slavery, conditions of success, 62.
Slobozii, 13.
Small farms, distribution, 240.
Small-holdings: demand for, 71 n.; laws concerning, 67, 68, 70, 71.
Social changes, caused by reform, 462.
Social effects of reforms: on the land, 470; on urban life and institutions, 463–70.
Social Psychology, characteristic stability, 531.
Socialism, agrarian program revised, 555.
Socialists, 95; join Liberal Party, 538; opposed to Peasantist policy, 113; support Peasant government, 558.
Sombart, Prof. Werner: on Rumanian industrialization, 575; on great social reforms, 101.
Souchon, on minimum size of holdings, 194.
Soutzo, N., 23.
Speculative leases, 80.
Srdinko, Czechoslovak Minister of Agriculture, 453.
Stahl, Henri, 225, 534, 585 n.
Standardization, of cereals, 452.
Stăpân, 13.
State, interference with economic life, 545.
State domains, sale of, 67 ff.
State farms, as outlets for social discontent, 534.
Statistics: land distribution, 227; farming and production, 332–4; exports, 355–6; co-operative movement, 412–3.
Steaua Română, 447–8.
Ştefan the Great, 14 n., 43, 107.
Stere, Constantin, on origins of reform, 97 n.; legislation on co-operation, 378; on class relations in Rumania, 536–7; on interdependence of land and political reforms, 553; leader of Populist current, 553; formulates Peasantist doctrine, 555.
Still-births, 498.
Stirbey, Barbu (1834–49), 33, 35; (1927), 443.
Stoicescu, C., 425.
Strat, Dr. G., 523.
Sturdza, Dimitrie, 85, 249; on expropriation, 91; on leases, 248; on peasant status, 95; on pellagra, 482; on reforms, 95.
Sturdza, Mihaiu, 584.
Subsoil: treatment in agrarian laws, 105; nationalization, 418, 545.
Subventions, to agriculture, 451.
Suicides, 500.
Şuman, 367.
Şupreme Agricultural Council, 88.
Survey: absence of, 114, 185–6; proposed bill, 455; work for reform, 195–7.
Sutzu, Mihaiu, and peasant grievances, 535.
Synadino, P. V., 203 n., 249, 328, 341, 417 n.
Syphilis: among peasants, 506; in the army, 506.

Szász, Zsombor de, 218–19.

Tabacovici, Nicolae, 503–5.
Tarla system, 589 ff. *See also Métayage.*
Ţara Românească, 5.
Ţăranul Român, 525 n.
Tariffs, 432 ff.
Ţarină, 583.
Taxation: allowances and rebates, 429; assessment for various categories of farms, 427; comparison between agriculture and industry, 429–30; low yield of land tax, 428; of agriculture, 427–41; of peasants, *see* Peasants.
Tea-houses, effects on alcoholism, 494.
Tenant farming: relation to size of farms, 263; extent after reform, 246–7.
Tetanus, at child-birth, 500.
Textor, Miss Lucy, 179 n.
Thünen, H. von, 254.
Timov, S.: on growth of capitalist farming in Rumania, 239; on differentiation among peasants, 401 n.; approves Rumanian reform, 425 n., 573; on agrarian tendencies in Rumania, 533–7 n.
Tocqueville, de, 541.
Town v. Country, 562.
Transport: conditions in Rumania, 441 ff.; effect on local prices, 442.
Transylvania: declaration of independence (Alba-Iulia), 111; distribution of land, 207–8; effects of reforms in, 206, 325 ff.; history of peasants in, 206; joint holdings in, 215–16; justification of reforms, 327; land tax in, 327; numbers of peasants entitled to land, 218; Rumanian grievances, 207; rising of 1784, 206; land tenure after 1848, 207.
Tschajanow, N., 461 n.; on objects of agrarian policy, 461.
Tuberculosis, extension of, 503.
Ţuica, 492.
Turkish corn monopoly, 25.
Turkish domination, influence of, 14, 567.

Unemployment, 447.
Union of Agricultural Syndicates, Memorandum of, 96.
Universal suffrage, 549, 551, 552–3.
Upper class, origin of, 11–12.
Urbarial system, 16–17.
Urbariu of Calimach, 21.
Urbeanu, Dr. A., 486–7, 489.
Urice, 11.
Usury, 84, 548.

Vaida-Voevod, Al., Coalition Government, 112; dismissal of, 115.
Vandervelde, E., xxv, 70.
Vatamani, 5 n.
Vatra satului, 585.
Vecini, 6, 15–16, 23, 583.
Viaţa Românescă, 553.
Vlachs, settled in Dobrogea, 212.
Vlad, Dr. Aurel, 222.
Vladimirescu, Tudor, 24.

Wages: inquiry into agricultural (1906), 82; lagging behind retail prices, 464; Ministry of Labour, index, 464; reasons for low, 465.
War: as cause of reform, 91–2, 569; effects on Rumanian farming, 307–8.
Weber, Max: on expropriation during French Revolution, 282; on slavery, 62.
Wheat: cultivation in Rumania, 296–7; limited cultivation by peasants, 297; decline after reform, 309; prices of, 350; effects of peasant diet on surplus of, 298; effect of export prohibition on, 300–1; expansion of cultivation, 289; imports of, 298; increased consumption of, 299; bonus for cultivation, 302; post-war production and exports, 338; prospects of export, 303; quality of, 321; cultivation and internal demand, 297.
Wheat exports, as standard of Rumanian prosperity, 295–6.
Women: conditions of labour, 41, 499; conditions of child-bearing, 499–500, 509; and drink, 494; franchise of, 551; inheritance of land, 585 n.; illiteracy among, 509–10.
Wright, Harold, 303.

Xenopol, A. D., 6 n.
Xenopol, N., 289 n.

Yield of crops, means of improving the, 323; Varjas experiment, 325.
Yeoman farming, decline not due to economic causes, 571.
Yeoman land, rape of, 22, 23, 583–4, 586.
Yugoslavia, paradoxical effects of reform in, 229–30.
Ypsilanti, Alexander, 20.

Zahareanu, Dr., 443.
Zeletin, Ştefan, 408 n., 545.
'Zemlia i Volia', 81.
Zileri, 207.

OTHER STUDIES IN THE RUMANIAN SERIES

Dr. Gr. Antipa (of the Rumanian Academy):
L'Occupation Ennemie de la Roumanie et ses Conséquences Économiques et Sociales. Paris, 1930.

Professor G. Ionescu-Sisești (Director of the Institute of Agronomic Research, Bucarest).
L'Agriculture de la Roumanie pendant la Guerre. Paris, 1930.

In Preparation:

Professor J. Cantacuzène (Director of the Bacteriological Institute, Bucarest).
Les Effets de la Guerre sur la Santé Publique en Roumanie.

PRINTED IN GREAT BRITAIN AT THE UNIVERSITY PRESS, OXFORD
BY JOHN JOHNSON, PRINTER TO THE UNIVERSITY

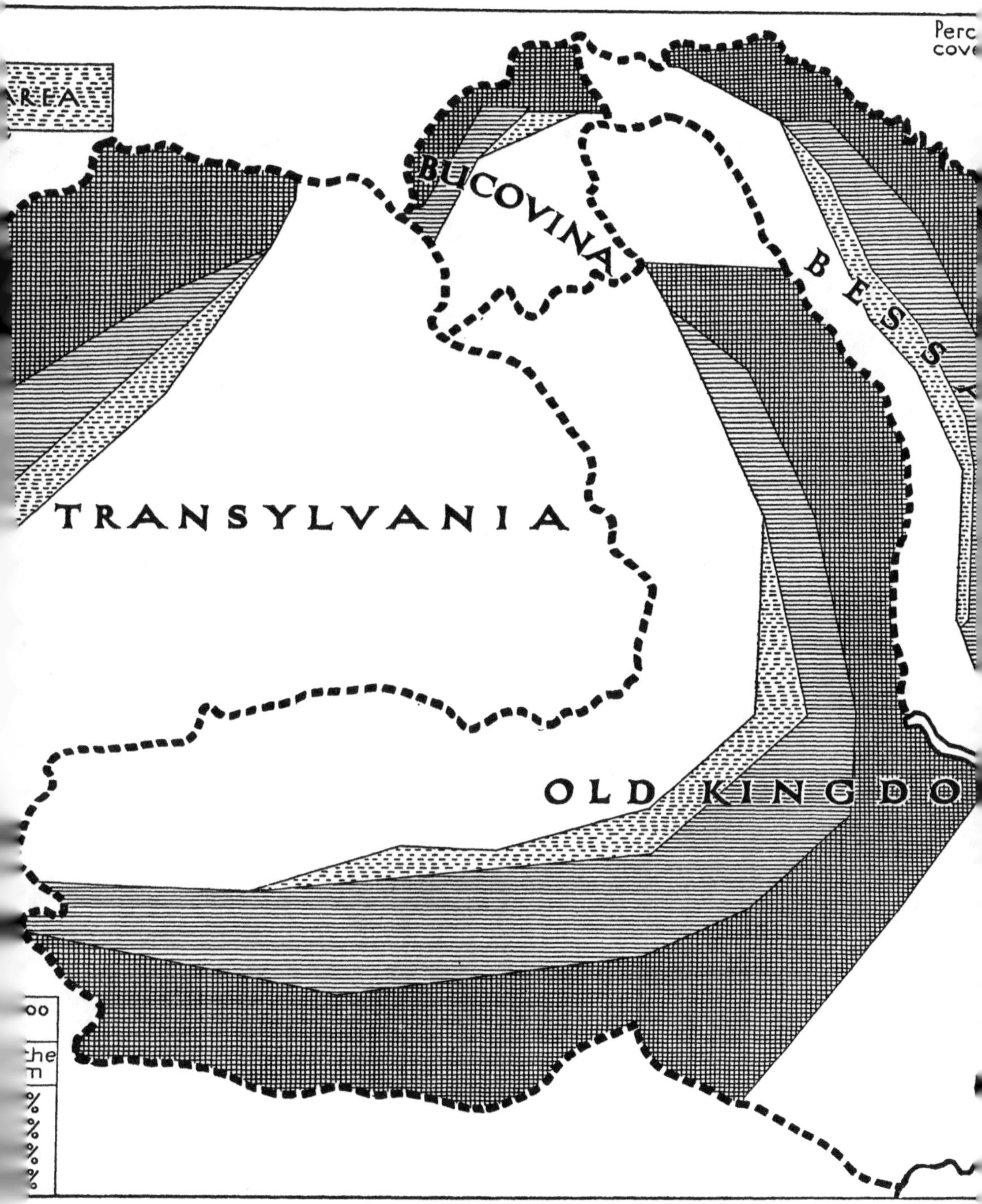

THE EFFECT OF THE REFORM ON THE DISTRIBUTION OF ARABLE PROPERTY

The extent of the arable land is drawn to scale, its situation is only approximate

www.ingramcontent.com/pod-product-compliance
Lightning Source LLC
LaVergne TN
LVHW052207200926
848481LV00001B/47

ACLS Humanities E-Book presents this volume as part of its Print-on-Demand (POD) program. This program offers a wide range of titles, across the humanities, that remain essential to research, writing and teaching. These titles are among the works chosen for digitization on our site in cooperation with ACLS's constituent learned societies for their continued importance to the scholarly community.

Part of the original plan for ACLS Humanities E-Book was to investigate the varieties of publishing formats that could be derived from single sources for both its retrospective collection and its new XML titles. Deriving multiple formats is essential for both publishers and scholars in today's rapidly evolving scholarly communications environment, and creating a production model that takes into account the multiplicity of access possibilities and audiences is an essential task of HEB.

For more information on our POD program and titles, visit http://www.humanitiesebook.org/pod.html .

ISBN: 978-1-59740-701-4

ISBN 978-1-59740-701-4
56600
9 781597 407014

AMERICAN COUNCIL OF LEARNED SOCIETIES, NEW YORK, NY